KU-320-247

Building Pathology

Deterioration, Diagnostics, and Intervention

Building Pathology

Deterioration, Diagnostics, and Intervention

Samuel Y. Harris, PE, AIA, Esq.

U.W.E.L.
LEARNING RESOURCES

ACC. No.
2382079

CLASS 622

690.
24

CONTROL
0471331724

DATE
27. JUL. 2006

SITE
HAR

John Wiley & Sons, Inc.
New York • Chichester • Weinheim • Brisbane • Singapore • Toronto

This book is printed on acid-free paper. ∞

Copyright © 2001 by John Wiley & Sons. All rights reserved.

Published simultaneously in Canada.

No part of this publication may be reproduced, stored in a retrieval system or transmitted in any form or by any means, electronic, mechanical, photocopying, recording, scanning or otherwise, except as permitted under Sections 107 or 108 of the 1976 United States Copyright Act, without either the prior written permission of the Publisher, or authorization through payment of the appropriate per-copy fee to the Copyright Clearance Center, 222 Rosewood Drive, Danvers, MA 01923, (978) 750-8400, fax (978) 750-4744. Requests to the Publisher for permission should be addressed to the Permissions Department, John Wiley & Sons, Inc., 605 Third Avenue, New York, NY 10158-0012, (212) 850-6011, fax (212) 850-6008, E-Mail: PERMREQ @ WILEY.COM.

This publication is designed to provide accurate and authoritative information in regard to the subject matter covered. It is sold with the understanding that the publisher is not engaged in rendering professional services. If professional advice or other expert assistance is required, the services of a competent professional person should be sought.

Library of Congress Cataloging-in-Publication Data:

Harris, Samuel Y.
    Building pathology :   deterioration, diagnostics, and intervention / by Samuel Y. Harris.
      p. cm.
    Includes index.
    ISBN 0-471-33172-4 (cloth : alk. paper)
    1. Buildings—Defects   2. Building failures   3. Buildings—Repair and reconstruction.
  I. Title.
TH441 .H295   2001
690′.24—dc21   00-063340

Printed in the United States of America.

10 9 8 7 6 5 4 3

Dedication To W.S.A. Harris, M.D., who taught me to ask questions, and to Nicholas Gianopulos, P.E., who taught me how to find answers.

Contents

Preface

∙∙

The term *building pathology* was introduced by James Marston Fitch in a course of that same title at Columbia University. When, in 1985, I proposed to teach a course under that title as part of the Program in Historic Preservation at the University of Pennsylvania, it was with considerable humility, then and still today, because of the high standard set by Jim in his course. In the intervening time, two things have happened, neither of which is attributable to an event or date as specific as the institution of a course: first, the term *building pathology* has become part of the common parlance of architecture and historic preservation, and, second, the field of historic preservation has become much more widely recognized as a design discipline, and not simply as an advocacy position.

What arguably began in the United States with Lee Nelson, Hugh Miller, Penny Batchellor, Henry Magaziner, and Nick Gianopulos as the application of design principles and analytical processes specifically adapted to aging buildings has merged with, then redivided, only to re-emerge and divide again in a sort of international morphosis of aging-building technology. Out of that collection of interests and efforts has emerged an idea that buildings in their end stages of deterioration can provide information and lessons as to the mechanics of those deterioration processes. Though this concept may seem obvious its application still is not altogether embraced by many practitioners. Until Jim Fitch articulated the idea, the presumption was that any informative aspect of deterioration was limited to the application of remedial intervention; in other words, the most we could expect to learn from a rotten building was what needed to be replaced or repaired. And that is still a prevailing attitude among many architects, engineers, preservationists, and conservators.

The incentive to remediate and to do so efficiently is certainly not inappropriate, but it is limited; and it is, at times, misleading in its consequences. The building design industry, which includes an ever lengthening list of specialists and professionals who directly or indirectly derive remuneration from dealing with the built environment, generally, gets paid for actively altering the existing condition, regardless of what that condition is. It follows, then, that, as practicing professionals, we have a vested interest in affecting alteration or intervention because that is how we earn our livelihoods. This subject is given further voice in Chapter 2, but the point here is that we have a potent impulse to advocate change, as opposed to maintaining the status quo.

The economic throttle on this impulse is the client's perception of value, which has two immediate components: first, the perceived value of the remediation itself, meaning the value added to the building or site as a consequence of the investment in the remedial measures; second, the perceived value of the guidance and presumed wisdom and skill provided by the practitioner. This, too, is an enticing subject well worth exploration, but not here. The impulse to alter and intervene in a building is not unlimited, but the client is ordinarily choosing among intervention options and, typically, has to raise the "do-nothing" option (called "abstention" in the book) on his or her own if it is raised at all. Consequently, the fees for designing interventions are more often set by externalities, not necessarily by the time or effort of exploration and analysis. Not only is there an impulse to remediate, but there is an equally strong impulse to minimize the time required to arrive at the intervention solution and design.

Some 20 years ago when I began teaching materials diagnostics at Penn, I labeled this phenomenon the "cookbook" approach to diagnostics and intervention, meaning that the self-interest and fiscality of professional practice impelled us to follow "recipes," rather than design unique solutions. In practice, I had the extreme good fortune to work with, and study, under my mentor, Nick Gianopulos, who was a master of many things, but none quite so apparent as his extraordinary ability to convince clients, whether they were owners, agencies, or other design professionals, to pay us to take a closer look at the problems and to explore additional options. In retrospect, I am aware that Nick's reputation was such that those clients were probably predisposed to do so, which is why they came to that office.

As I encountered students who were in a rush for to learn "the answers," I became increasingly aware that the impulse to solve problems and to solve problems quickly may have origins other than just cost and remuneration. Using devices of mild recrimination and disparagement were only partially successful in leading students to examine alternatives and to evaluate multiple options to a given set of signs and symptoms. Simply, deprecating the cookbook approach was not enough to carry the field.

Those first few years of teaching at a prominent university with classes of well-educated students powerfully demonstrated to me that we as a culture may be more or less proficient at imparting information to our scions, but we have uniformly failed to teach and train our successors to think critically or analytically. I was confronted with graduate students who seemed to be unaware of the later implications to their statement that they did not "do math or science." How could they aspire to solve the problems of physical nature in a material world if they did not? Hence, my job was and is to teach basic scientific principles to graduate students because it was acceptable to not "do science," in their undergraduate education.

If students are going to leave Penn or anywhere else with the ability to critically analyze and solve problems efficiently and profitably, it will come in one or two semesters of exposure to fundamental science and mathematics. The device that has emerged from my efforts to fill this gap is presented in this book: it is the intervention matrix. It is a compromise between the monocular prescription of cookbook intervention and analysis from first principles. In short, it is a tool. The premise is that just acknowledging there is always more than one way to solve a problem opens the mind to designing solutions appropriate to the facts at hand.

Much of the thinking and, therefore, of the language used to describe the development of this model is the result of my attending law school—which, I am convinced, is an experience everyone should have, but only after the age of 40. While the process is a bit harrowing, training in law is as close as current education comes to offering what might be called applied philosophy, meaning, simply, thinking with rules. Thus the study and practice of law is more the model for the analogue of this book than is medicine, despite the blatant pilfering of the medical vocabulary.

Law is the discipline more than any other that starts with a rule, the black letter law, and proceeds to prove that the rule rarely fits the facts. The practice of law, properly executed, is an exercise in knowing rules do not apply, because the reality at hand never fully conforms with the ideal. I knew for years that cookbook solutions were dangerous, but I had to go to law school to find the simple words to express why that is the case. As regards the subject of this book, the facts at hand in any building never match the ideal of the prescriptive solution to allow us to adopt a preconceived answer without analyzing the fact patterns. In the lingo of law, the answer always depends on the situation; and the situation is always fact-specific.

If the intervention matrix is to be supported, however, the designer will have to acquire at least a modicum of understanding of principles. The development of a device that provides multiple options means that there must be some way or ways to impart a set of minimum essential scientific principles in language and form that are palatable and useful to the audience. This, then, is the first and the primary objective of this book. If the notion of multiple options as an analytical and critical device is sound, then the principles underlying the options must be familiar, and the practitioner must be able to control the line of inquiry based on those principles.

To arrive safely at our destination in the land of intervention matrices, the voyager requires, not only a minimum understanding of how the machinery works, but some tools such as a compass or a map. Maybe some road signs would help. In this book, I have borrowed unashamedly from all over the teleologic map—a little symbolic logic here, some thermodynamics there, legal analysis over there. From these sources emerged the concept of the deterioration curve, which is not nearly so certain in its accuracy as the diagrams imply. The important idea, which the notion of a smooth curve conveys, is predictability. If deterioration is predictable, then the consequences of current intervention and the future of new installations are also predictable. A predictable progression allows for abstracting and globalizing ways to alter the curve, which are later identified as approaches to intervention. The deterioration curve is the map, which is better for showing there is a path than for indicating which path to take.

Once committed to using the term building pathology, it is tempting to follow the medical model further—and some authors have gone down that path. Certainly, in past semesters and years I have spent a fair amount of time in medical "lingo land." On one or two of these sojourns, I searched for analogues for disease and diagnostics; and for a number of years, I simply lifted the terminology, but it never quite fit. The better I became at explaining the medical model, the better the students and the clients were prepared to attend medical school, not to intervene in decaying structures. Therefore, in this book, the notion of disease has become the deterioration mechanism, and the notion of comparative diagnostics has been translated to the notion of necessary and sufficient conditions. Although the book introduces deterioration curves, deterioration mechanisms, and necessary and sufficient conditions before arriving at the intervention matrix, they are not stand-alone destinations; they are guides. I emphasize this point as a matter of encouragement to the reader. There may be times when the principles become opaque, when the exposition becomes confusing. It is at those moments it is important to remember that, while the understanding of fluid flow is not trivial, it is a means to an end, not an end in itself. If the reader finishes the book with no other lesson learned than that there are always six basic approaches to intervention, and what they are, then he or she will have invested his or her time and energy wisely.

As a practicing consultant to design professionals, as a colleague of teachers of students in those same disciplines, and as a founding principal in an architectural firm, I have come to realize that design is a process of resolution of conflicts. The difficulty in designing anything is the inherent impossibility of reconciling all of the competing interests represented in an object even as simple as a mouse trap, much less a building. For a variety of unflattering reasons, within the design professions, the operating definition of design has narrowed the lists of conflicting interests to ones that survive and manifest in a publishable photograph: the image of the object is the object itself. Keeping water out a building is not design. Intervening in a deteriorated condition is not a design. Restoring an interior while accommodating contemporary demands is not design.

If this blanket assessment seems a bit harsh, ask this question: Why are "design" awards given to buildings that have been judged only by photographs? Award juries never visit the building and have no idea whether the building leaks, whether the materials are holding up, or, for that matter, whether the building even works. The reward is for the image. Simply by changing the rules of design awards could radically alter perceptions of what constitutes design. I propose that no AIA chapter award a gold medal to any building that is less than 10 years old, and that it carry an affidavit that the owner is satisfied with the product and has no past or pending claims against the designers. This will not happen, but it is the secondary objective of this book to raise the query as to why that is the case.

The current state of design practice so deemphasizes responsibility for the future of the building that there is a disjunction on the part of the design mentality from the world of cause and effect. If the designer calls for a limestone façade in an acidic environment, the consequences are a predictable as when the preservationist restores the same limestone item in the same acidic environment. Whereas the preservationist is expected to understand the effects of acids on the ionic salts, for reasons related to the emphasis on novelty and entertainment, the architect can call the limestone art and decline responsibility for its otherwise predictable dissolution. For designers and educators of designers who want buildings to look good and to perform longer than is required for the initial photo shoot, read on. If the image is enough to satisfy your definition of design, put the book down now.

Acknowledgments

As with any venture, a book of this scope and breadth does not sprout full-grown from the head of any one person. I have been probing and staring at rust and rot for almost 30 years, and this book the compilation of the lessons of those 30 years. For almost 20 of those years, I have been teaching rust and rot to some very fine young minds, and some of the material in this book has been gleaned from learning from my own students, from their papers and theses, and sometimes, from just trying to stay—or to appear to stay—ahead of them. I am convinced of the virtue of the adage that the way to really learn some thing is to teach it. My first acknowledgment is, therefore, to my students at Penn and Goucher who over the years have taught me and who continue to teach me.

I have benefitted greatly from the direct instruction and training of a few superb teachers, notably Nick Gianopulos and Tom Leidigh, from my days at Keast & Hood, Co. Tom taught for many years at Temple University, and he is the author of the "ice column" story about which you will read in Chapter 3. Nick was and is far more than a teacher at Penn. For a generation of engineers, architects, and preservationists in Philadelphia and, increasingly, across the country, Nick is the mentor of mentors, *il maestro professore de edificio*. I worked with Tom and Nick for almost five years. Before then, old buildings were to me old buildings; since then, old buildings are rather more my life.

The connection from practice to teaching did not occur by accident, nor did it occur solely on my own efforts. It occurred because certain people provided me with the opportunities to bring the practice to the classroom. They took a chance on me when neither they nor I knew what the value would be. John Milner, FAIA, gave me my first lecturing opportunity in a course he taught, and that I inherited three years later. Peter McCleary was the Chair of the Preservation Program at Penn at the time, and he was the person who approved my first new course proposal, Material Diagnostics. Anthony N.B. Garvan succeeded Peter and he gave me my second course, titled Building Pathology. David De Long, AIA, succeeded Tony, and he has supported me and my teaching for as long as he served as chair; I now enjoy the continued support of Frank Matero, Chair of the Preservation Program; Richard Wesley, Chair of the Architecture Department; and certainly of Gary Hack, Dean of the Graduate School of Fine Arts. A few years ago, a long-time acquaintance and friend, Hugh Miller, FAIA, recommended me to Richard Wagner, AIA, the Chair of the Preservation Program at Goucher College, where I have been teaching for the past four years.

More immediately and specifically, no book gets into the hands of those holding it now without the efforts of many, many people. In particular, I very much appreciate the assistance of Alison Garwood, in editing and drafting sections of the book; and Diane Jackier, my teaching assistant during this period, for helping in innumerable ways, from tracking down references to developing diagrams to keeping my courses on track while I lived up to the stereotype of the absent-minded professor.

Every book has a publisher, and until I wrote this book, I did not know what publishers do. I still can't say I do, but I do know that without them, an endeavor such as this remains a distant idea. Because of the confidence and efforts of Amanda Miller at John Wiley & Sons, Inc., this book is real; and reality is always the first choice in any situation. For her assistance and support I will be forever grateful.

1

Introduction

On May 19, 1997, on the Lawn of the University of Virginia a tragic incident occurred: a suspended, second-floor porch that was attached to one of the original Jeffersonian pavilions collapsed. Twenty-three people were injured, one of whom died of her injuries. The structural element that failed was allegedly 175 years old at the time of the failure. The many lessons that can be learned from this incident are described throughout this book; the engineering details of the collapse are detailed in particular in Chapter 3.

According to the responsible agent of the university, there was no reason to suspect that there was a problem with the element that failed, for a variety of reasons, among which was that the element had, in fact, endured for all that time without exhibiting a problem—meaning that noone had been injured or had died prior to May 1997. Such a statement, presuming the condition of an element simply because it has endured, implies several things, one being the assumption that the original design was sound. In fact, the postincident analysis of the porch concluded that the maximum allowable imposed load for the porch, in pristine condition measured by today's standards of material safety, was 13 pounds per square foot. The weight of the porch was 15 pounds per square foot; in other words, by today's criteria, the porch was inadequate even for its own weight much less a superimposed live load. How is that possible? Why didn't it fall down sooner?

The answers lie in the seemingly innocuous phrase "by today's standards." When Thomas Jefferson designed the porches on the lawn, there were no standards, no widely known engineering techniques for analyzing the stresses in the members or for assigning load limits to structures.

Neither was there an understanding of the properties and ultimate strengths of the respective materials. The critical elements of these porches were designed with enough capacity to preclude their immediate collapse, but not enough to satisfy the most minimal of today's criteria of safe occupancy.

The critical elements of the assembly in question operated in what we describe as the *margin of safety* for a structural member. This margin or factor of safety is achieved in two general ways. First, we assign a reduction in the permissible stress, which we expect to incur when in service. If the ultimate stress is known to be 36,000 psi, we will arbitrarily limit the likely occurrence of stress to 24,000 psi and provide enough material to insure that this limit is not likely to be exceeded. This is the factor of safety, which in the above instance would be 36,000/24,000 = 1.5. This particular example is a common factor for steel. The applicable factor of safety varies depending upon the vagaries of fabrication and manufacture that can affect the reliability of a given material. In the case of steel and concrete, the factor of safety is relatively low because we have a relatively high degree of control over the manufacture of these materials, hence a relatively high degree of assurance as to the consistency of the materials involved. In the case of the designs of subsurface conditions, where variability and uncertainty are very high, the factor of safety may be 10 or more. In the case of timber members, the factor of safety is typically around 4. The values of allowable stress and, therefore by implication, the design criteria are derived primarily from industry standards, although some are imposed by the applicable building code, either explicitly or because the industry standard is adopted by and promulgated through the code.

The second general way we increase the margin of error in building design is through limits on the loads that may be placed on the structure. The most important of these load-limiting measures is the *safe live load occupancy* assigned by the building code. The code assigns an occupancy classification to every structure that is subject to code review. If the occupancy is likely to result in large crowds, then it is assigned an assembly classification that imposes a requirement that the affected area be designed to safely carry 100 pounds per square foot. Occupancy may also be limited simply by the number of persons allowed into an area based on the capacity of that area to be safely evacuated in an emergency. These two limiting factors are arguably conservative, maybe overly conservative; but the code is the only tangible article standing between a naive public and a careless practitioner.

For 175 years, the porches at the University of Virginia operated with little or no factor of safety. Eventually, materials deteriorated to a critical proportion such that when, on graduation day in 1997, in full view of the graduating class, a critical element was loaded to ostensibly modest levels by the weight of probably no more than eight people. The rod snapped, and those people fell. A grandmother of one of the graduates was killed; the 22 others sustained various injuries, some quite serious.

One of the assumptions underlying the writing of this book is that many of us value, even revere, at least some of our aging architectural inventory. We want to conserve and preserve older structures, generally and specifically, and we want to do so for a variety of reasons. It is important that we recognize that such reasons are preferences; there is no natural law that older buildings are better, nor that they are necessarily sacred. Those of us who take it upon ourselves to curate and conserve aging building fabric do so because we want to, not because it is either statutorily necessary nor required by nature. Unlike paintings or most sculptures, people occupy buildings, making those people vulnerable to the prospect that the building will collapse, either from under them or on top of them, or that they may be trapped inside the building in a fire or trampled in their effort to escape. It is for these reasons and others that people be able to enter into and occupy buildings with some assurance that they can do so free of unwarranted risk. Absent that assurance, people are entitled either to explicit warnings as to the hazard presented by the building or to be prevented entrance to the building. There is no law or requirement that people must enter a building regardless of the risk. Despite our apparently strong desire to have persons enter into and occupy old and, occasionally, historic buildings, there is no absolute requirement for them to do so. We can always lock the doors.

What we do not know about old buildings is profound, and the more we do not know, the greater the risk that our ignorance can result in profound damage to people in the building and to the building itself. When dealing with buildings—regardless of age—we must balance public safety and our comfort with our relative ignorance about how the buildings are performing and their condition. Our relative knowledge or ignorance about a given building depends on information, the accuracy of that information, and the reliability of that information remaining accurate over time. The sources of information about a building and its capacity are based on our awareness and knowledge about the criteria and standards to which the building was designed, the degree to which the actual design adhered to or varied from those standards, the properties and capacities of the actual materials and construction, and the actual condition of those same materials and assemblies as they have aged. The obligation to investigate, therefore, increases the greater the distance in time from the original design and construction.

In the case of the porches at the Pavilions at the University of Virginia, there was no surviving indication of the original design criteria, material properties, or construction details employed by Jefferson, meaning that even if the condition of the construction were the same as when it was originally constructed, there could be no way of determining the intended capacity of the decks or hangers. Of course, we know that over time materials and assemblies deteriorate, and do so inexorably. We know that the structural capacity of elements and assemblies is dependent upon the continued good condition of the materials; in short, we know that materials do not improve over time. Materials necessarily degrade over time, and capacity decreases; so while we may derive some reassurance from knowing the original design criteria, we cannot assume that the capacity has remained unchanged in the intervening years, in this case, 175 years. The magnitude of deterioration can and will eventually jeopardize any construction; so even if Jefferson's design was adequate, it was reckless to permit the continued loading of an exterior porch without investigation and verification of the condition of aged fabric.

The proposition of this book is that the incident that occurred in May 1997 at the University of Virginia was unnecessary and avoidable. The purpose of this book is to provide anyone with a vested interest in the condition and performance of aging structures with sufficient information to reasonably investigate, diagnose, and intervene in the deterioration of buildings. If the employment of this information is insufficient to determine the capacity of the building; if the this book does not lead in the direction of reasonable assumptions of risk; if the public cannot reasonably be protected from the deterioration of buildings, then lock the doors until you *know* the answers. Simply, if the fabric is so precious as to not justify the invasion of the requisite examination, then it is too precious to expose to the public.

1.1   Statement of the Problem

The essential problem is this: We cannot have certain knowledge of, nor can we exercise complete control over, the progress of deterioration. As buildings age, the lines of stiffness shift; the load-carrying distribution within the structure changes; at the macro-scale, nonload-bearing elements begin to carry load. At a colonial vintage stone house in Germantown, a section of Philadelphia, a kitchen wall collapsed. From the foundation to the second floor, the 12-inch-thick masonry fell into the side yard. The second-floor masonry, attic, and roof were being carried by the projecting ends of the second-floor joists. Those joist ends had been pocketed into the wall, which was prostrate on the side lawn. What was carrying the joists was the kitchen cabinetry, and it was doing a fine job of it. Over the years, however, as the wall relaxed, the cabinets were collectively thrown into compression. Despite the loads on the cabinets, there were no particularly compelling clues as to their distress nor to the movement of the exterior wall. Had an engineer been called in a year before the collapse, it is highly unlikely that he or she would have identified the kitchen cabinets as structural elements, although that is precisely what they were—or, more accurately, what they had become.

Our charge as building pathologists is to interpret the signs and symptoms of the aging process in order to assess the current condition and capacity of a the structure, to understand the aging processes and mechanisms to enable us to project the trajectory of that condition, and occasionally to intervene in those processes so as to alter the current and future capacity of the building. We will attempt to do all that, with the humbling realization that we will never completely comprehend the condition of the fabric, the distribution of the forces, or the precise state of deterioration of the materials. This means that the judgments and decisions we make will, therefore, be based on inexact information. Collectively, we accept that much of life, generally, is guided by imprecise and inexact knowledge. We accept that we must make calculated guesses to keep projects and society as a whole moving, to affect change, or to alter the course of events.

When we cannot rely, hence act upon, certain and/or complete information, we tend to substitute forms of knowledge called experience or wisdom. Experience is a form of information that may lack a theoretic structure, but that provides a sort of comparative basis for predicting future outcomes. Experience at its best is a way of saying, "I have no idea what is going on, but it looks like something I saw in '54 that turned out to be. . . ." Experiential knowledge is important, but there are severe limitations to practitioners relying on experience. First, not all experience is good experience; there is no way to verify that the recalled event from 1954 was not itself an anomaly. Second, experience is highly subject to interpretation and veracity of the person recalling the experience; some memories are inaccurate, other are simply self-serving. Third, we tend to place greater credence on an experience if it occurred more than once, particularly if it was experienced more than once by the same person. The older the person recalling experience, the greater the likelihood that he or she has experienced the same sort of event more than once. Replication is, indeed, a form of experimental verification, but repeated experience may simply be the repetition of the same fundamental mistake, not experience. The fact that someone has done something many times or for a long time does not mean the experiential value is high.

Wisdom differs from experience in that it tends to be less factually and more procedurally based. It may be experience that leads us to suspect freeze-thaw cycling, but it is wisdom that cautions us to consider other alternatives. Wisdom and experience are both subject to the authority of the practitioner. Authority, in turn, is always rendered by the observer. A person may be inherently wise or experienced, but that means little or nothing if he or she is perceived a fool. Credibility is a form of surrendering to the speaker the authority or his or her words. No one can demand or buy authority or credibility. But because much of what we do as practitioners of a far-from-exact science, wisdom and experience are valuable tools. The veracity and credibility to effect action based on that experience and wisdom is not the creation of the practitioner, but of the recipient of the information.

1.2 Problem-Solving Process

The fundamental process of solving problems is by itself a fascinating pastime. There are two basic reasons for deconstructing the process here: one is prospective, the other is analytic. The prospective motive goes to a common proposition that performance may be improved if we first dissect the object of inspection. The analytic motive is classically teleological in nature, that is, inspection for the sake of interest in the inspection process. It is the prospective motive that guides the analysis in this book. We want to know more about how problems are solved on the chance that we can improve the results of our efforts.

One notion about problem solving is that it is a so-called natural talent: some people are good at it, others aren't. And those who have it can improve on the talent with practice and training. There is a second view of problem solving, that is, the naturally talented problem solver is very much more dependent upon adherence to a learned discipline than to a natural predisposition. This is important as it relates to the issue at hand, because diagnosing problems associated with buildings and building pathology is rarely dependent upon natural talent. When it is done well, it is the result of rigorous application of discipline, learned skill, and hard work.

To that end it can be helpful to articulate the problem-solving process. The fact that the process can be articulated means that it is inherently rational. This is counterposed to another form of problem solving called intuition. We tend to equate intuitive and rational thought as coequal occupants of the same realm, primarily because we wish it to be the case.

The intellectual process of the rational problem solver is akin to that of the pyramid builder. The practitioner builds the pyramid from blocks of different sizes according to some blend of reasoned trial and error and theoretical rubric. When the pyramid is complete, the form is demountable and replicable. The rationality of the process is measurable by the ability of the practitioner to recount and to repeat the process. This is the essence of the rational problem solver.

In contrast, the intellect of the intuitive problem solver is a flat plane with many round-bottomed depressions. Rolling around on the plane are marbles of various but not necessarily unique combinations of diameters and colors. The intuitor tips the plane such that the marbles of his or her mind roll around until one by one they all nest in a depression and allegedly, therefore, come to rest at the point of solution. This is variably called the creative process, the intuitive process, the artistic temperament, even sometimes, the stroke of genius. It is also entirely possible that it is bunk.

1.2.1   Structured Problem Solving

There are as many problem-solving processes or sequences as there are people who have committed their processes to paper. And if a process has been recorded, there is already an excellent chance that its content is logical and, probably, sound (anyone able to commit a thought process to writing is past intuition and inspiration as the primary analytical devices). Table 1.1 outlines one such sequence of steps, which can lead a rational examination of the problem to a solution. But keep in mind, the sequence is not the solution; it is a tool. It is not a unique tool or even a particularly refined tool. We present it to demonstrate what a sequenced process is.

A common pattern of investigation begins with a question that is provable as either positive or negative, meaning simply that a statement or hypothesis stated in the affirmative is either true or untrue. This chapter, then can be considered the hypothesis. The next four chapters following the hypothesis in the sequence are methodology, which presents the process by which the hypothesis will be tested; data, which is the collection of known information and discovered information resulting from the application of the methodology; analysis, which is the assembly and interpretation of the data; and conclusion, which is the determination of whether the analysis supports or denies the hypothesis.

1.2.2   Establishing Necessary and Sufficient Conditions

This book establishes the necessary and sufficient conditions of the deterioration mechanism as part of the basic methodology of the analysis and, therefore, of the approaches to intervention. For many—presumably

Table 1.1   BUILDING DETERIORATION PROBLEM-SOLVING PROCESS

I. Define the problem internally.
   1. Identify the discrepancy.
      (a) Flaw announcement
      (b) Performance deficiency
      (c) Performance improvement potential
   2. Determine original objectives.
      (a) Document stated objectives.
      (b) Deduce undocumented objectives.
      (c) Measure current status relative to original objectives.
   3. Determine current performance standards.
      (a) Establish remaining serviceability.
      (b) Measure remaining serviceability relative to current demand.
   4. Specify performance objectives.
   5. Define performance measurement criteria.
II. Establish externalities.
   1. Define collateral issues.
      (a) Preservation/conservation priorities and objectives
      (b) Sustainability criteria
      (c) Environmental criteria
      (d) Socioeconomic implications
   2. Measure operating constraints.
      (a) Determine elements to be retrieved, and initiate test design.
      (b) Design test methodology and execute trials.
      (c) Execute tests and process data.
      (d) Verify data reliability.
III. Execute preliminary design solutions.
   1. Assess data compared to algorithm.
      (a) Assemble algorithmic models and initial design options.
      (b) Verify that options satisfy algorithm and externalities.
   2. Array options in order of preference.
      (a) Conduct feasibility/benefit analysis.
      (b) Synthesize to enhance strengths, minimize detriments.
      (c) Select preliminary preferred option(s).
      (d) Conduct life-cycle costs analysis.
      (e) Test for peripheral compatibility issues.
      (f) Confirm preferential selections.
IV. Present recommended solution(s) and elicit decision.
V. Proceed with selected solution development.
   1. Execute detailed design.
   2. Assemble cost and schedule data.
   3. Publish design documents and solicit implementation proposals.
VI. Execute implementation.
   1. Monitor compliance with design.
      (a) Resolve discovered inconsistencies in information.
      (b) Resolve design errors and omissions.

most—of the deterioration mechanisms discussed in this book, any given list of necessary and sufficient conditions will result in a unique deterioration mechanism. In other words, the combination of specific properties in a specific environment either will result in no mechanism or in a unique mechanism. If it is possible that from one set of conditions (meaning, from one set of material properties combined with one set of environmental factors) we may get two distinct and identifiable mechanisms, then we have a structurally logical dilemma. If this were a book on human pathology, the dilemma could be that, from the standpoint of medical diagnostics, we would be confronted with two possible diseases; therefore, we could not reasonably prescribe for one with the assurance that the treatment would not exacerbate the second. Fortunately, this is book about building pathology, and the dilemma is more apparent than real.

As we will discover in Chapter 2, one of the advantages of having multiple approaches to intervention—hence, multiple options for treatment—if we are confronted with a nonunique set of necessary and sufficient conditions, we can search the respective options for common options that will provide comparable intervention. We can also conduct modest diagnosis by treatment, using a distinguishing treatment to identify the specific mechanism. This is occasionally done in medicine as well; for example, one way to diagnose classic and common migraines is to treat them with migraine-specific medicines: if the medicine is effective, then the headache was, in fact, a migraine; if the medicine was ineffective, then the headache was of some other origin. We can occasionally use a comparable diagnostic approach with buildings, but not with comparable definition of results. We are more likely to spend time and money to the detriment of the building and still not have definitive diagnosis.

1.2.3 Establishing the Search Pattern

The collection of information is one step in all problem-solving sequences. Embedded in almost every experimental methodology is the equivalent of the conditional statement, "if A, then B." or "if and only if A, then B." Once we gather the known information regarding a set of conditions, we must construct the list of definitive unknowns. In other words, we will find in our search for conclusion that certain pieces of evidence will complete the set of necessary and sufficient conditions. Barring the double diagnosis dilemma, from that construction we can deduce the deterioration mechanism itself, hence the options for intervention. The search for the unknown information is one of the major flaws in the practice of building pathology because we are victims of our own preferences. We like diagnosing old building problems; we like designing structures; we like defending legal claims; we like collecting information about the objects of our preference;

we like knowing everything we can about old buildings; we enjoy the process of searching for the unknown information, we enjoy studying them and testing them and documenting them.

Unfortunately, we also have a tendency to overexamine the patient. Disciplined examination means, among other things, restricting the search pattern to those elements that will affect the outcome of the intervention matrix. One test of efficacy is to ask this simple set of questions: If we find that B exists, how will that affect the decision? And if we do not find B, how will that affect the decision? If the answer to those questions is that the existence or nonexistence of B is irrelevant to any decision, then the search for B is an extravagance, an indulgence. The design of the search pattern for information is, therefore, a critical element in the problem-solving process. It warrants emphasis because it goes to our professional vulnerability, namely our self-interest, and can put our preferences at odds with our clients' interests.

1.3 Statement of Objectives

The objectives in writing this book are synonymous with the audiences to whom it is addressed: For the practitioner, it is intended as a reference; for the non practitioner, it is a primer, meant to provide a bridge from the familiar to the unfamiliar; for the student, it is a basic text.

The material is immediately applicable to the remediation of aging buildings; and more than the designers are likely to admit openly, it is directly applicable to the design of new buildings as well, as too many architecture programs do not include a dedicated course in material properties and performance.

When designers draw a detail in the studio, it always works perfectly and is readily installed. But the detail is nothing more than lines on paper, and the laws of physics do not apply other than to the durability of the lines and the paper. Drawn details on paper never leak, and fabrication and installation are complete. The quintessential aspect of building design that separates architecture from the design of architecture is that the drawing is not a product; it is a *means* to a product. Architectural drawings are useful only to the extent that they assist in fabrication and assembly. That assembly will be done by professionals other than the designer, and those professionals have experience and preferences that are often contrary to those of the designer. When the preferences of the so-called trades collide with the preferences of the designer, the resolution of that conflict is a major determinant in the technical durability and critical success of the design.

To understand a tradeperson's concerns requires more than a faint familiarity with the materials involved. To know a material as an architect and to know the same material as a trade is to know the same thing twice, which is not the same as knowing the material from two perspectives. Among other things, that means that no amount of reading and studying about brick, can substitute for mixing the mortar and laying a brick in the mud. It is the only way to appreciate the reason that a trowel is the size and shape it is; why a brick is the size it is; why cavities behind bricks are the depths they are. It is not abstract theory; it is the size of a human hand and the way the wrist moves that makes a wall.

2 Mechanisms and Diagnostics

2.1 Building Pathology and Diagnostics

The science of building pathology and diagnostics addresses the deterioration and the demise of buildings and their component systems. The combination of building material properties and environmental conditions create the requisite components that perpetuate materials deterioration, hence building failure. It is the job of the building pathologist to aptly access and diagnose these sources of the decay; ideally, this assessment will be comprehensive, thorough, and circumspect. The object is to ascertain the condition of the building and the sources of the defects, and to prioritize them in an orderly and intelligible form. As evaluators, experts in the field, it is our job to meet all the technical objectives of the assessment and to present the assessment in a manner that is understandable and acceptable to the client or layperson.

There is, however, an unfortunate and ill-advised tendency in the field to divide deterioration into two definitive categories: materials deterioration and building deterioration. Insofar as this distinction is convenient for categorizing deterioration mechanisms, this book will respect it. The problem is that deterioration does not respect professional boundaries; rust and rot proceed apace, whether the observer is a chemist or an engineer.

Using metaphors and analogies as a teaching tool is done universally. The analogy in this case is of the deterioration of buildings to the human aging process. So commonplace is it that we hardly give a thought to its applicability or limitations. But it is essential that we do, for as much as we will use and rely on the aging process as an analogy for building deterioration, we reiterate here, at the outset, that buildings do not age the way people age.

It is even arguable that buildings do not age at all; they simply deteriorate, and there is a difference between deterioration and aging. The only inherent similarity is that both occur over time. Given the intangible nature of time, it is not surprising that we try to link time-related concepts one to the other.

Despite the analogical limitations, borrowing from the medical lexicon does provide us with a well-developed vocabulary and with a proven diagnostic methodology. And though the terminology may at times border on being glib, the invocation of the diagnostic methodology is sound and durable. Two aspects regarding aging versus deterioration are valuable and worth discussion because they are not fully appreciated by those of us who are stewards of old buildings. Aging and deterioration are inevitable and progressive. The evidence is inescapable, yet there is a tendency to discount the deterioration process of the built environment as episodic, idiosyncratic, and/or the result of error or accident.

Two prevailing forms of denial prevent us from recognizing and accepting the inevitability and progressive nature of deterioration. The first form of denial is characterized by designers who believe that chemistry and physics, gravity and reality, afflict only "other" designers; in short, they delude themselves into believing that deterioration will not occur in their buildings. The second form of prevalent is more likely prevalent among those of us who specialize in building geriatrics: preservation professionals and technicians. Among this group deterioration is indicative of an error or poor judgment, and if we exercise proper care and judgment in the remediation, deterioration will not occur again. This attitude is no less delusional than the first.

Let us dispel both forms of denial here and now: Deterioration affects *all* materials and, therefore, *all* buildings. The fact that the material is a recently developed, synthetic elixir does not mean it is exempt from the laws of thermodynamics. It, too, will deteriorate. That is not to say that science has developed and will continue to develop more durable, inert, strong materials. That is what materials scientists do. Teflon is a good example of such a material. It is so inert chemically we may some day have Teflon beaches because all that ever happens to the stuff is that it gets ground up into smaller and smaller pieces. But the grinding up is still an alteration. If the item started out as a Teflon-bearing plate in 1980 and winds up lasting for 200 years before it converts to dust, it still converts. Nature wins because she has all the time in the universe; we do not.

2.1.1 Deterioration as a Natural Process

Buildings may develop leaks, and fail, because the designer did not know a flashing from a raincoat, but that does not mean that all leaks are the result of human error. The best-designed details will eventually fail for the same reason that all materials must deteriorate. Deterioration is not an exception, nor is it synonymous with failure. As we will discover, there are times and situations where deterioration is preferred, even revered, for example to achieve patina. But, as is the case with most negative human conditions, there are always enough insidious aspects to failure to sustain a suspicion that all deterioration is the result of either human error or human frailty. The only human frailty that consistently contributes to or is the cause of deterioration is failure to recognize that it is normal, and will occur given time and exposure. This failure to recognize leads us to install particularly vulnerable materials and assemblies in hostile environments.

Deterioration is as inevitable as the passage of time because the two phenomena are not unrelated. Because the basis processes are not biological, in the sense that the building is not an organism, the progressive nature of deterioration is always and irreversibly negative. By this we mean simply that buildings cannot heal themselves. Although some of the environmental agents that act on the materials *are* organisms, and even though some of the deteriorating materials are inorganic, deterioration is not biological in character.

We as constructors of the built environment have known about deterioration for thousands of years. We accept it implicitly as a condition of habitation of this galaxy; but for reasons beyond reason, we find fault when certain select buildings deteriorate. This phenomenon is what we call the phenomenon of *altered impression*, a topic we address in detail in a subsequent section by that title.

2.1.2 Monitoring and the Time Budget

If this book teaches one and only one indelible lesson, it would be that nothing is static. From this lesson many things follow that are simple and profound at the same time. Most of the educational points about materials and buildings are static models, and as such represent little more than vignettes of reality, not reality itself, which is inherently dynamic. When we discuss systems and materials in the context of deterioration, we are discussing the very nature of change. Even in the course of discussing change, however, we will tend to look at it in terms of what is referred to as *steady state equilibrium*, which means that there is movement; the parts are moving, but there is no net alteration in the system.

Most of the content in this book is highly subject to alteration with time. It is not as simple as the a beam becoming more rotten with time— that is a fairly predictable linear function of energy dissipation; there is change in the material, but the rate of deterioration is linear which is to say that the steady state of the rate of rot is static. That is how we tend to view change, as sequences of a steady state of flow problems. Unfortunately, that is not reality; that is the abstraction of reality, which is necessary and functional if we want to convey the concepts underlying a complex reality.

Building pathology is an aspect of natural laws operating, the rates of which are constantly changing, which is one way of saying that the changes are changing and the changes in the changes are changing in nonlinear ways. The gibberish aside, we are not saying that reality is chaos. There is a great difference between chaos and complexity. Reality is complex, but it is also orderly, even if that order is extremely complex and constantly altering that complexity. So what does this mean? Among other things, the reader is cautioned as to the limits of this book to reflect the dimension of time as it affects every subject we will address. We will look at materials and at systems, and we will identify deterioration mechanisms that necessarily will result in changes to the fabric and to the building. As those changes occur, we will no longer be looking at precisely the same set of materials and systems we were previously, so while the same deterioration mechanisms are basically in effect, they are altered by the constantly evolving circumstances of prior deterioration.

2.2 Mechanisms and Degradation Model

A mechanism, as defined in this book, is analogous to building pathology as disease is to human pathology. Mechanisms are the consequence of the reaction and interaction of two independent variables: the physical object and the environment. The analogue of the body in medicine is the building, as the mechanism is to the disease; but that is about as far as the analogy can be useful.

The fundamental difference between buildings and humans is obvious: buildings are not organisms, and the deterioration patterns are unmistakably different. Buildings deteriorate from a point of maximum order at or near their completion, and they begin to deteriorate immediately. Initially, the type of deterioration is typically at the material and suboptical level. This is referred to as the *incipient period*, to sufficiently press the point that deterioration has begun despite the apparent absence of damage.

The incipient period is followed by a period of *accelerating deterioration*, in which the mechanisms initiated in the incipient period begin to coalesce and converge and become visible. In short, components begin to fail. Mathematically, the incipient period is not a linear deterioration, but a period of acceleration as well. The difference is merely a matter of recognition and scale; it is not conceptual. The acceleration progresses, and necessarily so, until the building as a total system fails and the building is abandoned or condemned. The final stage of building deterioration, the *deceleration period*, may or may not begin with abandonment. In fact, abandonment probably precedes the beginning of deceleration. At any rate, the building, absent intervention, will continue to deteriorate until it is nothing more than a pile of rubble, and that pile will continue to deteriorate virtually in perpetuity.

This description can be demonstrated in a smooth curve representing deterioration, which is the loss of ordered assembly, versus time. The general shape of the curve is portrayed in Figure 2.1. This type of decay curve is by no means unique, nor is it without pedigree. It is, in fact, rather typical of entropic disintegration of crystalline solids, which not coincidentally is typically the geneses of building deterioration. At no small risk to veracity, this curve is the sum of all the respective decay patterns of each and every element in the building. That said, it is necessary to point out that the curve is an educational tool, not an experimentally derived plot of data points. We will use the entropic deterioration model throughout the book to advance understanding, not to promote it as truth.

2.2.1 The Deterioration Mechanism

As stated, one of the fundamental premises of this book is the concept of the deterioration mechanism. This is the rudimentary basis for the notion of building pathology. To repeat, the deterioration mechanism is to building pathology as disease is to human pathology. There are, however, a few extremely important distinctions between a mechanism and a disease. A disease may or may not result in the demise of the patient, if for no other reason than the capacity of the human organism to contribute significantly to its own recovery. Buildings cannot spontaneously recover; a diminution of the fabric of a building is permanent without external intervention. For this reason, all mechanisms are inherently progressive. As long as the deterioration conditions remain in place, the deterioration mechanism will progress.

The components of the deterioration mechanism are the fabric itself and the erosive environment in which the fabric exists. The nature of deterioration mechanisms is that not only are they progressive, but the resulting damage is cumulative and irreversible. This means that deterioration mechanisms may begin and end temporarily, but, unlike disease, there can be no regeneration. This distinction is obvious, but it is also subtle, in its implications to this extent. A building-related deterioration mechanism may go into remission—meaning that the mechanism may be momentarily dormant—but when the mechanism regenerates, the revived mechanism begins where it left off and the resulting damages are additive.

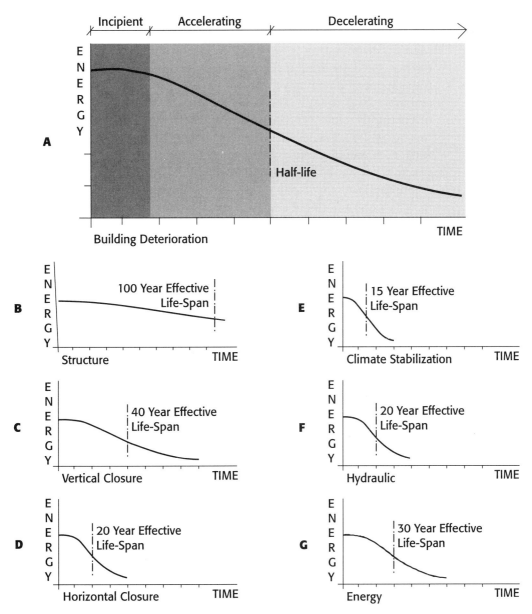

FIGURE 2.1 The composite deterioration curve (A) is the sum of the respective deterioration patterns of the subsystems.

The specific mechanism that affects an element or specimen is addressed in some books almost as if the mode of deterioration is random. Material deterioration is not random, nor is it a matter of chance. At its essence, the deterioration mechanism is a manifestation of the law of entropy in operation. Energy flows from a higher and ordered state to a lower and more random state. This is yet another way of stating the Second Law of Thermodynamics. The First Law of Thermodynamics is also relevant to this discussion. It states that in a closed energy system, there may be conversions from one energy state to another, but the total energy within the system is constant. This is also the Law of the Conservation of Energy. The second law is not inconsistent with the first law because the second law permits the same amount of energy to remain in the system, but in an ever increasing degree of randomness or chaos.

The deterioration mechanism is inherently an exothermic process, which is to say that deterioration is entropic. Viewed as an energy model, the building degrades over time along a declining S curve. While there are other basic decay curves, such as the hyperbolic decay curve corresponding to radioactive decay, the simple S curve is the most descriptive of building decay. The basic model assumes no reintroduction of energy from outside the system, such as maintenance or repair. The mathematical characteristics of the S curve are that the curve accelerates to a point of contraflexure, then decelerates. At no point is the slope equal to zero, nor is it ever negative. The upper and lower bounds are known, and the lower range of the curve is asymptotic to the ordinate.

Unlike the chance contact of an airborne virus and a human being, the forces requisite for the deterioration of fabric are omnipresent. The point is that materials and assemblies possess characteristics that can and will perform in predictable manners given certain conditions. If a building is constructed of timber we know to expect termite damage and fungal rot; if the structure is metal, we know to expect corrosion. The expectation is founded, however, not so much on the empirical accumulation of experience as on the scientific predictability of actions and reactions.

The nature of deterioration is, therefore, not a matter of whether a material will deteriorate; all materials deteriorate. The questions for which our studies are relevant are how and at what rate deterioration will occur. It is precisely because of the physical inevitability of deterioration that we are able to predict with some accuracy and precision that deterioration will occur and at what rate it will progress. The basis of this certainty is that the deterioration mechanism is irreversible and not subject to alteration internally.

The passive character of materials as compared to biological organisms means that absent intervention from outside, the deterioration mechanism, once established, will continue until one or more of the necessary and sufficient conditions requisite for the origination of the deterioration mechanism are violated. Only at this point will the particular deterioration mechanism cease. It is altogether probable, however, that another deterioration mechanism will ensue based on the properties of the residue and of the current conditions of the active environment. Buildings can respond to deterioration in one direction only, downward. If our intervention is not positive, the damage is cumulative on two counts: the original deterioration mechanism that was not abated, compounded by the ineffective intervention of the practitioner.

The deterioration mechanism, as mentioned, is analytically divisible into two factors that combine to produce a specific mechanism: (1) the inherent properties of the specific material or system, and (2) the atmosphere or environment in which those properties are operating.

The properties of a given material, element, or structure, when taken together, are not merely descriptive, they are definitive. Ordinarily, we deal with materials by nomenclature: this is steel, that is brick. The names of materials are neither descriptive nor definitive, only nominative. The word steel is communicative that the material is of a class of materials referred to by that name; but the word does not proffer enough information to enable us to predict behavior. Even if we add descriptive phrases to the word steel they cannot assist us in the analysis of performance.

In order for performance to be predictable and accurate, the defining attributes and properties must be *quantifiable*. When assembled into a list or table, these quantified properties provide more than a literary description of the material; they are the material. The word steel means that the table of properties defines not only steel, but a specific "flavor" of steel, such as A36 steel. Most of the terms we apply to materials are, in fact, generic brands of materials. Wood, as a term, encompasses a couple of hundred species; aluminum and steel, dozens of alloys each; granite, hundreds or even thousands of distinguishable rocks. In order to deal effectively with materials and material deterioration, it is necessary to move beyond the generic characteristics of materials and deal with the specific properties of the material at issue.

This table of values, then, is a profile of the material. As such, it is unmistakable for another class of material. Anyone familiar with the meaning of the terms reading the profile of wood, for example, could not mistake such a profile for that of steel. The profiles are linked to the material as surely as a photograph is linked to its subject. The potency of this concept begins to become apparent when we take away the name and deal only with the profile.

Imagine being confronted with a detailed list or table of material properties. The list is thorough and precise. We now know enough about material pathology to know that when exposed to an open environment, these properties will result in deterioration mechanisms. We may not know enough at this point to accurately describe those mechanisms, but we know that deterioration will necessarily ensue. As we learn more about the characteristics of properties and environments, it will follow that not only can we identify the deterioration mechanisms, but quantitatively predict the performance of the material without ever having to know what the material is called.

The point of severing the name from the profile is this: we have attached associations to materials that are not helpful in the analysis of performance. We attach to wood romantic notions of comfort and warmth; coldness to steel; solidity to granite. These associations are metaphors, not properties. They represent learned associations and attributions, which are derived from the symbols of the material, not from the material itself. It is not the purpose of this book to upset hundreds of years of symbolic associations except to this extent: symbolic associations can and, at times, do distort the perspective of the investigator. A preference for wood, for example, can prejudice an investigation of that material. The same is true of any other material, either in favor of or against the use or tolerance for the material. Such prejudices, while legal and under certain circumstances even justifiable, are not constructive relative to building pathology, diagnostics, and intervention. All materials are flawed in the sense that they all deteriorate. From that standpoint they are all equally flawed. Our task as building pathologists is to see them clearly, without the hype and lore, and to judge based on value and performance, not literature and history.

The second part of the deterioration mechanism equation is the operating environment. The operating environment essentially includes everything outside of the item at issue. In the case of a specimen of fabric, the operating environment is everything surrounding the specimen, including other materials in contact with the specimen—the air, the ambient temperature, the relative humidity, even the gauss value of the ambient magnetic field if it is relevant. Absent an operating environment, material or building properties are merely laboratory results and reports; there is no particular significance to properties per se. The vulnerabilities or merits of materials or systems require the description of the operating environment. As a matter of symmetry, the operating environment is by itself lacking significance in the absence of fabric. In the absence of fabric, the operating environment is only weather. In the absence of an operating environment, fabric is just stuff.

The interrelationship of material properties and the operating environment can be represented in a mathematical form as $M=f(P, E)$, meaning that a mechanism is a function of two sets of independent variables, properties (P) and the operating environment (E). It is the combination of the two sets that determines the deterioration mechanism. From the discussion above, we have concluded that the characteristics of M (the deterioration mechanism) are that, given P and E, M is inevitable. As long as P and E are present, M will operate and progress. M is cumulative and irreversible; and if P and E are known and understood, M is predictable and quantifiable. In terms of building history, the curve corresponds to three general phases of deterioration: incipient, accelerating, and decelerating deterioration.

As previously stated, incipient deterioration is that period immediately following construction when all the systems are simultaneously new. The building is in fine shape and the illusion is that nothing is happening in the form of deterioration. In actuality, the mechanisms of deterioration are active, albeit at the suboptical level. The microcrack networks are initiating; the corrosion cells are forming; and the weeps are gradually clogging. The rate and duration of the incipient period is dependent upon the inherent properties of the materials, the quality of detailing and construction, and the aggressiveness of the environment. The end of incipient deterioration may be as early as 15 years or as long as 50 years. The delusion of the incipient period is that it will last indefinitely, possibly justifying minimum maintenance or depreciation.

Accelerating deterioration begins with the failure of the first major subsystem and extends to a point of contraflexure, which is defined as the point of *zero utility*. Left relatively unimpeded, the building's subsystems will fail sequentially resulting in the catastrophic failure of the building as a whole. In reality, very few buildings go immediately to collapse; however, no building ever fully recovers to its initial state of simultaneously unimpaired subsystems. The condition of most buildings hovers precipitously between the end of the incipient phase and some point of accelerating deterioration.

Decelerating deterioration is that state that follows functional failure, and is relevant primarily for historic ruins. The rate of deterioration of a pile of rubble is relatively slow, and the condition is generally stable.

An important point in the life cycle of a building or component is the end of effective life. In other words, it is the point at which the functional aspect of the system fails. We have defined that point as the *half-life* of the system, because this is also the point of contraflexure in the decay curve. A more rigorous way of defining half-life would be that point at which half the original energy is dissipated; however, the system may have long since failed even with half the potential energy intact.

Most systems are rather predictable in terms of effective functional life. A 20-year asphalt shingle roof may perform for more than 20 years, but the incidence of leaks after 25 years is generally unacceptable, and the roof is consequently not worth repairing. Of all the variables required to construct a deterioration curve, one that is generally available is the *effec-*

tive life. The effective life of a given material should be deliberately set at the outset of service.

2.2.2   Necessary and Sufficient Conditions

An essential element of the diagnostic and intervention process is an understanding and application of the concept of necessary and sufficient conditions. Every mechanism has a specific set of requirements or conditions that must be met in order for the mechanism to exist. These are the necessary conditions. Closely related, but semantically distinguishable, are the sufficient conditions, which are those circumstances that, if present, *will* result in the mechanism. Any single necessary condition that is absent or defeated means the mechanism will cease; any single sufficient condition that is absent means the mechanism cannot initiate. When all necessary and sufficient conditions are present, the mechanism *will* activate and persist until a necessary condition is eliminated. This simple notion is powerful in the diagnostics of and, particularly, in the intervention of building pathologies.

Although it will mean getting a bit ahead ourselves, an example here might be instructive. Timber rot has five necessary and sufficient conditions: (1) timber relatively free of toxins, (2) fungi, (3) moisture, (4) heat, and (5) oxygen. Remove any one of the five and there is no timber rot. Bring all five into proximity and rot will ensue. It will ensue each and every time the necessary and sufficient conditions are met, regardless of who designed the building. There may be many factors that will influence the rate at which and the degree to which the fungi will destroy the wood cells. These factors are extremely important when we get to the design of the intervention package. It is, however, important not to confuse mitigating factors with the necessary and sufficient conditions. The presence of the necessary and sufficient conditions does not suggest that the fungal rot will progress rapidly or slowly, only that it will occur.

The clear and specific articulation of the necessary and sufficient conditions for a particular mechanism provides a structure both for accurate diagnosis and for efficacious intervention. Given a suspect mechanism, the citation of its necessary and sufficient conditions provides a reasonable checklist against which to test at least the feasibility of the candidate. The use of the necessary and sufficient conditions for this purpose has a latent defect of which the practitioner must be aware: the list may not be unique or exclusive; in other words, more than one mechanism may have several of the same necessary and sufficient conditions as the set for another mechanism. For example, as just established, the necessary and sufficient conditions for timber rot are precisely five conditions. These five conditions, when present, will result in timber rot at some rate or another. The presence of these five conditions, however, do not distinguish between brown rot and white rot.

This appears initially as a limitation to the application of necessary and sufficient conditions as a diagnostic concept. Necessary and sufficient conditions, are, however, conceptually valid for the intended purpose of establishing the prerequisites for the initiation and progression of a deterioration mechanism where and when the consequences of the mechanism are measurable and definable. The distinction in the differences between white versus brown rot are appropriately included in the elaboration of the condition that rot be present. The purpose of using necessary and sufficient conditions as a diagnostic tool is primarily to establish the conditions such that a deterioration mechanism will arise and progress, not to distinguish between two subtly distinguishable groups of fungi.

2.3 Complex Mechanisms and Diagnostics

It is by no means simple to determine a specific deterioration mechanism responsible for an observed manifestation of decay of a particular element or material. It is even more complicated when two or more deterioration mechanisms are operating at the same time. Nothing in the description of the deterioration mechanism model or the notion of necessary and sufficient conditions precludes two simultaneously occurring deterioration mechanisms. As long as the requisite conditions of each deterioration mechanism are met, then each deterioration mechanism will occur. This makes the unraveling of the pathology particularly difficult.

If, for example, the only manifestation of the operating deterioration mechanism is a crack or a network of cracks, there is a distinct prospect, even the probability, that the cracks were precipitated by more than one simultaneously operating mechanism. In fact, given the complexity of the general environment, it stands to reason that multiple deterioration mechanisms are always operating at the same time. It is a major complicating factor of diagnostics to recognize distinct and addressable mechanisms in the maze of simultaneously operating sets of mechanisms.

The complications caused by a simultaneous occurrence is, however, rather more a result of the manner of analysis. As we will discuss in the section on crack formation, the device we have developed for analysis is an abstraction. The forces we assign to the formation of the crack are abstractions. The only reality is the crack. Although we may assign multiple sources or origins to mathematically assignable force vectors acting on the specimen simultaneously, the response of the specimen is the simple sum of the forces, or, more accurately, one indistinguishable force. The reality of building pathology is simple: our ability to comprehend that reality is extremely limited. If we are to intervene in the deterioration mechanisms that are the objects of our interest, we need tools and devices that allow for the proper prognosis to ensure the desired effect. This, unfortunately loops us back to the fact that analyzing complex deterioration mechanisms is tedious and difficult.

2.3.1 Typical Mechanism Combinations

When confronted with a complex array of potentially interlocking mechanisms, it is helpful to think diagnostically in terms of commonly occurring sets of mechanisms. As we present the catalogue of deterioration mechanisms throughout this book, we will also identify the common concomitant mechanisms. This is not to suggest that such common sets are invariably coincident, as much as they are a suggestion for further analysis and speculation.

We have elected to arrange the presentation of types of mechanisms as structurally, thermally, hygroscopically, or chemically induced. That implies that such categorizations are functional. They are not; they are instructional. Functionally, any and all mechanisms that can occur simultaneously, will occur simultaneously regardless of the category we may assign it to in this book. We categorize solely to assist our learning and understanding of the variously identified deterioration mechanisms.

An example here, too, may be helpful. We discuss stress cracks in Chapter 3, freeze-thaw cycling in Chapter 4, and corrosion in Chapter 5. In a simply reinforced masonry wall, typically, all three phenomena will occur for the fundamental reason that the necessary and sufficient conditions for each are present. Arguably, a more accurate way to analyze the condition is to categorize problems by construction type rather than mechanism type.

Organization by material or construction type is not, in fact, an unusual way of assembling the subject matter in this type of book. The primary reason for not doing so in this case is that such assemblies typically suffer from two problems: one of scope, the other of pedagogic origins. The scope problem is, simply, that the potential list of materials and assemblies is endless. There are a number of ways of organizing materials such as metals, stones, woods, and so on. If pathologies were sufficiently general to apply to all metals or all woods—which, to a limited extent, they are—then we could organize the literature for materials along those lines. As long as we are content to deal with materials in general ways, there is much to be said for such an organization. Indeed, many books are organized along such lines. Naturally, the problem with such an approach is compounded when the inventory grows to include assemblies and whole building types. The specifics associated with buildings alone is so site-determinate that generalities quickly become platitudinous and cumbersome. The problem for the student and the general practitioner is finding a balance between comprehensive scope and definitive detail. The problem with surveys that address materials and assemblies in encyclopedic order is that they are voluminous and daunting.

The second problem, conceptual in nature, is more important. Addressing materials and building pathology issues in encyclopedic order tends to educate by prescription. In this way, an encyclopedic exposition implicitly says that there is an answer for each and every item. This leads to "cookbook" type solutions as well as to prescriptive thinking. The reality of building pathology is that there are few situations that are sufficiently similar to allow prescriptive solutions without the considerable risk of oversimplification. The descendants of oversimplification are misidentification and misapplication. Cookbooks are valuable when the object is to replicate the same recipe. They are not designed to solve, nor are they particularly effective at solving, even simple culinary problems. It is a fundamental objective of this book to provide the principles and methodologies with which to analyze and diagnose uncatalogued problems, and to challenge the generalities for those that are catalogued.

2.3.2 Variable Environmental Effects: A Dynamic Approach

Another complicating factor related to materials and building pathology is the variability of the environment. As stated, if a deterioration mechanism is a function of properties and the operating environment, it follows that when the environment varies, the deterioration mechanism will vary accordingly. This means that the rate of deterioration is subject to many variables, some of which are constantly changing. The only certainty involved in the process is that the net change in material and building integrity, absent intervention, is negative, and that the nature of the as-yet undeteriorated materials is constant.

A caution is in order here: The presentation of deterioration mechanisms suggests that the functions are not linear and univariable in their response. In fact, deterioration mechanisms are multivariable functions and nonlinear in response. Again jumping ahead, let us look at the mechanism identified as freeze-thaw cycling (see Chapter 3). The necessary and sufficient conditions for the origination and progression of freeze-thaw cycling are (1) porous, permeable materials; (2) surficial moisture saturation; (3) relatively low modulus of rupture; and (4) temperature declining through freezing point. The completion of the cyclic notion is that the material subsequently thaws, resaturates, and refreezes; but the minimum essential elements that account for the deterioration mechanism are as presented.

The chronology of the freeze-thaw mechanism is that water, generally in the form of rain, runs down the vertical surface of an exposed porous, permeable surface, such as brick masonry. The water is absorbed into the brick; it freezes and thereby results in the expansion of the water and, subsequently, the formation of ice. The expanding water and ice exert pressure inside the brick; if the pressure exceeds the modulus of rupture, a crack is initiated. The water subsequently melts, and the process occurs all over again, but this time the crack fills with water as well. Pressure inside the crack extends the crack, and the cycle begins again.

At this simplified level, the deterioration mechanism is relatively clear and uncomplicated: freezing water results in a cracked brick. There are, however, several interesting details about the process that are not as simple. We know the thermal resistance property of the brick; we know the temperatures of the extreme outside and inside surfaces, hence the temperature gradient of the brick (see Chapter 3). Based on that, we can compute the location of the line within the brick where the temperature is 0° C. This then is the depth to which saturated brick will freeze. But that is too simple. A brick that is saturated to some depth is not a thermally homogeneous material; it is thermally *two* materials, saturated brick and unsaturated brick. Each has a different thermal transmission resistance, and the location of the line where freezing occurs must be analyzed as a two-layer situation, not one. As the ice forms, however, there is a third layer of thermally distinct material—saturated and frozen brick. This means that, as the ice forms in the brick, the line of potentially freezing water is shifting. The computation of the depth of penetration of frost within a saturated brick is a multivariable, dynamic problem. As the phenomenon occurs, the variables change, which alters the phenomenon itself.

This particular problem will be discussed in detail in Chapter 3. It is relevant at this point only to emphasize that many deterioration mechanisms are, in fact, dynamic equilibrium problems, and precise simulation of the phenomenon is difficult, if not impossible, to calculate manually, other than by crude approximation.

2.3.3 Integrated Systems and Complex Pathogenesis

The problem of simultaneously occurring deterioration mechanisms is especially vexing at the building systems level. The components of a building are responding to and deteriorating because of accelerating materials-related problems, but they are simultaneously deteriorating because of systems either unrelated or only partially related to materials pathogenesis. Consider the building as a set of interrelated subsystems.

It is important to stress that the subsystems should not be looked upon as isolated components, for the deterioration and demise of one subsystem critically affects the function of the entire system. A building's division into subsystems is done simply to assist the practitioner in grouping portions of the structure for evaluative and assessment purposes. There are many approaches to doing this, among them various forms of directional method. Some use a checklist of 16 division specifications, while others implement a top-to-bottom or inside-to-outside assessment. The primary objective of all these approaches is that, if done thoroughly, all the pieces of the building will have been inspected and duly noted. For analytical purposes here, we have organized building-related issues into six primary subsystems.

As practitioners, the motives for developing a theoretical model is to be able to explain to others the connection between what we see and what we conclude based on those observations. The value of a systems approach has less to do with qualitatively improving the analytical process and more to do with the communication of that process to an audience. The process provides a framework for the practitioner as well. The point of the model is more to explicate what we already know and do, as opposed to reconstructing the process.

Typically, it is not necessary to articulate the analytic mechanism of an assessment, as long as the results are valid; however, there are two general conditions when the intuitive; or at least unspecified, techniques are at a disadvantage. One of those conditions occurs when a client or other lay audience member asks how we arrived at our conclusions. The other occurs when we attempt to teach the subject of building analysis and assessment to others. In the first case, a lay person often has a vested interest in contesting our conclusions, in questioning our methodology, whereas the "student" group has a legitimate need to re-create the experiment with some prospect of reliably achieving similar results. Unfortunately, most building pathologists acquired their skills through experience, without much theory, and consequently may be less than articulate about the connection between inspection and conclusions.

For assessment purposes, we can divide a building into two functional systems: passive and active. These can be further broken down into six subsystems: structural, horizontal, and vertical systems define the passive systems; utilities, environmental, and protection systems are categorized as active systems. Each of these six subsystems is functionally based, which means that each has the characteristic of being evaluated based on performance. Independent of age or appearance, the quality of the subsystem depends on the degree to which it satisfies its functional requirement.

The primary difference between the two systems is that the passive system contains few moving parts, whereas an active system generally embodies numerous moving parts. Replacement occurs much more regularly in an active system for two principle reasons: the elements wear out and fail, or a continual change of parts is necessary before they or the whole system wears out. Doing the latter is called *preventive maintenance*. The common denominator of these subsystems and their respective components is that deterioration is inherently progressive; unchecked, their subsequent failure will jeopardize the integrity of the building.

While we recognize that any of these systems can fail, and when they fail, they jeopardize the macrosystem, the bulk of this book addresses passive systems. Together, the structure and the closure systems comprise the envelope and stiffening elements. This is not to imply that the active systems cannot be every bit as problematic; indeed, they can threaten the building in ways that failures in structure and envelope rarely approach.

STRUCTURAL SYSTEMS

The function of the structural system is to minimize deformation. Ninety percent of all building failure occurs during the first 18 months of construction, primarily as a result of incomplete construction. If the building passes the 18-month test, most likely it will remain standing from 20 to 400 years, barring any intervention. As we will discuss in more detail in Chapter 3, the test of a structural system, once in service, is rarely whether the building actually collapses, rather whether it remains rigid enough to preclude damage to other systems and surfaces. The demise of a building can occur from the catastrophic failure of the structure or because the structure flexed to a degree that caused cracks in the skin, permitting water to penetrate the substrates of the walls. The consequential damage could be every bit as devastating as the fracture of a load-bearing member.

The principle components of the structural system include, but are not limited to, the foundations, footings, compacted earth below the foundations, foundation walls, excavated backfill, earth-retention structures, columns, beams, girders, joists, roof rafters, trusses, and wind bracing.

CLOSURE

The overall closure of a building is composed of two systems: the vertical closure system and the horizontal closure system. These two systems are the essential elements in providing protection to the occupant and the building contents.

Vertical Closure System

Essentially, the vertical closure system is designed to provide a barrier, but it may also perform a structural role. For example, a wall that is structural may also serve as a barrier to unwanted intrusions from weather, pollutants, noise, critters, and people. Primarily, the vertical system elements include the nonstructural components of the exterior wall. The integral parts are siding, furring, wall ties, weeps, coping, interior surface substrate, windows, doors, louvers, sills, and wall flashings.

The inherently difficult aspect of designing a vertical closure system is not only the number of items *for* which or *against* which a barrier must be presented; it is that the barrier must be variably selective. In other words, it would be ever so much simpler to design to exclude all outside air, but occasionally we want to let outside air inside the building. Likewise, we want to be able to selectively admit people and animals, as well as variable amounts of light and heat. We want to be able to look through the wall, while prohibiting bug entry. To accomplish all these competing objectives, a vertical closure system must be a remarkably complex system involving both static and moving parts.

Horizontal Closure System

The horizontal closure system provides for the collection and transportation of precipitation. There is an important barrier function, not dissimilar from the function of the vertical closure system, but the specific and peculiar function of the horizontal closure system has to do with water. A horizontal closure system may or may not be an ideal barrier for heat, but it must control and direct water or it is in a state of failure, by definition. The system operates from the time water enters the site, by whatever avenue, until that same water is *appropriately* discharged from the site.

This system of course includes the roofing material and substrate, which is the largest component, but it is important to not stop at the roof. Other elements include flashings, gutters, downspouts, boots, subgrade conductors, French drains, retention ponds, sump pits and pumps, and storm sewers.

FIGURE 2.2   Although located at the roof, according to the systems approach presented in this book, the skylight header and rafters are part of the structural system.

UTILITIES SYSTEMS

The power and hydraulic supply systems fall under the overall umbrella of utility systems. These types of systems have the potential to wear out or become outdated every 7 to 15 years. To the detriment of the building, the utilities systems are sometimes embedded and/or difficult to replace, thereby increasing the chances of demolition rather than overhaul. (Note:

This book does not address utility systems in a dedicated chapter; a summary of the implications of failures of these systems is given in Chapter 6.)

Power Supply

The power supply distributes energy and collects its by-products throughout the building. It includes fuels of all types that travel to the motor, furnace, or fixture, as well as the return or waste system. The actual motor, furnace, or fixture itself, however, is not included in this system. Primary and secondary electrical services, transformers and switch gears, breakers and fuses, wiring and junctions, conduit, grounds, lines and meter, oil tanks and pumps, and pressure regulator and controls are among the primary components related this system. External combustion sources, such as flues and chimney, fall within the power supply subsystem. It also includes any auxiliary and emergency generators, as well as firewood storage and protection.

Hydraulic Supply

The hydraulic system's primary function is to collect and distribute the water intentionally brought into the building for direct use and to collect its waste. Specifically, direct use includes domestic water use and fire protection. The domestic water service includes the water main, meter, hot water heater, the well, pump, storage tank, filtration tank, supply lines, drains, traps, control valves, bibs, vents and back vents, sanitary sewers, septic tanks, tile fields, grease traps, and disposals. (Note: plumbing for the hot water system belongs to the heating system; and fire protection will be discussed in greater detail in the subsequent emergency protection system section.)

ENVIRONMENTAL MODIFICATION SYSTEM

The principle role of the environmental modification system is to maintain the balance of heat, humidity, and oxygen. Systems are incorporated into the building to manipulate and maintain a copasetic environment, which includes the heating, air conditioning, and ventilation systems. More specifically, primary components include the ducts, steam risers and runs, controls and thermostats, dehumidifiers, humidifiers, filters, fans and air handlers, condensers, chiller, pumps, motors, gauges, and valves associated with those systems. Wood-burning stoves, fireplaces, and dampers, as well as thermal and moisture control items in walls, particularly vapor barriers and insulations, are also components of this system.

Heating and Cooling

The heating and cooling systems assist in controlling the temperature and humidity within a building. Both systems must meet the needs of the occupants but avoid damaging the building itself or the contents within.

Air conditioning uses a mechanical plant to clean and deliver air. The system may have both a negative and positive effect on the life of the building. An increase in condensation is possibly due to the development of negative or positive pressures. On the other hand, air conditioning may reduce the hazard of pollutants and dust ingress.

Radiation, convection, and, in small part, conduction are the primary vehicles for which heat is dispersed. (These principles of heat transfer will be discussed in Chapter 4.) Like air conditioning, the implementation of artificial heat throughout a building can also produce unwanted condensation in unheated spaces. How a room is heated or cooled is dependent upon the relationship between the heating and cooling services, which in turn is dependent upon the energy efficiency of the building's skin.

Humidity Control

Humidity can impart devastating effects upon building materials. Too much can facilitate mold growth, whereas too little can cause shrinkage, brittleness, and cracking. In the summer months, air conditioning can control the amount of humidity in the building. If reliant on electricity, as most systems are, the loss of that source can shut down the air conditioner and upset the controlled environment, which can be extremely damaging to the building or the contents within. In the winter months, a heating system could introduce and control the humidity. Continuous and severe changes within the internal environment can be particularly damaging to the building; therefore, the humidity control systems must run consistently.

Ventilation

Ventilation can be provided naturally and/or artificially. Artificial ventilation can be executed through an air conditioning system. Natural ventilation in a building depends on wind pressure, the stack effect, or a combination of both. The stack effect can be described as the rising and escaping of warm air through high-level vents, to be replaced by cool air distributed through lower-level vents. Natural ventilation is related to numerous variables such as the immediate weather, building exposure, building size, and general air tightness. Reliability on such precarious variables can be a disadvantage for maintaining an even and comfortable temperature for occupants, as well as for the building materials.

PROTECTION SYSTEMS

Protection systems include emergency and security systems, and they have the annoying characteristic of failing in arguably the most dramatic ways of all the building systems. Failures of protection systems are rarely partial; almost always they are total failures. Moreover, these failures often go unnoticed until a crisis. Although we will not expand on these systems as we will structure and closure, they are extremely important, and only vaguely understood by laypeople because the they are presumed to be a collection of sophisticated mechanical devices beyond their comprehension.

Security

The primary function of a security system is to protect life, property, and terrain, in that order, without exception or hesitation. A mechanical security system typically consists of a detector box, a control box, the wires, and an alarm signaler. As with many of the systems, it is common to identify the security system as consisting of the mechanical elements alone. Consequently, the inspection and evaluation of the whole system is based on the operation of these devices.

The security system is, in fact, far more than its alarms and detectors. Physical security is a state of mind and attitude as much as it is of hardware. Experience shows that, of all the systems in a building, the security system is the least understood, appreciated, or tested. Chapter 6 directs attention to the common-sense aspects of security.

Emergency

The largest hardware component of the emergency system is the fire emergency system. As with physical security, the evaluation and assessment of the entire emergency system often consists of the testing and evaluation of the hardware components of the fire emergency system. This again indicates failure to factor in the human element to emergency response, or to appreciate all the other types of emergencies that can occur to a building and its occupants.

The fire protection systems can be divided into two types: passive and active. Both, should be implemented simultaneously to maximize protection. Encasing and subdivision characterize the passive system, in that areas are divided into sections bounded by fireproof divisions. The active systems entail the warning alarms and automatic sprinklers. More particularly, the system includes the risers and branch lines, pumps, sensors, and alarms.

2.4 The Impact of Material Standards over Time

A popular material of the latter nineteenth century was what we call brownstone, a weakly cemented sandstone. (We will discuss the material in more detail later in this book.) As a consequence of its properties, in combination with the hostility of the environment—particularly the freeze-thaw action that is characteristic of the Mid-Atlantic region of the United States, which was a region of immense brownstone popularity—brownstone façades today are in high states of disrepair: flat surfaces are peeling away and carved details are eroding. A common response among conservators, preservationists, and architects is to question why designers and/or constructors of brownstone façades used such an inferior material.

Ironically, the same material is also the object of acclaim and admiration. It is also ironic that those most critical of brownstone are those who stand to gain from that choice, making their livelihoods repairing and replacing the "inferior" material.

When the owners commissioned and architects originally designed brownstone façades, they did not consider that a hundred years later those same façades might be criticized for not enduring better than they have. Do the architects and owner today? Occasionally, an institution may take the long view of a design decision, but it is rare for an owner to take a hundred-year view of construction; it is not even good business to do so. The decision to use brownstone in 1880 was reasonable. What changed more than the condition of the material was the standard to which the material was being held. In other words, we changed; and when humans change, we expect—unreasonably—the building to change with us. The building is the same as it was before it became historic, deteriorating in precisely the same manner as it was deteriorating before it became historic.

Not only do standards change over time, but so does the building itself. Obviously, there is the deterioration of fabric and systems. There are also deliberate alterations in details and in the gross. Furthermore, buildings have been maintained overtime, so, for example, the replacement pointing may be more problematic than the original pointing; the rotting timber may be white pine, which lasted 10 years, having replaced the heartwood red cedar, which lasted a hundred years. The point is that interventions may now be more problematic than the original fabric, and these interventions are often less well documented than the original design and details, and they often trigger compatibility issues. As we will demonstrate throughout this book, changes in material properties within the body of the fabric produce an inherently suspicious condition that is the source of several identifiable deterioration mechanisms.

Buildings are also altered in use and function. Original openings are closed; new openings are created. Stairs are moved with remarkable frequency. When changes of these types are made, the signs and outlines of the original items are carefully obscured. That is the whole point of making such changes. As pathologists, our concern in this regard is that virtually every opening in a building is a point of discontinuity, a hole. If an opening is filled and covered with an opaque finish, it is still a hole and, therefore, a point of discontinuity. The perforation of the structure is cumulative. Each opening, whether it is subsequently filled or not, adds to the sum of the perforations in the structure.

Assessors and pathologists must be particularly aware of vestigial holes that are not readily visible, meaning those modifications that have been deliberately disguised as the consequence of past modifications and alterations. Assessors and investigators should routinely expect and suspect that stairs have been moved, that windows have been covered up, that partitions have been relocated, that shafts have been installed and abandoned. Buildings originally constructed for industrial or agricultural use are particularly likely to have trap doors and floor penetrations.

A rule of thumb for old buildings is that what is there now was not always there, and, conversely, what is not there probably was at one time. Floors are not level, walls are not plumb; right angles are either greater than or less than 90°, and parallel lines eventually will converge.

2.4.1. The Myth of Modern Materials

Using the term "modern" to describe recently developed synthetic or modified materials is inappropriate; nevertheless, it has come into widespread use. The boundary between "modern" materials and "not quite" modern is predicated only indirectly on the date of introduction into the market, but generally speaking "modern" refers to materials that are approximately 40 or fewer years old.

Occasionally it is suggested that "modern" materials are inherently different from their predecessors, and are therefore special. To the contrary, new materials and new modifications of existing materials are not special; they are simply new, and undocumented. All materials perform in accordance with the same set of rules.

Newer materials also deteriorate. Whether we understand the deterioration mechanisms or not may well be debatable. We are still waiting, for example, for the results on EPDM single-ply roofing, not because we suspect that EPDM will last interminably, but because we do not have a basis for analyzing the modes of end-stage failure. A new material simply cannot enter the market with full information about how it will perform in the future. This acknowledgment creates one of the inherently difficult dilemmas of materials science: How can we assess the performance of a material that has not existed long enough to deteriorate in any environment however hostile or benign? Obviously, we cannot.

But what if we rephrase the question as follows: Are there principles of physics and chemistry that provide us with some guidance as to the performance of newly developed materials? The answer is a qualified yes. The qualification is that predictions based on currently available property analyses are *indicators* of long-term performance, not *proof* of long-term performance—it takes time for time to pass. Weathering machines and other methods of artificial aging of materials are just that, artificial, which usually means not nearly as complex as natural environments.

Despite the qualification, however, we are justified in questioning hydrocarbon-based material response to ultraviolet light, for example. Hydrocarbon-based materials often exhibit relatively high coefficients of thermal expansion, which triggers issues of shear gradients and stresses. Thus, we can ask questions about homogeneity and hygroscopic expansion. If the values and analytical values of these investigations correspond to the necessary and sufficient conditions of known deterioration mechanisms, then we can certainly speculate as to performance.

To be sure, promotions for "modern" materials suggest that they are different, that some new set of laws of physics are operative. It is more accurate to say that newer and, therefore, less familiar materials may posses properties such that the modes and forms of deterioration mechanisms are not familiar. The fact is, the principles of science that will ultimately have an impact on them are the same as those affecting materials since the beginning of time.

2.5 Intervention Methodologies

If for every flaw there was one and only one response, then performing diagnostics would require only a list consisting of two long parallel columns. The left column would itemize conditions, and the right column would detail the corresponding solutions; for example, if paint is cracked and/or peeling (left column), remove all paint to bare wood and reprime and paint (right column). But what if the paint is cracked only a little? What if the removal process would damage the substrate? What if the reason the paint cracked and peeled was because the wood was and still is wet? What if . . . ? Reality, as we all know, is too complex, and the parameters too highly variable, to allow for solutions in the form of lists of causes and responses. Even establishing the conditions is subject to interpretation and variation, much less the responses.

Despite the obviousness of this observation, much of the literature surrounding building pathology and diagnostics and intervention, whether it addresses "modern" materials or not, is of the cookbook variety: find the right page and follow the recipe. There are, for example, reputable texts that state plainly and simply to use a 1:1:6 lime-to-cement-to-sand mortar for repointing, the implication being that any other mix is wrong. In fact, a 1:1:6 mix is a reasonably safe mix most of the time, but notice the qualifiers "reasonably safe" and "most of the time." A better approach is one that accepts complexity and allows for variation.

2.5.1   The Client as Commander

Before adopting a general intervention strategy, much less a specific technique, it is essential to gain the concurrence of the owner. As obvious as this may seem, too many practicing professionals leap to prescriptive conclusions, then implement them without the advice and consent of the client. We move quickly from symptoms to prescription, sometimes without considering less obvious possibilities. And because we are so confident in our diagnosis, we assume that any right-minded client will agree with us.

It is essential that the client remain the commander of the team, and that the professional not circumvent the authority of the commander. The professional's job is to help the client make well-ordered and informed decisions, by providing complete and accurate information. After arraying the options and their respective attributes, the professional then recommends a course of action. It is at this point that the merit of the practitioner as the professional agent of the client is most likely to be tested, because it is precisely at this point that a client may disagree with the practitioner and insist on an approach not recommended by the professional.

There are two basic responses available to a professional with a "non-compliant" client, both of which have advocates. He or she may withdraw, the rationale being that the client is determined to engage in a behavior that is contrary to the advice of the professional. The client is free to exercise his or her ownership prerogatives, but the professional is not obligated to endorse and advance those decisions. A professional may adopt certain tenants of practice that, if violated by a client, render further service untenable. The other general approach is for the professional to carry out the decisions of the owner in the most prudent and providential manner possible. The professional is bound within legal limits to advise and counsel the client as to options and the consequences of each of those options. Once an informed decision is made, it is the obligation of the professional to assist the client in the execution of that decision in the most judicious and responsible manner practicable.

> The most articulate version of professional withdrawal or termination is explained in the Model Rules of Professional Conduct §1.16, propounded by the American Bar Association, "Declining or Terminating Representation." It specifically addresses the conditions and manner of unilateral termination by an attorney. The American Institute of Architects (AIA) has no such misgivings regarding the subject, and the Model Rules promulgated by the National Council of Architectural Registration Boards (NCARB) do not address the subject.

The first approach is the position of the practitioner as the principled idealist; the second, that of the professional as the ethical pragmatist. Regardless of the position or positions held by the practitioner, the point is, professionals should clearly state their preferences and prejudices to clients. Too often practicing professionals do not disclose their preferences, and instead manipulate their clients into agreement and compliance, sometimes to avoid disagreement, and other times to exert personal control over the process on the grounds that as experienced practitioners we know what is good for our clients whether they agree or not.

Regardless of how we rationalize such behavior, it has the effect of circumventing the prerogatives of the client. Such professional manipulation becomes manifest at the end of the project when the client complains that we have not produced what he or she wanted, and demands an explanation. The only defense at this point is that we used our best judgment of what we *presumed* to be in his or her best interest.

The simplest way to avoid such situations is to establish from the beginning of any project the objectives for the project, forgoing "insider" lingo and jargon. If it becomes apparent that you and the client have irreconcilable differences of what is reasonable and proper, you can quietly withdraw or persevere with open reservations as you see fit.

In the event that this important conversation does not occur at the beginning of the project, there is another opportunity to recoup the situation, typically, following the assessment and analysis phases. As a matter of self-discipline, and to preserve the authority of the client, it is wise to disclose and discuss the following six approaches to intervention. Together, they comprise a checklist of tactical approaches to intervention; taken separately, each is emblematic of various schools and attitudes toward old buildings generally. By discussing these approaches with your clients, you simultaneously open your mind to alternatives while engaging the client at the policy level.

2.5.2 Approaches

In the field, clients or other practicing professionals may expect specific answers to problems; even in the face of a disclaimer, occasionally someone will demand a one-to-one correlation of mechanism to solution. If, for example, the problem is rising damp, we are expected to insert an interdiction layer. If the problem is differential settlement, we are expected to underpin.

This approach to building pathology is common; it is part and parcel of the "cookbook" method of diagnostics and intervention, mentioned earlier. The theory behind such an approach is that every problem is uniquely diagnosable and uniquely rectifiable; for example, all walls wet above-grade

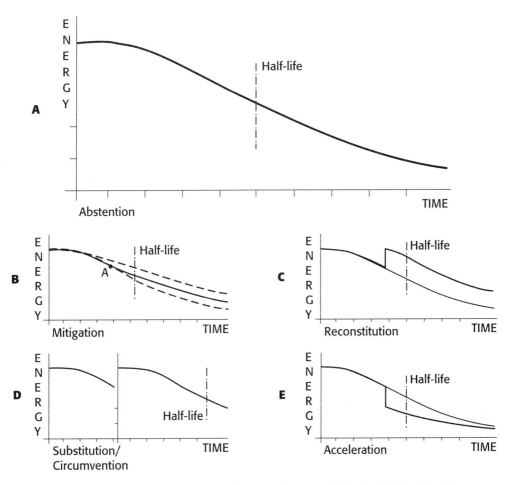

FIGURE 2.3 The composite deterioration curve (A) is alterable in distinctive ways depending on the intervention approach (B–E).

suffer from rising damp, and the "cookbook" says this is preemptively remedied by inserting an impermeable layer between-grade and the upper reaches of the wall. This is weak thinking and weaker practice. In reality, though there may be one-to-one correlations of symptoms to prescription, they are rare and risky.

There is no singular guaranteed technique for ensuring diagnostic accuracy, but the use of the six general approaches given in the following subsections at least assures that the diagnostician explored alternative interventions, if not alternative mechanisms. But a word of caution is in order, before introducing these six approaches: They are general, meaning that within each category there may be, probably are, more than one specific application. Some of the specific interventions will prove to be unfeasible, but the value and purpose of developing the complete set is not only to ensure the inclusion of the atypical intervention, but to rationally evaluate the obvious interventions. The six approaches are: abstention, mitigation, reconstruction, substitution, circumvention, and acceleration.

ABSTENTION

Abstention simply means that there is no intervention in the existing mechanism. Abstention is, simultaneously, the most obvious and least respected of the six options. It is the most apparent option because it is the one already operative; it is in place and it is working. It is the least respected for two reasons: The first, is the reasonable presumption on the part of the practitioner that the abstention option was considered and already discarded by the client. If the do-nothing option were preferred, the client would not have engaged a professional. The second, and not so evident, reason dovetails with the first. The abstention option is the practitioner's blind spot. After all, the professional is generally remunerated and rewarded for doing *something*. In short, leaving ethical compromise aside, professionals have a vested interest in positive intervention.

The client's initial observation of a suspected deterioration mechanism is done so in a comparative void. The client calls a professional, not because he or she has evaluated the do-nothing option and rejected it, but because he or she is concerned about the building condition generally. If the professional fails to include the abstention option with the other available options, it is, at a minimum, self-serving and, at a maximum, fraudulent. The National Environmental Policy Act of 1966 requires that, as a basis of comparing and evaluating qualifying alterations to the environment, the evaluator must include the option of doing nothing. In the evaluation of alternative methods of intervention, we are likewise obligated to consider abstention.

A counterargument is that alterations to the environment are assertive acts of potential net destruction, and that the status quo is inherently acceptable. That is, deterioration mechanisms are inherently and irreversibly destructive, hence all interventions are a net positive result and preferable to abstention. But this argument first discounts the possibility that the deterioration mechanism in place, while inherently destructive, is more desirable than any alternative, and second, discounts the possibility that an intervention may, in fact, accelerate deterioration. Not all interventions are preferable to the existing condition.

Moreover, doing nothing is initially less costly than doing something. But the fact that abstention has the lowest immediate cash outlay is not necessarily a convincing argument. It is the role of the professional to explain the advantages and disadvantages of *all* the alternatives, including short- and long-term costs, to the client.

MITIGATION

Mitigative interventions are defined as acts that alter the environment that supports the deterioration mechanism. Thus, the intervention is necessarily and exclusively external to the fabric. Mitigative interventions do not address the fabric directly; the fabric remains undisturbed as a matter of definition. This means, among other things, that mitigative interventions cannot or rarely can completely arrest the deterioration mechanism. Mitigation of the external conditions constituting the deterioration environment will result in a change in the rate of deterioration, not the cessation of the deterioration mechanism.

A change in the rate of deterioration can be reflected in a change in the slope of the basic entropic curve initiating at the point of intervention. If the selected mode of intervention results in a net reduction in the rate of intervention, the curve will be shallower than the original curve—which is, by definition, the same as the abstention curve. The objective of a mitigative intervention is to effect a net positive change in the entropic curve. Unfortunately, this is not always the case. A net positive change in the entropic curve means that deterioration may continue; however, at any future point, the absolute level of deterioration given the intervention will be less than had the intervention not occurred. The significance of this statement is that deterioration will continue, but that the rate and resulting damage will be reduced.

RECONSTITUTION

Reconstitution is the exact inverse of mitigation. Whereas mitigation acts on the environment at the exclusion of the fabric, reconstitution acts exclusively on the fabric; for example, a deteriorated wood sill will be replaced in kind, size, and location. The degree of precision of the match of the replacement to the original is a matter of economics and advocacy. This book does not advocate one level of precision over another; it states only that as an approach reconstitution is a valid intervention approach.

The advantage of reconstitution is often more than the satisfaction of complying with standards of authenticity; it is often an economical solution as well. If the rate of deterioration of the original was a hundred years, and the cost of replacement is nominal, recycling the process may be the optimal solution. Reconstitution is often the approach of choice for wood members, because the costs of material and fabrication are relatively low, while the feasibility of demolition and reinstallation are relatively high.

Even though with reconstitution the deterioration mechanism is unaltered and, as such, the deterioration clock is, essentially, reset to zero, the benefit of simply restarting the mechanism maybe, and often is, an acceptable decision. This is the approach implicit in repainting, repointing, and many other simple maintenance replacements. More than any other intervention approach, reconstitution demonstrates the basic concept of intervention generally. Because, as stated repeatedly, all materials deteriorate, all that we do as preservers, conservators, curators, and stewards is to manipulate the rates, consequences, and alterations of the deteriorated material. Reconstitution is the essence of what we do in our respective roles relative to building pathology; we can do no more.

That said, note that the disadvantage of reconstitution is the same as its advantage: the basic deterioration mechanism is still in place and operating at the same rate as prior to the intervention. Nothing suggests that the rate of deterioration in the future will be any less than the rate experienced to date. Within a comparable period of time, the replacement element will and should be in the same state of deterioration as the item that was replaced. However, the rate of deterioration is known, not speculated. It is no small comfort to know what the future holds, even if it is less than ideal. There will be no surprises either as to the rate of untoward developments that sometimes occur when we substitute or treat materials. Predictability offers the opportunity to program and budget for intervention.

As to the entropic curve, reconstitution is a vertical step upward at the point of intervention: the curve is reset at or near its original value. Once in service, the material will proceed to deteriorate along an identical path as the original material.

SUBSTITUTION

Substitution is the direct replacement of a material with another material for the purpose of enhancing its performance; in other words, it increases the efficacy of the overall deterioration curve. It is the common presumption of material substitution that the new material will outlive the replaced material. But this sometimes ignores the fact that all materials deteriorate.

A disadvantage of substitution is that the rate and deterioration mechanism of the replacement material is something of a gamble. Despite best efforts to predict performance, the peculiarities of the substitution condition are idiosyncratically specific, meaning that the substitute material brings with it a level of uncertainty as to performance.

It is the purpose of this book to reduce the uncertainty of future performance, and it is the purpose of analysis and design of remediation measures to minimize the negative consequences of materials substitutions.

CIRCUMVENTION

Circumvention is substitution not only of the material, but of the manner in which the material functioned. Whereas substitution is the replacement of the deteriorated material with another material matching the original in placement and configuration, circumvention disregards the original material and its performance altogether. The easiest way to explain circumvention is by way of example. Imagine the common situation of ground-level floor joists pocketed into an exterior foundation wall. The wall is sufficiently moist as to wet the end-grain of the joists, which have rotted. Reconstitution would replace the timber joist with an identical timber joist; substitution would replace the timber joist with a cold-rolled metal joist, or inject the existing joist with reinforced epoxy. Circumvention would install a line of screw jacks and a collector beam under the ends of the rotted joists and shore them from the basement floor.

Thus, circumvention means going outside of the original envelope altogether. It is readily seen in structural applications where a parallel or redundant system or element has been installed that effectively obviates or relieves the original material. The original material may remain in place or not; nevertheless, it still has been circumvented. In the place of the deteriorated material is a new system, operating in an altogether new manner. The materials involved are different; the forces on the members are necessarily different; and the deterioration mechanism is likewise different.

With circumvention, no residual information from the original system remains to assist in the prediction of the performance of the new material or systems, though the original material is still in place and the original deterioration mechanism may still be operative relative to that material. In the example of the rotting joist ends, the installation of a line of shoring will not arrest the rot mechanism; in fact, it is possible that the rot will progress to a point that the shoring no longer collects load from sound joist material and eventually will have to be reset further inboard or abandoned.

There is no broad generality regarding cost or predictability with circumvention. It may be permanent or temporary, cheap or expensive, predictable or risky. Probably it is more often applicable to structural members or systems, but even that is less than certain. If the deterioration of mortar has progressed to a degree and in a manner that prevails against repointing, then covering the wall with an applied cladding is arguably a form of circumvention.

ACCELERATION

However distressing, there are times and circumstances that demand partial or total demolition or dismantlement. Professionals resist acceleration not because of a vested interest, but because, as engineers, architects, preservationists, or conservationists, our instincts and motives are constructive. Destruction is counter to what we do. But we must recognize and acknowledge situations that are beyond reasonable redemption, when our skills and talents must serve to economically and efficiently reduce the clear and present danger.

Acceleration is doing in a controlled manner what will happen in an uncontrolled and potentially catastrophic, dangerous manner. Because of the irreversible nature of demolition, acceleration is not nor should be an easily determined alternative, but it is an option.

2.6 The Intervention Matrix

The six intervention approaches and the notion of necessary and sufficient conditions can be combined to form a matrix, with the approaches on the horizontal axis and the necessary and sufficient conditions along the vertical axis (see Table 2.1). At the top of the column of necessary and sufficient conditions is a box titled "General Mechanism," which is the sum of the necessary and sufficient conditions. Each intersection represents a specific intervention tactic that satisfies the criteria established by the combination of the general or specific condition and the intervention approach.

The intervention matrix is a tool that addresses several perplexities of preservation and conservation. One of the most powerful messages of the matrix is that deterioration mechanisms are divisible, meaning that it is not necessary to attack a deterioration mechanism on a broad basis or from all angles. If even one necessary and sufficient condition is affected, the entire mechanism is affected. The intervention matrix graphically reminds us of this powerful condition of logic.

The intervention matrix also helps us to avoid formulaic responses to commonly encountered deterioration mechanisms. In the case of solving

Table 2.1   INTERVENTION MATRIX

Mechanism Title: Timber Rot

Necessary and Sufficient Conditions		Intervention Approaches					
		Abstention	Mitigation	Reconstitution	Substitution	Circumvention	Acceleration
	General Mechanism	Accept condition of rotted member and rate of continuing rot.					Demolish rotted member or members, and do not replace.
1	Toxin-free timber		Inject toxin into or on surface; apply toxin-bearing material onto timber.	Remove and replace timber in kind.	Remove and/or replace with another material; inject with epoxy, or replace with steel.	Install parallel member for structural purposes, and relieve rotted member of load.	
2	Select fungi		Sterilize situs locus.				
3	Moisture content, approximately 28–30%		Reduce moisture available to the wood.	Dry timber to lower moisture content.			
4	Oxygen		Reduce atmospheric oxygen content; store in oxygen-depleted vault.			Encapsulate member or element and evacuate volume.	
5	Temperature not less than 40°F nor more than 120°F		Place member or element either in artificially high or low temperature.				

problems associated with building pathology, we tend to simplify the process to stylized solutions, gradually resorting to a few reflexive pre-scriptions as soon as we identify the deterioration mechanism. Sometimes we solve the problem before we even examine "the patient." The interven-tion matrix reminds us that the real world is complex, and that there is al-ways more than one way to intervene in a deterioration process.

The matrix presented in Table 2.1 follows on the example from earlier in this chapter of the necessary and sufficient conditions associated with timber rot. The six approaches, the general mechanism, and five necessary and sufficient conditions are arrayed as described above. At the various intersections of conditions and approaches, there may be an entry that is a specific intervention tactic. Certain of the intersections are empty, which suggests that there is no specific tactic that applies to the respective neces-sary and sufficient condition and, simultaneously, to the corresponding approach. This may be the case, or it may be that no specific tactic comes to mind. One advantage of this tool is that it posits the implicit question of a blank intersection and the challenge to fill it.

Even at intersections in which there is only one entry, there is a distinct prospect that there are other tactics that also would satisfy the conditions. The intersection of substitution and the basic fabric (toxin-free timber), for example, includes two possible substitutes: steel or injected epoxy. With some time and imagination, it is highly probable that the list of potential substitutions would grow even longer. It is important to remember that, at this point in the diagnostic process, it is counterproductive to preclude any possibility, however impractical it may seem.

Having assembled the basic intervention matrix, we begin the process of evaluating the various options. There are many ways of going about an evaluation, yet they all share the common objective of finding the solution that offers the greatest advantage with the least disadvantage. Each of the ad-vantages and disadvantages has several components that warrant analysis.

To some degree or other, options have an element of feasibility, mean-ing offering the potential to practically and reasonably implement the op-tion. To clarify: An option may, indeed, satisfy the corresponding criteria for reducing or eliminating the deterioration mechanism, but that does not mean the implementation of the option is feasible. There is always the issue of costs versus benefits. There may be a question of how closely the option conforms to some external standard of performance, such as the Secretary of Interior Guidelines. There is a highly important criterion of whether a particular option conforms or digresses from the client's objectives, and if so, to what degree. These considerations, while certainly common and im-portant, are by no means exhaustive. The process of evaluation warrants a sizable book of its own, but we offer a few suggestions and cautions below.

We begin with the caution that any evaluation technique is corruptible by the prejudices of the evaluator or the designer of the evaluation process. As a rule of bureaucratic control, manipulation of the evaluation technique is right up there with control of the communication channel. We, as evaluators and evaluation designers, need to recognize our prejudices and preferences, and to be aware of how easily we can influence decision making. If we do not, we allow our preferences to contaminate the process. In the example presented below, we will first look at a rather common evaluation technique, and then at how such a tool can be skewed.

The technique demonstrates an approach to evaluation that assists in initially the sorting of options. The first step is to articulate the criteria against which the options will be evaluated. In this case, the evaluation is the combination of the implementation feasibility and accessibility of the basic necessary and sufficient conditions (vertical axis) versus the relative conformance of the general approaches to established historic preservation criteria (horizontal). Each label is awarded a numeric value between 0 and 10 to represent the weight or degree of conformance of the condition or approach with its respective ideal. For example, among the general approaches, all other things being equal, abstention is the most highly preferred option based on a preservation criterion of retention of original fabric and minimum intervention.

Having established the combination of external criteria, and having weighted the conditions and approaches accordingly, the next step is to engage the algorithm, which, in this case, simply means to multiply one numeric value by the other. The highest value of these products is an indicator of the option that is the optimal solution based on the combination of these two sets of criteria. In this case, the option that calls for a reduction in the moisture content of the wood comes as no surprise to anyone familiar with the practicality and cost of the intervention options of timber rot intervention (see Table 2.2).

The result of this simple evaluation is neither particularly sophisticated nor contrary to our presumptions prior to the exercise. There is often remarkable resistance to invoking such a simple aid. One reason reflects the manners and modes available to distort, if not to corrupt, the results. We often do not trust numeric evaluation techniques precisely because we instinctively, if not specifically, know how easy it is to twist the results. In this simple example, there are two general ways to do so.

One way, obviously, is to adjust the weighting. The assignment of weighted values is an inherently subjective exercise, thus we can justify virtually any defensible set of values. Although the values assigned in this example are far from unreasonable, they are by no means inviolate. In order to accommodate debatable variations in values, an added level of

Table 2.2 SORTING INTERVENTION OPTIONS

Mechanism Title: Timber Rot

Necessary and Sufficient Conditions		Intervention Approaches					
		Abstention 10	Mitigation 8	Reconstitution 6	Substitution 4	Circumvention 6	Acceleration 2
	General Mechanism 6	Accept condition of rotted member and rate of continuing rot. **60**					Demolish rotted member or members, and do not replace. **10**
1	Toxin-free timber 7		Inject toxin into or on surface; apply toxin-bearing material onto timber. **56**	Remove and replace timber in kind. **42**	Remove and/or replace with another material, e.g., epoxy injection. **28**	Install parallel member for structural purposes, and relieve rotted member of load. **42**	

Table 2.2 SORTING INTERVENTION OPTIONS (continued)

Mechanism Title: Timber Rot

Necessary and Sufficient Conditions		Intervention Approaches					
		Abstention 10	**Mitigation** 8	**Reconstitution** 6	**Substitution** 4	**Circumvention** 6	**Acceleration** 2
2	Select fungi 1		Sterilize situs locus. 6				
3	Moisture content approximately 28–30% 9		Reduce moisture available to the wood. 72	Dry timber to lower moisture content. 54			
4	Oxygen 3		Reduce atmospheric oxygen content; store in oxygen-depleted vault. 24			Encapsulate member or element and evacuate volume. 24	
5	Temperature not less than 40°F nor more than 120°F 3		Place member or element either in artificially high or low temperature. 24				

49

Table 2.3   COMPARING A RANGE OF VALUES TO A SINGLE VALUE

Mechanism Title: Timber Rot

Necessary and Sufficient Conditions	Intervention Approaches					
	Abstention 10–8	Mitigation 8–4	Reconstitution 6–5	Substitution 4–3	Circumvention 6–2	Acceleration 2–0
General Mechanism 9–1	90–8					18–0
1 Toxin-free timber 7–6		56–24	42–30	28–18	42–12	
2 Select fungi 1–0		8–0				
3 Moisture content, approximately 28–30% 9–7		72–28	54–30			
4 Oxygen 3–1		24–4			24–2	
5 Temperature not less than 40°F nor more than 120°F 3–2		24–8				

sophistication is to run the same analysis with a *range of values* rather than a *single value*. In Table 2.3, the intersections for itemized options are simply shaded, and the resulting ranges of values are entered at the intersections. Not surprisingly, the results are *not* radically different. Indeed, if the original values were fairly and reasonably established, and they remain so, the results should at least be comparable. And they are. There are, however, some instructive lessons to be learned from this step.

The two shaded intersections tell an instructive story. In the prior analysis, mitigation of moisture content outscored "do nothing" by 72 to 60. In the second analysis, abstention had a higher absolute value than moisture mitigation, 90–72; but at the lower ends of the respective ranges, moisture mitigation was higher than abstention, 42–8. What, if anything, does this tell us? Under ideal circumstances (low associated uncertainty), abstention may be both more desirable and more feasible than moisture mitigation. The range of possible outcomes, however, is such that the outcome of abstention might be decidedly less efficacious than the worst-case scenario for moisture mitigation. While abstention may be more promising, it is simultaneously more risky.

The numeric values of the numbers, therefore, reflect efficacy, and the spread, or range, of the values reflect the associated risk. In this case, the potentially more efficacious option is also the more risky option. Such a description of viable options is certainly not uncommon. If to the client, or the property, the risk associated with an abstention option is tolerable, then that is the reasonable choice. If the loss of the property or element is less acceptable, then risk reduction prevails, and moisture mitigation is the reasonable choice.

This is sometimes referred to as the Theory of Second Best. The general situation is an interaction between two parties, each with two options that are the same for each of the two parties. Option A is high in efficacy but concomitantly high in vulnerability or risk; option B is decidedly inferior in terms of efficacy, but it is safe. Both parties recognize that the total sum of satisfaction will be maximized if both parties choose option A simultaneously; but each is vulnerable to the chance that his or her counterpart will choose option B instead to exploit the other party's vulnerability. As a consequence, both parties choose option B even though it is neither's first choice, but because neither can risk being exploited. Both parties wind up with their second choice, hence the name of the theory.

The Theory of Second Best and its innumerable variants is routinely exhibited as a demonstration of a frailty in human nature: "If we were not so suspicious, we could achieve more; we could be happier; the world would be a safer place; there would be peace in the world. . . ." There is a distinct absence of experimental proof for the Theory of Second Best. Is it really second best? The critique of the model is invariably based on a skepticism about human nature and trust. The presumption is that, but for our cynicism, we would get our first choice. But this may not be the only interpretation. Suppose human nature is not that we value happiness or peace over unhappiness or conflict, but that we value survival over all else. As long as we survive, there will be opportunities for efficacious interactions at other times. If survival is always the first choice, then second best is the preferred option not by default but by design.

This simple example illustrates one of the most important points of the intervention matrix: the preferred option may vary depending on externalities, namely the values and preferences of the decision maker. Even for the options that initially appear far less desirable than either of the more highly rated tactics, there may be a set of circumstances for which that option is appropriate and desirable. When a practitioner emphatically prescribes one way of intervening in a situation as "the way," he or she is short-circuiting the range of options to that which is her preference, regardless of the circumstances, thereby executing the classic device of controlling the choices to unity.

Let us now add another dimension to the matrix—cost. The preceding analysis weighed feasibility, meaning technical feasibility, and preservation efficacy. It said nothing about cost. If we add cost directly to each intersection, the content of Table 2.4 may be representative of the results.

Now the picture is more complicated, but still quite manageable. By adding cost to the abstention option, we learn that to do nothing may cost nothing, but the risk of doing nothing may cost a great deal if the building falls down. The costs of abstention are directly related to risk, because the cost is incurred not in the actual abstention, but in the failure of the approach and the potential loss of the building altogether.

The costs of moisture mitigation are inverse to risk, as are most options other than abstention. Certainly they are not free, but even a lower-risk solution costs significantly less than the loss of the property. Higher risk in this case is defined more in terms of failure of the process to arrest the problem than in the loss of the property.

We also learn that when we add cost as a variable, other options appear to be worth consideration. The message in these cases is that arguably less efficacious, even more risky, options may be sufficiently inexpensive as to warrant either accepting the risks and repeating the process later if necessary, or accepting the risks and opting for another tactic if the first does not work. Fabric mitigation and fabric reconstitution both are so inexpensive as to warrant consideration on these bases. In fact, based on the information contained in this matrix, assuming it is accurate, the cost of one reasonably effective installation of whatever will effect moisture mitigation could pay for 50 to 200 applications of whatever will accomplish fabric mitigation.

Table 2.4 ADDING COST TO THE EQUATION

Mechanism Title: Timber Rot

Necessary and Sufficient Conditions	Intervention Approaches					
	Abstention 10–8	Mitigation 8–4	Reconstitution 6–5	Substitution 4–3	Circumvention 6–2	Acceleration 2–0
General Mechanism 9–1	$0–1 million 90–8					18–0
1 Toxin-free timber 7–6		$400–100 56–24	$1,000–600 42–30	$4,000–2,000 28–18	$8,000–4,000 42–12	
2 Select fungi 1–0		6–0				
3 Moisture content, approximately 28–30% 9–7		$20,000–8,000 72–42	$4,000–1,000 54–30			
4 Oxygen 3–1		24–6		24–2		
5 Temperature not less than 40°F nor more than 120°F 3–2		24–12				

Table 2.5 ADDING DURABILITY AS A FACTOR

Mechanism Title: Timber Rot

Necessary and Sufficient Conditions	Intervention Approaches					
	Abstention 10–8	Mitigation 8–4	Reconstitution 6–5	Substitution 4–3	Circumvention 6–2	Acceleration 2–0
General Mechanism 9–1	N/A $0–1 milllion 90–8					18–0
1 Toxin-free timber 7–6		0.5–0.1 yrs $400–100 56–24 $800–1,000/yr	10–4 yrs $1,000–600 42–30 $100–150/yr	20–10 yrs $4,000–2,000 28–18 $200/yr	20–8 yrs $8,000–4,000 42–12 $400–500/yr	
2 Select fungi 1–0		6–0				
3 Moisture content, approximately 28–30% 9–7		20–10 yrs $20,000–8,000 72–42 $1,000–800/yr	0.25–0.1 yrs $4,000–1,000 54–30 $16,000–10,000/yr			
4 Oxygen 3–1		24–6			24–2	
5 Temperature not less than 40°F nor more than 120°F 3–2		24–12				

We can continue to increase the number of variables that affect the decision matrix. Each time we do so, the model comes closer to matching the complexity of the reality and, of course, the difficulty of sorting and evaluating the options. Keep in mind, the tool is intended to assist in the analysis not obfuscate it. Even with that said, let us add one more variable, namely durability. What we want to know is how long a given option will remain effective, or the time when the deterioration mechanism cycles to the same point on the entropic deterioration curve as at the point of initial intervention.

The durability variable is inserted into the matrix preceding the cost range as a corresponding range measured in years until the intervention cycles back to a comparable state of disrepair (see Table 2.5). At the bottom of the box in bold type is a range of values that represents the cost range divided by the durability range. These values represent a range of values of costs per year of durable performance.

What a difference a few numbers can make. What looked like interesting and tempting options—that is, moisture mitigation and moisture reconstitution—based on efficacy and cost, are far less attractive when viewed on an investment basis. Even if we were to increase the level of sophistication another notch by incorporating the time value of the money, it is unlikely that these figures and their respective differences would change enough to close such dramatic gaps. Even in analyses where the differences are not as dramatic, such an evaluation may be quite revealing, and affect the decision at issue.

As we have progressed though this analysis, we have seen the options wax and wane as we added each level of complexity. As the model increases in complexity, it approximates the complexity of the basic issue and becomes increasingly cumbersome. The purpose of the model is, indeed, to systematize and to simplify our understanding of reality. At its simplest level, the distortion brought on by the model was so great as to skew the results. As the model increases in complexity, the reliability of the proffered option increases, but so does the risk of corruption of the model. Every level requires the application of a certain degree of judgment and discretion to distribute the weighted values and to assign the categorical values. As discussed earlier, these assignments of value can be manipulated, and are, therefore, inherently vulnerable.

Although we mentioned it earlier, we did not discuss another aspect of potential vulnerability and manipulation. Recall at the very beginning of the exercise, the algorithm for the calculations of values within the matrix. In the first instance, the value relating to objective performance was multiplied by the value of technical feasibility. Subsequently, the high end of one range was multiplied by the high end of the other. Subsequent to that, the cost was divided by the durability expressed in years. Although there is a sort of intuitive logic associated with each of these mathematic operations, there is no firm correlation between the operation and tested outcomes. In short, the algorithm is untested and unconfirmed. Though this may be sufficient for the illustrative purposes of this chapter, it may not be for decision-making purposes when someone's money and property are at stake.

3 Structural Systems

3.1 Functional Definition

The function of the structural system of a building is to minimize distortion. To anyone who has studied structures, this definition will appear far too weak. We all learned that failure of the structural system results in failure, meaning collapse. From this not inaccurate observation, we inverted the observation and concluded that the function of the structural system is to prevent collapse. This, too, is accurate, but still not enough. If we determine that the purpose of the structural system is to prevent catastrophic collapse, then our expectations are far too low.

One of the consequences of defining the structure in terms of collapse is that we measure success and, therefore, design the systems from the outset, in terms of strength. If success is avoidance of fracture, the principle design criterion is preclusion of any critical member achieving ultimate strength. This is by no means enough. We cannot simply accept an unfractured structure if in operation the system is distorted to the point at which façades crack, floors sag, walls bulge, or foundations slide.

The design of structures is less an issue of strength than it is of stiffness and, therefore, rigidity. Likewise, the requirement for rigidity is less a matter of aesthetic preference than of compatibility of members, materials, joinery, and components.

3.2 Components Description

The names of the structural components of the structural system of a building are among the more familiar terms in architecture: column, beam, truss. The design of these elements is the subject of structures courses in both architecture and engineering curricula; and certainly, the accurate and adequate design of structural elements is a necessary and proper condition of building design.

Of concern to us as building pathologists, however, is not so much the adequacy of the original design as the gradual devolution of the structure to a point at which unacceptable distortions, much less outright failures, occur. This chapter studies the gradual disintegration of what were once adequate and proper structural elements and systems. After all, inadequately designed structures are unlikely to survive long enough to be the subject of building pathology.

3.3 Materials Pathology

If building pathology is the study of the failure of structures after prolonged service, then materials pathology is similarly the study of the failure of specific materials after prolonged service. The origins of building pathology typically are found at the materials level, thus materials pathology. At some level or other, the basic mechanisms of mechanically related materials pathology eventually devolve to form cracks. The sources of the stress that precipitate such fractures are varied, but the class of materials deterioration mechanisms that qualify as mechanical in character are exercises in crack formation.

3.3.1 Differential Deformations

Fundamental to the study of building pathology is the concept of stresses induced by differential movements across homogeneous sections and differential stresses occurring along the boundaries of two different materials. Both conditions are common, and together are the quintessential mechanical deterioration mechanism. The details of these two basic mechanisms are explored in the following sections; but we begin with an overview to guide the discussion.

The fundamental mechanical mechanisms for crystalline solids are either fractures or elongations of intermolecular bonds. The former is a condition requisite for cracking; the latter, for distortion. The forces that account for such deformations or fracture are readily apparent when the sources of the stress are more or less physical, such as gravity loads or wind loads. Certainly, such forces can and do stress materials to the points at which distortions and fractures occur. At the materials levels, however, it is as likely or more likely that the sources of the internal stresses that initiate small fissures and cracks are not the gross external loads, but changes in dimension resulting from differential movements either relative to different materials or within a single material.

The former is the more readily envisioned. Imagine any two materials being bonded together, such as the aggregate and cement paste in concrete or the paint on a wood clapboard. It is theoretically possible that these two materials have identical material properties, but unlikely. The consequence of these differences is that, among other things, when the materials are exposed to heat or moisture, they will change dimension. If each is unaffected by the other, the degree of movement will be dependent upon the specific property of the material and the magnitude of the stimulating environment. It is readily apparent that in an unbonded condition, the two materials will change dimension by different amounts in their respective properties, relative to the change in the surrounding environment.

The two materials are, however, bonded together such that at the immediate boundary plane, the two materials will and must change to the same degree. The only way for them to change by different amounts is for the bond to break, which is precisely what occurs when the stress along the plane exceeds the ultimate strength of the bond. As long as the bond remains intact, however, the dimensional change of one material must be identical to the dimensional change of the second material at the interface between the two. Each material applies a sort of restraining influence on the adjoining material. One of the materials is, for example, trying to expand more than the other. The former will effectively stretch the latter; the latter will effectively compress the former.

Not as obvious is what happens internally within a single material that is subject to differing dimensional change across its section. Imagine a homogeneous material consisting, in fact, of many layers of the identical material. The particle is exposed to an environment on one face, which causes that face to attempt to change dimension. The next layer inboard of the surface is less affected by the environment, and is, therefore, less prone to change dimension. The result is a sort of mutual restraint similar in effect to that where the materials were totally different. In this case, the material is the same, but the degree of actual change is different.

In either case, the consequence is that one material or one layer is being restrained by the adjacent material in one case or layer in the other. The stress along the boundary lines may be sufficient to exceed the ultimate strength of the material, resulting in a small fracture.

The primary sources of such dimensional variation in crystalline solids are as follows: change in temperature, change in stress, change in moisture content, and change in chemistry. There may be others, but these are the four basic sources of dimensional alteration, and they are described in detail below.

It is beyond the scope of this text to completely explore the behavior of crystalline solids exposed to stress; however, such is the basic level at which all mechanically related failures begin. The properties and characteristics of physical chemical construction of various compounds pay out in the deterioration mechanisms at the macroscopic level. The difference in behavior in the presence of an acid of calcium carbonate (lime) and calcium sulfate (gypsum) has relatively little to do with the property differences of carbon and sulfur, but everything to do with the packing arrangements of the two crystalline compounds, their respective bonding strengths, and the physical dimensions between adjacent ions.

In high school chemistry, most of us were exposed to the notion of valence balancing (a chlorine ion has a negative valence of 1; calcium, a positive valence of 2. Calcium chloride, therefore, must be $CaCl_2$. What we were not told was that balancing valences alone does not mean the compound actually exists, because the packing arrangement of the calcium and chlorine must also allow a closely arranged regular packing pattern. Imagine calcium chloride represented by a large cookie tin full of two sizes of marbles. Imagine that the calcium and chlorine ions are of different sizes. We know that the ratio of calcium to chlorine must be 1 to 2, but how exactly are they arranged? Classic chemistry suggests that each calcium matches up with two specific and discrete chlorine ions; but if that is the case, how are these little triangular arrangements packed into the cookie tin? Even if we could maintain the little triangular molecules of calcium chloride, the packing would be loose, and the resulting compound would be very loose, lacking density with attenuated and weak bonds between molecules.

In fact, such compounds are laid down in layers. Each layer is densely packed generally in a hexagonal pattern, in the case of calcium chloride, beginning with a chlorine layer. The single layer of chlorine spheres will easily arrange into a dense hexagonal pattern with six equally spaced spheres arranged around a central sphere in tight contact; bond distances are minimized. The next layer is a repetition of the first chlorine layer, and will nest very nicely on the first layer. Bond distances will be minimized if the second layer nests on the interstitial voids of the first layer. This tight, close packing produces the minimum volume and most efficient packing arrangement of two layers of same-sized spheres. If we added a third layer of chlorine ions to the first two, we would have two options for nesting the third layer. We could repeat the exact location of the first layer; or if we look carefully, we would see another set of interstitial voids in the first layer that are not filled by the second layer.

Such packing sequencing of layers of like-sized atoms offers some interesting and important consequences. If the first pattern is extended ad infinitum, the sequencing of the layers is a-b-a-b-a-b-a-b . . . , repeating every other layer. If the second pattern is extended ad infinitum, the sequencing of the layers is a-b-c-a-b-c-a-b-c . . . , repeating every third layer. Playing with this idea offers countless layering patterns among like-sized, closely packed spheres: a-b-c-b-a-b-c-b-a-b-c-b-a-b-a . . . , repeating every fourth layer; or a-b-a-c-a-b-a-b-a-c-a-b-a-b-a-c-a-b . . . , repeating every sixth layer. The phenomenal aspect of this simple layering exercise among atoms of the same size is that each variation can and probably will exhibit different physical properties even though they are all chemically identical, pure monoatomic crystals. This is why graphite and diamond are so very different. Both are pure carbon, but the packing patterns of the carbon atoms are different, with the consequence that the two minerals have drastically different physical properties.

When there are two elements of differing spherical radii, the packing problem may be complicated significantly. In the case of our example of $CaCl_2$, if the Ca and the Cl are of the same diameter, then the issue becomes much a matter of the just-described sequencing. If, however, the diameter of Ca and that of Cl are significantly different, the packing problem is dramatically different. If one is much smaller than the other, it is possible that the smaller atom will actually fit into the voids between the larger atoms without disrupting the dense, close packing of the larger atoms. This does occasionally occur; but it is far more likely that the smaller atom is only enough smaller than the larger to result in some spacing of the smaller atoms as they nest into the voids formed by the packing of the larger atoms. Depending on the degree of variation in size, the pattern may be compromised, and the only manner to accommodate the two sizes at all is to shift the packing pattern to some other geometry, such as an orthogonal packing, as is the case with sodium chloride.

THERMAL VARIATION

With a few notable exceptions, materials expand as they are warmed, and contract when they are cooled. It is those exceptions to the rule regarding thermal expansion and contraction that are particularly important to the field of building pathology. Among the class of compounds that do not rigorously follow the general rule is water (another is ammonia). Water is at its densest at 4° C (39° F) which means that as water increases in temperature above this point, it does, indeed, follow the general rule for liquid expansion. As the water cools from 4° C to freezing, the liquid actually expands. When it reaches freezing, and converts to a solid, the solid water (ice) is less dense than the surrounding liquid, which is necessarily at the same temperature. If the surrounding water were, in fact, at any other temperature, energy would necessarily flow back into the liquid until equilibrium were achieved, thus defeating the formation of ice. The fact that ice expands in volume relative to the surrounding water is of profound consequences. For the world of building pathology, this means that when the freezing water is confined in a surrounding rigid vessel, the expanding ice exerts pressure against the walls of the vessel. The further consequences of this phenomenon are discussed in greater detail in Chapter 5.

The consequences for the rest of the world are even more profound. Not only did floating ice allow the species to avoid becoming primordial pesto, ice has a substantial insulative value. Thus, a blanket forms over the lake, inhibiting the prospects of freezing the lake all the way to the bottom.

If a block of stone, for example, is sitting in the open without anything touching it, and it increases in temperature, it will expand in all three directions. Each dimension will increase by the same percentage in proportion to the magnitude of the temperature change. This means that the percentage change in dimension is the same regardless of whether the temperature change (ΔT) is from 10° C to 20° C or from 910° C to 920° C. This is not quite the case. The rate of dimensional change due to a change in temperature does vary depending on the magnitude of the initial temperature; however, for the range of most materials operating in common conditions, the variation from a linear relationship to the actual curvilinear relationship is minimal. The total magnitude of the dimensional change is, therefore, a function of, one, how long the item was to begin with, two, its inherent propensity to expand or contract due to changes in temperature, and, three, the magnitude of the change in temperature. The greater the temperature change, the greater the dimensional change.

The inherent propensity to change with temperature varies from material to material, and is experimentally calculable. The resulting number can be multiplied by the number of degrees of temperature change and the original dimension of the element to determine the actual change in dimension. If the original length (L) is 100 cm, the change in temperature (ΔT) is 15° C, and the percent change per degree is 1 percent, then the total amount of change in length (ΔL) is $L \times \Delta T \times 1\% = 15$ cm.

In the world of thermal induced dimensional change, this example would be immense. In fact, actual changes due to changes in temperature are quite small, but as we shall see, very powerful in their effects. A more realistic percent change per degree centigrade is between 1/500 percent and 1/1000 percent. This means even for a 100° C temperature change, the dimensional change is between 1/5 percent and 1/10 percent. For a 12-meter long beam made of steel undergoing a 40° C temperature change, the actual dimensional change is less than 6 mm (approximately one-fourth of an inch).

The characteristic percentage change per degree is called the *coefficient of thermal expansion*. Values for materials are published, and they vary considerably. Metals are typically fairly high; stone, fairly low. The coefficient of thermal expansion for steel is two times that of brick. If the same 12-meter steel member were encased in a brick wall, and both underwent a 40° C change in temperature, the steel would increase by one-fourth inch. The brick, however, would increase only one-eighth inch. If the steel were literally inside the brick, there are only a limited number of outcomes. One possibility is that the steel would overwhelm the brick and expand to its full length. For this to occur one of two other things must happen: either the bond between the brick and steel must fracture, or the brick must fracture in cross-section at several locations, with each segment of divided brick remaining bonded to the steel.

In composite construction such as described in this example, either, or even a combination of both such failures, can occur. In reinforced concrete design, once the magnitude of the maximum tensile force is accommodated by the addition of tension steel, the designer checks the shear stress at the boundary surface of the steel to avoid *pullout*, which is the term used to described the failure at the boundary surface. He or she then checks the balance of the steel in the total section of concrete to prevent the concrete from merely traveling with or following the tendency of the steel, which would otherwise result in the second type of segmenting failure.

In a properly designed composite beam of steel and brick, a more likely scenario is that neither the bond nor the brick fractures. The brick inhibits the steel from achieving its unrestrained expansion, and the steel simultaneously elongates the brick more than it would from thermal expansion alone. The bond between the two remains intact; however, both materials are stressed, each by the other. The steel is effectively compressed by the restraining action of the surrounding brick; the brick is effectively elongated, hence is put in tension, by the encased steel. This means that at some arbitrary point along the bonding plane between the steel and brick, the brick is pulling on the steel, which means the bonding material itself is in state of shear. Ultimately, the capacity of any combination of two materials bonded to act compositely depends on the capacity of the bond along the boundary plane to transfer the requisite shear stress generated by the relative movements of the two materials.

In the case of reinforced concrete, the specific material that provides the "glue" that bonds the steel to the concrete is a complex hydroxide formed by the relative alkalinity of the cement acting on the iron. The material formed by this chemical reaction at the time that the concrete is poured can be disrupted or damaged by a number of intervening events at this critical point in the process. If the surface of the reinforcing steel is corroded, the formation of the hydroxide is deterred; and for what hydroxide does form, the bonding surface is compromised by the rust. If the reinforcing steel is coated with mud or oil, the surface of the reinforcement is effectively coated and thus denied to the chemical deposition of the hydroxide. Subsequently, when the reinforcement is placed into service and stressed, the shear transfer contact area between the steel and the concrete is reduced, and an initial flaw in the contact plane is already in place. The consequence is that the entire bonding plane is jeopardized, and debonding and pullout may occur.

It is important to note that the term relative movement refers to the relative direction and magnitude of otherwise unrestrained movement. Seldom is the case so clear as one material moving in one direction and the other in the opposite direction. The typical scenario is that both materials, if unrestrained, would increase or decrease in dimension. The stress between the two is generated not by opposite tendencies, but by similar tendencies of different magnitudes. The result is that one material is constrained compressively and the other is effectively elongated tensilely.

HYGROSCOPIC VARIATION

Whereas thermal expansion and contraction is a general property of solids and liquids, the propensity of materials to absorb moisture and, subsequently, to increase in volume is specific. Some materials, such as metals or glass, simply do not absorb moisture. Others, such as many thermoset plastics, technically absorb water, but the amount is negligible. Still others absorb water and expand, but at different rates depending on the axis of orientation.

In general, a material will absorb moisture and, as a consequence, expand equidimensionally in all three axes. Such materials are invariably isotropic and homogeneous, such as brick and concrete. Granite and other igneous rocks also behave in this manner although the degree of expansion may be less than for the former two materials. In such cases of triaxially equal hygroscopic expansion, the consequences are similar to thermal expansion. If the material is unrestrained, the material expands or contracts volumetrically without internal stress. If the material is restrained, then the restricted expansion or contraction will result in internal compression or tension, and such internal stresses will be evenly distributed through the material and equally in all restrained axes.

The situation is complicated if either the material is anisotropic or nonhomogeneous. As if that is not enough, anisotropic materials may be anisotropic in two or all three axes. Similarly, materials may be nonhomogeneous in either two or three axes. These complications do not alter the basic premises, but they certainly complicate predictability and computability.

The significant difference between thermal differentials and hygroscopic differentials is their relative degree of predictability and reliability. Within the limited range of temperature changes that we experience in service, the relative differentials associated with temperature change are nearly linear, which means that we can assign to thermal change a simple coefficient to each material and multiply that number by the number of degrees involved in the shift to compute a dimension change. The calculation is simple, accurate, and reliable. All we need to know is the proper coefficient for the material at issue and the temperature differential.

The expression of hygroscopic expansion is typically expressed as a change in dimension as a percent of original dimension. For example, the hygroscopic expansion of a brick may be 1 to 2 percent, meaning that the difference in dimension between a desiccated sample and a saturated sample is in a range between 1 to 2 percent. This means that the material may expand as much as a couple of percent if it were to go from a completely dry condition to a completely saturated condition. This range between completely dry and saturated is often referred to as the *moisture content* of the material, and it is a useful concept.

Achieving an absolutely desiccated state is practically impossible for most building materials. American Society for Testing Materials (ASTM) standards for dry weight require subjecting the sample to heat for protracted periods of elevated temperature, which theoretically drives off excess moisture. This, however, will not release the bound water in many common materials. For example, the bound water in the calcium carbonate crystal is an integral part of the crystalline structure. To release the bound water in limestone requires subjection of the sample to temperatures in excess of 700° F for protracted periods of time. Having accomplished desiccation of the limestone sample, the calcium carbonate crystal disintegrates and converts to calcium hydroxide (quick lime).

The moisture content at absolute desiccation is, by definition, 0 percent. At total saturation, the moisture content varies among materials, and is, therefore, a measurable material property. The moisture content is the weight of the water as a percent of the total weight of the material, including the water at that point. The terms *maximum moisture content* and *moisture saturation* are synonymous. In the case of a common sample of wood, the saturation point, hence the maximum moisture content, is approximately 30 percent. For brick, the maximum moisture content may be only as much as 4 percent, keeping in mind that such a measure is typically based on weight, not volume. The same specimen may, in fact, contain as much as 8 percent water based on volume if every void were in fact filled with water.

Such values for the maximum moisture content of a given material are readily available in the literature. Also generally available is the percent change in dimension over the range of moisture content from 0 percent to maximum moisture content (MC_{max}). For example, a typical piece of timber will exhibit an 8 percent change in dimension in the tangential axis over a range of MC from 0 percent to MC_{max}, which is typically 30 percent. Such information does not, however, allow us to do other than linearly interpolate the change in dimension if the MC increases from 10 to 15 percent, which is a common enough event. By interpolation, we may conclude that a 5 percent change in MC ($\Delta MC=5\%$) represents approximately one-sixth of the total potential moisture content fluctuation (5%/30%); therefore, the dimensional change (Δl) is in proportion to the change in MC, hence one-sixth of 8 percent or approximately 1.33 percent of the original length.

While such an interpolation is logical, it is an inaccurate representation of the phenomenon. The actual change in dimension of materials as a result of hygroscopic absorption is far more complex than the interpolation suggests. Despite the imprecision of the analysis, the resulting computation from interpolation is sufficiently accurate to allow us to function in most building conditions. The difference between the approximation and the actual dimensional change is, in the end, not enough to warrant development of a more precise tool for calculation.

Isotropy

In the case of laminates, which are biaxially anisotropic materials such as thin-layered sedimentary rocks, the performance perpendicular to the laminars is inherently different from the performance parallel to the layers. A simple example will illustrate one such difference. Imagine a bar of sandstone in tension. This particular specimen consists of alternating layers of materials of different tensile properties. One layer is predominated by silt in a clayey matrix; the other, by well-graded quartz granules in a ferrous matrix. Such situations are, in fact, rather common among brown sandstones. The quartz layer is competent in tension relative to the silt layer. If the direction of tension is parallel to the layers, each layer will resist the elongation in proportion to the modulus of elasticity and area of each contributing layer. If the direction of tension is perpendicular to the layers, each layer is subject to the full value of the tensile stress without contribution from or distribution to other layers; hence, the bar is only as strong as the weakest layer. Similar consequences follow when the bar is in bending or shear in the respective axes.

In the case of hygroscopic expansion, the two layers may absorb water at different rates and amounts, and they may respond differently dimensionally. The degree of such differences combined with the strengths and elastic properties of the two materials can and will generate internal stresses. For example, again imagine the hypothetical sandstone: the sand layer expands hygroscopically more than the clayey layer. The relatively greater expansion of the sand layer perpendicular to the layers otherwise unconfined has no particular consequence to the internal stress state of the specimen. Parallel to the layers, however, the sand layer will attempt to expand more dimensionally than the clay layer, thereby dragging the clay layer along as it, the sand layer, increases in laterally. Depending on the its elastic limits, the clay layer may, in fact, be elongated to a point of fracture. This very simple and equally common condition is a major weathering mechanism in such laminate structures.

In triaxially isotropic materials, the situation is quite similar, except that the performance varies in all three axes. This means that the there is the potential for internal stress variations in all three directions, along with different degrees of hygroscopic expansion in all three axes. The most common example of such a state exists with most, if not all, timber species. The nature of wood formation is such that the material properties of a sample of wood vary, depending on the specific axis; and all three axes perform differently hygroscopically, mechanically, and thermally.

Homogeneity

The effects of homogeneity are no less significant than isotropy. If a material is a composite of one or more component materials, then it is nonhomogeneous at some scale or other. At a structural level, such nonhomogeneity may be significant only at a rather large scale, such as the reinforcement steel in the reinforced concrete. At a materials level, nonhomogeneity is significant at any scale where there is a boundary or interface between two component materials.

At some level or other, many materials are nonhomogeneous. In fact, true homogeneity is rather more the exception than the rule. Because the absence of homogeneity is only occasionally of structural significance, and virtually impossible to preclude, it is typically discounted as a significant material property. Ironically, however, homogeneity is arguably a major, if not *the* major, origin of material pathology. Many of the mechanisms discussed in this book have their origins in the essential difference in properties of the component materials in nonhomogeneous composites and agglomerates.

One manifestation of the homogeneity issue relates to hygroscopic expansion. Take, as an example, concrete, which consists of a mixture of cement, sand, and aggregate. The terms alone disguise some of the problems associated with concrete deterioration. Sand is generally, but not always, quartzite beach sand or bank run; however, at times it is crushed quartz. The significance of this seemingly minor difference is that surfaces of bank run and beach sand are relatively smooth; the surfaces of crushed quartz are sharp and angular. The former presents a relatively high ratio of volume to surface; the latter, just the opposite. The consequence is that the bonding strength of the former is relatively lower than the latter.

Aggregate is a simple term that belies the obvious fact that it is a generic reference for gravels of many types and sizes, as well as for crushed rock of just as many types and sizes. The origins of the aggregate may vary among dozens of petrologically identifiable rocks, ranging from granites to diorites to sandstones. The properties of these rocks vary considerably; it would not be reasonable to expect otherwise. The same is true of different sources of cements. It follows from all this that there is no reasonable prospect of material property identity, even similarity, among the cement, sand, and aggregate.

Each of the component materials comprising concrete is absorptive, and each is hygroscopic. By virtue of the disparity of material properties, the absorption and expansion rates are different; therefore, the dimensional changes will vary. Suppose that the aggregate is relatively absorptive and expansive compared to the cement paste—a not uncommon condition. This means that, as the stone and the cement each absorb moisture, the stone will expand at a greater rate than the surrounding cement matrix. The stone will exert an outward pressure on the confining cement.

The effective confinement of the stone by the cement results in a relative compression of the stone, and a relative stretching of the cement to wrap around the more rapidly expanding stone. The stretching of the cement means that the cement surface in direct contact with the stone is in tension. If and when the value of this tensile stress exceeds the fracture limit of the cement, either or both things may occur. A fracture may occur in the cement perpendicular to the interface of the cement and the stone. This would be a result of the stretching force exceeding the ultimate strength of the cement in tension.

Another possibility is that a crack forms between the cement and the stone surfaces. In effect, the cement debonds from the stone. This type of crack is a failure in shear of the contact surface of the two materials, and is the result of the dragging along of the less expansive material (the cement) by the more expansive material (the stone). If and when the shear limit is reached, a shear failure occurs along the contact surface. While it is possible to imagine a shear crack (debonding) occurring without a perpendicular tension crack, it is geometrically impossible for a tension crack to form without some degree of adjacent debonding occurring simultaneously.

MECHANICAL VARIATION

Variation of mechanical properties is particularly significant as a source of deterioration as it relates to stress distribution. This subject is discussed in detail later in this chapter under the topics of axial and shear stresses and in the section addressing stress gradients. The fundamental point here relates to materials within which the mechanical properties vary within or across the section of the member. Such variations are also normally associated with changes in chemistry, but not always.

It is possible to develop mechanical properties variation due to density variation as well. For example, a brick is manufactured by pressing clay into a mold or extruding it through an orifice. In the case of molded bricks in particular, it is likely that the pressure applied to the plastic clay will not be perfectly uniform. As a consequence, the clay that is more highly compressed will have a lower initial moisture content and a higher density when fired. In materials such as brick and concrete, density is a significant factor in their elastic behavior. In our example, then, the variation in density may and probably will result in a variation within the body of the material in the modulus of elasticity, hence the bricks' response to load.

Variation across the section of the member in elastic properties means that internally the member has soft spots and stiff spots. For reasons explained more extensively later, stresses will be concentrated in the stiffer zones, which as a result may fail at external load levels below what would otherwise be predicted based on more uniform properties. The reason is that, though the elastic property may change, the ultimate strength of the stiff spots may not increase proportionately. Imagine two bricks made of the same sample of clay and fired in precisely the same manner, but one is perfectly uniform in its density whereas the other varies. The one that varies includes pockets of softer brick surrounded by a matrix of denser brick. The average value of the variable brick is such that, in theory, both bricks can support the same compressive load before crushing. When both bricks are compressed, predictably, the uniform brick crushes at the predicted load. The variable brick is placed in the same test stand and crushed. As the load is imparted to the variable brick, the stiffer portions resist deformation more than the softer portions; the consequence is that those stiffer portions absorb more of the load. The softer portions deform relatively easily; the stiffer portions, not as easily. This phenomenon is the source of the axiom that "load flows along lines of maximum stiffness."

Load does not actually flow, but the sense of the axiom is correct inasmuch as stiffer elements deform proportionately less, and therefore act to resist deformation more than relatively elastic components.

The internal stress in the stiffer segments is greater than the softer segments as a result of the disproportionate amount of load in the stiffer portions. Although it may seem counterintuitive, it is altogether possible that the ultimate stress of the soft and stiff portions is the same. The internal stress in the stiffer elements will reach ultimate strength first, and fail, followed by an abrupt redistribution of load into the as-yet uncrushed softer material, which will necessarily fail because it is no stronger than the now-crushed stiffer material.

The summary of this somewhat sad tale is that the variable brick failed at a lower level of total load than the uniform material because the dissimilarity in stiffness triggered a sequential failure of the stiffer material, followed almost immediately by a failure of the softer material. Even though the ultimate strengths of any of the materials involved are essentially the same, the variable material will perform less well than the uniform material.

If this conclusion is less than convincing at this point, it is probably because it is difficult to imagine materials of differing density that are not also of differing ultimate strengths, meaning fracture points comparable in value to the density of the material. It is often the case that denser brick is also stronger brick, but this is far from universal.

To illustrate the point of strength versus density, let us examine that class of materials called metals. We associate iron and its close cousin, steel, with relatively high density (490 pcf) and high ultimate strength (60,000 psi or more). Taking steel as a basis for comparison, aluminum is one-third as dense, but comparably strong. Aluminum also has an elastic modulus one-third that of steel, which means that under similar loading conditions, the aluminum will deform approximately three times as much as steel; and it is not unusual to associate deformation alone as failure. Deformation, however, is not the same as fracture; and while a sample of aluminum may well deform three time as much as the steel, it will not actually fracture at one-third the stress. In fact, many alloys of aluminum have ultimate strengths as high or higher than some grades of steel.

Lead is a good example of a material actually denser than steel with a significantly lower ultimate strength. The point is that material density and material strength may be loosely related for brick and concrete, but the correlation is not strong.

A class of materials that comes closer to our brick example is the general class of materials of stone. The density of stones is remarkably close, typically around 150 pounds per cubic foot (pcf). There are of course some extreme examples, such as loess or tuff, which are very light, but most construction-grade stone is close to the same density. Despite the similarity in density, stones vary considerably relative to modulus of rupture, which is a measure of ultimate strength. In the case of an agglomerate, the variation in type and stiffness is fairly apparent; but the same phenomenon may be at work in a granite that consists of several identifiable minerals of varying moduli of elasticity. The very condition described above may result in the fracture of the stiffer minerals first.

In actual service, for this phenomenon to become a deterioration mechanism, the stress must increase over time, otherwise the failure mechanism would have ensued immediately, and the specimen would not have survived long enough to be the subject of pathological investigation. Remember, building pathology is quite simply the study of failures in materials and building systems *after* the building went into service. In other words, the materials and building systems were serviceable at one time, and subsequently failed due to some change in conditions. The most common changes in condition are those resulting from changes in the materials and the building due to deterioration mechanisms. Change may also be due to changes external to the fabric, such as changes in loading. Later in this and subsequent chapters, we will investigate various forms of material and load fluctuation.

COMBINED VARIATION

In service, materials and building systems are rarely subject to a set of conditions such that one and only one of the three types of variation discussed above are affected unilaterally. As materials increase in volume or attempt to increase in volume because of hygroscopic variation, it is almost necessarily the case that they will do so uniformly across the entire cross-section of the material. As the material begins to absorb water, for example, the outer surface is wetted first, and the moisture content of the outer surface increases at a greater rate than the interior of the specimen.

The same phenomenon occurs as the material is heated or cooled. The outer surface is affected first, and the change in temperature is initially greatest at the outer surface. Gradually, heat is transferred through the body of the material. If the external thermal condition is maintained for a sufficiently long period of time, the material will achieve a state of thermal equilibrium. Similarly, if the humidity of the surrounding atmosphere is sustained for sufficiently long period of time, the moisture content of the material will stabilize and will achieve a state of equilibrium.

Until these states of equilibrium are established, however, there exists within the material a variation or gradient, either a moisture gradient or a thermal gradient. The significance of these gradients will be discussed in detail, but for now it is important to recognize that they produce variations in stress, mechanical variation, within the material. Variation in moisture content may also affect stiffness directly, and extreme temperature variation may embrittle certain materials.

The point to keep in mind as we progress through this text analyzing deterioration mechanisms, is that such mechanisms operate purely and simply on one and only one material property.

3.3.2 Crack Formation, Extension, and Coalescence

The cracks causing immediate concern are small, generally suboptical in width and length. Optically identifiable cracks are important, but more so the building or systems level, less so at the materials level. The cracks that form and that account for the weathering of exterior materials are measured in microns and angstroms, not inches and millimeters. In subsequent sections, we will discuss the physics and mechanics of crack initiation and progression.

As the cracks grow and coalesce, they form a mat or network of merging and intersecting cracks. The cumulative effect of the sum of all the small cracks is a larger crack. The growth and coalescence of small cracks into larger ones is one of the two fundamental methods of materials deterioration, the other being chemical alteration. When these basic mechanisms operate on the built environment, they form or are at the root of what is the subject matter of this book, namely deterioration mechanisms. Many texts on the subject of deterioration of the built environment treat the progression of deterioration as a phenomenon unique to buildings and the materials we use in those buildings. Other texts describe deterioration mechanisms as peculiar to materials once they are assembled, unrelated to any prior existence. The truth is, materials are simply materials, nothing more. Fabric deterioration in buildings is a response to precisely the same environment factors as produce weathering in materials not installed in buildings. There is no special rule of nature that applies to materials in buildings and not to materials not in buildings. Whether we call the results of these environmental factors deterioration or simply weathering, the causes are the same and the results are very much the same. Only the professionals studying the problems use different labels.

3.3.3 Mechanical Mechanisms

There are several mechanically related deterioration mechanisms, of which two account for much of the deterioration that we assign to the mechanical properties and performance. These two general mechanisms appear in many guises, but they are derivative of the same two simple concepts: the differences between two adjacent materials, and the change of stress across the section of a material. The former is related to and a part of what is called *composite action* at the structural level; we will refer to this phenomenon as *composite materials behavior* at the materials level. The latter phenomenon we will refer to as the *stress gradient mechanism*.

COMPOSITE MATERIALS

The behavior of composite materials refers to the general situation where two materials are adjacent to each other. They may be bonded to each other or not, although the bonded situation is the more general class, and

FIGURE 3.1A   The bracket above the caryatid is disengaged from the beam, above which, in this instance, is a plaster encased steel beam.

FIGURE 3.1B   The separation demonstrates the rule of compatible deformations. If the bracket were being pushed down by the beam, there would be no separation.

FIGURE 3.1C   The wood armature holding the cornice below and behind the plaster is vulnerable to the same moisture that is causing the visible damage to the plaster. One mechanism suggests the presence of the other, because the necessary and sufficient conditions exist.

the debonded case, a special class. The phenomena associated with the situation of two proximate materials are independent of scale; they are present whether the specimen under investigation is a composite bridge girder or a speck of sand in a mortar joint.

With that said, virtually every material found in the field is a composite material at some scale or other, which is another way of saying that all materials are nonhomogeneous at some scale even if we have to examine them at the atomic level. It is possible to achieve pure homogeneity in a laboratory specimen of a monoatomic crystal, but even monoatomic crystals (e.g., pure metals) are rarely that pure in construction applications. Far more common is a wide array of materials that are relatively homogeneous at the optical level, such as construction-grade metals, glass, and thermoplastics; and an equally wide array of materials that even at the optical level are nonhomogeneous, such as concrete, most bricks and stones, and wood. This phenomenon also occurs at a structural level, for example, in reinforced concrete, and at even larger scales such as in geologic layering of rock formations.

The general situation common to all these examples is that two different materials are proximate to one another and generally bonded to one another, such as the cement paste bonding to the aggregate in concrete, the early wood bonding to the late wood in timber, or the sodium layer bonding to the chlorine layer in simple table salt crystal. In any and all such situations, the individual substances comprising the overall material will have individual characteristic properties aside and distinguishable from the gross material. It is, for example, possible to distinguish the properties of the aggregate and the properties of the cement paste as distinct and identifiably different from the concrete as a whole.

Where there are distinguishable materials with identifiable properties, it is highly probable that the properties of the two materials are different in value. It is statistically possible that the aggregate and the cement have the same modulus of elasticity, but it is far more likely that the respective values of the moduli of elasticity are different. It is also more likely that their respective coefficients of thermal expansion are different, as well as their respective rates of hygroscopic expansion. Taking those three properties alone, it is quite apparent that the statistical likelihood of coincidence of all three values is negligible. Different materials have different properties.

There are some, albeit quite few, nominally different materials that are virtually indistinguishable among their respective properties. For example, quartz and amethyst are identifiably different; we can and do refer to each as a separate material (and their respective prices reflect that distinction). Their material properties, however, are virtually, if not actually, identical because they are essentially the same material regardless of the visual distinction between the two.

When two materials immediately proximate to one another are subject to change in temperature or moisture content, each responds in a predictable and measurable manner *if unrestrained*. But typically, one or the

other or both are restrained. The concept of restraint is one of the fundamental issues associated with the deterioration of materials and structures,
and it is often misunderstood or not understood well. Because unrestrained conditions are easier to explain, we will start there and work to restrained conditions.

If subjected to changes in temperature or moisture content, unrestrained materials and systems will change dimension. If we know the
properties of the materials and the degree of change, we can calculate the
total magnitude of the volumetric change of the material. As discussed previously, the coefficient of thermal expansion and the rate of hygroscopic
expansion based on moisture content allow us to accurately predict the
change in dimension, hence volumetric changes. This is true for any material for which we know the properties and magnitude of thermal or moisture shift. If the same specimen is restricted in any direction from achieving the full predicted volume at the time of the thermal or moisture shift,
then the material will not achieve its full predicted dimension. At the instant
that the specimen incurs a restriction, the motive force to achieve the full
unrestrained dimension does not cease; the specimen continues to attempt
to expand. The fact that it is restricted does not eliminate the impulse to
change dimension. The inability to attain the unrestricted dimension is
not the consequence of a cessation of the effort to change; it is the result of
the confinement imposed by whatever is the source of the restriction.

To succeed in confining the specimen to a restricted dimension requires that the source of restriction exert a force on the specimen that is
sufficient to accomplish the confinement or reduction in dimension. The
force required to accomplish such a confinement is the effective equivalent
of subjecting the same specimen to a force sufficient to accomplish the
same change in dimension as a result of elastic deformation due to externally applied stress. Put more simply, this means that as the material tries
to grow, for example, it is squeezed by some other surrounding material. If
the surrounding material were infinitely rigid, it would restrict the expanding material from growing by a dimension equal to the magnitude of
otherwise unrestricted growth. The external material is, therefore squeezing the specimen by an amount equal to a force that would accomplish the
same amount of elastic deformation in an unexpanded situation.

It may help to picture the situation as a two-step process. Step 1 is that the
material in question expands due to a change in temperature or moisture
content. It grows 1 millimeter in length. At this point, there is no internal
stress generated by the growth because there was no restriction on the mil-
limetric growth. Step 2 is that the now-longer specimen is squeezed back to
its original dimension. To accomplish this reduction in dimension, the
specimen is compressed elastically, back to its original dimension. The
amount of compression necessary to achieve this reduction in dimension

is predictable based on the elastic properties of the material and the amount of deformation.

Imagine a 40-foot-long steel beam spanning two massive concrete abutments (infinitely rigid). The temperature of the beam changes from 40° F to 90° F ($\Delta T = 50°$ F). The coefficient of thermal expansion for steel is 0.0000065 inches of change per inch of original length for each degree of change in temperature (F). The total unrestrained change in the length of the beam would be equal to (40 feet × 12 inches/feet) × (50° F change in temperature) × (coefficient of thermal expansion for steel of 0.0000065 inches/inch/° F) = 0.156 inches, which is approximately $\frac{5}{32}$ of an inch. This does not seem like a very great change given the length of the beam and the change in temperature; indeed, as a dimension in its own right, $\frac{5}{32}$ of an inch is no more than the diameter of a pencil. As such, it should not require a major effort to squeeze the beam back from its assumed length of 40 feet − 0.156 inches to 40 feet − 0 inches.

Suppose the beam is a W27 × 94. The area of steel of this section is 27.7 square inches, which means that when the concrete abutments begin to impart compression into the length of the beam, the force will be resisted by a piece of steel 27.7 square inches in area. When the concrete abutments resist the expansion of the steel, and effectively exert a compressive force to the steel, that force will be distributed over the cross-section of the beam, and each square inch of the steel will resist the compressive stress.

The modulus of elasticity (E) of steel is 29,000,000 pounds per square inch (stress) per inch of change in length per inch of original length (strain). Inasmuch as it is a unitless expression, the last part of the expression of the modulus of elasticity is often deleted, but it is conceptually important to remember that the modulus of elasticity is a numeric expression of the characteristic ratio of stress to strain of a particular material. The modulus of elasticity is a measure of the stiffness of the material. Specifically, it is a ratio of the magnitude of stress to a unit of deformation or strain. The higher the modulus of elasticity, the stiffer the material and the more resistant a material to deformation.

The force (F) induced by the inflexible and rigid abutments is distributed over the area of the beam (A = 27.7 inches ^{2}), resulting in a level of internal stress (σ): σ = F/A = F/27.7 inches^2. The total change in length (ΔL) of the compressed beam will be − 0.156 inches, which we know because that is the amount of growth restricted by the abutments; therefore, the strain (δ) is equal to the change in length divided by the original length (L = 40 feet = 480 inches): $\delta = \Delta L/L = -0.156/480 = 0.000325$ inches/inch. This is a very small number but, as we shall see, it is far from insignificant.

We know that the modulus of elasticity (E) is a measure of the relationship of stress to strain; E = σ/δ. We know the value of E (29,000,000 psi/inches/inch) and the value δ (0.000325 inches/inch) from which we can compute the stress: $\sigma = \delta \times E = 0.000325 \times 29,000,000 = 9425$ psi. This means that the pressure on both ends of the steel beam exerted by the concrete abutments is almost 5 tons for every square inch of area, which as we know is 27.7 square inches. The total force $= F = \sigma A = 261072.5$ pounds. That's over 130 tons!

What does all this mean? When materials are restrained, even though the dimensions involved may be small, the forces involved can be huge.

STRESS GRADIENTS

Earlier, we introduced the notion of gradients, specifically, moisture and temperature gradients. Is this section, we will discuss stress gradients. A stress gradient exists when the value or magnitude of internal stress is not uniform across the cross-section of the sample. This can occur because the source of stress, the load, is not uniformly applied at the surface of the material. In fact, a very common source of stress gradient occurs in simple bending. As an introduction to stress gradients, we will examine the bending situation from this new perspective.

When an element is in simple bending, the bottom surface of the element or member is in tension, and the top surface is in compression. Consider the notion of a neutral axis of the member, which is that plane within the member that is neither in compression nor tension and that, in symmetrical rectangular sections, runs along the midline of the member. For this discussion, we are interested in the two zones between the extreme edges and the neutral axis.

In this zone, the stress changes from the zero at the neutral axis to a maximum value at the edge. If we imagine the member as being composed of an infinite number of thin layers of material, then we can imagine that each layer is at a slightly different stress level from that of the adjacent layers, corresponding to the change in stress between the extreme fiber and the neutral axis. This is necessarily the case if the value of stress varies across the cross-section of the member. If each layer is at a slightly different stress level, and the elastic properties of each layer are identical, then each layer is slightly different in length from the adjacent layers .

One possibility to account for such a dimension shift is that one layer must be able to slide relative to the adjacent layer. A graphic example of this is to take a telephone directory and bend it. The pages slide slightly across one another while the free edge of the directory shifts in such a way as to display these slight variations in dimension. But note that the directory, despite its thickness, is not a stiff beam.

Our experience, as well as our training, informs us that solids are not composed of layers that are free to slide relative to one another when the member is subjected to bending. If the elements are not free to change position relative to adjacent layers, then two phenomena must be present. The layers must be bonded together, and the bonding material must be in a state of elastic shear in order to return to its original position when the bending load is removed. This means that, even if the basic layered material remains in pure tension or compression, the bonding material is in shear, and it is deformed. It is also necessarily the case that, upon release of the external forces, the material returns to its original shape and dimension, which is the essence of elastic behavior.

As we know, isotropic crystalline solids are not composed of finite layers, imaginary or otherwise. Thus, the transfer of differential shears does not occur in discrete layers, but continuously. Transfers across layers are analogous to going down steps; transfers across homogeneous solids are analogous to sliding down a slope. In either case, there is a change of stress across the cross-section, and the difference in deformation from layer to layer means that there is necessarily a continuous, albeit changing, state of stress across the section. This changing state of stress is the stress gradient. The stress gradient (which is to say, anytime there is a variation of stress across the cross-section of a material) is a source of stress in and of itself. The external stress may be vertical in the form of gravity load. The resulting stress gradient results in horizontal shear stresses between actual or theoretical layers inside the member. The stress gradient, regardless of source, signifies an internal shear stresses.

COMBINED STRESS

Materials deformed in shear are typically described as being subject to a completely different set of variables than materials in tension or compression. The implication is that shear is an inherently different type of stress than tension or compression. Discussions of tension and compression often suggest that they are inherently different from one another, when, in fact, tension is little more than the mathematic negative of compression, and vice versa. A similar mathematic relationship exists with shear: shear is actually transformed tension and/or compression.

Imagine a small piece of material, square in shape and subjected to an external shear stress. This means that equal and opposing forces are present at two opposite edges of the specimen. This is the definition of shear: opposing forces operating at a distance from one another. Unless opposed by some restraint, the shear forces will generate a rotation of the specimen, and it will begin to spin. Obviously, a material specimen inside the body of a larger crystalline solid is not spinning into the universe. It must, therefore, be restrained, which means that there must be more going on than simply two opposing forces at a distance. In fact, given that the specimen is not spinning off into space, the shear forces *must* be opposed in both magnitude and direction as the original shear pair or couple. The source of such opposing forces is the restraining effect of the surrounding material; and the specific opposing forces are analytically another set of shears acting on the faces of the specimen at right angles to the original shear couple.

If the shear couples are analyzed as vectors, the graphic representation of the two shear couples is such that a shear force is present at each on the four edges of the small sample. Because one couple must oppose the other in magnitude and direction, each force vector is opposed by a counterpart 90° to it, and meeting head to head at the corner of the specimen. Consequently, there are then two sets of opposing vectors meeting at diagonally opposed corners of the specimen.

We can merge two shear vectors into two equal and opposite vectors acting along one of the great diagonals of the specimen. Therefore, relative to the major diagonals of the specimen, the specimen is not in a state of shear at all, but in a state of axial compression or tension. In other words, shear is nothing more than tension or compression viewed on a diagonal. Conversely, any state of tension or compression can be converted to a state of shear by rotating the ordinates. In fact, any set, however complicated, can be converted to a simple set of principle stresses acting perpendicular to some plane.

Principal Stresses

The phenomenon of transformation of principal stresses was most eloquently articulated by Professor Otto Mohr in Germany in 1895. Mohr developed an elegant graphic transformation tool now reverently called Mohr's Circle. Mohr's hypothesis was that the deformation of a specimen is a response to the sum of all imposed stresses. The sum of all stresses can be expressed as a set of simple vectors. Using the geometry of the circle, Mohr demonstrated that the sets of shears and axial forces that resolve to a single pair of principal stresses lie on the perimeter of that circle. By transforming the coordinates on Mohr's Circle, we can calculate the plane on which only axial (principal) forces will result in the deformation of any other set of shears and forces that will accomplish the same deformation. One of the implications of Mohr's transformation principle is that, for any given stress state, there is *not* a unique set of contributing external stresses. In fact, for any given internal stress state, there is an infinitely large set of external stresses that may result in the same resultant or principle stresses.

If this is the case, namely that there is not a unique set of external forces for any given set of principal stresses, it follows that there is no set of unique forces that explain an observed deformation. Imagine a square face of solid material deformed to a trapezoid. Such a deformation can be explained in three basic ways:

- Compression is exerted along a great diagonal, squeezing the square into a trapezoid.
- Tension can be exerted along the other great diagonal, pulling the square into a trapezoid.
- Shearing forces are exerted along the edges of the square, wedging the square into a trapezoid.

The results are identical: the square is deformed into a trapezoid. The only observable phenomenon is the deformed shape.

On an even broader level, this discussion of stress transformation highlights the fact that we tend to be too literal in our interpretation of forces, stresses, and deformations. For analytical purposes, we use concepts of tension, compression, and shear to explain deformation. These are all analytical and mathematical abstracts that have no verifiable independent reality. The only verifiable reality is deformation (physical change); forces and stresses are theoretical abstractions, the manipulation of which coincide with observed deformations.

Deformation and Crack Formation

In the course of stress development at the crystalline level, there is typically some point or magnitude of stress at which the material fractures; that is, the material parts, and is discontinuous across the fracture. For a variety of reasons, the material along either side of the fracture may remain proximate, such that the now-severed fragments remain more or less in the same relative position that the materials held prior to the fracture. This is the functional definition of a crack: the material is fractured, but the now-rendered pieces remain in close proximity and, generally, in contact.

The single most common reason for the enduring proximity of the cracked material is that the crack did not progress completely through body of the gross element, that there is some surviving zone of the original body that remains intact and continuous despite the presence of the crack somewhere within the element. We regularly see these types of cracks around us: bricks with cracks across half the face, or cracks in plaster that end in the center of the wall. These are partial cracks, in that they do not extend completely, but only partially, through the body of the material.

Whether we are describing cracks at the suboptical or the gross structural levels, partial cracks are divisible into two categories: stable and unstable. Stable partial cracks are fractures that are not appreciably progressing in length, which means that the forces that gave rise to them originally may have dissipated. This is the case with cracks that are precipitated by impact or a similar trauma, that does not completely sever the specimen. The impulse is great enough to produce internal stresses sufficient to initiate a crack; and if such a force is sustained, the crack would progress until the specimen fractures completely, but if not, the precipitating force dissipates instead.

An impulse is defined as (Force) × (Time) or FT. The concept is particularly applicable to high levels of force imparted over particularly short periods of time, such as an impact of an object on a surface, or a collision of objects. The units that result from multiplying force (mass) × (distance/time^2) by time become (mass) × (distance/time), which are identical to the units of momentum (mass) × (velocity) or (MV) = (mass) × (distance/time). This identity (FT = MV) is useful in analyzing resonance and vibration resulting from impact. Imagine a truck (24,000 lbs) moving at 60 feet per second (fps) (MV = 24,000 × 60 lbs feet/second). The truck strikes a corner of a building and the momentum decreases from a value of 1,440,000 lbs feet/second to 0 in the course of one-hundredth of a second. The equivalent force (F) at the point of impact is MV/T = 140,000,000 lbs.

The far more common type of stable, partial crack is, effectively, a strain relief crack, better known as a shrinkage crack. The initial internal stress was the result of a change in temperature or moisture content, resulting in a change in dimension, which we will assume resulted in a reduction in dimension (shrinkage). The reduction in dimension was resisted by an edge restraint, such as an intersecting floor or foundation that is not changing dimension at the same rate. The resulting internal tension is relatively evenly distributed throughout the wall or panel; however, there is some idiosyncrasy in the panel or initial flaw that results in a local concentration of stress that is higher than the average value of stress in the panel. The tensile stress at the initial flaw reaches the ultimate strength of the material even if only at this single point, and the material cracks ever so slightly at the site of excessive tension. Once a crack, however small, is initiated, the crack may and probably will progress relatively easily and rapidly, at least for a short distance because of a phenomenon associated with nature of geometric changes in a material under stress.

In samples of uniform cross-section that are subjected to external stress, the actual stress level at one site is assumed to be the same as at another similarly situated site. This assumption is not literally correct, but is sufficiently accurate as to allow the simplification. Let's say, for example, that a square bar of steel is subjected to axial elongation; and let's also say we limit our investigation to the middle third or so of that bar. We will find upon close examination that the internal stress is not precisely even across the section. This is primarily because microcracks form at the surfaces, which relieve the "skin" of the sample of some of the stress generated by the gross elongation. Such cracks are deeper and more closely spaced at locations of stress concentrations, specifically abrupt geometric changes such as corners. Keeping in mind that the microcracks are very shallow, the consequence of these slight, but discernible variations, is that the core of the specimen is at a higher stress level than that which is computed simply by dividing the area by the applied force. The variation from computed to actual average stress is less among specimens of ductile materials with simple geometries than among brittle materials of eccentric geometries.

If, however, we were to cut a notch into the side of the sample, the stress at the reduced section would, of course, increase due to the reduction in material adjacent to the notch. Of greater significance, however, is that the stress would not be uniformly distributed across the body of the material at the location of the notch. In fact, at the tip of the notch the stress would be significantly higher than the computed average at that section. Such geometric changes are often referred to as *stress raisers* or *stress concentration points*. The magnitude of the stress increases as the stress raiser is related to the abruptness of the geometric change. Generally, the sharper the change, the greater the increase in stress concentration.

Square corner notches, for example, have a higher multiplying effect than rounded corners; acute angles, higher than obtuse angles. Typical stress-raising geometries or shapes have been analyzed and assigned multiplier values such that, when encountered, the average stress can be multiplied by the appropriate factor to account for the stress concentration generated by the particular change. The tip of the common crack is about as sharp a stress raiser as can be designed. The abruptness or sharpness of the geometry is very great, to say the least. The application of almost any axial load across an existing crack will generate disproportionately high stresses at the crack tip.

Stress concentration is often explained as stress "flows" or stress trajectories through the specimen. These analogies are helpful if the we are somewhat familiar with Bernouli's observations regarding fluid velocity and pressure at changes in pipe or channel geometries. Our common experience of observing fluid flow assists in understanding stress concentration, if we accept the analogies. As is often the case with teaching science by analogy, the reality suffers in the translation. Stress does *not* flow; there is no movement, velocity, or mass associated with stress. Furthermore, while fluids may increase in velocity at points of restriction, there is not a local increase in pressure at the point of restriction. The analogy is not only inaccurate in the portrayal, it is even inaccurately portrayed.

The classic test for this phenomenon is the Charpy V-notch (CVN) test. The specimen is placed in three-point bending by placing the specimen on two support points, one at each end of the sample. At the midpoint of the sample, and opposite in direction to the supports, a load is imposed so that the sample is in flexure, with tension occurring on the face of the sample opposite the point of load application. At the midpoint of the tension side, which is to say directly opposite the point load, there is a notch cut into the sample. The load is increased until a fracture occurs and the specimen fractures. Among the variables at issue and that can be tested using this simple technique include the inherent resistance to crack formation depending on the material. This property is referred to by the misleadingly simple term "toughness." Toughness, in turn, is severely and often catastrophically associated with temperature. Crack initiation and progression can also be investigated relative to the geometry of the notch and the presence and depth of the initial flaw.*

If the material is ductile, the material may deform plastically at the crack tip, which enables nonlinear deformation, and may preclude further crack progression. If the material is brittle, however, relaxation at the crack tip is extremely limited and further fracture is a more likely consequence. Certainly, even in ductile materials, the ultimate strength of the material may occur at the tip and progression will, therefore, continue.

It is important to note that not every crack will progress simply because of the reapplication of external force. The mere reopening of an existing crack will allow strain to occur, which may be enough to accommodate the deformation without necessarily reintroducing stress at the crack tip. Because many deformations are strain-limited—meaning that the expanding material is limited by the degree of thermal shift or hygroscopic expansion—it is only necessary for the deformed material to accommodate a certain dimensional change. If such deformation can be accommodated by reopening the existing crack without significant stress at the crack tip, then the crack will not progress. Such cracks are, in fact, quite common in masonry walls and plastered surfaces.

An example of a structural scale crack of this first type—namely stable, partial cracks—that is strain-limited, is the typical differential settlement crack. When a foundation subsides, the exterior wall be "rack" or distorted, to a degree that it may crack. The foundation eventually consolidates the compressible soil, and the subsidence arrests. When the movement stops, the racking in the wall stops, and the cracks stabilize. Such cracks are also

---

*Note: Other variables affect crack formation and progression, including initial flaw size, notch shape, specimen thickness, and notch depth relative to specimen width among others. For specific stress concentration factors, see *Formulas for Stress and Strain*, by R.J. Roark and W.C. Young (New York: McGraw-Hill, 1975). For a general discussion of stress concentration, see (in order of increasing density and authority but also difficulty) *Elements of Strength of Materials*, by S. Timoshenko and D.H. Young (New York: Van Nostrand Reinhold, 1968); *Introduction to Mechanics of Solids*, by E. Popov (Englewood Cliff, N.J.: Prentice Hall, 1968); *Structural Engineering: Behavior of Members and Systems*, by R.N. White, P. Gergely, and R.G. Sexsmith (New York: Wiley, 1976); and *Fracture and Fatigue Control in Structures: Applications of Fracture Mechanics*, by S.T. Rolfe and J.M. Barsom (Englewood Cliffs, N.J.: Prentice Hall, 1977).

quite common, and while some may appear quite alarming, they are often simply large stable strain-limited cracks.

The third most common reason for crack stability is that the otherwise completely severed pieces of the original body are externally and independently held in position by some other surrounding material. Take, for example, a microcrack occurring across the cross-section of a piece of aggregate embedded within a matrix of concrete, or a check within a section of timber. These cracks are characterized by two general factors that explain their stability. First, the stresses that initiate this class of cracks are internal. Residual stresses within steel and glass are examples of such internal, residual stresses, as are many examples of thermal- or hygroscopic-related stresses. If such stresses result in cracks, the stress is effectively relieved, and the crack reaches a point of equilibrium within the fractured material. Second, such materials, once cracked, must be encapsulated within surrounding material that is not fractured and that will, effectively, splint the fractured material.

A fourth type of stabilized crack is actually an arrested crack that, absent the arresting device or bar, will continue. These cracks are not strain-limited; they will continue to progress until arrested. Unlike the third type, they are initiated by external forces and are, therefore, not dissipated by internal stress relief. Such cracks occur, for example, in steel members composed of multiple steel sections bolted together to form a girder or beam. A crack forms in one of the individual pieces and runs through the cross-section of that element. When the crack reaches the edge of the element, naturally it stops, because there is no more material through which to propagate. Such cracks often run out from bolt or rivet holes to the edge of the plate and are arrested by the discontinuity of the element. In order to bridge the interface between the cracked element and the adjacent, proximate material, the crack must reinitiate in the second element.

One reason this may *not* occur is that it is easier to propagate an existing crack than to initiate a crack in the same material. Recall that the stress raiser at the tip of an existing flaw is extremely high; to initiate a crack in an otherwise unflawed material requires that the average local stress must exceed the ultimate strength of material, whereas the stress needed to progress an existing crack need only reach ultimate strength at the tip of the crack. It is for this reason that cracks, once initiated, sometimes accelerate through materials. During World War II, at Scappa Flow, Scotland, while unloading grain from a Liberty ship in extremely cold weather, a crack formed that almost instantaneously ran through and around the entire hull of the vessel. The ship broke cleanly into two halves and sank at dockside. Subsequent analyses demonstrated that a crack, once initiated in the Liberty ship could and would continue to travel until arrested at some break or discontinuity. Inasmuch as the Liberty ships were all-welded structures, there were no (or very few) such breaks, and cracks would travel uninterrupted or arrested through the entire body of the ship. The problem had not manifested itself in riveted ships because the breaks between riveted plates acted as crack arresters. This phenomenon is present in any all-welded construction, such as pressure vessels and welded, plated girders, because the weld metal is fused with and continuous with the base metal. There is no discontinuity, hence no crack-arresting potential.

Glass is notable for this phenomenon. Take an apparently stable piece of glass struck by a small projectile. The damage done by the projectile is a small initial crack, which rapidly or, sometimes, slowly migrates across and through the body of the glass, resulting in a crack vastly larger than would otherwise be warranted by the initial impact. The reason is that glass often includes substantial residual stress, which is "released" once a crack is initiated, combined with the phenomenon that propagation requires less force than crack initiation.

One technique for arresting a migrating crack is to "drill out" the crack. This is accomplished by drilling a hole at and through the crack tip. The theory behind this technique is that the crack tip—hence, the high stress raiser associated with the tip geometry—is blunted by the now larger and smoother drill hole. The theory is sound, but the results are mixed. Close examination of the Liberty Bell reveals that its crack was drilled out several times, only to continue on the opposite side. Although the Liberty Bell performed reasonably well beforehand, on July 8, 1835, the great bell cracked while tolling the death of Chief Justice John Marshall. It has never rung since. At this point, the crack is rather more a succession of drill holes than a true crack.

In summary, stable cracks, regardless of their origin, size, or type, are just that, stable. They may be disfiguring, even alarming, but the fact that they are not progressing means that there is stress equilibrium. If the element or sample has not catastrophically failed because of the crack, and the crack is stable, then the member or element will not fail, at least not from the stabilized crack. This is not to say that the section is not weakened and, therefore, vulnerable to other sources of stress; but a stable crack is a stable crack.

Unstable cracks are altogether different, simply because they may continue to catastrophic proportions and jeopardize the integrity of the element, member, or structure. Unstable cracks may originate for the same reasons discussed above and simply not have reached equilibrium; they may never reach equilibrium. Because of the uncertainty of the future of unstable cracks, all are of concern regardless of width or length; they are suspect until proven otherwise. Given the risks associated with crack instability, it is obviously a priority to ascertain stability or instability. If the determination is that the crack is unstable, then, among other things, we want to know the rate and degree of instability.

A relatively accurate, albeit unsophisticated, technique for determining crack progression is visual inspection of the exposed crack, more specifically, the exposed faces of the fracture plane. Obviously, many cracks do not expose a crack face, much less one that is visible to the unaided eye. Those that do are of such a width as to risk having reached structural proportions. Allowing for these limitations, however, crack faces can be inspected for clues as to age and progression. As soon as a crack forms, even a suboptical crack, the face is exposed to conditions radically different from the precracked condition. The availability of moisture and oxygen alone irrevocably alter the micro-environment of a crack face. As soon as the environment changes, the deterioration mechanism of the material at the crack face changes. If the products of the altered deterioration mechanism are detectable and reasonably measurable, then the crack age can be approximated, and the progression estimated, based on the degree and magnitude of the deterioration process.

Basing a determination of crack progression on visual inspection of the exposed crack face is of little real value other than as a quick indicator. Even a practiced observer of weathered materials will use other means to definitively corroborate his or her preliminary findings. The techniques and devices available for such determinations, though many and varied, are generally assignable to groups or types.

If circumstances permit, it is quite simple to mark and date the location of the crack tip. At a subsequent observation, the previous mark allows the observer to determine whether the crack has advanced, the magnitude of the progression, and the time required to achieve such a change in length. This is a great deal of information for the price of a simple pencil mark. A slightly more sophisticated version of this approach involves mounting a transparent, plastic grid over the tip of the crack and using the gridded plastic cover to track the progression of the crack tip. There are a multitude of variations on this theme, all of which are methods for tracking and dating the advancement of the crack tip.

Another general approach is to monitor crack width. The manner of tracking the width of opening of a given crack can be as simple as fixing a board to one side or the other of the crack, such that it lays across the crack itself. The board will move with the crack edge on the same side as that to which it is fixed, and relative to the opposite side of the crack. By striking a line on the edge of the board corresponding to the free edge, we can record the relative location of the free edge.

More sophisticated and precise gauges can be used in a similar manner to measure crack width and movement. Some such devices are fixed to both sides of the crack and have sliding vernier scales to measure distance.

Measuring crack width alone, while quite common, is not very informative. Cracks may open and close, but width may or may not correspond to progression. The theory behind the notion that crack width corresponds to crack progression is based more on geometry than on material properties and performance. The geometric assumption is that a simple crack is shaped like an extremely long isosceles triangle. The sides of the triangle correspond to the opposing faces of the crack plane, and the base of the triangle corresponds to the crack width. In order for the sides of the triangle—hence, altitude of the triangle—to increase, the length of the base must increase. Euclid would be proud, because the geometric logic is flawless. Unfortunately, reality is somewhat more perverse than this. The presumptions underlying the geometric analogy are, one, that cracked bodies are absolutely and perfectly rigid, and, two, that the forces that precipitated the crack are uniformly distributed along, and perpendicular to, the crack.

The first presumption accounts for the notion that an increase in the crack width at any point will necessarily be transmitted though the material in such a way that all other widths across the crack along the entire crack length will also increase proportionately, and that the resulting wider crack is geometrically similar to the original crack. This is possible if and only if the material translates rotationally about the crack tip but is otherwise is absolutely undeformed. To disprove both basic presumptions simultaneously, imagine that the material at issue is a board of wood, and that we are trying to split the board along its long axis. We place a wedge into the butt of the board and drive it into the wood with a sledge. As the wedge advances the, a crack forms and the wood begins to split. We replace the original wedge with a wider wedge and continue the process. After a bit, the split is well advanced. We will notice, however, that the faces of the split wood are not straight, but curved, and that as the wedge advances, the tip of the crack may or may not advance proportionately with each advance of the wedge.

What is happening is that the material is flexing; the degree to which the force responsible is advancing the crack tip is heavily dependent upon the varying degrees of resistance to fracture at the tip, combined with the varying degrees of rigidity of the material between the tip and the wedge. This is, in fact, a multivariable problem much more complex than the Euclidean model suggests. Cracks can and do advance with very little, if any, general increase in crack width. Similarly, cracks may increase or decrease in width without necessarily advancing the length. In order to effectively monitor cracks using these general approaches, both tip and width should be monitored simultaneously. But even when both tip and width are tracked, there are certain limitations to this combined approach. Monitoring requires time, which may or may not be available.

Even when time is available for a reasonable duration, it is also necessary to make frequent observations. It is altogether possible for cracks to increase and then decrease in width between observations, with or without crack progression. This undetected information may have been relevant for an accurate assessment of crack activity, but was never recorded simply because of the frequency of observation. Another limitation of this approach is that it is generally limited to optically sized cracks, which is to say cracks that can be observed without much, if any, magnification. This not so much because we have to see the crack to measure it, as much as it is that we have to have gauges that can respond to and measure the movements of the crack. At crack sizes below the optical level, the devices described above are inadequate and dysfunctional.

Provided that time is available, observations are frequent, and that the crack is of sufficient size, however, the general approach to crack analysis of monitoring width and tip progression is informative and quite inexpensive.

Where the crack is either not visible or there is some question about incipient progression, it is possible to invoke strain gauges. These ingenious devices are used quite often in determining linear elastic deformations, which, in turn, can be translated into to stress within the member. A strain gauge works on the principle that electricity incurs resistance as it passes through a simple wire proportional to the length of the wire. If a given wire is stretched, its resistance to electrical current increases. If we measure and calibrate the resistance of a known current in a known length of wire, we can calculate the increase in length of that wire if we observe a measurable increase in the resistance of that wire.

The strain gauge consists of a very fine wire applied to a mylar backing, such that the wire loops back and forth along the strip of backing many, many times. The total length of a single loop may be only an inch, but that inch may include dozens of loops of the wire. If the backing is rigidly adhered to a surface, and if that surface moves, the wire on the backing will be stretched the amount of the elongation of the material times the number of loops of the wire. A current passed through the wire as it is elongated by the deforming substrate will incur added resistance, which can be measured on a volt meter. We can calculate the amount of elongation in the wire, hence in the substrate. If we know the modulus of elasticity and area of the material, we can calculate the force in the member and the internal stress of the material. The calculation is quite direct. Wire length (L) is proportional to resistance (Ω), and a change in resistance is proportional to the change in length ($\Delta\Omega/\Omega = \Delta L/L$). We can, therefore, solve for ΔL. The change in wire length divided by the number of turns or loops of the wire ($\Delta L/n$) is equal to the change in length of the substrate over the length of the strain gauge (L/n): $\delta/d = (\Delta L/n)/(L/n)$. If we know the modulus of elasticity (E), the force in the specimen (F) = $E\delta/d$ and stress (σ) = F/Area. In terms of crack analysis, this means that we can apply a strain gauge ahead of an advancing crack tip and monitor the stress in the region essentially at the crack tip. Whether the crack advances or not, we can monitor the stress condition. The device that supplies the current and measures fluctuation can be attached to a continuous recorder; this frees us from the observation task as well.

Strain gauges are both accurate and precise, but also have certain limitations. Unless they are applied in a biaxial array, we can determine only the change in dimension in one axis. Fortunately, biaxial gauges are available, but they are not cheap (the wire is typically silver), and they are not easily retrievable. Because they are applied with super adhesives to the surface, not only can they not be retrieved, they remain permanently visible. The leads can be cut, but the gauge becomes part of the ambient decor. Furthermore, the application of the gauge to the surface requires experience and skill. If the surface is not properly prepared, the gauge may actually measure the behavior of a surface coating, or it may simply disengage.

As desirable as it may be to monitor cracks, the time required can have debilitating consequences when the crack is advancing rapidly. The need to ascertain crack stability must be balanced by the urgency and risk associated with allocating significant quantities of time to the analysis of the situation.

PERMANENT SET AND PLASTIC DEFORMATION

Implicit in much of the discussion about cracks is that the materials at issue are brittle, elastic solids. This is not an altogether unwarranted assumption. Building materials are, on the whole, far more likely to be rigid solids than they are to be liquids or gases. They are, as a matter of construction economy and appropriateness, also quite commonly classified as brittle. Some materials, however, are not technically brittle, elastic solids, although that statement requires an immediate qualification: this does not mean that the world of solid materials can be divided simply into two classes: brittle and elastic, and other.

To review briefly, being elastic means, one, that a material will deform under load, and, two, that the material will return to its original shape when the load is removed. Such deformation and recovery will recur virtually without limit to the number of cycles of loading, and without change in the amount of incident deformation upon subsequent equivalent load. The magnitude of load versus deformation is a property of the material, and is not, generally, a linear function. Brittleness is a measure of the fracture or cracking behavior, which is characterized as being rapidly accelerating, through-body cracking and without significant collateral deformation.

When materials do not perform elastically, they will usually deform plastically; that is, they may deform predictably enough under load, but they *do not return to their original shape or position when the load is removed*. This lack of recovery is the essence of plastic behavior. Put another way, the original deformation of the material is not the key to plasticity; it is the failure to bounce back. Common modeling clay is an example of a nearly perfectly plastic material: it will deform under load or pressure; it will even resist deformation. When the load is removed, however, the deformation remains imprinted in the clay.

It is important to remember that plastic materials resist deformation; they are not to be confused with liquids. Clay will carry load, albeit in a deformed condition; and it is likely that the deformation will arrest, which is to say that the deformation will not necessarily continue until failure occurs. This statement, too, requires immediate qualification; in this case, that clay, as well as other plastic materials, will display some elastic behavior. When the clay is depressed, but before is begins to deform plastically, it will, however briefly, exhibit some slight elastic deformation. When the pressure is removed, there will be some elastic rebound though virtually undetectable. A somewhat more accurate description of such plastic materials as a whole is that they are predominantly plastic or that they are characterized by their plastic properties, but technically they are elastic-plastic materials with rather short elastic ranges and significant plastic ranges. Soils, particularly clayey soils, are much more accurately described and engineered as plastic materials. As the sand content of the soil increases, the soil will behave more elastically, such that pure sand is an elastic soil type.

The use of soils as examples of plastic materials has some potential simplification problems, as does the linear elastic issue discussed above. Clays, for example, if confined, will behave more elastically. When we press on clay with our fingers, the material immediately below the point of the application deforms by displacing other material, which "flows" up the sides of our fingers. The clay is not so much compressed into its deformed shape as it is shifted into its deformed shape. The volume of material remains relatively constant. This is in contrast to the behavior of foam rubber, which is an elastic material that may be compressed into a smaller volume. Repositioned clay will not "flow" back to its original shape if the load is removed, hence its plastic property. On the other hand, if the clay were confined in a cylinder and compressed via a piston apparatus, and the clay could not reposition, or "flow," around the piston, then the confined clay would behave very much as a true elastic material.

Another detail regarding clay soils worthy of mention has to do with long-term deformations. It is a characteristic of clay soils to continue to deform over time even though the load is not increased. This phenomenon is not to be confused with plastic deformation; it is *consolidation*. Clay soils typically contain quantities of water that are interspersed among the clay lamina at the molecular level, and are not easily removed or drained because of the hydrophilic character of the clay particles. Assuming there is someplace for the water to migrate, when pressure is applied to the clay, the water is gradually squeezed out of the clay. The result is a net reduction in the volume of the clay mass, which, in turn, may result in further subsidence of the structure. This process, consolidation, is a function of the inherent density of the soil, the amount of entrained water in the clay, the magnitude of the applied pressure, and the confining pressure, which may inhibit the migration of the water.

A note regarding sand and silts, which are very fine-grained sands: When sands are fully saturated with water, and they are vibrated, the sand will disperse under applied load. Anyone who has jiggled his or her feet in the surf has observed the phenomenon. This is not plastic behavior; it is called *liquefaction*. It occurs when granular materials are supersaturated, under pressure, and vibrated. The liquid/sand slurry is mobile and easily displaced by a load that alternately pumps the slurry away from the site of contact, followed by the reoccupation of the voided volumes by the piston applying the pressure.

This action is used deliberately to drive certain types of piles into granular soils. The pile is a perforated pipe into which water is injected, then forced out through the perforations. At the same time, an eccentric, oscillating weight inside the pipe vibrates the entire pile. The mere weight of the pile, combined with the liquifaction of the surrounding soils, "drives" the pile into the soil.

Another class of materials that exhibits an elastic-plastic behavior is, of course, metals. This entire group of materials behaves elastically up to a certain limit, at which point they behave plastically. One of the consequences of this elastic-plastic deformation is that when the deforming load is removed, the metal will elastically recover with some permanent residual plastic deformation. This plastic deformation is called the *permanent set*. A common occurrence is when we bend a coat hanger to the point at which the wire remains permanently bent. Initially, the bending effort is resisted elastically, followed by a period of plastic deformation. When we release the hanger, it springs back only partially, not to its original condition. The amount of residual "bend" in the wire is the permanent set.

As a matter of building pathology, plastic deformation of materials is not a major source of long-term failure; however, it is associated with a phenomenon that is relevant. When metal sheets, bars, or plates are bent deliberately in order to fold or shape the material, at the line of the fold the material must exceed the elastic limit, otherwise the sheet or bar will spring back to its original shape.

At the precise line of the fold, several things may occur that will affect the long-term performance of the material. When the metal is folded, typically, the material to the inside of the fold will be put into compression, and the material to the outside of the fold will be in tension. Even though the material may be quite thin, somewhere within the modest thickness of the metal sheet is a neutral axis. As in the typical bending problem, the stress varies across the section of the material; but unlike the typical elastic bending problem the stress gradient is no longer linear. The extreme fibers of the material must exceed the elastic limit of the material or, upon release, the sheet will spring back. Having achieved the elastic limit, the extreme fibers will continue to elongate as the folding process continues. The actual stress at the extreme fibers that are behaving plastically will continue to increase, but the increase will no longer be proportional to the elongation.

On the compression side of the fold, the material compressed plastically may actually be distorted into long linear pleats of metal on the inside face of the fold. On the tension side, it is possible, even probable, that the very outside surface will exceed ultimate strength and actually tear along the fold. The depth of the tear will arrest as soon as the material at the base of the tear is plastically deformed at a stress level below ultimate strength. All of this is a long way of describing what most of us have observed in bent plates and sheets.

From a pathological standpoint, this simple act of folding forever changes the properties of the material at the site of the fold. The inside face is deformed in such a way that, when the fold is reversed even slightly, the "pleated" material will not unfold to its original flat condition. Instead, the ridges of pleated metal will constitute relative concentrations of material, and the material in the "valleys" between or at the base of the pleated material will immediately go into tension, not simply the reversal of some sort of residual compression. At the same time, on the opposite face, the tears in the surface will close but not mend. The net effective cross-section is forever diminished. Any reversal of the fold will have less effective material to resist opening of the fold. The net result is that plastically folded plates and sheets are normally severely damaged in the folding process, such that working the fold can and will result in tears in the previous compression face and in pleating on the previous tension face.

When a bent metal is worked back and forth through its plastic range, and then reversed and folded in the opposite direction, the tears quickly coalesce, and the material fractures. Anyone who has ever broken a coat hanger by bending it back and forth is familiar with the phenomenon. In buildings materials, complete reversal is rare; but the important point is that even the slightest reversal is sufficient to advance the fracture plane.

Strain Hardening

Once a metal has exceeded the plastic limit and the external force has been released, the material will rebound along a path that is parallel to the slope of the modulus of elasticity. When the stress reaches zero, the offset on the strain axis from the origin is the permanent set. This then establishes a "new" elastic constant originating at the offset or permanent set. When the metal is reloaded, the stress/strain relationship will follow the relocated but still constant slope of the modulus of elasticity up to the point where it was previously in the plastic zone. The difference in the second loading is that, effectively, the elastic limit has been extended to a value previously in the plastic range. This process can be repeated several times, with each iteration resulting in an added permanent set and an increase in the elastic limit. This process is called *strain hardening*, and is occasionally done deliberately to enhance the yield point of the metal. An important note regarding strain hardening is that the ultimate strength is unaffected; thus, theoretically, a metal can be "worked" to a point at which the plastic range is eliminated and the metal assumes the fracture properties of a common nonductile material.

Historically, strain hardening has been employed at weld metal to bring it to a yield point at least equal to or in excess of the base metal. The process for achieving strain hardening of weldments was through peening, whereby the weld was beaten with a peening hammer. At the point of contact, the impact of the hammer raised the stress level of the weld metal into the plastic range. As rapidly as the stress was imparted, it was released, leaving behind an indented spot of strain-hardened metal. The typical peening hammers were named for the shape of the impact head of the hammer—for example, a point peen hammer, a chisel peen hammer, and a ball peen hammer. Of these three, the ball peen hammer remains a common item in the typical toolkit even though the owner of such an item probably has not the faintest idea as to its origin or purpose.

Creep

A particular and not well-understood phenomenon associated with solid materials, notably wood and concrete, is called *creep*. Creep manifests by irreversible, long-term deformation of structurally adequate members under constant load. The member deforms in bending, for example, and develops the characteristic deflection. After a period of time, the load is removed but the deflection remains. Another example is that of a cantilevered retaining wall, which deforms elastically, precisely as engineered, and continues to slowly deform more over a long period of time.

For the most part, creep is a predictable and not terribly difficult problem to solve, as long as it is recognized for what it is. Most structural engineers will suspect creep whenever dealing with concrete in bending, and probably whenever dealing with long-term deformations in timber, but many practitioners may mistake creep for elastic deformation. Even if the deformation is mistaken relative to its origins and causes, there is still no particular consequence to the error until the practitioner effects an intervention. When load is relieved from a member in elastic bending, the member will rebound and gain its undeformed shape. A member deformed by creep will rebound some elastic portion of the deformed shape; but even after all load is removed from the member, there will be a residual deformation. In terms of the resulting permanent deformation, creep resembles the permanent set of plastic deformation. Some theories regarding creep are based in plastic deformation theory, but it is probably more accurate to say that creep remains incompletely understood.

When practitioners attempt to reverse the effects of creep, often they achieve untoward results. A member deformed by creep is not in a state of internal stress, so reversing the deformation will not reverse the strain of elastic stress; instead, it will induce stress in the member. To the extent that the member may have accommodated the deformed condition, reversing the deformation may not only induce negative stress into the individual member, it may also disrupt an entire system.

3.3.4 Biological Mechanisms

The actual topic of this section is considerably narrower than the title implies. It addresses timber and two general timber-related mechanisms, timber rot and termites. The reason for focusing specifically on timber is one of economics. There are other biologically based deterioration mechanisms that we will outline, but not examine in detail, because the magnitude of all other biologically derived deterioration combined is insignificant compared to deterioration of timber, simply because there is such a huge inventory of the material. Moreover, there is no foreseeable prospect that the dependence on timber will diminish. After choosing timber as the dominant topic of this section, the decision to concentrate on rot and termites was obvious: they are the dominant mechanisms associated with timber deterioration.

But rot and termites are well documented and treated in the more popular preservation-related texts, so why replicate the effort? The reason is different for rot than for termites. In the case of rot, the reason is depth of subject matter. The reason for discussing termites is that there have been some changes in the termite world since some of the more popular texts were published.

Before turning our attention to rot and termites, however, we should discuss a few items regarding other materials and other biological mechanisms. Simply, we must acknowledge that many organisms view our buildings and the materials with which they are constructed as either food or shelter, and occasionally, as both. To do otherwise is naive. Our concern in these regards is generally limited to the damage these organisms do to our constructions and materials, and so it is this aspect of the relationship we share with biological organisms that we will discuss. However, there is another aspect of biological infestation that is arguably of equal or greater concern, and that is human health. We tend to discount the health risks of some infestations.

More people die every year from diseases carried by pigeons than have died or ever will die from the effects of asbestos, lead paint, and other materials hazards combined. For reasons that defy logic, we will spend billions to rid schools of asbestos, but allow pigeons to roost where they will. As professionals in building pathology, we have a special interest in such health-related issues because we often must enter buildings heavily infested by potentially infectious birds, bats, rodents, and other disease carriers. Or we employ others who enter these areas on our behalf. We have an obligation to our staffs and clients to warn them of this class of biohazard.

Infection aside, there is physical damage associated with the habitation in buildings of larger species such as birds, rodents, reptiles, and others. These classes of animals almost always treat buildings as shelter. The chewing and/or clawing damage they cause is more modification of the fabric than it is consumption—though there certainly are exceptions. Mice and rats, for example, will eat almost anything at one time or another. There are cases of mice eating insulation from around electrical wiring, deer chewing on house siding, and rats eating caulking compound; still, the collective damage caused by these instances is small. Thus, the occurrences are not considered pathological because they do not conform to the notion of necessary and sufficient conditions. Of course, damage may result from such activity, but we are limited in what we can do to prevent it or to mitigate future damage, other than eliminate the intruders.

We respond similarly to infestations of certain insects, even as we acknowledge the damage as undesirable. As a class, for example, we accept damage from carpenter bees and wasps, because the prevention and cures are as expensive and unreliable as the infestations are unpredictable. We tend to respond aggressively to infestations by carpenter ants, but more from a distaste for the large, black insects than because of the damage they cause. Carpenter ants rarely jeopardize a structure, although, arguably, they may remove reserve capacity in a structure. We get rid of them, if we can, because we do not like them, not because they are particularly damaging to fabric.

When a building and its constituent materials shift function from shelter to food source, our concerns legitimately elevate as well. But even then our concern is mitigated to some extent by the rate of damage versus the cost of eradication. Powder post beetles, for example, use wood both as shelter and as food. They are very difficult to eradicate because they inhabit the wood well below its surface, and most treatments are not particularly effective and are fairly costly. Tenting and fumigating the building, for example, is an excellent technique for ridding the house of flies and moths, but the prospect of propelling the toxic fumes into the tissue of wood, thereby killing the beetle larvae, is dim. Fortunately, the rate of powder post beetle damage is slow enough to accept without much, if any, intervention. The same is generally true of death watch, old furniture, and other beetles, which feed on and live in wood. There are bacteria that exude oxalic acid and digest stone, and lichens that exude chemicals that accomplish the same results. The consequences are, however, even less significant than those caused by beetles. The damage is real, but the significance is minor. See Figures 3.2a and 3.2b.

FIGURE 3.2A Vegetation, particularly rapidly growing woody species, take hold in almost any condition that allows anchorage and moisture. Note the dislocation of the stone.

FIGURE 3.2B   Trunk and root structures extend into and through the wall with enough lateral force to generate a local collapse. Removing the entire root structure can result in as much damage as caused by the original growth.

FUNGAL ACTION AND OTHER MICROBIAL ACTIVITY

In many respects, timber is far from an ideal material. In fact, it has only two redeeming qualities: it is plentiful and it can be fabricated by hand with sharp-edged tools. Beyond that, timber is a design and construction disaster. The highest and best use of timber is arguably as a structural support for oxygen-producing leaves. If is must be consumed, then it may as well be consumed by termites and fungi, which indirectly recycle the nutrients back into other trees. Although we will focus on only two food cycles that rely on cellulose, a major component of timber, there are thousands of species that rely on living and dead trees as a food source.

There are two primary reasons for the popularity of wood as food and shelter. One is that cellulose is chemically closely related to sugar, which is an ingestible food by numerous species. The reason for that development leads to the second reason for the popularity of wood: cellulose is a very old molecular form. It has been on the planet millions of years, hence biological organisms have had all that time to evolve enzymes and ecosystems capable of exploiting the energy potential of the molecule. Perhaps when rayon is as old as the wood in trees, insect and fungi may consume it as rapidly and thoroughly as they currently do wood. In the meanwhile, we have to deal with the deterioration mechanisms we have, and those that affect wood are particularly important.

The part of the tree that is the target of all this interest is located in the wall of the common wood cell, the tracheid. To understand fungal rot and termite infestation, we will begin with a review of the structure of trees, generally, and the cell wall construction, specifically. We will examine the chemistry of the cell wall fibers, and the conversion of the cellulose polymers from wood into food.

Trees are classified into two extremely broad categories based on the manner in which the seeds are formed. Angiosperm species produce covered seeds; they are the broad-leaved, flowering-tree varieties. Gymnosperms, the older and more primitive of the two classes, produce naked seeds, which are exposed in cones produced by coniferous trees. The former are popularly called hardwoods, and the latter, softwoods. Unfortunately, these terms, while pervasive, are very misleading in their connotations of durability and strength.

The terms "hard" and "soft" are not characteristic of the density, strength, or even hardness of the wood, but primarily refer to milling terminology. They do, however, vary distinctly in their behavior toward deterioration, chemical cell structure, and growing patterns.

Most wood, both hard and soft, consists of five distinct layers. Beginning from the external layer to the center, the five layers are the bark, cambium, sapwood, heartwood, and the pith. Bark, technically called phloem, has an outer layer of dead cells with a thin layer of live cells. The cambium layer is the living tissue responsible for the production of both bark and the new wood, or xylem. It produces both varieties of cells through cell division. The live wood cells to the immediate interior of the cambium layer are collectively called the sapwood. Some of these cells are active in sap conditioning and the transportation and storage of food. The production of sapwood cells varies somewhat depending upon the time of year. In temperate zones, most softwoods have two distinct growing intervals. Not surprisingly, earlywood is the portion of growth developed in the early part of the growing season. The cells are usually large, thin-walled, and can be light in color. Latewood, on the other hand, is formed during the slow growth periods, typically in late summer. Its cells are usually smaller, thick-walled, and primarily dark in color, which in turn, produce a stronger and denser wood. It is this variation in an alternating pattern of light and dark wood that constitutes the annual ring that represents one full year of wood growth. The variation in seasonal growth is not as distinct in hardwoods, and, not surprisingly, is virtually indistinguishable in many tropical species where there is little variation in the seasons.

Helpful in the classification of wood species are the variations in cell reproduction that can be visually identified through growth rings. Identification becomes a critical issue when trying to characterize and understand the fungal attack of a particular wood species, because specific types of wood are more or less resistant to infections. Visual identification of species based on figure and ring conformation can, however, be tricky. To illustrate, in a cross-section of red oak, the largest vessels are seen in the earlywood growth period, which is referred to as ring-porous. Vessels in black cherry, maples, and birches are identical in size, and exhibit an even distribution pattern. This type of wood is described as diffuse-porous, because it is difficult to differentiate between the beginning of the latewood period and the ending of the earlywood period. Yet a third group, termed the semi-diffuse porous hardwoods, generally have smaller vessels in the latewood, but the distinction is not clear. Black walnut is a good example of this type of wood.

In softwoods, the size and distribution of a tracheid, the primary cell, assists in the identification between earlywood and latewood. Still, identification can be complex. For example, in southern yellow pines, the diameter of tracheids is large in the earlywood phase, but changes abruptly when entering the latewood stage. Conversely, the cell sizes in the eastern pine do not change in the same manner, making it complicated to differentiate latewood and earlywood. See Figure 3.3.

Physiologically inactive older cells comprise the center of the trunk or branches, commonly referred to as the heartwood and pith. Their principle contribution is as structural support; however, the heartwood is of particular economic significance. The sapwood, which consists of younger cells, is the zone of fluid transportation up and down the trunk. As the tree grows, the trunk increases in diameter. The cells that transported fluid when the tree was smaller are no longer proximate to the active portions of the limbs and, hence, the leaves. The older cells are, in effect, buried alive in the interior of the trunk. The fluid transportation function is obviated by the burial, and the cells become inactive. The inactive cells are potential sites of disease, so they are often "treated" with infusions of spontaneously produced chemicals, which, in many cases, "preserve" the inactive

FIGURE 3.3 This roof-rafter condition shows three levels of rot. Member A is present but has lost all structural integrity. B is absent altogether. C is present with significantly diminished capacity.

cells that lend structural support to the trunk without developing fungal disease.

Wood itself consists of a complex mixture of large, heterogeneous organic polymers that comprise the majority of the cell wall (90 to 95 percent). The three primary chemical constituents are cellulose, which imparts the skeletal strength of the wood cells; hemicellulose, which exists as a matrix material; and lignin, which provides stiffness within the cell. The highest concentrations of cellulose are found in the inner wall layers of the cell wall, whereas the greatest amount of lignin is found in the outer regions. There are even higher concentrations of lignin at the corners of the cell hexagon, about 85 percent.

Chemically, cellulose is defined as a monosaccaride, to describe the fact that it is a simple sugar that cannot be further hydrolyzed to produce a simpler sugar. In contrast, hemicelluloses and lignins consist of various polymer or polysaccaride types. It is believed that the hemicellulose and the lignin may form an interpenetrating polymer complex, with covalent bonds between them.

The remaining substances that make up the dry weight of wood consist of resins, fats, waxes, and simple phenols, which are called extractives and which are concentrated in the heartwood. It is composition and concentration of these extractives that typically accounts for the resistance to disease in wood. The fact that they are more highly concentrated in the heartwood accounts for its relative resistance, and, therefore, for its relative economic value.

Cellulose, the principle structural component in all plants, is the main source of mechanical and hygroscopic properties. Of the 90 to 95 percent of the dry weight of wood, cellulose makes up 40 to 45 percent of that weight. The cellulose monosaccaride is formed through photosynthesis of atmospheric carbon dioxide. It is polymer of D-glucose, a six-carbon sugar. Each cellulose molecule consists of 7,000 to 12,000 glucose residues ($C_6H_{10}O_5$). The linkage of any two glucose molecules is accompanied by the elimination of one molecule of water; and each glucose molecule added to the chain is rotated 180 degrees. When this process is repeated, it is called polymerization, hence, a polymer chain.

The molecules, which are 3 to 5 micrometers long, are organized in linear bundles called elementary fibrils, which have strong intermolecular hydrogen bonds. Two or more elementary fibrils assembled into a flat ribbon form a microfibril, which is considered the smallest structural unit of the cell wall. Each of these units contains approximately 40 cellulose molecule chains, and has a diameter ranging from 2 to 4 nanometers (1 nm = .000001 mm). Between the microfibrils are intermicrofibrillar spaces about 10 nm wide. Also, within the microfibrils are intermicrofibrillar spaces about 1 micrometer in width. These spaces are primarily filled with encrusting surfaces such as hemicellulose, lignin, and pectic substances. The cellulose content of the cell wall is very high in the secondary walls, but decreases toward the intercellular regions.

The tensile strength of the wood comes from the high percentage (70 percent) of crystalline glucose molecules in the fibrils. Some experts say 30 percent of these molecules are amorphous, as opposed to a crystalline form, which means that they are in a random state. Others believe only 5 to 10 percent are amorphous. These molecules are generally arranged lengthwise with regard to the microfibril axis, but are parallel to each other only in portions. The regions that remain are oriented cellulose chains called crystallites, each containing approximately 100 cellulose chains. The length of a cellulose chain is approximately 100 times that of a crystallite, so one cellulose chain may pass through many crystallite and amorphous regions. The highest degree of cystallinity is in the core of the microfibril, with less-ordered polysaccharides near the perimeter.

Hemicellulose, which comprises approximately 20 to 30 percent of the tissue matter, is a low-molecular weight polysaccaride. It tends to be highly soluble and is likely to decompose by hydrolysis, which is the splitting of the chemical bonds involving the addition of a hydroxide anion of water. Hemicellulose contains fewer than 200 sugar units, which are less ordered than cellulose, and unlike cellulose, are branched rather than linear. Whereas cellulose consists of a repetition of like glucose units, hemicellulose consists of several different sugars: polymers of D-glucose, D-galactose, D-monnose, l-arabinose, D-xylose, and 4-O-methly-D-glucuronic acid. The composition of hemicellulose differs between hardwood and softwood. For example, softwoods contain galactoglucomannon and arabinoglucuronoxylan, whereas hardwoods contain glucuronoxylan and glucomannon.

Lignin is a cross-linked phenolic polymer that is unique to vascular plants. It is an amorphous chain in that it does not have a long-range order. Lignin constitutes approximately 20 to 30 percent of the wood weight. Unlike cellulose and hemicellulose, the exact structure of the lignin is not well defined. It is not a carbohydrate like cellulose or hemicellulose, nor is it an aromatic material produced by nature. Lignin is a three-dimensional amorphous polymer consisting, to the best of our knowledge, of pheynyl-propane units of three cinnamyl alcohols: p-coumaryl, coniferyl, and sinagpyl. During synthesis, the monomers are converted to free radicals that randomly join to form a structurally complex three-dimensional polymer. Because the bonds are held together by a variety of ether and carbon-carbon bonds, lignin is strong and largely resistant to hydrolysis. Lignin polymers in hardwoods contain a lower structural integrity, and in comparison to softwoods, chemically degrade more easily. The amount of lignin within the cell walls varies based upon its location within the wall itself; in the middle lamella between the cell wall there is a high percentage, which then decreases toward the inner cell wall layer.

Despite their small percentage, extractives perform a significant role in determining numerous properties. Although they are not part of the wood substance, extractives are deposited in cell lamella and cell walls. Given the species of the wood, the extractives can govern the color, odor, permeability, resistance to decay, and the specific gravity. Furthermore, extractives can determine the resistance of wood to fungal attack; for example, heartwood usually contains a small percentage of extractives that assist in its resistance to microorganism attack. Depending upon the species, extractives can make up 1 to 20 percent in some tropical species. Extractives include, but are not limited to, wood resin, fatty acids, sugars, terpene, polyphenols, and tannins.

There are six primary cell types within a softwood: axial (longitudinal) tracheids, axial parenchyma, epthelial, transverse (radial) tracheids, ray parenchyma, and ray epthelial. A softwood cell is predominately composed of axial tracheid cells—elongated, tubelike, conductive cells that are the primary source of water and particle transportation. Tracheids run parallel to the stem and are usually 3 to 5 mm in length, with a diameter ranging from .02 to .04 mm. As previously stated, earlywood tracheid fibers exhibit a wide diameter and thin walls, whereas the latewood cells have a narrow diameter with thick walls.

Parenchyma cells have living cell contents in sapwood or organic inclusions, sometimes in crystalline form. They are bricklike cells with simple pits that assist in food storage. Ray parenchyma cells can make up the entire wood ray, or may combine with ray tracheids. Although this cell variety occurs in most species, these cells are not found in pine or spruce species.

Thin-walled, parenchymatous, epithelial cells serve a secretory function by producing resins, gums, latex, and other materials. As such, the epithelial cell, with the exception of a few rows of young cells in the cambium and parenchyma cells in the sapwood, is the primary live cell within the wood, and contains a nucleus and protoplasm.

The center of the typical tracheid cell exhibits a warty-type membrane, which lines the inner cell lumen or cavity. Surrounding the lumen are the three secondary layers, or lamella: inner, middle, and outer. The middle secondary layer contains the bulk of the cell wall material, and is composed of approximately 50 percent cellulose. Microfibrils are oriented within 20 degrees of the fiber axis, which causes tangential and radial swelling when moisture enters in between the chains of hydrated wood. It is the absorption of water in this layer of the cell wall that ultimately accounts for most of the shrinking and swelling of timber at the gross level. In addition, the inner and outer layers of the secondary wall serve the purpose of constricting the amount of swelling permitted in the middle layer, and contributes to the dimensional stability of the wood.

A pit is the recess or gap in the secondary cell wall with an external membrane that enables the flow of water and small particles between microfibrils. Typically, the pit is located at the tapered end of a tracheid where they overlap. A pit's auxilliary function is to connect adjacent and contiguous axial and ray cells. There are several types of pits located within a cell: pit pairs, simple pits, and border pits. Pit pairs are complementary pits that connect cells. Simple pits connect parenchyma cells, and are relatively uniform in diameter along their lengths. Border pits, on the other hand, vary in diameter along their lengths and are typically located in the radial walls of the axial tracheid, but they can also be found in other cell wall types, depending upon the species of the wood. Pits assist in the decay of wood, as they provide one of several access points for the hyphae, a thread-like substance of fungal rot, which will be discussed in greater detail in a subsequent section.

The cell functions in hardwoods are more specialized than those of softwoods. Similar to softwoods, hardwoods have six types of cells, but are all axial in orientation. They include vessels, tracheids, fiber tracheids, libreform fibers, axial parenchyma cells, and epithelial cells.

Hardwood contains three primary cellular components. The prosenchymatous tissue includes the vessels (axial section) and the pores (transverse section). Fibers encompass both the tracheids and the libriform. Because of their thick walls and closed ends, both the tracheid and libriform fibers are unsuitable for fluid conduction; their primary function is to lend strength to the wood. The long and narrow, finely tapered ends of the libriform fibers can exceed 1.5 mm in length and 20 to 40 mm in diameter. They have a thick wall that surrounds a narrow central lumen. Tracheid fibers are elongated, yet shorter and less pointed than the libriform fibers. Furthermore, adjacent libriform fibers have simple pit connections, whereas fiber tracheids have slitlike bordered pits with a central pit chamber.

The transportation of water and particles primarily flows through the open vessels traversing the pit pairs to the ray cells, vertical parenchyma, and fibers. These vessels account for 50 to 60 percent of the volume of hardwoods. The thin-walled vessels tend to act as capillaries, and pits lead from the vessels to surrounding fibers, tracheids, or parenchyma cells. The vessel membrane is continuous across the entire chamber, and consists of primary wall material with a random pattern of microfillibars. It is incapable of aspiration due to the absence of torus, the point at which the floral leaves grow, but the flow path exists between the microfilbrillar strands.

Vessels tend to be short and perforated, and align end to end with perforation plates. Perforation plates vary in size and number of openings, but their relative resistance to conduction is low. Simple perforations are single, large, rounded openings. Multiple perforation plates create scalariform plates, which have bars that form a ladderlike configuration in the vessel opening. Flow can vary within the heartwood and the cell type of various woods.

Like softwoods, the hardwood cell embodies axial parenchyma cells; however, there are two forms: paratracheal and apotracheal—those associated with vessels and those not in contact with vessels, respectively. Lastly, tracheids include vascular, which do not have openings and resemble small latewood vessels, and vasicentric, which have tapered ends. Similar to softwood, radial cells are either parenchymatous or epithelial. Epithelial cells secrete gums and resins. Other components of hardwood are the intercellular spaces known as gum canals, and parenchyma elements.

Fungal attack results in changes of color, structure, chemical composition, and properties of wood. Most critical are the changes that occur during the incipient period, because once in the advanced stages of decay, the wood is useless. In order for wood-destroying fungi to infiltrate the wood, certain environmental conditions must be in place. These conditions can be divided into two primary groups, those that relate to its establishment and those related to its continued survival and growth. Three primary elements are needed to foster infection:

- The source of the infection, that is, the introduction of spores or invasion by fungal mycelium from an established source.
- The presence of a suitable substrate, suitable temperature, and adequate moisture supply.
- The absence of poisonous or inhibiting substances including preservative chemicals and toxic heartwood components from the substrate.

For continued growth, fungi require: food, oxygen, moderate temperatures, and moisture. Ph and competition among microorganisms are also factors. Although most fungi do not need a light source to grow, a few types require light for continual reproduction.

Generally speaking, the optimal temperatures for fungal growth range between 70° F and 80° F (20° to 25° C). Fungal activity is reduced at temperatures below 50° F (10° C) and higher than 85° F (30° C). Fungi are killed, and growth ceases, at very low or high temperatures, such as below 32° F and above 104° F. High temperatures in concert with very high relative humidities (steam) are more effective in killing fungi than just dry heat alone. These types of temperature-related control mechanisms, do not, however, protect wood from new attack.

Favorable combinations of moisture and air are requisite for fungal activity and growth. Air occupies the cavities of the wood mass that are not occupied by water. To initiate fungal attack, generally wood must be at or near fiber saturation, that is, contain 28 to 30 percent moisture, based on the oven-dry weight. Different species do, however, have varying moisture content requirements. As moisture reaches the fiber saturation point, fungal activity increases, but cell cavities are still empty of free water. Optimal conditions for fugal decay are at moisture levels slightly above that of the saturation point, namely moisture contents between 35 percent and 50 percent. At these moisture levels, the cell walls are saturated and there is a thin layer of free water on the inner surface of the inner wall layer, which directs the diffusion of enzymes from hyphae to the cell walls. Wood tends to shrink with a relatively low moisture content, and swell when the moisture content increases. Increased water absorption in the cell wall will result in an increased density of bound water and increased specific gravity of the moisture in wood.

At low moisture levels, the fungus is not killed, but remains dormant for long periods of time, and can be revived under conducive conditions. Some species have thick-walled spores that are resistant to drying. Others can actually transport water to decaying wood to increase the moisture content by releasing water during decomposition. Because the high moisture content deprives the fungi of oxygen, it can have a detrimental effect on its growth and infiltration into the wood.

There are two types of fungi: parasites and saprophytes. Parasites exist on living plants and animals, whereas saprophytes feed on dead matter, such as the cell walls of deceased trees. Our concern is obviously with saprophytes. Because fungi are unable to produce their own food, they must derive nourishment from organic materials synthesized by other living organisms. There are four groups of wood-inhabiting microorganisms: wood-destroying, wood-staining, surface molds, and bacteria. Wood-destroying fungi are the most devastating because they disintegrate the cell walls and change both the physical and chemical composition of wood. In living trees, fungal rot, though rare, usually attacks the heartwood, and sometimes, in advanced stages, extends to the sapwood.

Spores act as the main reproductive agent in wood decay, which has two life-cycles: asexual and sexual. In late spring, summer, and early fall, the spore reproduces asexually; the sexual life-cycle occurs during the late fall and early winter months. These microscopic spores are dispersed by wind, insects, and rain, and enter the wood through wounds in the trunk, roots, or branches. When in contact with wood, the spores secrete enzymes that attack wood cells and create threadlike filament called hyphae. Hyphae grow by the elongation of their tips, taking nourishment from the cell walls or cell contents. In an advanced stage, they can appear as a cottony mass on the wood surface, branched out from aggregates called mycelium.

Interactions between microorganisms can have an effect on fungal activity, related to the cohesiveness of the fungi to the substrate. On a macro level, this relates to the wood species as a whole; on a micro level, it refers to the specific elements of the wood, such as cellulose, starches, and soluble sugars. The competitive nature of saprophytes is determined by the following:

- The speed of spore germination and the growth rate of hyphae in the presence of soluble nutrients from the substrate.
- The ability of the fungus to produce enzymes necessary to decompose both soluble and insoluble wood components.
- The ability of the fungus to release antibiotic substances that inhibit the growth of other fungi and bacteria and its ability to resist fungistatic substances released by other microorganisms.

Wood-destroying fungi have a preference for a pH that is acidic, somewhere usually between 4.5 and 5.5. Optimum growth for both white and brown rot fungi are within this range. In addition, both of these fungal types increase the acidity of wood in the process of metabolizing, especially brown rot.

Cellulose and hemicellulose are broken down into simple compound sugars by the enzymes secreted by the fungi, which are then absorbed and metabolized by the degrading organisms. Extractives determine the resistance of the wood. A higher content imparts more resistance, whereas as lesser amounts of extractives put the wood at a higher risk for fungal attack.

Lignin is generally not affected by hydrolysis or the enzymatic attacks because of its three-dimensional amorphous polymer make-up. White rot, however, has the ability to degrade even the lignin. Typically, in most rot species, after the enzymes attack, lignin is the only element left, because the extractives—cellulose and hemicellulose—are removed by the infection.

A primary concern of fungal attack is the loss of wood strength through the breakdown of cellulose in the secondary cell wall, particularly crystalline cellulose, of tracheids in softwoods and axial fibers in hardwoods. This decomposition process involves both hydrolysis and oxidation, which is the chemical bond-splitting through dehydrogenation by oxygen. Wood decay fungi possess enzymes called cellulases (endocelluses and exocelluses), hemicelluloses, B-glucosidases, oxidases, and lignin-attacking enzymes (white rot only), which break down isolated wood polymers. To be effective in the degeneration of the cell wall microfibrils, the enzymes must work together simultaneously. Cell-wall-degrading enzymes are generally too large to penetrate and move freely in wood cell walls. Oxidative free radicals act as deterioration mechanisms (in brown rot) to allow for enzyme penetration.

An important factor in controlling the rate at which cellulose enzymes can access the cellulose microfibril core are the characteristics of the encrusting substances, hemicelluloses, and lignin. For example, while the hemicellulose is generally hydrolyzed by decay fungi, lignin is much more difficult to break down. The lignin type and quantity in a particular wood species is a significant factor with respect to soft rot and is, to a lesser extent, a factor of white rot. Other factors to consider in the deterioration phase are the degree of crystallinity of the cellulose and the degree of cellulose polymerization.

The specific enzymes produced by fungi that degrade the cell wall are polysaccharide degraders and lignin-modifying or lignin-degrading enzymes. The polysaccharide degraders, cellulases and hemicellulases, are both facilitated by coincidental hydrolysis and oxidation reactions. Otherwise, the enzymes are too large to attack the cellulose molecules directly. In a multistep process, hydrolic cellulases break down cellulose polymers into saccharide monomers. It requires synergistic action by two types of cellulase enzymes and B-glucosidase. Exocellulase hydrolyzes crystalline cellulose into cellobiose in conjunction with the endocellulase that is hydrolyzing the amorphous cellulose into oligosaccharides. Cellobiose is then converted to glucose by the action of glucohydrolase, the hydrolyzing form of B-glucosidase.

Oxidative enzymes reduce cellulose through the cellobiose and glucose levels. For the reduction of cellulose to occur, all enzymes are needed in the degradation process. To illustrate, oxidases serve to prevent an accumulation of the end products of hydrolysis, which, if too highly concentrated, will cease the chemical reactions, thereby depriving the fungus of glucose monomers, its main nutrient.

The hemicellulose cells in the matrix region of the middle wall layer are of great concern with regard to cell-wall degradation. These hemicelluloses, surrounded by lignin in the microfibril core, are interdispersed with cellulose in the amorphous regions, and surround the perimeter regions of crystalline cellulose. To access the crystalline substrate, the enzyme mechanisms must break down and loosen this particular area. Specific enzymes, however, are required to break down certain hemicellulose types. For example, hemicellulose forms of xylans, mannans, and galacatans are reduced, respectively, by xylanaxes, mannanases, and glacatanases, which in simple terms means that the enzymes required to break down a particular saccaride are very specific. Hemicellulose composition and, in some instances, lignin content and types will dictate which wood rot fungi will attack certain wood species. If a particular fungus does not produce a complete set of enzymes, the decay process may progress, but either not as rapidly or not as completely as when a complete set is present and active.

The degradation of lignin remains only partially understood, although it is known that only white rot fungi will degrade lignin in large amounts. Brown rot and soft rot have the ability to demthoxylate, or modify, lignin. Some soft rot is able to degrade lignin, but at a nearly undetectable rate. Lignin deterioration is achieved by an oxidative radical mechanism involving lignin peroxidase, orligninase, and hydrogen peroxide.

Fungi are classified by the type of damage they produce and by their visual appearance. For example, there are brown rot, white rot, and soft rot, which are the most common; within those broad classifications are numerous specific types of fungi. There are, for example, more than 100 types of brown rot. Brown rot is usually identified with the fungus subdivision basidomycetes, though more recently it has been associated with dacrymycetales. The most common types found in North America are cellar fungus, scaly letinus, oligoporous placenta, gloephyllum trabeum, maze gill, and dry rot fungus called meruliporia incrassata.

Two forms of brown rot deserve particular attention: wet rot (*coniophaora cerebella*) and dry rot (*merulius lacrimans*). To prevent confusion from the beginning, we point out a misnomer inherent in the term "dry rot": dry rot refers to fungi that have the ability to transport water that attacks actual dry wood; the rot does not lack moisture, as the name implies. The term derives from the fact that the actual wood appears dry in the advanced stages of decay. In the early stages of brown rot, the wood lacks luster and appears dead. As the rot continues its infiltration, the wood develops an abnormal brown color. In the final phases, a cross-grained, checked mass occurs, followed by abnormal shrinkage, crumbling, or collapse. At this point, a small amount of pressure can reduce the wood to dust. Filaments of the hyphae usually are not visible on the wood surface, although in certain ideal conditions, a fluffy mycelium may be apparent. In fact, when hyphae are well established, they can produce long, pencillike growths called a rhizomorphs. At times this growth can extend 33 to 66 feet (10 to 20 m). Furthermore, hyphae strands can also penetrate through damp brick or stone, traversing mortar joints or moving behind plaster in search of another wood source. The property that brown rot attacks and degrades in the wood is its strength. Brown rot primarily attacks softwood, although it can penetrate some hardwoods.

Wet rot is a less serious affliction than dry rot. Because wet rot does not have the ability to transport moisture, its attack is more localized. To facilitate growth, wet rot needs a moisture content above 25 percent. When conditions are suitable, wet rot produces fine, dark-brownish strands and a green, leathery, fruiting body that produces masses of white hyphae.

Brown rot enzymes feed on cellulose and hemicellulose, the carbohydrate properties of the wood, thereby leaving a chemically unaltered lignin residue. Hyphae grow mainly in the wood cell lumina, and in the early stages of infection invade nearly every cell. As the cell walls are being consumed in the later stages of decay, the lumen hyphae decrease due to autolysis, which is the destruction of cells or tissue by their own enzymes. Brown rot colonizes the longitudinal surfaces of wood via the rays, and the hyphae penetrate into the axial cell systems of the tracheids and fibers. Pit tori, the central membrane of softwood border pits, are easily destroyed, but the hyphae may also attack by means of bore holes, which are direct transverse penetrations through the cell walls. In later stages, cell walls begin to collapse, and small splits occur near one of the secondary layers (S2). The order of cell wall loss is S2, S1, then S3, which shows that brown rot has a highly diffusible decay system. Only a lignin skeleton remains in the last stage. In brown rot, different layers of cell walls are attacked at the same time, whereas, in white rot, the layers are removed sequentially. The warty layers seems to be more resistant to brown rot and most susceptible to white or soft rot.

The bleaching of wood is the result of white rot fungi (*phanerochaete chrysosporium*) attack. This type of deterioration is associated with the subdivision of basidiomycotina, as well as higher forms of ascomycetes. The wood does not change drastically until it is well into the final stages of decay, when it appears as a spongy or fibrous mass with white pockets or streaks, separated by areas where the wood remains strong. Cuboidal cracking that is characteristic in brown rot does not occur in white rot. Surface toughness is quickly degraded, whereas there is a gradual deterioration of the strength properties. White rot is a common deterioration factor in hardwoods, but it can attack softwood, particularly those located at the ground level.

Unlike brown rot, white rot removes the lignin or the lignin and cellulose in tandem. In the beginning phases of decay, numerous white rot hyphae grow in the cell luminae. Later stages exhibit the lumen hyphae in autolysis, but they remain more abundant in white rot than in brown rot. Early colonization in hardwoods occurs through ray parenchyma and vessels, whereas in softwood the attack is primarily limited to ray parenchyma and resin canals.

Ray parenchyma may be nearly destroyed in hardwoods before there is substantial colonization in the vessels. Similar to brown rot, hyphae colonization from cell to cell occurs via pit pores and, eventually, bore holes. Recently, the presence of bore holes, erosion troughs, and the gradual thinning of the cell wall have been observed. The middle lamella is dissolved only in the final phases of deterioration. The corners with the highest lignin content remain intact until the absolute last stage. Both holes are created by finely tapered canals, which enlarge, then break open through the cell walls.

As previously stated, the decay mechanism in white rot has a preference for the degradation of lignin. Hyphae gain access to the lumen cell wall inner surface, then progress through the secondary wall layers S3, S2, and S1. Eventually, the compound middle lamella (the middle lamella and primary wall) is degraded, resulting in the separation of the wood fibers.

Soft rot is a relatively new term, coined in the 1950s to reflect its obvious softening effect on wood. It is a member of the subdivision ascomycotina and deuteromycotina. Differing from brown and white rot, soft rot has a superficial softening effect that makes it easy to remove by sanding or planing. Substantial weight loss can occur, however, but there is no appreciable effect on wood strength except for the loss of surface toughness. As the rot progresses, it superficially stains the wood into shades of black, green, orange, and other colors. When dried, the surface layers become checked and brittle, yet the underlying wood remains firm and sound. Soft rot typically attacks softwood, but hardwood is also susceptible.

Soft rot fungi affect timber exposed to very high moisture conditions, such as woods placed in water or woods placed in contact with moist soil. Common decay symptoms include cavity formation in the S2 layer, some erosion of the luminal surface of cell walls, and cell wall decay. In the early stages of colonization, the hyphae gain access through the rays of softwoods, the rays and vessels of hardwoods, and the axial parenchyma. In softwood, once the fungi has entered into tracheids, hyphal colonization proceeds rapidly; but unlike brown and white rot, the spread is passive in nature between cells. A principal feature of soft rot attack is the formation of biconically shaped cavities formed around the hyphae.

Although not considered a decay fungus, the soft rot can present a significant aesthetic problem. The microorganism usually attacks the sapwood of softwoods, but rarely affects the heartwood. Blue stain or sapstain is produced by hyphae or hyphae-producing by-products, of which the generic forms include ceratocystis and graphium. Starch that is stored in the ray parenchyma cells and sometimes in the tracheids is the primary food source. The staining fungi pass from cell to cell by way of pits or by boring holes, similar to white rot. This can cause minor problems in wood toughness, which can be reduced by 15 to 30 percent. Unlike soft rot, planing or sanding are not effective remedies because the staining fungi penetrate particularly deep into the wood. Sapstain primarily affects timber under adverse conditions and, as a general rule, does not affect wood in service.

Molds cause a discoloration on moist wood surfaces. Generic types include penicilium, trichoderma, and gliocladium. Usually, the discoloration is superficial and easily removed. In rare situations, the mold may penetrate deep into the timber and possibly engender the erosion of cell walls.

Probing the wood with a small pick and performing a visual inspection of the microbial phases are simple methods for identifying fungal attack. In the latter, the existence of molds, staining fungi, and soft rot on wood surfaces can be considered a precursor to wood-destroying fungi, hence they sound warnings for further attack. If, when probing with a pick, the wood breaks in long splinters, the wood is sound, but if decayed even slightly, the splinters break abruptly.

Visually inspecting heavy timbers for decay can be particularly deceptive and should be done using borings. The problem is that timber tends to absorb water along the long axis of the tracheids. Locations that present a cut across the cells, so-called end cuts, are particularly open channels for water absorption. The ends of timbers buried in pockets of masonry walls are prime candidates for absorption of water at considerable distances into the timber. The exterior surfaces may remain dry and sound while the core of the timber rots. This condition is called tunnel rot. Tunnel rot can easily escape visual inspection and even probing because the rotten core may be protected by a shell an inch or more thick of sound wood. Sounding, or tapping, the wood with the expectation that the rotten core will resonate a hollow sound is futile inasmuch as the rotten wood may be saturated and every bit as dense as sound wood.

The tool of choice for inspecting heavy timbers is a forester's auger. The device has a cutting tip and a hollow bore, and is screwed into the timber. A core of wood is cut and pushed into the hollow bore, from which it is retrieved with a long "spoon." The retrieved core provides an accurate cross-section of the timber, and can be microscopically inspected for fungal damage.

As stated earlier, the chemical make-up of the wood and its natural resistance to fungal decay are important considerations. For example, spruce timber is highly susceptible to heartwood decay, whereas black locust is extremely tolerant. The variation is vexing when considering wood-destroying basidiomycetes. Methanol, an extractive in wood, can resist fungus, as well as be used as a preservative. Other decay-resisting extractives are hydrolysable tannins, lignins, courmarins, alkaloids, terpenoids, and steroids, all produced as secondary metabolites. Some biocidal activity is effective in controlling wood-destroying insects, but varies according to concentrations and fungal decay types. Natural resistance, however, is not restricted to nonnutrient extractives. Wood density, nitrogen content, starch content, and lignin quantity and type contribute to the fungal resistance of both heartwood and sapwood.

In addition to the environmental and physiological requirements, there are necessary conditions for fungal activity that directly relate to the wooden components of structures at or near ground level. Near this level, maximum moisture and oxygen conditions exist for fungal growth. Bacteria, molds, staining fungi, and brown and white rot basidiomycetes are attracted to this region. In dry areas above this region, molds and staining fungi will be active if the post is exposed to the weather. In addition, all horizontal surfaces of wooden components exposed to weather are subject to direct rainfall, and the intermittent moisture conditions created will promote staining fungi, soft rot, and, occasionally basidiomycete attack. Exposed wood vertical members are the most susceptible, because of the capillary action of water in the axially arranged cells.

A final factor related to moisture conditions is the premature decay of external trim joinery on structures. A moisture content profile in a window sash extends from the lower joints upward. The breaking of protective coatings, such as paint, provides means for the infiltration of water; and the joints where this occurs provide suitable substrates for decay fungi.

There is a somewhat misunderstood or mythical logic that states that wood submerged in water, because it is devoid of oxygen, is not susceptible to decay. To a certain, but very limited, extent this is true. It is a fact that the submersion of the wood precludes its contact with airborne organisms that facilitate the types of decay discussed above. There is, however, a devolution of the wood within its aquatic environment that is typically the primary mechanism to determine the type and degree of decay the wood encounters. Many other deterioration mechanisms can induce the decay within waterlogged wood, such as the chemical nature of the water, the depth of the object within the water, the speed at which the water is moving, clays and silts in the water, water temperature, the organisms and microorganisms present in the environment, pH and saline contents, alkaline level . . . the list could continue indefinitely.

Waterlogged wood deteriorates very similarly to fungal rot as it breaks down the chemical composition of wood. The extractives are the first to deteriorate, because many are soluble and easily decomposed by hydrolysis. Next, the soluble, less stable hemicelluloses, which include the pectins and pentosans, decompose. The sugar-based chemical structure of the pentosans ($C_6H_{10}O_5$) is the same polymer chain carbohydrate found in cellulose. Next, the more stable hemicellulose sugars are attacked by enzymes of microorganisms that pass through the large openings in the wood polymer matrix due to swelling. Cellulose decomposes simultaneously with the hemicellulose. It has been suggested that this process can occur with both microorganism and hydrolysis attack, but evidence of enzymatic process seems to prevail.

Once the cellulose, hemicellulose, and the extractives have decomposed, lignin is the remaining substance. As previously discussed, because it is a three-dimensional amorphous polymer consisting of phenylpropane units, it is not affected by hydrolysis or enzymatic attack. It can, however, be removed in the presence of ethanol or organic solvents.

There are two forms of decomposition in the cell walls of waterlogged wood. The initial stage affects the entire axial length of the cell. It begins at the interior of the cell with the degradation of the tertiary wall, followed by the secondary wall structure. The second form begins at the outer layer of the secondary wall, and shifts with the dissolution of the middle lamella, to adjacent cells dissolving the cell bonding. This process represents a radial deterioration that opens passageways between cells for microorganisms.

The removal of the elements of the cell walls allows for the increased penetration of more liquid into the wood structure, which decreases the specific gravity and the relative strength of the wood. The additional water also tends to act as a cushion and buoys up the structure, keeping it from collapsing upon itself.

The added water also bonds to the hydroxyl groups of the remaining cellulose, which prevents the cellulose molecules from bonding and forces the cell walls to shrink inward. The hydrolysis of the crystalline regions of the cell wall and the resulting depolymerized cellulose chains make more polar terminal groups available for bonding. Thus, the water again acts as a stabilizer to prevent the internal structure of the wood from collapsing.

In badly degraded wood, the water content can exceed 100 percent based upon the dry weight and swelling of the timber. If the wood is retracted, the water is removed from the wood, and it is no longer stabilized. The polar groups on adjacent cellulose molecules tend to bond together and are unavailable for future water absorption. Additionally, the surface tension of the receding water is greater than the strength of the cell walls. Both of these factors can cause collapse of the cell wall structure, evident in shrinkage, cracking, twisting, and warping of a dried piece of wood. Once the cell wall structure has collapsed, even if the timber is reimmersed, it will not expand to its original size. The wood could essentially disintegrate.

A certain amount of water will always be a component of wood. In its waterlogged state, the water is only one of the mechanisms that causes deterioration. The access water facilitates the hydration process of the cells, allowing for enzyme attack and the decomposition of the extractives, the primary component in the defense of attack. If wood were submerged in pure water under ideal conditions, deterioration to the point of collapse of cell wall structure may or may not occur. Even with the enzymes present, total deterioration of the extractives, cellulose, and hemicellulose may not be evident for hundreds or thousands of years, hence the notion that submergence is beneficial. It is not.

TERMITES AND OTHER INSECTS

Wood-attacking insects fall into three classes: *Coleoptera*, which generally includes various types of beetles; *Hymenoptera*, which includes wasps and ants; and *Isoptera*, which includes termites. Termites are social insects, meaning that they are reliant on one another for their individual as well as collective survival. Because of their body make-up, they must live in a closed environment, which is the nest, with regulated temperature and humidity. This requirement limits their mobility. Likewise, their diet is restricted to cellulose obtained from wood, leaves, and humus. That said, termites have been able to radiate adaptively into various food niches, such as books and other paper sources.

The soft-bodied termite is a six-legged insect with a body divided into three parts: the head, thorax, and abdomen. But the size, color, and uses of these parts vary from species to species and even within the various families.

In most instances, situated on the head of the termite is a pair of bead-like antennae. In addition, there are usually a pair of mandibles used to devour wood and for some defensive activities. Two black, compound eyes complete the head. Some species, not all, exhibit a fontanelle, a needlelike protrusion used to excrete a milky-type substance. Typically, this is used for defensive purposes. The legs and, in some cases, wings are attached to the thorax.

Though a large body of literature exists on the termite in regard to its preference for eating wood in service, most discusses the insect and its habits as if it were one species exhibiting uniform characteristics and eating habits. Termites are much more complicated. Within the United States alone, there are more than 45 species; and throughout the world, there are approximately 2,000 to 3,000 known species, as well as many thought to be still unidentified. Species differ in size, color, nesting habits, formation of nest, reproduction behavior, and countless other ways. Each species brings its peculiar style of destruction to a building, and it is important that professionals recognize these differences and identify each species of termites properly before employing a control method because no one method will work for all species. For example, the same control method is not as effective for drywood termites as for subterranean or underground termites.

This leads us to types of termites; there are three general categories: subterranean, dampwood, and drywood. All demonstrate differing living and wood-destruction habits. The most common subterranean termite within North America includes two families: *Rhinotermitidae* and *Termitidae*, with numerous genera under both. *Heterotermes, Reticulitermes*, and *Coptotermes* make up the *Rhinotermitidae* family; *Microcerotermes, Macrotermes, Odomtotermes, Microtermes*, and *Masutitermes* comprise the *Termitidae* family. The drywood species are much more limited. The two families include *Masotermitidae*, with one genera termed *Mastotermes*; and *Kalotermitidae*, with three genera called *Incisitermes, Cryptotermes*, and *Neotermes*. The primary dampwood species is the *Zooterrmopsis* family, in which three genera are represented, the *Nevadensis, Laticeps*, and the *Angusticollis*.

Most North American termite species are found in the Nearctic region, which is generally defined as the area lying north of the Tropic of Cancer and south of the Arctic Circle. The majority of the more than 2,000 identified termite species occupy the hot and humid tropical environments in other parts of the world, such as Africa, South America, and Asia; but there are a number of termite species that can survive in the extreme temperature ranges of Nearctic region.

According to the Building Research Advanced Board, the heaviest concentration of the termites within North America reside in the southeastern states and California. Eastern, midwestern, and southwestern states host a moderate population. Generally, the farther north one travels, the lower the concentration of termites. In fact, termites are rarely found in parts of northern Michigan, North and South Dakota, Montana, Wyoming, Maine, Vermont, and Canada.

Among the most widespread species within the United States is the arid land subterranean *Reticulitermes* species. These termites pose a significant economic threat and are responsible for the bulk of the wood damage done to wood structures throughout the United States. Many can be found in the bulk of the lower 48 states. Within the *Reticulitermes* genera are five species spread throughout various parts of the United States: *Reticulitermes flavipes* is common in the eastern states; *Reticulitermes viginicus* most commonly occurs in the Midwest; *Reticulitermes hesperus* appears on the Pacific Coast, from British Columbia to lower California; *Reticulitermes tibialis* is found in the West, but this species is not nearly as common as the *hesperus*; finally, *Reticulitermes humilis* prefers Arizona and New Mexico.

Compared to the drywood and dampwood species, these subterranean insects are small; they nest in the soil, wood, or vegetable material that is in contact with the ground, and need a source of constant moisture to maintain themselves. Through the use of earthen tubes they construct, they can gain accessibility to wood or cellulose material above the ground. These tubes, or galleries, are made of "frass," which consists of triturated wood and earth cemented together using saliva and liquid feces to form a mud-like cement. (Unlike the drywood and dampwood termites, the feces of subterranean termites consists of liquid drops, not pellets, which they deposit upon the excavated wood, clear evidence of infestation.)

The tubes are the most conspicuous sign of infestation by subterranean termites. They can easily be seen, as typically they are of contrasting color to concrete walls, wood, and metal; moreover, they may be hanging or freestanding. However, these tubes may be indistinguishable if they are located on a dirt-streaked foundation or hidden behind a sill or foundation or between two adjoining pieces of wood.

Underground termites gain access to buildings through adjacent earth-filled steps, porches, terraces, patios, breezeways, and planters. And because the aforementioned constant supply of moisture is mandatory for their survival, any potential water source is a target for all subterranean termites as well. Leaking wooden water tanks, condensation on cold water pipes or air conditioners, clogged gutters or drains, leakage from showers, and plumbing or other faulty drains are all potential vehicles for the underground termite. Even buildings on concrete slabs are at risk, as they offer an easy access to the termites via the soil to wooden framing joints and trim. Door jambs and grade stakes frequently penetrate the concrete as well. Heating ducts, insulation around the slab, and expansion joints also allow for termite passage through the soil to the building.

Subterranean termites, typically, eat the soft spring growth and summer growth, but leave the late wood half eaten. Their intricate galleries tend to form concentric circles around the annual rings in the cross-section of the wood, working along the grain, leaving paperlike pieces of wood divisions.

In contrast, drywood termites establish themselves in wood that is not decayed, not in contact with ground moisture, and which, to all appearances, seems perfectly dry. Unlike the dampwood or subterranean termite, the drywood termite can live on as little as 2.8 to 3 percent moisture content. These types of termites are commonly found in attics of buildings, where they can survive without any soil connection. Damaged wood infested by the drywood termite exhibits clean, smooth cavities, which look as though the surface had been gone over with sandpaper.

Drywood termites, sometimes called the "white ant," because of their close resemblance, feed just below the surface of the wood, and in most timber, attack the sapwood. Not found in the colder northern states and in Canada, the drywood termite prefers California, southern Florida, and Hawaii. Their feeding produces a granulelike dust that is substantially coarser in comparison to the furniture beetle. When a piece of attacked timber is broken, dust pours out of the interior even though the skin of the wood appears sound.

Typically, the only evidence of a drywood termite infestation is the mounds of dust that are pushed out through the exit holes of the nest, though heaps of excreta pellets resembling small, light-colored seeds are also a clue. These heaps are generally found below the infected timber. The pellets usually fall to somewhere just below the exit hole, and the shape of the deposit is a guide to the distance they fell. The pellets are regular in size, fluted longitudinally, and vary in color from nearly white to light tan to reddish brown and, occasionally, to dark brown or black. The differing colors help to ascertain the age of the pellet. Similar to the wood from which they originate, their colors are bright when fresh but become dull with age.

The drywood termite produces smaller colonies than the ground termite, thereby making its destruction much slower. Similar to subterranean termites, drywood colonies are begun when a reproductive pair enter a building though a crack or crevice. Typically, the drywood species *Kalotermitidae* maintains the most simplistic nests. Construction activities are limited to the erection of a few partitions. They do not employ the intricate gallery system of the subterranean termite.

Another economically important termite species located within the United States is the dampwood insect *Zootermopsis angusticollis*, which is a part of the *Zootermopsis* genus. They are present on the Pacific Coast, particularly in Oregon, Washington, and southern British Columbia; evidence of their presence is the appearance of pellets. These insects are the largest of the termite species and, because they require a large amount of moisture, are often found in logs and damp or decaying wood. They are able to withstand relatively low temperatures, even though they are confined to the wood in which they are working.

In younger colonies, these termites locate in the cells just below the bark of downed or standing dead logs. From the initial cell, passageways are developed perpendicular to the wood surface. Transportation of this species is generally through lumber shipments that have originated from the Pacific Northwest. They are typically termites from mature colonies rather than young colonies.

Most termite species have an annual cycle of activity and the amount of damage caused over a year is not high. The list of materials that termites damage does not stop with wood stumps and structures. In their quest for food and shelter, termites are seeking cellulose, and they do not care whether it comes from plywood, paper, books, cartons, or any other of a great variety of materials containing cellulose.

In many modern structures, both residential and commercial, termite attack is not limited to the structural elements. The insects may feed on flooring, paneling, window and door frames, even when the structure is of a concrete or steel frame. Wallpaper and wallboards are at risk, as are furniture and fabrics. Termites may not be able to digest synthetic carpeting fibers, but they have been known to cut channels in its backing or fibers as they seek food.

Termite environments typically include a wide range of fungi, as both organisms feed on wood and plant remains. Such competition may lead to habitat partitioning and, subsequently, to the evolution of symbiosis, parasitism, or pathogenic relationships. Termites usually must adapt their feeding habits and digestive processes to partially decomposed food, which is generally a gradual process involving mutual adaptions. Cellulose and lignin are among the least easily digestible materials, as a great deal more energy is required to break them down, as opposed to sugar, starches, and proteins. Therefore, the presence of fungi has often proven beneficial to termites; however, there are differing levels of dependency between families and generas.

The *Termopsidae* and *Rhinotermitidae* are typically found in wood that shows signs of decay from fungi. There is evidence of better growth and viability when termites fed on rotted, as opposed to sound, wood. The fungi might break down harmful extractives in the wood. For example, as discussed in the section on wood rot, white rot fungi break down lignin into the readily digestible carbohydrate. Termites are unable to digest lignin, and welcome its conversion into a carbohydrate.

Several authors have noted that termites are attracted to certain types of rot fungi, which has sparked the notion of using the attractants as a means of control. One author has discovered that the *Retculitermes* species, common in North America, tends to go straight to wood that is decaying.

Attempts are being made to identify the specific attractants. On the other hand, some fungal breakdown products are poisonous to the termite, and so can be used as a repellent. To illustrate, another study suggested that the deleterious effect of the wood rot *Lenitnue lepideua*, and its metabolite methyl cinnamate could control the mortality of the *R. flavipes* and *R. viginicus* species because of its apparent toxicity to the termite.

There are also termites that produce fungi; however, these species are typically located in the more tropical regions. A symbiotic relationship exists between termites and *Termitomyces*, which grows on the fecal fungus combs of the *Macrotermitinae* species. The combs are an integral part of the normal food cycle of the colony. Again, the function of the fungus seems to be the breakdown of the lignin, but it probably also supplies the nitrogenous materials and possibly other nutrients, such as vitamins.

Termites live in complicated colonies, and rely on one another to maintain their living, working, and eating habits. Typical of most termite colonies, the king and queen are the founders and rulers in the caste system. The queen has a very large, long abdomen and, in an established colony, can lay upwards of 1,000 eggs a day.

Scientists disagree about the foraging activities of the termite. One school of thought says the termites forage randomly; that the paths they create are determined purely by chance. Other experts believe that foraging may have an instinctual structural underpinning. One experiment suggested that tunnels were split and branched in new directions at statistically measurable intervals, to maximize the colony's the use of space.

In the caste, below the king and queen are two types of reproductive termites: primary, or imago, and supplementary. Alates, or swarmers, are the primary reproductives, meaning they leave the colony to establish new ones. The supplementary reproductives are blind, nonwinged, lighter in color, and do not leave the colony. In most termite colonies, the supplementaries take over reproduction if the primary king or queen dies or becomes separated from the main colony.

Wings are useful for identifying the types of species in an area, but the appearance of a few termite wings in or near a structure is not in and of itself evidence of infestation in that structure. Scattered wings found outside the structure may have been airborne, and not related to the particular building. A large amount of wings found inside the building generally does indicate termite infestation.

Another caste comprises the soldiers, whose job is to defend the colony. Soldiers have long, powerful, mandibles (similar to pincers) or other structural modifications to defend the colony against predators. They are usually wingless and blind, as are the worker termites, described next. If the colony is disturbed, the soldiers will remain around the breakpoint until the workers have repaired the area. Soldiers will also protect the open-flight-slits during swarming season. In some species, there are three types of soldiers, major, intermediate, and minor.

The workers, the lowest members of the caste system, are the most abundant; they are typically small, white, blind, and quick-moving, and live approximately five years. In some species, there are two types of workers, the major (large) and the minor (small). As the name implies, workers are in charge of maintaining the colony. They forage for food, feed the others, monitor the eggs, build tunnels and carton nests, open and close the flight-slits for swarming, and bury or cannibalize abnormal or injured colony members. Generally, the workers are responsible for all the damage the Formosan species imparts upon wood.

Also in a termite colony are the nymphs, young termites that have not been formed to fit into one of the castes. They are capable of developing into members of any one of the castes through the application of pheromones and hormones by the queen and king.

Though relatively little of termite behavior is understood by scientists, it is known that termites rely on chemical codes to guide their activities. Termites exchange food and chemicals called pheromones, which are distributed throughout the nest. The pheromones determine the social status of the termite and guide the formation of its distinct body type. The king and queen emit the pheromone, called juvenile hormone, which suppresses the sexual development of the termites. Depending upon the dosage, a termite will become a worker, a soldier, or reproductive. The highest dosage produces a soldier.

Termites are one of most pervasive wood-destroying insects, but their destruction can usually be controlled by various preventative treatments, such as mechanical barriers or toxic chemicals. But one, the Formosan subterranean termite (*Coptotermes formosanus* Shiraki), has proven to be a fairly invincible pest. Usually found in southern and southwestern regions, where the temperature and humidity are high, this particular species presents an extremely difficult problem to homeowners, government officials, preservationists, and entomologists alike.

The Formosan Termite

In contrast to the typical native termite colonies, which number in the thousands at their largest, the Formosan is known to form supercolonies, averaging some 5 million, and covering as much as 6.5 acres of land. Furthermore, the species is exceedingly more aggressive; they will eat wood nine times faster than their native counterpart. Formosans are known to sacrifice their workers, eat through corpses, or go through chemical barriers to get back to the wood source. With this insatiable appetite, a Formosan colony can collectively consume 1,000 pounds of wood per year, making it one of the most devastating wood-destroying insects known. Consequently, it is no longer believed that the Formosan termite can be completely eradicated. New chemical pesticides and stricter building codes can only help to control the destruction caused by this species.

The destruction caused by this insect in New Orleans, primarily in the historic French Quarter district, has garnered the most recent attention, because the damage there has been the greatest in the United States, excluding Hawaii, which has been battling the Formosans for more than 100 years. To illustrate, the Algier Public Library in New Orleans had undergone 79 treatments within a period of three years when a subsequent inspection of the ground beneath revealed a colony of 70 million Formosan termites, the largest recorded in the world. The effect of the termites feeding on the building was analogized as equal to that of a 500-pound animal feasting for 24 hours a day. In another instance, a façade in the French Quarter fell into the street after termites had eaten through the connections between the floor joists and the walls.

The Formosan termite is considered so invasive and destructive that, in 1997, it was placed on the National Trust for Historic Preservation's list of the top 11 preservation threats. Since 1990, the insect has caused an estimated $2 billion in damage in New Orleans alone, which is more than that caused to the city by hurricanes, floods, and tornadoes combined during the same period. Annually, the Formosan termite causes approximately $300 million in damage to private properties in the New Orleans metropolitan area. The city government incurred an additional $15 million in damage. A survey has also proven that 85 to 90 percent of the homes in New Orleans are infested with Formosan termites. Needless to say, the insect is responsible for severe damage throughout the southern and southwest regions. In the South alone, it causes upwards of $1 billion in damage annually.

New Orleans has been so hard hit because, in addition to having the optimal subtropical climate for the insect, the city has a vast number of historic buildings, constructed primarily of wood. These buildings offer the insect countless entry points from the soil to the timber, an ideal food. Also, bricks retain the moisture necessary for the termite's survival, and wooden lintels are ideal places for the pest to hide. Once their destruction of the beams, floors, and other structural elements has begun, the damage is nearly impossible to undo.

The Formosan termite is subterranean, and a native of China, Formosa (now Taiwan), and Japan. Probably the termite immigrated to the United States in the 1940s via ships whose cargo contained infested wood crates, pallets, and other packing materials. Following the offloading of the cargo in New Orleans and Lake Charles, Louisiana, and Houston and Galveston, Texas, the wood containers were buried in landfills, a highly conducive environment for the establishment of Formosan colonies.

Grossly misidentified as being the less destructive dry-wood species, the Formosan had more than 20 years to establish, unchecked, their characteristic massive colonies. The first colony was discovered in a Houston, Texas, shipyard in 1965, where a warehouse was entirely infested. By 1966, New Orleans and Lake Charles, Louisiana, and Houston and Galveston, Texas, were harboring well-established colonies. A year later the species was found in Charleston, South Carolina, and by the 1980s the insect had built colonies throughout Florida.

Fortunately, the spread of the insect is slow. Though swarming is the primary method of spread for most termites, the Formosan is a weak flier and does not spread rapidly by itself, as witnessed by the fact that it took approximately 20 years for it to infiltrate Florida from Texas and Louisiana. Today, however, the insect can be found throughout the South, as far north as North Carolina and as far west as southeast Texas, with a small contingent located in a neighborhood in San Diego. With the exception of San Diego, the common denominators in the majority of these locations are heat and humidity, ideal living conditions for the Formosan. The primary reason for their survival in San Diego's hot, but dry, environment is the effects of suburbanization. Landscaping in the area includes exotic plants, which require large quantities of water to keep the soil consistently soft and wet. This creates a very lush, pseudo-tropical environment for the Formosan termite.

Because of its success in San Diego, some experts believe the Formosan could infiltrate coastal areas as far north as Washington, D.C., and Boston. The termite seems able to adapt to most climates, soil types, and settings, though it prefers a subtropical or temperate environment within the latitudes of 35 degrees north and south of the equator, with a line going through the middle of the North Carolina. At present, the insect is not able to withstand freezing temperatures; however, during the winter months in northern cities, buildings become heat sinks, in that they warm the soils beneath them, creating a conducive environment for the Formosan termite. Furthermore, numerous cities have large networks of underground tunnels—subways, for example—that remain warm throughout the winter months, thereby offering a suitable home in which to establish colonies. Consequently, a slow migration northward is possible. It is difficult to predict their northern spread, and ultimately it will be many years before the Formosan will impart destruction in these regions, but it is an all-too-real possibility.

As in all termite colonies, the Formosans are based on a hemimetobolous caste system, meaning that their caste members are usually bisexual and have no known subsocial group. There are four primary divisions: royals, reproductives, soldiers, and workers, plus the nonclass nymphs.

The winged Formosan reproductives are yellowish-brown, 12 to 15 mm (0.5–0.6 in.) in length. In contrast, the native species are shorter, at

10 mm (0.4 in.); and the *Reticulitermes viginicus* and *Reticulitermes flavipes* species have black bodies. But the *Reticulitermes hageni* native species exhibits the same yellowish-brown color as the Formosan. The Formosan's skin is thick, enabling it to survive in dry environments when swarming. Also, the ocelli, the eyespots, of the Formosan termites are larger than those of most native termites.

The Formosan soldiers have a shorter, egg-shaped head that is hard and brown. Its jaws resemble pincers, which are used strictly for fighting. If disturbed, the Formosan will excrete a white substance from the fontantelle, a conspicuous tubelike gland located on the top front of the head. In native termites, the fontantelle is small and barely visible.

As previously stated, it is possible to mistake the winged forms for the common native drywood species, because they are similar in size and color. To positively identify the Formosan, it may be necessary to inspect the heads or wings under a microscope, or closely examine the damage caused by the insect. For example, the wings of a native termite have few or no microscopic hairs, whereas the Formosan's hairs can be seen in relative abundance. Furthermore, Formosan wings are 10 mm (0.4 in.) in length and have two largely pigmented veins near the front edge. The middle vein is visible, and may or may not be branched.

Formosan swarms can number in the billions when winged reproductives leave the colony toward a light source. A mature colony can produce up to 50,000 swarmers each season. Even if only 1 percent of the population survives, which is probable, the density of the Formosans in an area can increase rapidly. The insects settle into piles of rotting leaves in the crevices of roofs, or they burrow under the eaves. Swarming typically occurs from April to July, generally in weekly intervals. In between, however, smaller swarms can take off.

To begin the swarming process, the Formosan alates require certain environmental conditions. Windy and rainy weather keep them at bay; they wait for balmy, calm nights, when they will leave the nest in massive droves, usually at dusk, unlike the daytime swarming of many native species. (Some native species also swarm in the evening, such as the *Prorhinotermes simplex*, of Florida, and some drywood species.)

Formosan worker termites clear a path from the nest in preparation for the alates' flight. (If an alate misses its flight opportunity, it is killed and eaten by the colony and its nitrogen and protein are recycled.) Once in flight, the alates are attracted to any type of light source, and converge upon it. Following an approximate half-hour flight, the winged reproductives ground themselves and drop their wings; the males and females pair off, with the male following close behind the female as they run in circles, a behavior termed tandem running. If the female accepts the male as a mate, the couple, now repelled by the light, burrow into the nearest area that supplies moisture and food, where they mate.

Despite their seemingly sophisticated colony organization, the Formosan is considered the least advanced of the 2,400 known termite species. Still, its colony is a cohesive unit governed by instinct. It is believed that the species came into existence more than 200 million years ago.

As stated at the beginning of this section, the Formosan devours wood at a rapid pace, feeding on the cellulose of wood and excreting the lignin; it prefers dead wood. With the exception of some species of beetles, the termite is the only insect that can break down cellulose and use its energy and elements to create new life. Lignin, when passed, is used to fortify the nest material and shelter tubes. In this way, termites serve a specific and necessary niche within the ecosystem. Without them, dead wood would eventually clutter the landscape. Unfortunately for humans, they cannot differentiate between dead wood in service or in a log.

Whereas the native subterranean termite generally feeds along the grain of the wood, feasting on the earlywood and leaving the latewood behind, the Formosan eats both types of wood, forming a hollow within the attacked wood. Within the hollowed-out wood, they build a complex series of galleries, called cartons, in place of the consumed wood. The species has been known to eat away huge swatches of wood in a structure, but leave the paint on the surface untouched. Visually, the wood seems intact, but if touched it readily collapses.

Generally, the Formosan termite is attracted to the bases of poles, old tree stumps, or other wood in contact with the soil. It has been known to bore through brick, cement, fiberglass insulation, and rubber seals to reach the wood or moisture source located on the opposite side. The aggressive soldier has a gland located on its head, which secretes a solution that enables the insect to penetrate a variety of organic and inorganic materials. In buildings, they will enter through cracks or holes in the materials, targeting leaky roofs, faulty plumbing, and any holes that allow moisture to enter and collect in the building.

Formosans are also unpredictable, which can be a major problem; if for example, a colony is disturbed, within a week's time that group could take up residence elsewhere. In fact, in general, the Formosan is somewhat transitory. It may be sighted in one area and then move six months later. At present, it is not clear why they relocate.

The Formosan is the only termite species known to attack living trees. Though repelled by the phenol and alkaloid chemicals in the live wood, it feasts on the dead part of the trees. It cannot chew through the outer layers, so it enters through a dead root, or will travel up the tree's exterior, constructing shelter tubes along the way, until a wound or rotting piece of bark is encountered. The crook of a branch where bark has accumulated and has begun to rot is a favorite entry point. The termite will burrow into the crook and eat the branch until it weakens and falls. Upon entry, the termites generally build a nest underneath the tree and burrow upward through its core, where the dead wood cells are located. Replacing the tree's core with its nests is a trademark of the Formosan termite. It may eat to the farthest edge of the tree, leaving only one or two inches of the wood.

Urban trees, because they are more susceptible to disease, have weaker resistance to termites. And nicks and other damage provide entrance points for the aggressive termite as well. Conversely, the work of the termite can also make the tree vulnerable to disease and other maladies. In New Orleans, it is estimated that nearly 30 percent of the live trees are infested with Formosan termites. Its victims include 47 species, including citrus, sugar cane, avocado, wild cherry, cherry laurel, legustrum, hackberry, cedar, willow, tallow, wax myrtle, sweet gum, mimosa, cypress, red bud, Chinese elm, white oak.

Generally speaking, most subterranean termites have a soft cuticle, hence are easily desiccated if exposed to harsh temperatures. For this reason, the termite must live in a closed environment with a stable microclimate. The Formosan species builds nests of the aforementioned carton, which is composed of chewed wood, saliva, soil, and fecal matter. The nests are homeostatically regulated units whose primary function is to serve as a barrier to predators, much like the shell of an animal. The primary nest is usually constructed near a food source in or on the soil or in wall voids. Soil is their usual mandatory moisture source, though they can obtain the necessary moisture from leaking pipes or roofs as well. Foraging galleries lined with carton material are then constructed to the food source. The use of carton is exclusive to this species.

Formosans are not restricted to subterranean living. Once the carton has been incorporated into the wall voids or hollowed trees, they can exist for longer periods of time above ground because the material traps moisture. This is another exclusive behavior of the Formosan. Nests have been found as high as the attic in a three-story house, or reaching the top of living trees. The nests serve as a self-regulating, optimal environment for the insect's existence and survival and are, to some degree, independent of the external environment. To illustrate, in approximately one year Formosan termites have been known to burrow through four feet of concrete and steel, then carton themselves off, to preserve moisture in a wooden beam of a newly constructed porch addition. In another example, termites cartoned themselves within a tree that had been removed from its natural environment, and survived for nine months without a direct food source.

Needless to say, this type of colony system is extremely effective for, first, penetrating structures, then defending from human and other enemy contact. A colony must be attacked at its core, the location of the king and queen, to achieve somewhat positive results. If any other spot is attacked, the colony will simply replace their dead. With that said, it may not even be effective to kill the king and queen, because some reproductives have the ability to become the new royals.

As noted previously, the Formosan depends on the chemical signals and sensory organs to guide their foraging, eating, and defending activities. The worker termite has numerous tiny hairs on its body called sensilla. These are utilized to detect certain chemicals and sensations that guide the insect in feeding, defense, and reproduction. It is thought that the Formosan may be able to detect stresses in the wood, thereby redirecting it from a beam that is near collapse

Workers departing the nest to forage emit the chemical pheromone by dragging their hindquarters and creating a trail the other worker termites can follow. If the workers discover a good food source, they will lick it and emit a pheromone that other termites seek to feed upon.

It has also been discovered that the carton material consists of a substance called naphthalene, a chemical repellent that is component of mothballs. The Formosan can tolerate this chemical, but its enemies, particularly ants, nematode worms, and microbes, are repelled by it. In effect, the termites have their own pesticide to deter interlopers from their nests.

The Battle against the Formosan

Humans have been battling Formosan termites for years, to varying degrees of success. Ten years ago, the primary insecticide used to keep the pest at bay was chlordane. It succeeded only in keeping the pests from coming into the structure, but it killed relatively few, and the Formosan just sought another food source, primarily trees. Ultimately, the chemical was eliminated from the pesticide inventory because, in 1988, it was proven to be a health risk to humans, animals, and the environment. Since its elimination from the inventory, scientists and entomologists have been struggling to develop an effective and environmentally safe alternative for Formosan eradication.

Tenting and ground treatments commonly used to eradicate the native termite are only marginally successful in combating its Formosan cousin. Efforts are in full force in Hawaii, Louisiana, Florida, and Texas, but, currently, no single mechanism is available to solve the Formosan termite problem. However, numerous research projects are underway. In the meantime, the goal is to keep the control efforts going, and to bring greater public awareness to the problem. The downside to greater awareness is that focus will be directed to declining property values, rather than to devising a concerted action plan to eliminate the pest.

Federal policy initiatives have been launched. In 1998, the U.S. Congress established Operation Full Stop, and appropriated $5 million to fund the fight against the termite problem, especially in Louisiana. The USDA Agricultural Research Service is leading the campaign, which is underway in New Orleans in a 16-block pilot area located in the French Quarter. The organization's tactic is broad-based. It is not sufficient to target a single building or tree because, as explained, the termite will just move and grow in number. The attack will encompass the use of baiting systems as well as chemicals.

Operation Full Stop has also established a collaborative effort with the National Aeronautics and Space Administration (NASA) and the U.S. Department of Defense, with the hope of using sensing techniques and technologies to locate termites within walls or in the ground. Presently exterminators must rely on visual inspection. In the future, they hope to incorporate electrical current, acoustic sound, infrared light waves, and other means of technology.

At present, two baiting systems with a chemical component are being used in the 16-block federal initiative in New Orleans. Dow AgroSciences and FMC Corporation will supply the baiting systems, and the Bayer Company will supply the liquid pesticide. The basic method for both baiting systems is to drill holes into the pavement or the ground at intervals around the target structure or the general area, then to insert two pieces of thin wood within a plastic tube into the holes, and finally to monitor for activity. If activity is present, the wood is retracted, bait is inserted into the tubes, and resubmerged into the ground.

Here is how the two systems differ: In the Sentricon baiting system, from Dow AgroSciences, a poison called hexatulmuron is injected into the wood which, when ingested by the insect, inhibits the production of chitin, the material that makes up the termite's outer skin. As such, it prevents the worker termites from molting, or shedding their skin, which is fatal. The theory is that, because it is the job of the worker termite to feed the colony, if the worker termite is not able to shed its skin, causing it to die, food distribution will cease, and the colony will eventually die. In contrast, the FirstLine baiting system from FMC Corporation uses the poison sulfluramid, a slow-acting stomach poison that disrupts the insect's metabolism and its ability to transform the energy contained in food into nerve impulses. Basically, the termite becomes immobile and dies. FMC Corporation also encourages the continued use of chemical barriers, whereas Dow ArgoSciences does not.

The different approaches reflect the companies' underlying understanding of the foraging habits of termite. Dow scientists believe the insects forage in a random state, while the scientists at FMC believe they are guided by moisture gradients, objects, or imperfections. Consequently, the installation methods differ. Sentricon baits are placed in the ground at 10- to 15-foot intervals. FirstLine, on the other hand, recommends the strategic placement of baits in areas where termites are likely to be found.

A third bait system, not yet approved, is from Ensystex. Its bait system, Labryrinth, uses another chitin inhibitor called diflubenzuron. Four wooden strips are placed around the hollow monitoring station. Upon evidence of termite activity, a pest-control operator places a cardboard bait containing the poison into the hollow. The theory is that this approach will lessen the accidental disturbance of colonies. One study, however, has indicated that the diflubenzuron poison is not effective against Formosan termites.

Also on the market is Bayer's liquid Termitidae Premise. This remedy seems to bridge the gap between the baiting systems and chlordane by creating the traditional barrier and working somewhat like the baits by killing them slowly. Similar to chlordane, Premise disrupts the insect's nervous system, but unlike chlordane, it does not harm other animals. The termites stop eating and feeding others in the colony. The first application of the chemical is strong enough to kill the termites outright. Over time, as it breaks down, Premise's behavioral effects cause death by fungus and disease. Unfortunately, this type of attack is not guaranteed, as are other baiting systems.

Because the Formosan termite is chemically oriented, many of the methods being studied for its eradication seek to modify the behavior of the insect. The goal is to discover its internal chemicals and try to use them against the insect through feeding stimulants and attractants. For example, researchers are hoping to synthesize the chemical luring signals used by the termites to attract fellow termites to food sources, and turn it against them. Another study is investigating the types of wood species that the termite finds undesirable. The idea is that the undesirable woods may exude certain chemicals that the termite knows to avoid. The goal is to identify these elements and incorporate them into a new pesticide. Studies are also underway to identify the natural predators of the termite, then transplant the insects, parasites, or fungi into termite colonies.

In New Orleans, public policies may be able to assist in the eradication of the Formosan termite. It has been suggested that a new state law be passed to require that wood beams used in construction and repairs be treated with chemicals that can kill the pest. Another possible treatment is to pressure-treat building materials with the chemical Borate.

Hawaii, which has been battling the Formosan termite longer than mainland states, is several steps ahead in its control efforts. The insect was introduced to the island in the mid-nineteenth century, and thus Hawaiians have nearly 100 years of experience with it. Hawaii has implemented various policy measures in its attempt to control the termite. Building codes require that new homes constructed of wood be treated with a ter-mite repellent; concrete slabs must be treated with termiticide. In addition, all new government buildings and military compounds are required to be termite-proof. This is accomplished through the implementation of the environmentally safe, Basaltic Termite Barrier (BTB), a barrier made of special granules that the termite cannot move, chew, or penetrate. The bar-rier is formed by placing a four-inch layer of the granule material between the building and the ground. A minimum of 60 percent of the granules in the barrier must have at least one dimension measuring between 1.7 to 2.4 mm. The hard and dense material must be graded precisely to the proper depth before the slab is poured over it. The gravel is also poured into the vertical space between building's rear wall and the hillside. This method is relatively inexpensive and retreatment is not required.

Another preventative measure used in Hawaii is the integration of Termi Mesh, a fine stainless-steel screen that termites are unable to pene-trate. It is placed along the seams of the concrete and around pipes and other break points. Remarkably, despite these concerted efforts, the For-mosan manages to infiltrate the materials, but at a much slower rate. Any-one interested in buying property in any of these locations should have a thorough inspection done before purchase, followed by annual inspections of the premises. The interior and exterior must be examined for tunneling, especially those that disappear into cracks in the masonry, in and around doors and window frames, and along siding. Because tunnels are not al-ways evident, probing the wood ensures the discovery of infestation. Hol-low sounds within walls indicate the presence of termites; mud trails near doors and windows also suggest infestation. It is also a good idea to keep shrubs and trees from touching buildings.

3.4 Systemic Pathology

The essential difference between the structural failures typically investi-gated in academic structures classes and actual examples of deterioration in the field is shown when a member fails. Structures classes emphasize the proper sizing and design of structural elements in order to preclude cata-strophic failure, hence to ensure safe occupancy and service. Implicit in the textbook cases is the notion that such failures will occur immediately upon application of a predictable critical load.

The problem associated with structural pathology is that the element of the system at issue presumably was structurally sound when installed, but has somehow become unsound over time. Or the element was not structurally adequate upon installation and has deteriorated to an even lower level over time. The initial design inadequacy is a modern problem, but this situation is particularly likely among precode and preengineered structures, complicated by the fact that many admirers of older structures assume that older buildings were designed and built to higher standards than modern buildings.

It is true that we do not design or construct buildings the way we did in 1825. And it may be arguable that some modern materials and details are not as fine as, for example, hand-planed chestnut interior trim; but in most other respects, modern buildings are safer and structurally stronger than buildings of comparable construction that are over a hundred years old. We can attribute this general improvement to two reasons, both of which are really a cluster of associated notions. The first has to do with advances in scientific and engineering understanding of structures and materials. The second has to do with an increased appreciation of the vagaries and risks associated with construction and use.

Historically, erecting buildings for the most part has been an exercise in trial and error; construction technology was very much a matter of copying patterns and details. In some cases, the suitability of construction and design to physical reality resulted in reasonable and durable details and assemblies, particularly when the materials used were relatively inert and massive, such as thick masonry walls. In other cases, the margins of safety and durability were so narrow that the building or detail disappeared early on. Viewing history as represented by surviving remnants gives a distorted picture, because those things—in this case buildings—that have endured show us only the best of a particular age. The poorly crafted or deficient creations have deteriorated or failed, leaving few or no clues as to what went wrong and why. We are left with older buildings that, in almost every case, were better built or maintained, and so are not truly indicative of the general state of construction of a given era. A more balanced view of the history of construction is that, in any given era, a great deal of the construction was of modest or even low quality.

Of course, we still build junk today. But we also build tough, durable buildings that, if durability is the only criterion for their survival, will be here for hundreds of years. In other words, there is nothing remarkable, one way or another, about the mix of construction quality in this era, versus any other era, except that contemporary buildings are probably safer than those of prior generations.

Buildings are generally safer today for two main reasons; but before explaining those reasons, we will define what is meant by safer. In general, the hazards to which occupants of buildings are exposed have remained unchanged throughout history: building collapses, explosions, fires, releases of noxious fumes or vapors, as well as the risk of injury incurred when trying to exit a building in jeopardy. There are hazards associated with projections and other features of buildings that increase the prospect of our falling or tripping, thereby hurting ourselves.

In every one of these risk categories there has been a reduction of the risk to life and health over the past hundred years, with the possible exception of the danger posed by noxious fumes or vapors, in particular, the release of such fumes and/or vapors in the context of open combustion. Although this text is primarily focused on those building pathologies associated with progressive deterioration, as opposed to traumatic incidents, some discussion of these types of conditions and incidents is warranted, because their consequences are mitigated for the same two basic reasons as are the consequences of progressive deterioration.

The two basic reasons why buildings are safer today are, one, advances in technology and engineering science, resulting in a lower risk factor associated with the basic designs and materials; and, two, an increased application of margins of error and risk analysis. The former, improved design, is much the result of advances in the understanding and measurement of material properties and structural analysis. Most of the engineering analysis we take for granted today when designing even a modest building is less than 300 years old.

Augustin Cauchy presented his landmark paper on the relationship of stress and strain in 1822. The introduction of this concept, which we now call *elasticity*, into the technical schools took some additional time, and the general application of Cauchy's work took even longer. Prior to the conventional quantification of stresses and strains, building design was strictly a matter of repeating what appeared to work and avoiding what did not. The only thing remotely resembling an engineering axiom was that more material was better—that is, more durable—than less material. When an attempt was made to minimize material or to stray from convention, the methodology was one of trial and error, and somewhat risk-laden trial and error at that. Remember, we quantified gravity only relatively recently. (Sir Iassac Newton died in 1727.) The pyramids, the great cathedrals, the Roman viaducts, the Great Wall of China, and the Welsh castles were all built before we knew how to quantify loads or stresses. What those builders did know, however, was that straightness and plumbness were important, that sandy soils could carry more than clayey soils without moving, that water was generally detrimental, that bugs ate wood, and so forth. What they could not know was when shear failures would occur, how fractures progress, or at what stress level a material would rupture.

The consequence of our ancestors' comparative ignorance was that many structures were built that, even if they did not fall down, were by no means safe. Recall the example used in the Introduction of this text, in which a porch at the University of Virginia collapsed, killing one person and injuring 22 others. That porch and the others like it are examples of structures that were never adequately designed for reasonable loads, which included the weight of the porches themselves. They survive without causing harm or damage for as long as they do primarily because of the reserve capacity of the building members, combined with the relatively intermittent and noncritical loading, supported by the occasional repair to obvious deterioration. The elements and materials of many older structures operate in ranges of internal stress that, while not immediately high enough to precipitate a fracture, are well above what is now considered safe or prudent practice. As the materials and assemblies deteriorate, that margin of safety is reduced ever further. Deterioration can, of course, occur in a newer building as well. The difference is that, in a newer structure that is subject to code requirements and standards of common practice, we know the magnitude of the allowable stresses and what the applicable standards of safety were at the time of construction. As the building deteriorates, we can more readily assess the reductions in safety, because we can rely on the initial standards and factors of safety.

The combination of increased understanding of the behavior of materials and of engineering principles has significantly improved our ability to predict building performance, allowing us simultaneously to design more economical and more reliable buildings. Of course, there are famous episodes, such as the John Hancock Tower in Boston, that give lie to this assertion. Although recognized by other architects as an exemplar of architectural design in 1974, the tower had the disturbing predilection for shedding it windows onto the plaza and streets below. (See *Why Buildings Fall Down*, by Matthys Levy and Mario Salvatore, New York: W.W Norton & Company, Inc., New York, 1994, for this and other twentieth-century examples of less-than-adequate design and/or construction.)

The fact that a structure has remained standing for a long period of time is in no way logically connected to its state of repair or structural capacity. It does not follow that because a given building has remained erect for a protracted period it will continue to do so. Simply put, an old building is, as a matter of physical science, in a lesser state of repair than it was when originally constructed. No reasonable practitioner can or should take comfort in the structural capacity of an element simply because the building or an element of a building has not yet failed. In fact, he or she should take the opposite view. In the absence of specific information to the contrary, meaning direct investigation and observation confirming proper condition, advanced age is prima facie grounds for suspicion of a failing element or structure.

Fortunately, structurally related catastrophic failures of newer buildings in service are relatively rare, simply because construction and engineering practitioners rarely miscalculate the design requirements by such a magnitude as to result in collapse. They do, however, occur. On July 17, 1981, a suspended walkway in the lobby at the Regency Hyatt Hotel in Kansas City, Missouri, failed, resulting in the deaths of 111 people and the injury of 186 others. The failure was attributed to a connection that, though reasonably designed, was altered in the shop drawings, a fact that went undetected during review. Two years earlier, on June 4, 1979, also in Kansas City, the Kemper Arena roof blew off in a storm. And during a particularly infamous storm in January 1978, the Hartford (Connecticut) Coliseum roof collapsed, followed three days later by the collapse of the C.W. Post Auditorium in Brookville, Long Island. These failures were less obviously the result of error. Unusual conditions that result in relatively sudden and extreme loads will test the marginally designed structure or the structure with a latent flaw well after occupancy. In retrospect, all three of these failures appear to have been avoidable, but in at least two of the cases, the causes were fairly exotic, albeit predictable.

By far the most common type of structural failure of our era is typified by the collapse on August 13, 1979, of a stadium roof in Rosemont, Illinois, which killed five construction workers. This type of structural failure is distressingly common, and accounts for the majority of structural collapses, namely those that occur during construction. Generally, construction failures may be the result of inadequate design; however, the majority, as was the case at Rosemont, are the result of inadequate staying and stabilization during the construction sequence when the building is structurally incomplete. A variation on this theme is typified by a collapse during the construction of a cooling tower at Willow Island, West Virginia, on April 27, 1978, attributed to concrete that did not meet specifications. More recently, the New York City construction community has experienced a rash of collapsing construction towers and cranes. The failures were not directly connected to building failures, but rather attributable to the same laws of physics or, more accurately, the ignorance and/or violation thereof.

Of far greater concern to us as building pathologists are the gradual changes in the building fabric and/or the external operating environment of a building that result in a failure of some part or portion of a building. These are the mechanisms defined in Chapter 2, namely progressive alterations in the structural materials and elements that, if unchecked, will result in the failure of the system.

As suggested above, in the case of structural mechanisms, there are two broad classes of mechanisms: those internal to the fabric of the structure and those that are external to the fabric of the structure. Those internal to the fabric result in the deterioration of materials such that, at some point, the structural integrity of the member is reduced to a point of failure, either by collapse or by deformation. Those external to the fabric are typified by changes in the forces acting on an otherwise intact element or structure.

3.4.1 Differentiation between Deformation and Failure

The term failure is commonly used in connection with structural elements and members to denote inadequacy, but, unfortunately, it is less than precise as to the manner and characteristics of the inadequacy. For example, the failure may result from a through-body fracture of the member, which is a result of exceeding the ultimate strength of at least some critical area of the member. The failure may be a result of the formation or occurrence of geometric instability, wherein the individual members are intact, but the combination of members forms a mechanism leading to collapse.

Yet a third definition of failure may not involve a collapse or a fracture. A member has arguably failed if it can no longer serve its original purpose, and such a condition may occur well shy of fracture or collapse. We are all familiar with buildings that are noticeably deformed: bulging walls, leaning parapets, sagging lintels. It is equally evident that, while such members may eventually collapse, they may require replacement simply because the distortion alone is unacceptable.

FIGURE 3.4   The exterior wall was never bonded to the interior structure. As the exterior wall was pushed laterally by freezing water and debris accumulation, there was no restraining force to counteract the movement. When the wall reached geometric instability, it collapsed.

An important corollary of failure by distortion is the equally apparent observation that many members that are distorted are still serviceable. Whereas collapse is a matter of fact, failure by distortion is a matter of degree. A wall façade less than one-fourth of an inch out of plane is not problematic; the same wall four inches out of plane is problematic.

3.5 The Pathology of Common Structural Elements

Although many deterioration mechanisms originate at the material level, they are not manifest until they reach at least the optical level. Even though the underlying pathogenesis may be the equivalent of a cellular dysfunction, in many cases the first manifestation of a problem is at the scale of a building element, such as a deflecting beam or bulging wall. In addition to material-level pathogenic deterioration mechanisms, attributes of the configuration and geometries of structural members can precipitate progressive deterioration.

The following sections of this chapter comprise, essentially, an inventory of the components of the structure of buildings, and identify the characteristic deterioration mechanisms associated with those elements. The discussion begins at the foundation and moves upward through the superstructure of the building. The subsections address characteristic deterioration mechanisms associated with the structural components at the various locations in the building. For deterioration mechanisms where intervention may be complex or where there are multiple options for intervention, we will use the intervention matrix to organize those options and to introduce those complexities. In many instances, we will discuss the design, manufacture, or installation of the member, because the origins of pathogenesis in many cases lie, not in the concept nor in the material, but in the process of production. Understanding the pathogenesis can be of great help in understanding the appropriateness of a particular intervention option.

3.5.1 Foundations and Retaining Structures

Problems associated with foundations are, as a class, good examples of external pathogenesis. The progressive deterioration is, generally, traceable to conditions outside the element at issue or, at least, originating outside the element. The source of those external conditions and changes in those conditions comprise the subject matter of the disciplines of soil mechanics and geotechnics.

When it comes to foundations, the fundamental problem is that soils simply are not the ideal platform upon which to place the relatively dense mass of a building. Soils tend to be very unpredictable, variable, and unreliable. To explain the progressive building problems associated with soils, we begin with a short course in soil mechanics.

Foundation materials may be divided into rock and soil. Generally speaking, soil is the deterioration product of rock, either in situ or as transported and redeposited. Rock is a complex subject that we address in greater detail when we discuss walls and cladding materials. For now, it is sufficient to describe rock as a firm and more reliable basis for construction than soil. Rock is quite variable relative to bearing capacity, but most undecomposed rock provides considerable bearing capacity, but as a group, rocks provide good bases for foundations. There are, however, some notable exceptions. For example, certain classes of shale expand when the overburden is removed. Sedimentary rocks are formed under considerable compression, so when pressure is removed during excavation or quarrying, the material expands elastically. This behavior is compounded by the fact that shale derives from clays, which at the time of deposition were very probably saturated. Subsequent compression of the bedded clays squeezed the water out of the clay, a process called consolidation. The fully consolidated clay was compressed into rock, which now is not only elastically compressed but hygroscopically desiccated. When the material is exposed, not only does it rebound elastically, it immediately begins to rehydrate. The expansive characteristic of certain shale can be enough to lift and crack a building.

Even when the expansive property is more limited, excavating in shale can be an adventure. In 1981, while I was serving as the structural engineer for a Robert Venturi-designed building at Princeton University, a site known for its shale deposits, as expected we hit undecomposed shale and had to jack-hammer the shale out. In the evening after the excavation crews left and the site was quiet, while standing in the pit we could hear the shale popping as it expanded and cracked. Occasionally, a small piece snapped off and was projected some distance through the air. The "cure" was to get the building back into the hole as rapidly as practical to effectively recompress the shale.

Soils are classified in a variety of ways. One common method, based on terrain and vegetation, was developed by the National Forestry Service (NFS). Although useful as a designation device and for identification, the NFS system is not particularly helpful in classifying soils relative to bearing capacity. Soil mechanics tends to analyze soil based on three constituents: clay, granular material, and organic material. Granular soils include, as a large category, sands and gravels; technically this category also includes silts, which are very fine-grained materials, but which do not behave the same way as larger-grained materials. The organic material derives from the vegetation that grew on the and in the soil.

Clay, also referred to as cohesive soil, consists of fine particles that tend to be quite thin, so thin that water that gets in between the particles is effectively bonded to the particles by electrostatic forces. It is important to note that the clay is not dissolved in the water, even though the particles are so fine and of such low molecular weight that they are easily suspended in water and, therefore, are transported by moving water. The response of clay to water is what distinguishes it as a bearing material for buildings.

Water can migrate through clay, but the process can be extremely slow. This fact means clay can be and is used as a liner in reservoirs or to protect structures from migrating water. That said, it is also true that clays will gradually absorb and hold water. The presence of water between and among the clay particles increases the volume of the clay and, generally, the plasticity as well. When pressure is applied to the clay, the water is squeezed out, but the process can be amazingly slow. Naturally, as the water dissipates, the clay volume is reduced. The relevance to buildings is that the weight of the building may be sufficient to consolidate the clay below it, but the rate of consolidation is such that the building will have long since been completed and occupied. As the clay consolidates, its volume decreases, and the building subsides. The subsidence of a building built on clay is not automatically a problem if the subsidence is nominal and uniform; when it is not, long-term problems can develop.

Granular materials, notably sands and gravels, are by definition well drained. They cannot hold water longer than it takes for the water to pass through, unless the materials are in fact submerged, which is technically a supersaturated condition. Sands and gravels tend to react elastically to applied loads. Any subsidence or soil deformation occurs as the load is applied. Granular soils can consolidate because there is no particular reason for the granules to be as densely packed as possible when the load is applied. For the granules to compact, however, they have to be rattled or disturbed in a dynamic manner. The static load of the building is generally insufficient to achieve compaction of the granular soil below it. If, however, the granules are vibrated—in particular if they are vibrated while saturated—they can and will shift to a more compact configuration. Buildings on top of loosely compacted granular soils may incur subsidence but, for all practical reasons, only if the soil is vibrated. Such vibration can derive from earth tremors, passing traffic, industrial operations, mining and quarrying, or in some cases from machinery within the building.

Organic soils are the least desirable foundation bases, so much so that they are avoided; and, if they are unavoidable, they are removed and replaced with inorganic soils. The reason is simply that the organic compounds decompose and lose volume, hence major occurrences of subsidence are likely. The initial capacity of organic compounds to carry load without mechanical crushing is very limited and highly variable. Despite efforts to avoid organic soils, we occasionally encounter them, rarely with happy endings to tell. For example, two neighborhoods in Philadelphia, Logan and the Great Northeast, were developed in terrain originally traversed by streams. The developers leveled the sites by cutting down the higher elevations and filling the drainage courses. In some cases, the organic materials, such as trees and undergrowth, were piled into the water courses and covered with the soil. The water, continuing to follow the old water course, although below-grade, caused the organic materials to rot. Several houses in the areas had to be condemned due to settlements that fractured their structures beyond redemption. In another case, a developer in the Andorra section covered and built houses on a landfill. Within a much shorter period of time than in Logan or the Great Northeast, the houses subsided and fractured.

When constructing buildings on soils, the goal is to avoid subsequent building movement as much as is practical and economically feasible. The problems are relatively obvious: clays consolidate, granules compact, organic materials decompose. The solution to organic material is to get rid of it. The solutions to cohesive and granular soils are a bit more complex, and there are not many options. The basic objectives are to determine how much compressive stress the soil can endure and to avoid undesirable subsidence. For both types of soils, not to mention all of the blends and combinations of the two, this value—which is called the *allowable bearing capacity*—is established by laboratory analysis conducted by geotechnical engineers. The results of the analysis produce a value measured in tons per square foot (Tsf). While such values as 2 Tsf sound large and impressive, note that 2 Tsf is less than 28 psi, which is not a particularly high value compared to other familiar materials.

Soil-bearing capacities are among the lowest allowable stress levels encountered in the building design business. The example using 2 Tsf (200/144 = 27.8 psi) is not unusually low for soil-bearing; 28 psi reflects that this is a lower allowable stress than Styrofoam. Keep in mind that soil-bearing capacities typically invoke very high factors of safety, sometimes reaching as high as 10. The justification for such high factors of safety is easily given. Sampling of soil is basically done blind; there is no reasonable way to know with certainty that a given set of samples is typical or just coincidentally similar. Soils can vary tremendously both horizontally and vertically. There is little assurance that the soil two feet in any direction away from the sample location is the same or similar to the sample. It is, for example, possible to miss large boulders, lenses of clay, underground streams, or archaic vegetation pockets (peat).

SHALLOW FOUNDATIONS

One very common method of keeping a building where we want it is, essentially, to balance the weight of the building against the bearing capacity of the soil. If the bearing pressure of the building load remains sufficiently below the bearing capacity of the soil, even if it is clay, and consolidation occurs over time, then the building should move only nominally, if at all. This general approach results in what are classified as shallow foundations, though the classification can be a bit of a misnomer in that, sometimes, the depth of the shallow foundation is far from shallow. All shallow foundations are predicated on a form of load and soil capacity balancing. Once the bearing capacity is determined, there are a variety of techniques available to achieve the desired balance; however, they all share the common characteristic of limiting the pressure between the soil and the contact surface of the building to an acceptable level.

Probably the most common of these shallow approaches is the *spread footing*, which involves placing the weight of the building on pads or footings. The combined area of the footings is sufficient to distribute the load over a large enough area so that no single spot exceeds the allowable bearing capacity. The options available for manipulating the design, hence the size and distribution of the footings, include controlling the weight of the building. It may be more economical, for example, to build a shorter building with larger floors, than to build a tall building requiring an expensive foundation. For a given building weight, it is possible to increase or decrease the area of the footings to limit the bearing pressure.

Another technique is to search for better soil. This does not necessarily mean abandoning the original site, although that is sometimes the solution. Remember that soils are generally consolidated or compacted depending upon the amount of overburden that has been compressing them for geologic ages. The weight of that overburden varies with depth; therefore, as general rule, the bearing capacity of soil increases with depth. If suitable bearing capacity is at or below six feet below-grade, it is economical to go ahead and finish the area below the building and call it a basement. Basements are, or should be, the result of proper footing design, not the predetermination to have a basement. By themselves, basements are two or three times more expensive than the same volume built at grade. Homebuyers often mistakenly believe that houses must have basements. It is the responsibility of the engineer to optimize the footing the design.

Variations of the shallow foundation theme include raft and mat foundations, which are different from spread footings only in configuration, not concept. Mats place the bearing pad under the entire area of the building and, perhaps, beyond the perimeter of the building. This has the advantage of averaging the weight of the building over the entire excavation area. Rafts are employed in particularly wet and/or soft soil and, basically, turn the basement of the building into a reinforced concrete ship's hull. The soil is displaced, and the building floats in the soil rather than bearing on it. Once again, the common denominator is the balancing of the weight of the building with the capacity of the soil.

All shallow foundations are subject to the same set of risk factors and, therefore, to predictable pathogenesis. By their nature, shallow foundations are subject to inaccuracies in the determination of the design values. This is particularly true of structures erected prior to the development of geotechnics and soil mechanics. An early form of spread footing was simply to thicken the bottom of the wall slightly and bear the wall directly on the soil. Such rules of thumb for determinations of depth and wall thickness were developed mostly by trial and error, and tended to be conservative and generally valid even by current standards.

The problems associated with walls bearing directly on soil were often the result of conditions that continue to plague shallow foundations generally. For example, there were and continue to be problems with the validity of the assumed bearing capacity. In older buildings, the designers did not have the advantage of borings and soil samples. It was altogether possible to put up a wall on a sound granular soil only to see it sink into an undetected clay layer a few feet below excavation depth. There is a somewhat notorious development called Fox Hollow in Marlton, New Jersey. As with much of South Jersey, the soils are generally quite sandy, and this was ostensibly the case at Fox Hollow. Unfortunately, the development was in an area that has thick lenses of clay at various depths below the excavation of the houses. In time, many of the houses settled into the clay, with disastrous consequences. Remedies were so expensive as to rival the value of the property. Even today we incur the risk of underground conditions not being what was predicted. The more common pathologies associated with shallow foundations go to the following issues: differential settlement, heaving, sliding, and moving water.

Differential Settlement

Despite our knowledge about consolidation and load distribution, it is still not unusual for a building to settle in an unpredictable manner. There are two general classes or types of settlement problems, both of which are lumped under the rubric of *differential settlement*. The first is technically not differential settlement, it is *absolute settlement*: the building subside evenly, but more than was expected. In Mexico City, buildings have sunk so far that entrances are now on the former second- and third-floor levels. In many parts of the United States, there are instances of settlements measured in multiples of whole inches.

Far more common and devilishly irritating is the building that may settle far less, but unevenly. When a building settles evenly, there is not as much of a problem. This is true differential settlement. One place in modern buildings, particularly those that are slab-on-grade, where uneven settling is especially annoying is at exterior doors at-grade. This happens where doors must swing out for egress reasons, and the designer placed the door directly at the edge of the exterior foundation wall. If the building settles even as little as one-eighth of an inch more than the slab outside the building, the door will bind on the lip of the sidewalk and not open smoothly.

Other examples that may be the result of differential settlement include: some part of the building settles more than some other part, and a crack runs right up the wall; roofs tear; windows and doors rack; corners droop; sags appear in the masonry coursing; porches pull away from the main building. In fact, many building movements are labeled as differential settlement whether their behaviors fit the requisite conditions of foundation movement or not. In order for differential settlement to occur, there must be a local vertical displacement of the foundation that varies relative to an adjacent portion of the foundation. Above-grade distortion or cracking alone is not enough to categorically diagnose a problem as differential settlement. The investigation must identify the location and magnitude of the vertical foundation movement.

There is good reason to be conscientious about making rigorous differential settlement diagnoses. Intervention in true differential settlement problems can be extremely expensive, so expensive as to preclude intervention at all. When this is the case, intervention is either undertaken at great expense or suspended because of unacceptable expense. Either way, a misdiagnosis can have profound consequences, the latter because the true pathogen goes unabated, the former because the wrong pathogen is abated at great expense.

Applying the principles of intervention developed in Chapter 2, we begin with the necessary and sufficient conditions for differential settlement, which are (1) compressible soil below the footings, (2) varying compressive stress on the soil below the footings, and (3) relative flexibility of the footings to subside with the soil. According to our definition of necessary and sufficient conditions, when these conditions are met differential settlement will occur.

Also recall from Chapter 2 that there are six general approaches to interventions, from which we can construct an intervention matrix, to which we add, in this instance, three necessary and sufficient conditions for differential settlement (see Table 3.1). This will enable us to consider a wide range of options relative to intervention. Note: The options described within each matrix intersection in the table are for purposes of example

Table 3.1 INTERVENTION MATRIX: DIFFERENTIAL SETTLEMENT

Necessary and Sufficient Conditions	Intervention Approaches					
	Abstention	Mitigation	Reconstitution	Substitution	Circumvention	Acceleration
General Mechanism	Accept settlement as optimal solution.					Demolish building and redesign foundations.
1 Compressible soil	Ditto	Shim between top of footing and foundation wall so as to level building.	Elevate building to original position and reinstall original soil.	Elevate building to original position and install alternate soil.	Underpin building to lower bearing stratum of higher compressive resistance.	Dewater site for cohesive soils in order to increase settlement until achieving equilibrium.
2 Variable bearing stress	Ditto	Reduce or balance loads so as to alter settlement rate and/or total subsidence.	Demolish footings and add soil back to original elevation, and reinstall footings.	Demolish footings and soil below footings; backfill with denser material; reconstruct footings.	Inject soil with consolidating material, resulting in less compressible material and higher bearing capacity.	
3 Footing flexibility	Ditto		Add reinforcement to continuous footings to bridge settlement zones.	Demolish existing footing and redesign for continuous mat or raft.	Drill piles adjacent to existing footings and needle foundation wall to bear on piles or piers.	

only; they are not intended to be the definitive list of differential settlement options. Within an intersection there may be several options; an intersection that is blank offers the opportunity to think of an option that meets the criteria. As a formalized brainstorming tool, the matrix is meant to be used to consider all the possibilities. It is recommended that the exercise continue until at least one option exists for each of the six general approaches. And keep in mind that the abstention approach applies primarily to the general condition inasmuch as it subsumes all of the necessary and sufficient conditions.

The relative feasibility of the various options range from the practical to the seemingly absurd. An apparent absurdity must, however, be taken in context, because what may be absurd for a differential settlement of a small fraction of an inch may be preferable if the differential settlement is four inches or four feet. One of the most famous of all differential settlement cases is the Leaning Tower of Pisa. The fact that this tower has survived at all and was not demolished at an early age is more the mystery than the reason for its leaning. A recently completed project to "stabilize" the tower has every prospect of success, but it was not reached before exploring and exhausting several of the options included within the Intervention Matrix, which at first glance appeared ridiculous.

As a general approach, circumvention is probably the most common and feasible group of options to deal with differential settlement, which this is not surprising inasmuch as the options that address fabric alone or environment alone are not as accessible as those that go outside the fabric and the existing environment. In no small part this is because of the physical proximity of the foundations and the soil below the foundations. Access to the locus of the issue is sufficiently tedious as to invite options that are physically more available.

In fact, the options of injection grouting, underpinning, and needling have all been employed in settlement situations. In extreme cases, buildings have been partially or completely demolished; and certainly, there are many cases for which diligent investigations led to the decision to do nothing.

Heaving

Heaving is a phenomenon familiar to residents of colder climates—in particular colder climates where the underlying soils are clayey, such as much of inland New England. Heaving is dependent upon the properties of water, which we will discuss in greater detail in Chapter 4. It is sufficient here to our understanding of heaving to accept as fact that as water freezes, it also expands.

If the freezing and, therefore, expanding water is trapped inside soil, the freezing water will expand the soil-water mixture in all directions. Inasmuch as the material is confined laterally and from below, the primary direction available for expansion is upward. If an object of some mass were placed on the vertically expanding soil, that mass, too, would exert a confining force locally on the expanding soil. If the force exerted by the mass of the object were less than the upward force of the expanding soil, the soil would lift, or heave, the object. The remaining indeterminants in this scenario are the potential upward pressure of the expanding soil and the pressure exerted by the counterposed mass.

The potential expansive pressure of confined freezing water and/or ice is 10,000 psi, a very high value. Such a value of pressure can easily crack concrete or brick. Because we typically evaluate soil pressures in tons per square foot, the value of 10,000 psi converts to 720 tons per square foot. To appreciate how much this is, imagine a rather typical office building with columns 30 feet on center and 10 stories high. Now imagine each of the interior columns of that 10-story building being supported on a footing of only one square foot—that is, 720 tons per square foot.

Several factors significantly affect the likelihood of frost pressure ever achieving 10,000 psi. This value is a theoretical, not a practical, limit. In actuality, and for reasons we will examine in great detail in Chapter 5 when we discuss freeze-thaw cycling, as the water begins to expand vertically, it also expands laterally. If there is anywhere at the perimeter of the footing to which the expanding water can release, it will dissipate to that zone, thus relieving some of the pressure. It is also a simple fact that while freezing water may be able to exert a confined pressure of 10,000 psi, an unconfined specimen of ice will fracture at values substantially less than such stresses. If the bearing pressure of the building were remotely close to such a value when there is freezing water below the footing, that same bearing pressure will exist when there is no frost pressure below the footing; and 720 Tsf is well above the capacity of almost any known soil to support without failure. For all these reasons, the values of frost pressure reasonably likely to occur below a footing are well below the theoretic limit, but still significant. Much of the actual value of the frost pressure depends on the relative ease or difficulty of the water gaining release.

In order for frost to form below a footing, and thereby exert upward pressure, the water must be relatively confined. If, as mentioned above, there is a way the water can escape, it will do so. In other words, as long as the water is fluid, and the capacity exists to equalize pressure in any direction, the fluid will flow to the point of lowest pressure. One way for water to escape is through the soil itself. In fact, soils that are porous and permeable, such as sands and gravel, are virtually impervious to frost heaving, simply because it is virtually impossible to maintain the water in a confined condition long enough for it to freeze *and* to exert vertical upward pressure. No sooner will the water potentially collect below a given footing than it will dissipate into or through the soil as the pressure increases. In most cases, granular soils will drain so rapidly that accumulation of water below the footing is only likely if the water table or perched water table is above the bottom of the footing. In short, heaving is an unlikely event in granular soils.

The potential for heaving is the generating condition that gave rise to the general requirement to bury bottoms of footings below the level of likely frost penetration into the soil. In most temperate zones, this depth is three feet. In some jurisdictions, there may be an amendment to the building code requiring that footings be at or below a certain depth typically three feet. Such a requirement is not entirely ill-founded; however, from the above discussion, it follows that such a requirement is unnecessary if the soil is granular and the water table is below the alternative footing depth. It is, in practice, a waste of good foundations to bury a footing three feet into granular soil if the sole reason for doing so is to preclude heaving.

Even in soils where the potential for heaving exists, such as clays and silts, the frost must penetrate to the bottom of the footing. A rather simple way to minimize heaving potential in newly designed construction is to place a layer of gravel below the bottom of the footing, especially if such a layer is proximate to a perimeter drain. Any water that might collect below the footing dissipates to the drain; and any water that might be in the soil below the gravel layer either dissipates longitudinally along the length of the perimeter footing, vertically downward through percolation, or laterally into the drainage line. It is extremely difficult to concoct a set of circumstances given this construction detail that would result in heaving, even if frost were to reach the depth of the footing or below. Another consideration in the placement of footings at regulated depths of three or more feet are the circumstances required to actually achieve frost penetrations to that depth. During the winter of 1991–1992, the Mid-Atlantic region of the United States experienced five consecutive nights of below-zero temperatures contiguous to over 20 additional consecutive nights of below-20 degree weather. The frost penetration for that record-setting cold spell was 14 inches. Even if we were to experience a 36-inch frost penetration in the Mid-Atlantic region, heaving footings would be so far down on the list of disaster-related problems that we would probably not get around to dealing with them until the following year.

The problem of foundation heaving is largely an issue of the perimeter footing, as that area is exposed to the harshest temperatures. The heat from the building will generally be sufficient to protect the footings of slab-on-grade constructions, and certainly foundations at lower elevations. This brings up one of the multitude of reasons that abandoned buildings are at such disproportionate risk. When a building is allowed to drift with the ambient atmospheric temperature, the interior column footings may be subject to temperatures never experienced in an occupied building. Assuming the beneficial effects of interior heating, it is not unusual to place interior footings at shallower depths. But, when the heat is terminated, the shallowness places the bottom of the footing close to potentially freezing conditions. This is true even in buildings with basements and footings generally well below even the requisite frost lines. Without heat, the temperature in the basement may dip to below freezing. The exposure to heaving in such circumstances is not limited to interior footings, but perimeter ones as well.

Still, water below the footing is a necessary and sufficient condition for heaving to occur. Therefore, in abandoned buildings, heaving does not necessarily occur simply because freezing temperatures penetrate to the bottom of footing. It is a coincident disadvantage of abandonment that the potential for undetected accumulations of water to occur in the lower levels of buildings increases, particularly as other systems begin to fail and atmospheric moisture begins to penetrate the protective enclosure.

A location that at times is overlooked as a potential site of heaving is in concrete frame buildings at the perimeter-grade beams. The footings in these buildings are rarely a problem because of the depth of burial common in such construction—if no other reason than that the weight of the construction dictates higher bearing strata. The bottoms of the grade beams may have been poured directly on earth at relatively shallow depths with gravel-layer frost breaks. Uplift of a grade can have devastating effects on the first-floor spandrels and glazing. A similar phenomenon can occur at turned-down slabs used as perimeter foundations. Again, it is important to remember that *all* of the necessary and sufficient conditions for heaving must be met in order for the mechanism to initiate. The shallowness of a foundation bottom alone is not enough to condemn a structure to heaving.

Certain classes of construction are particularly vulnerable to heaving, namely pavings, roads, retaining walls, and other hardscape construction. All exterior construction is, by definition, exposed to the elements and, therefore, without the benefit of mitigating heat from the interior of a building. That alone places the materials at greater risk of freeze-thaw cycling and other thermally related mechanisms. Horizontal placements such as walkways and roads also are oriented to receive maximum water and sun exposure. Roads and some walkways endure substantial rolling loads, which can be especially devastating.

To understand the basic deterioration mechanisms associated with roads and railroad beds requires a modest understanding of their construction. The basic principle is to accept a fairly high and concentrated rolling load at the surface or rail and transmit that load to the bearing strata. The trick is to accomplish this with enough dispersion so as not to precipitate a local soils failure or subsidence, and enough resilience so as to regain the same elevation and shape of the road or rail after the load passes. The basic process of building a road consists of scouring the surface of organic matter by grubbing the top 12 or so inches of soil and removing it from the site. Depending on the desired finished surface elevation relative to existing grade, more material may be removed. Then the earth is compacted, usually with a sheepsfoot roller or similar device. Controlled fill is added and compacted. Then various layers of gravel are placed over the compacted earth. In railroad construction, this layer is the ballast.

The depth of the ballast depends on the magnitude of the design load. The greater the load, the thicker the gravel layer. Gravels and most granular materials, generally, are rigid and elastic, and they will distribute loads over a fan initiated from the point of the load through the gravel to the base with a fan slope of 1:1. If the load is an 8-inch wide truck tire, and if the gravel is 1 foot thick, then the same load is distributed over 32 inches laterally at the top of the base layer. Assuming that the tire imprint axially is also approximately 8 inches, a similar distribution is occurring fore and aft of the rolling load. The sum of this biaxial distribution is that a load imparted at the surface over an area is 16 times as great. If the original load were 16 tons at the surface, the pressure at the base would be slightly less than 2 tons per square foot.

If, in addition to the ballast, the surface itself also has a thickness, that too contributes to the distribution. And if, in addition to depth, the surface also has bending capacity, such as a concrete slab or a steel rail, the distribution is even greater. The composite of relatively stiff surfaces and thick ballasts is an effective roadbed or railbed. The basic deterioration mechanism associated with such constructions is that the ballast gradually becomes contaminated by the fine particles of the base layer. These "fines" are literally pumped into the ballast when the base is saturated and are relatively easily transported at the same time that the rolling load applies pressure to the surface of the base layer. The fines, now collected in the interstices of the ballast, act as a lubricant between the granules of the ballast, allowing them effectively to be squeezed laterally as successive rolling loads pass. The result is a depression in the path of the rolling loads.

If the surface is brittle, cracks may form in the bottoms of the depressions that permit even greater quantities of water to percolate through the ballast to the base layer. The depression accelerates to become a pothole in a highly fractured surface. The whole process is exacerbated in the extreme if, during the process, freezing temperatures acting on the saturated base layer heave the entire roadbed. This reverses the curvature of the depression, which accelerates the cracking—particularly given that heaving occurs at reduced temperatures—hence embrittling the surface even more. When this process is advanced, and the road surface is essentially destroyed, we are likely to see road signs warning us of frost heaves.

Sidewalks and terraces are no less likely to be subjected to heaving because of temperature conditions, but the loads are so nominal compared to those on roads and rails that there is scarcely ever enough of the prolonged pumping required to contaminate the granular layer. When walks and other pedestrian pavings are subjected to vehicular traffic, they deteriorate very rapidly simply because they are not designed with a deep-enough granular ballast or rigid-enough topping to prevent depressing the surface past its elastic limit, thereby compressing the substrate into the fines of the soil and initiating gross contamination.

Retaining structures, particularly cantilevered retaining walls, are vulnerable to lateral heaving. As the soil/water mixture expands behind the retaining wall, the wall is put into a state of bending or general rotation. When the soil melts, the wall rebounds, but as the melting generally proceeds from the top of the retained soil downward, there is still substantial lateral pressure, as the soil above is melting. The self-weight of the soil compacts and consolidates behind the flexed wall, providing permanent lateral restraint to the complete rebound of the wall. Over time and multiple iterations of the process, the heaving progressively rotates the top of the wall, potentially to failure.

Prevention of such a condition is simple and fairly common in retaining-wall construction. The hydrostatic pressure of the water behind the retaining wall can be effectively relieved by installing a gravel collection layer behind the wall, and drain lines through the wall to the low side. The gravel provides a porous conduit for the water, and the drains provide points of relatively lower pressure. The combination not only will eliminate accumulated water immediately behind the wall, but also will provide a pressure-release zone for water accumulated in the soil beyond the gravel layer. A commonly added detail is a compacted clay cap above the gravel backfill. The cap is typically sloped with a high edge at the contact line with the top of the wall. This acts as a relatively impermeable layer above the gravel backfill, which diverts water away from the zone of vulnerability.

Despite long-standing awareness of the value of such low-tech details, many earth-retaining walls were not built to observe them, and they eventually fail for these very basic and simple reasons associated with freezing water.

Heaving and other freezing-water-related mechanisms can be extremely destructive simply because of the forces that can develop. As discussed above, even though the actual forces are rarely equal to the theoretical limits, the forces are significant. Freezing soil can and will lift the perimeter of a building if all the necessary and sufficient conditions are met for heaving action. The actual magnitude of one cycle of such lifting is quite small, particularly if the heaving is opposed by counteracting pressure. An unopposed cycle or modestly opposed cycle, such as with a light paving, may lift some number of millimeters, but significantly opposed heaving lifts are typically on the order of a few micrometers. As we have observed and will observe over and over again, small movements can, however, induce stresses in materials and elements, which, despite the seemingly nominal magnitude of the strain, will exceed elastic limits, moduli of rupture, and ultimate strengths. But it is also true that freezing-water-related mechanisms are inherently strain-limited phenomena. This means that the expanding water, whether it is confined or unconfined, will assume an equilibrium dimension. Once that expanded dimension is achieved, the expansion stabilizes; expanding water does not continue to expand without limit.

In many mechanisms involving expanding water, the damage is progressive, as described with lateral expansion behind retaining walls; heaving, however, is not necessarily a progressive deterioration mechanism. (See also Chapter 5 regarding freeze-thaw cycling.) It is an inherently progressive deterioration mechanism resulting primarily from the expansion of freezing water. That said, freeze-thaw cycling is not technically progressive in nature simply because of the cyclic freezing and thawing of water; for the mechanism to be progressive, the material must crack, and the subsequent intrusion of water must fill that crack in order to progress the crack length. If the water filled only precisely the same volume in the pre- and postcracked conditions, then the expansion would assume the same strain limit. The existing crack would open to the same degree as when it was originally initiated, and there would be no motive force to further widen the crack. In vertical heaving, soil does not crack; there are no newly initiated capillaries to subsequently fill with water. In fact, the opposing force of the footing tends to reconsolidate the soil to or near its original elevation as the water reliquifies. It is entirely possible that a heaved footing, particularly one where there is no collateral cracking or fracturing, will return to or very near its original elevation when the soil thaws.

Intervention in instances of heaving is primarily an exercise in mitigation; however, a review of the necessary and sufficient conditions is informative. The necessary and sufficient conditions for the formation of heaving are as follows:

- *Cohesive soil.* The soil must be rich either in clay or silt.
- *Saturation.* The soil must have sufficient water content as to effectively eliminate air pockets in which expanding water might otherwise escape (see discussion of air entrainment in Chapter 4).

- *Freezing temperatures at or below the bottom of the footing.* The water can and will expand between 4° and 0° C; however, if it does not continue to freezing, the hydrostatic pressure is likely to dissipate.
- *Compressive load less than expansive pressure potential.* The pressure generated by the expanding water and/or ice must be greater than the confining pressure of the footing; otherwise, the footing does not move.
- *Vertically mobile foundation.* The foundation must be capable of vertical translation even though the movement may trigger substantial resistance, hence opposing pressure. The source of the resistance may be rigidity in the structure or some other strain-limiting device.

From the previous discussion, it is reasonably apparent that, depending upon the specific application, namely whether we are investigating a paving or a retaining wall, the relative importance of these conditions will vary; but these are the minimum essential conditions requisite for the initiation of frost heaving.

Specific circumstances may alter the relative efficacy of the various options, but the general approach with the most apparent promise, as just mentioned, is mitigation. The message of the matrix in Table 3.2 is that other options exist, but they are rather expensive and/or infeasible given the existence of the phenomenon. Heaving is certainly one deterioration mechanism that is far less expensive and, usually, simple to prevent than to definitively eliminate. By default, what is left is a set of mitigating options that, while not definitive solutions, can cost-effectively reduce the damage.

Lateral Sliding

Lateral sliding is a term applied to foundations or footing that, for one reason or another, are displaced sideways from their original positions. The most common occurrences are where an external motive force has the capacity to overcome the frictional resistance below a footing, such as at the bottoms of basement walls and earth-retaining walls. This is not to suggest, however, that lateral sliding only happens in earth-retention locales.

Before any footing or foundation moves in any direction, there must be a force acting in the direction of translation or movement. The source of the force, while typically lateral earth pressure, can be from any external source that acts horizontally on the footing or foundation. Indeed, a possible source of lateral foundation displacement can be freezing water acting in an identical manner as described in the section on heaving except the line of force is horizontal, rather than vertical, at the level of the footing. In

Table 3.2 INTERVENTION MATRIX: FROST HEAVING

Necessary and Sufficient Conditions	Intervention Approaches					
	Abstention	Mitigation	Reconstitution	Substitution	Circumvention	Acceleration
General Mechanism	Accept heaving as optimal condition.					Demolish construction.
1 Cohesive soil	Ditto			Remove and replace soil with granular material.		
2 Saturation	Ditto	Reduce the amount of water reaching the soil in the vicinity of the footing.			Remove soil, install a drain line, and backfill with granular material.	
3 Freezing temperatures	Ditto	Install heat-resistance tapes or other linear heat source in ground.			Underpin foundation to elevation below frost line.	
4 Compressive load > expansive potential	Ditto	Deliberately increase load on footings by adding weight to the building.				
5. Vertically mobile footing	Ditto					

fact, horizontal translation is a potential component of the heaving that occurs behind a retaining wall.

The basic condition involved in the mechanism is that a foundation or footing bears on soil with some pressure. The combination of the pressure and the contact plane of the foundation creates a frictional plane between the bottom of the foundation and the bearing surface. The amount of force required to break the bond between these two surfaces in contact is a function of the inherent frictional properties of the two materials and the magnitude of the applied force. Where lateral pressure is a foreseeable design consideration it is not unusual for the designer to install a lug or cleat on the bottom of the footing, to backfill against the passive side of a footing, or to install some kind of lateral brace to the footing.

Common examples of the cleat approach are found on many cantilevered retaining walls. On the bottom of the horizontal leg of the cantilever, the designer calls for construction of a lug to be continuous with and part of the basic footing. The lug is placed in a trench as the footing is poured (most cantilevered retaining walls are poured concrete). If and when the wall as a whole is pushed in the direction of the low side, the lug or cleat pushes against the short, but effective, wall of soil to the low side of the construction. This action has the effect of using the passive resistance of the soil into which the foundation is buried to balance the active pressure of the retained soil to avoid lateral translation or sliding. Moreover, the passive edge of the footing itself also serves the same function as the cleat, so the cleat is going to occur only when the passive footing edge and the friction on the bottom of the footing together become insufficient to balance the active earth pressure.

Such an imbalance is more likely to occur in connection with a modern cantilevered retaining wall than in a traditional gravity retaining wall because of the significant differences in relative masses of the two approaches. Gravity retaining walls are generally wedge-shaped masses of masonry or concrete that place the vertical or slightly inclined face (the *battering*) against the retained earth. This side of the retaining wall is the active pressure face of the construction, meaning that the pressure of the earth acts upon or is actively bearing against the retaining wall. The horizontal surface of the footing is placed on the soil below the wall, and is extended far enough in the low-side direction. This provides sufficient bearing area for the weight of the construction, to enable the frictional area to resist lateral translation, and to resist overall overturning of the construction. The exposed face of the wall on the low side may be vertical or battered, and most of this surface will be exposed above-grade as a result of the change in grade from the active side to the low or passive side. The balance of the passive side extends to the footing or the bottom of the wall, and some of the lower portion is likely to be buried below-grade.

FIGURE 3.5A   A crack extends completely through the wall and the terrace. Although the lateral separation is small, the return wall to the right has structurally released from the front wall. The problem originated from a foundation loss at the return wall.

The portion of the wall below-grade on the passive side is an important component of the overall resistance to lateral sliding. In addition to the frictional resistance of the footing, the vertical face of the footing, or the wall itself on the passive side, will press against the earth, which, in turn, will resist displacement. The resistance of earth to compression is called the *passive earth pressure*, as opposed to the pressure this same earth may exert on the active side of the wall. This point sometimes causes confusion: active earth pressure is the force exerted *by the soil* in repose against a surface, in this case the retaining side of the wall; passive earth pressure is the reactive force of *a surface compressing the earth* and the soil resisting that com-

FIGURE 3.5B The disengaged wall is restrained by a vertical brace, called a strongback, and six star bolts. The strongback is anchored through the wall and into the floor plates, which act as horizontal diaphragms and engage the entire structure. The underlying problem at the foundation remains unaddressed.

pression. The distinction is important because the values of these two potential forces can be and usually are rather different.

Most soils, in particular, cohesive or clayey soils have a tendency to stick together. If they are mounded up and released, they will slump into a lower but still distinguishable mound. If this same mound were placed against a vertical surface such that the surface were to interfere with the shape of the mound in repose, then there would necessarily be some force exerted by the soil on the surface and an equal force exerted back by the surface against the mound. If the surface, in response to the force of the soil, relaxed ever so slightly—which it must do as an elastic body—only part of the active pressure of the mound would continue to press the wall because of the inherent tendency of the soil to cling to itself rather than flow, to follow the now slightly moving wall. Granular soils exhibit a lesser tendency to do this, but they, too, will not simultaneously follow the moving wall as if they were fluid.

When a soil is itself being compressed, the tendency of the soil to stick to itself or to resist flow is much less relevant. The dominant factor or property in such a case is the elastic or plastic response to compressive force. In most cases, the resistance to deformation, the passive resistive property, of soil is greater per unit volume than the capacity of that same soil to flow against a surface and exert a force against that surface. The modest burial of the wall into the soil and, therefore, the development of even a modest passive bearing surface, significantly increases the resistance of that same footing to lateral sliding.

That said, it is notable that such burial is not essential, and its absence is not necessarily an indication of impending failure. The development of friction on the bottom of the footing may by itself be sufficient to overcome a tendency to slide laterally. In the case of cantilevered retaining walls, the situation is a bit more severe, in that the weight of the wall itself is significantly less than that of a comparable gravity retaining wall. Consequently, the frictional resistance at the bottom of the footing is lower than similarly sized footing below a gravity retaining wall. In order to compensate for the relatively lower frictional capacity, cantilevered retaining walls tend to rely more on passive earth pressure to avoid lateral sliding. Cantilevered retaining walls are more likely to include cleats or lugs on the bottoms of the footings, which increase the contact surface for the development of passive earth pressure.

Cantilevered retaining walls are typically L-shaped constructions, with the horizontal leg turned either to the low side or the high side of the construction. There are also cantilevered retaining walls constructed by burying a straight piece of construction into the ground and allowing the upper portion of the wall to act, quite literally, as a vertical cantilever with earth banked against one side. One of the more common applications of this buried-plank approach is sheet piling. In excavations, one way of holding back the earth is to drive planks or sheets of steel into the ground, then excavate on one side of the piling to a depth that leaves enough of the piling imbedded in the ground as to provide cantilever bending resistance against the retained earth. When the footings or foundation walls are installed, the sheet piling can be pulled out or simply left in place and buried. Regardless of the configuration, they all share the common denominator that they are forms of cantilevered beams, and, therefore, rely on the bending capacity of the section to resist the lateral load of the retained earth. The ability of the cantilevered retaining wall to develop bending means that the wall need not rely on the bracing provided by the mass of the gravity retaining wall. This means that cantilevered retaining walls are typically lighter and thinner than gravity retaining walls. The reduction in volume and weight offers certain economical advantages, but there are some trade-offs relative to resistance to lateral sliding.

Foundations walls, generally, are braced retaining walls, and, as such, exhibit many of the same problems as any other earth-retention structures. Bracing the top of the foundation wall tends to counteract the overturning force against the wall, but the concern about lateral sliding of the base still exists. Very often, the bottom of the wall is also effectively braced against translation by the basement slab, but there is not always a basement slab. In fact, in older buildings, the basement floor is very often exposed earth. Even where a basement floor is present, it may be little more than a thin mud slab, capable of very little compressive resistance.

Where there is a basement floor and that floor is of measurable thickness, and that thickness is reasonably intact, the slab works as a plate in being compressed along one edge in the plane of the slab. If the slab is very thin or the compressive load is particularly high, the slab may buckle locally along the edges or globally over its entire width. Very often, fractured basement slab perimeters are misdiagnosed as vertical settlement of the foundation that has dragged the edge down and fractured the slab in bending. In fact, the bottom of the wall has moved laterally and compressed the slab edge until it has buckled and fractured. Although the broken pieces of basement slab will appear somewhat similar, the distinction is not difficult to see.

If the deterioration mechanism is, in fact, lateral sliding and edge compression, the elevation of the line of contact between the slab and the wall will be the same as it was originally. If the deterioration mechanism is differential settlement, the line of contact will be lower than it was originally. The test is easy to conduct by finding some reference to the original elevation of this line of contact. If the slab was fairly level, then contact between the slab and an interior column may provide a valid reference plane. If the floor joists are original and undisturbed, they may provide a parallel plane to the original basement slab. If the joists slope downward toward the wall, then differential settlement is more likely; if they do not, lateral sliding is more likely.

Foundation walls in basements with earth floors are not necessarily pending cases of lateral sliding. The bottom of the wall may be buried sufficiently into the ground below the basement floor level to engage enough passive resistance to preclude lateral sliding. The foundation wall may be installed on a spread footing that has enough frictional resistance to preclude lateral sliding. For all the ways that lateral sliding can be avoided, it is still relatively common in residential structures with earthen basement floors, particularly in eighteenth-century and nineteenth-century masonry structures with relatively light foundation walls, where it was not unusual for the foundation wall to be on one course of brick or the equivalent dimension in stone thicker than the upper wall. The bottom of the foundation wall was placed directly on soil without a spread footing, and the elevation of the bottom of the foundation wall was at or only slightly below the elevation of the dirt floor. These are precisely the conditions to anticipate lateral sliding of the bottom of the foundation wall.

When lateral sliding in a foundation wall does occur, the form of the deformed wall is dictated by the boundary conditions. The first floor will brace the top of the wall, so *that* line will remain relatively straight. The intersecting foundation walls at the ends of the wall in question will brace the corners on the building and, therefore, on the panel of foundation wall at issue. This means that the translating bottom edge will be greatest at the midpoint between the return walls. From that point of maximum translation, the foundation wall will slope back toward the outside of the building until it engages the straight edge of the ground-floor plate. In advanced cases, a vertical crack will form on the inside of the foundation wall, initiating the point at the bottom of maximum translation and running upward, growing thinner and thinner until it dissipates as it approaches the ground-floor plane. There may be other, finer, vertical cracks starting at the base and diminishing as they traverse the wall.

Lateral sliding is similar to heaving to the extent that each increment of translation is somewhat strain-limited, because the soil that is actively exerting the translational force to the wall is somewhat cohesive. Thus, at the instant that the wall relaxes slightly and slides laterally, the pressure against the wall will decrease. The movement will stabilize simply because the motive force is momentarily relieved. Over time, the soil will be recompacted against the wall, and the active pressure will gradually return to its former magnitude. If that magnitude was enough to initiate lateral sliding, there is no physical fact that suggests that it will not be sufficient to continue the progression of the lateral sliding.

End-stage lateral sliding is manifested by a horizontal rupture of the lower portion of the foundation wall, followed by a vertical collapse of the upper wall above the point of maximum translation. Eventually, the base of the wall slides so far into the basement that the line of force of the wall above is outside the middle third of the foundation. The bottom of the wall will slide into the basement, and the upper wall will slide downward and to the outside of the foundation. The primary vertical crack, described above, may rupture, and the now-unbraced edges of the fracture wall plane may buckle simultaneously or independently. The main point is that lateral sliding is inherently progressive, and results in an unstable compression condition.

Structurally, end-stage buckling as a result of sliding is somewhat more understandable if the problem is turned upside down. The term unbraced basement foundation wall bottom means that the foundation is an open box turned upside down. The centers of the unbraced edges will collapse inboard first, and the collapse will progress toward the corners of the structure, which are generally the stiffest portions of the box. If the same open-ended box is turned right side up, these same unbraced edges will collapse first. Such a collapse mode is associated with the unzipping of a building along the roof line as the connection between the roof and the wall deteriorates, and the bracing effects of the roof disintegrate.

The buckling length of an unbraced base foundation wall bottom is two times the height of the wall from the basement to the ground level. This may seem surprising, but since the bottom of the wall is free to translate, the mode of buckling is such that k=2 in the formula for critical buckling load:

$$P_{CR} = \pi \, E \, I \, / \, kL^2$$

From the preceding discussion, we can conclude that the common element necessary and sufficient for lateral sliding is inadequate lateral restraint at the base of the wall, meaning simply that the potential restraining resistance is less than the actual lateral force. The restraining resistance is typically supplied by some combination of frictional resistance at the interface of the footing and the bearing material—passive earth pressure against the footing, and/or physical restraint of the footing by an intersecting slab or other bracing structure.

Among earth-retention systems, the inventory of such physical restraints is long, varied, and often quite ingenious, far beyond the scope of this book. We will, however, mention a few. If excavation allows, it is possible to employ what are called *deadmen*. A deadman acts much like an anchor. It is a log or bar buried well below-grade on the active side of the retaining wall. Cables are tied from the deadman to the retaining wall. In order for the wall to move, the deadman must plow its way through the ground. When rock is remotely convenient to the active side, rock anchors are a possibility. Rock anchors perform much the same way as deadmen, except the far end of the cable is attached to what amounts to a large cinch anchor sunk into the rock. Two factors significantly exacerbate a lateral sliding mechanism: inclination of the bearing surface and hydrostatic pressure on the active side.

If the bottom of the footing or foundation is actually sloping downward toward the passive side, there is a fairly obvious tendency for the footing to slide downhill. In addition to the lateral force of the retained soil, the force of gravity is also pulling the entire structure down the slope. A more likely situation is that the retaining wall is benched, or cut into an existing slope, where the bearing strata are layered, and the layers are parallel to the slope. If the boundaries between the layers are weakened planes, the weight of the retaining wall may be enough to precipitate a slump failure of the slope. Extreme cases of slump failures are exhibited in the mud slides of southern California; less extreme examples occur with regularity in far less vulnerable terrain. Such conditions may also occur at road cuts and other steep excavations, where the angle of the cut exceeds the angle of repose of the soil. In the case of whole structures such as buildings, sloped bearing strata may result in lateral sliding of the entire structure.

Hydrostatic pressure on the active side can occur either because the footing or foundation is below the permanent, or a perched, water table, or because even as water is passing through the soil, it exerts hydrostatic pressure against the retaining wall. The only essential difference between a perched water table and the migrating water passing through is a matter of rate. In the case of passing or migrating water, the prospect exists that the hydrostatic pressure is sufficiently transient to challenge the wisdom of investing much in the way of intervention. With permanent and perched water, the time factor weighs against doing nothing. Regardless of the duration, hydrostatic pressure can be a severely exacerbating condition. The discussion of heaving explained how water trapped behind a retaining wall could result in lateral displacement. Here we evaluate the effects of trapped water on the lateral stability of a retaining or foundation wall. The pressure of water against a surface increases linearly with depth, at a rate equal to the density of the fluid, which is approximately 64 pounds per cubic foot. At the bottom of a swimming pool six feet deep, the pressure against the bottom of the pool, as well as at the very bottom of the side walls is approximately 384 pounds per square foot. Soil is similar to water in that the lateral pressure increases with depth; but, as we explain in the next section, the rate of increase is significantly less than with water.

When water is added to soil, the pressure of the water begins at the top of the water level, and the pressure increases with depth from that point downward. The pressure of the water is added to the pressure of the soil. There is some compensatory effect because of the buoyance that water adds to the soil, but the computation of the magnitude of this effect is best left to a geotechnical engineer. For common exterior retaining walls, basement walls, and similarly scaled retaining structures, the straight additive effect is sufficiently accurate and appropriately conservative. An adequately designed dry retaining wall can fail utterly when the soil is saturated.

The added disadvantage relative to lateral sliding is the effect that the water has on the soil below the footing. If the soil is permeable—and all soils are permeable at some rate or other—the water retained behind the wall will necessarily seep under the footing. In earthen floor basements, this is a primary source of dampness. Even in basements with slabs, moisture can seep through the slab in both liquid and vapor form. As described in the section on heaving, water working in the interstices of the soil below the footing, combined with the fines in the soil, can effectively act as a lubricant and reduce the coefficient of friction between the footing bottom and the soil. For a footing on the verge of sliding, the combination of the added lateral pressure of the retained water and the reduction in sliding friction can initiate lateral sliding in a footing, which, when it was dry, was secure and stable.

Neither inclined bearing nor hydrostatic pressure is a necessary and sufficient condition for the initiation of lateral sliding; however, they can significantly reduce the capacity of a wall to resist lateral sliding. From that standpoint, they are notable facilitating factors, whether essential or not. In the evaluation of intervention efficacy, inclination and hydrostatic pressure can, if they are applicable, influence the choice of tactics (see Table 3.3).

Intervention in deterioration mechanisms involving lateral sliding are typically tedious and expensive because the sources of the problems are below-grade, hence inaccessible, and large in scope, because they typically affect substantial portions of the building or larger retaining structures. Among the general approaches, those that are most feasible and of long-term benefit tend to be mitigative, although under certain circumstances some of the other options may be desirable.

It is important to note with regard to lateral sliding that any approach that arrests or mitigates the translation is worth consideration, whether it addresses the source of vulnerability or not. This is true of many of the deterioration mechanisms that we will discuss throughout the text. For example, the basic weakness of the retaining structure may be the minimal nature of the passive pressure-contact surface. This does not mean that we are limited to the intervention listed in that column to affect alteration of the mechanism. If, in this example, there is no floor slab, and that will solve the problem just as well as installing a turndown at the back edge of the slab, it hardly matters that the slab is a bracing device, as opposed to one that will increase the passive pressure-bearing area. The slab may well be more cost-effective because it also has significant benefits, in addition to stabilizing the lateral sliding.

Another example of multiplication of benefit is to completely excavate the active side and reinstall an engineered backfill. If the problems associated with a particular structure are limited to lateral sliding alone, such a tactic is probably rather extreme, but if the problems also include moisture

Table 3.3   INTERVENTION MATRIX: LATERAL SLIDING

Necessary and Sufficient Conditions	Intervention Approaches					
	Abstention	Mitigation	Reconstitution	Substitution	Circumvention	Acceleration
General Mechanism	Accept the rate of sliding.					Demolish structure.
1.a Inadequate resistance: low frictional capacity		Reduce the surcharge of soil; reduce the amount of water getting into the soil.	Demolish the footing and install the same footing in the original location.	Demolish the footing, and install a similar footing in the same location; remove the old soil below the footing and replace with gravel.	Remove the surcharge entirely and replace with designed backfill; remove and replace or modify footing for greater bearing area.	
1.b Inadequate resistance: low passive-pressure capacity		Add soil or other surcharge to greater depth on passive side, and/or same as above.	Ditto	Remove existing soil on passive side and replace with more resistive material.	Modify footing to increase passive pressure-bearing area.	
1.c Inadequate resistance: insufficient lateral bracing		Same as 1.a.	Ditto	Add buttresses or counterforts.	Install slab or other horizontal brace.	

162

seepage through or under the wall, other structural deformations in addition to sliding, splashing of water at-grade against the wall above the foundation, failure of perimeter drainage systems, or any of a number of other collateral problems that could be mitigated or eliminated coincident with complete replacement, then what may seem like a rather drastic intervention may become quite feasible.

Other Lateral Retention Problems

Lateral retention structures fall into two general classes: those restrained only at the bottom of the structure and those braced at both the top and the bottom of the structure. The most common example of the former is the earth-retaining wall; of the second, a basement foundation wall. In the preceding section we discussed some of the parameters associated with these two classes of structures. In this section we will amplify these two classes relative to two other specific variations of the basic themes described above: foundation wall bulging and the undermining of foundations.

We have mentioned several times that soil lying against a vertical surface exerts pressure against that surface. The magnitude of the active earth pressure increases with depth as a result of the cumulative compression of the soil above, effectively squeezing the soil below. The consequence of the compressive force is that the soil is squeezed in all directions, including, obviously, against the vertical plane of the wall. A diagram of these increasing forces would be a triangle, with the tip of the top angle of the triangle at the top of the soil surface representing zero force at-grade. The triangle widens with depth, representing the gradually increasing magnitude of the force of the soil against the wall. The widest portion of the of the triangle is at the lowest portion, or extension of the wall, even if that extension is not coplanar with the primary exposed surface, such as a cleat or lug set back from the face of the wall. Though the cleat is set back from the active face of the wall, the total lateral force acts on it even though that portion of the active force is seemingly offset by passive resistance on the opposite side of the cleat or downturn.

For design purposes, particularly when designing for overturning and lateral sliding of the wall, the triangularly distributed load can be simplified to a single vector. The magnitude of the force vector is the same as the sum, or total value, of the triangular load. The direction is the same as the triangular load. The point of application is at a point one-third the height of the triangular load measured upward from the base of the load. It is a small but important detail that the point of application is the third point of the triangular load, not the third point of what we think of as the height of the face of the wall. Most of the time these two dimensions are coincident, but not always. Any cleats or turndowns, even if they are on the opposite side of the footing, away from the active face of the wall, add to the exposed active pressure surface.

When a triangular load is applied to a basement wall that is braced at the bottom and at-grade, the wall will bulge inboard in section. If that same wall is restrained by the return walls, this same bulge will also dissipate toward the corners of the room or building. The point of maximum distortion is approximately a third of the way up the wall between the basement floor and the floor above. This is what distinguishes this deformation from lateral sliding. Foundation wall bulges are also exacerbated by hydrostatic pressure, and for much the same reasons. End-stage bulge failure is also similar to lateral sliding end-stage failure.

A common and rather interesting consequence of bulging, and to only a slightly lesser degree of lateral sliding, is upper-level bulging. When the basement wall is deformed inboard, the portion of the wall above the bulge is inclined, generally on a slope from the inboard point of the bulge to the outboard edge of the floor at-grade. This incline often continues past the floor at-grade, which means that the wall above the floor level at-grade will lean away from the building. If the wall is not positively restrained at the second level, the wall of the building will pull away from the second floor. The same pulling away may also occur at other upper levels. At some point, typically the roof line, if not before, the wall is tied back to the building, with the consequence that the wall above-grade bulges outboard over the height of the building from grade to cornice. In a three-story building, the point of maximum lateral deformation will be between the second and third levels.

This façade bulge is quite common among row houses and other light masonry construction, because of the combination of relatively thin masonry and the fact that the floor joists typically run parallel to the front façade. The thickness of the masonry is significant because the stiffness, hence the inverse—namely, flexibility—is influenced by the thickness of the wall. The direction of the floor joists is significant because, as a result of running from party wall to party wall, they may not be positively attached to the front wall of the building.

Stiffness is a function of the moment of inertia. The maximum deflection of a uniformly loaded beam is as follows:

$$\delta = \frac{5}{384} \times \frac{wl^4}{EI}$$

In this and similar equations for maximum deflection, the variable I, which is the value of the moment of inertia, is inversely proportional to the deflection or distortion. Moment of inertia (I) is a measure of the geometric distribution of a given amount of material. The greater the value of I, the greater the efficiency of that distribution, and the greater the stiffness of a beam of element. In beams with rectangular cross-sections, the value of I is a function of the breadth (b) and the depth (d) of the beam in accordance with the following proportion:

$$I = b\ d^3\ /\ 12$$

This means that the beam will be more resistant to deflection as it becomes wider, but it will become stiffer as it becomes deeper, by the cube of the depth. In walls, the depth of the wall relative to lateral bulging is the thickness of the wall. In a solid 8-inch-thick wall, $I = 512$ in^4 per liner foot of wall. In a solid 12-inch-thick wall, $I = 1728$ in^4. Thus, for a 50 percent increase of thickness from 8 to 12 inches, the stiffness of the wall increases 337.5 percent.

The prospect of such bulging was so evident to the original row house builders that they often installed positive anchors to tie the front wall back to the floor with devices generically called star bolts. (Although star bolts are often shaped like stars or, somewhat more precisely, like pentagrams, they come in many shapes, including disks, squares, diamonds, even initials.) The principle is the same regardless of the shape of the exterior plate. The plate is attached to a rod or bolt that passes through the wall and is anchored into the joists such that any exterior migration of the wall will engage the rods, which, in turn, will be restrained by the sum of the flooring and joist plate.

Even after the façade begins to pull away from a building, a similar device can be retrofitted to a façade to stabilize upper-level bulging. When star bolts are installed initially, the rod is commonly a strap, which is let into the tops of the first three joists, and hooked down at the extreme inboard end. Solid blocking or ridging is installed between the joists, and subflooring and flooring run over the straps. The assembly is integrated, and the restraint is positive. The assembly need not be adjustable because any slack in the straps could be taken up at the inboard end with shims or packing.

The retrofitted star bolt is more likely to be a true bolt with at least one threaded end. A hole must be drilled through the masonry and the first three or so joists. The ceiling below where the bolts come through the joists must be removed, and solid blocking must be installed between the joists. The rod is passed through the wall with the plate or star bolt attached to the extreme end. The take-up nut can be on the joist end, the star bolt end, or both. Whether star bolts are installed initially or as a retrofit, they are positive and effective mitigation of upper wall bulging.

While star bolts or other specific wall ties will restrain the wall at the levels where they are installed, the motive force of the basement-level bulging is still present, consequently, the tendency to deform the façade in an outwardly bulging deformation is still present. What has changed by the installation of tiebacks is that the effective buckling height of the wall is significantly altered by the added number of nodes. Whereas the unrestrained wall had a buckling length of the full height of the wall, once restrained at each floor with star bolts, the buckling length may be halved or even quartered. Critical buckling capacity is adversely affected by the square of the unsupported length. In the case of the unrestrained façade, the effective buckling length is the full height of the façade from the grade-level floor to the point of restraint, probably the roof line. In even a short, two-story building, restraining the wall at the second floor reduces the buckling height in half, which increases the load-carrying capacity of the same wall by four times.

When the buckling length is reduced, the effective capacity of the wall increases; and although the load on the wall has not changed, the resistance to distortion for the current load is increased, and the rate of distortion is reduced.

As is the case with all retained-earth situations, the wall-bulging phenomenon is exacerbated by water saturation of the soil. In urban areas where wall bulging is quite common, the immediate area in front of the bulging façade is typically paved with sidewalk and street. Even in imperfect conditions, exterior paved "aprons" are very effective barriers to precipitation. The source of water in urban areas is very often a leaking sewer or water main. A front wall that has performed without distortion for many years may in such circumstances develop significant problems because of a below-grade source of water.

Another type of problem altogether, but related to the phenomenon of lateral sliding and bulging, is undermining of an otherwise stable footing or foundation. The most common source of the undermining is from careless excavations adjacent to footings or foundations. It is a rule of thumb to avoid excavating adjacent to an existing foundation anywhere within an imaginary cone emanating downward and away from the bottom of the foundation. The slope of the cone is ordinarily accepted as 1:1, meaning that if the excavation is to be one meter below the bottom of the footing, then it must be at least one meter laterally away from the footing as well. If not actually a rule, this is a reasonable guideline for short-term excavations. But it is somewhat risky for permanent excavations below an existing footing or any other surcharging load on soil. The premise behind the 1:1 rule is the conflation of two separate notions about the behavior of soil under load. One is the notion of repose, which is the angle which a pile of soil will assume if left undisturbed except by its self-weight acting under gravity. The other is a notion of a shear plane or plane of failure below a footing that is actively compressing the soil.

The angle of repose of many common soils is in the range of 30° to 32°. This means that the slope of an uncompressed pile of common soil is 7:12, meaning a rise of 7 meters to a horizontal run of 12 meters. According to the rule, if 7:12 is a sort of natural zone of influence, then anything outside that cone will not affect the position of the footing. Unfortunately, there is a serious flaw in this logic, namely that an excavation of 1:1 will, in fact, cut into a 7:12 cone emanating from the footing. Fortunately, the concepts of angle of repose and active zone of influence are totally disconnected, so the error in logic is irrelevant. The apparent logical leap from angle of repose and zone of influence simply are unrelated.

The concept of a shear plane or failure plane developing below a footing is more descriptive of the actual behavior of soil under load. If soil is loaded locally, as in the case of an isolated footing, until the soil fails, the form of that failure will be as though the footing scooped out a large spoonful of soil. The plane along which the soil will fail is a form of shear fracture or failure. The shape of the line along which the failure will occur resembles the section of a spoon. Picture sinking a spoon into a large bowl of sugar until the rim of the spoon bowl is just visible at the surface, and the handle is angled up and away from the surface so that the rim of the spoon is horizontal. The point of load of this imaginary footing is at the intersection of the handle and the bowl of the spoon. If we press down on the handle hard enough, the bowl of the spoon will slide along its bottom until even as the heel of the bowl is moving down in the direction of the pressure, the spoon is scooping up and out of the sugar dish a spoonful of sugar.

In soil failures, the plane of fracture is shaped very similarly to the shape of the bowl of a common spoon. It begins at the point of application of load and descends at a relatively steep angle. In a smooth curve, the trajectory of the shear plane shallows out and begins ascending at a shallower angle than the descent angle back to the surface. When soil fails, the result is a large scooping of earth moving down at the point of application while laterally pushing ahead of it a "spoonful" of earth up and out of the ground.

The part of the 1:1 rule that corresponds with the geotechnics of the situation is that by staying outside the 1:1 cone, a great deal of the actual shear plane is confined within the boundaries of the rule. The limits to the 1:1 rule are manifest in rather deep cuts, even if they conform to the rule adjacent to footings that are heavily loaded. The closer a foundation is to developing a failure along the shear plane, the less reliable is the 1:1 rule. If soils are lightly or nominally loaded, the permissible angle of the cut may actually be much closer to the edge of the footing than would be allowed under the 1:1 rule.

Another important variable is time, meaning the duration of open cut. The development of a failure along the shear plane in soil takes some time to develop, particularly if the actual load is close or equal to the critical failure load. Because of the plastic characteristics of soil, the deformation along the shear plane prior to release is considerable, and occurs over measurable amounts of time. The consequence is that the time required to develop a shear failure in soil is related to the magnitude of the load, as well as to the degree of interruption of the shear plane by the excavation. This is why grave diggers stay in business. A grave digger can in ideal soil conditions, meaning very clayey soils, dig an almost vertical grave without the sides collapsing into the hole. The keys to grave-digging success are to get the hole filled fairly soon, avoid rain between excavation and refilling, and pick cohesive soil to begin with.

If an excavation is close to, much less in contact with, an existing footing, and extends vertically from that point deep enough, and if the load on the footing is great enough, that footing *will* slide into the excavation. Observing the 1:1 rule will avoid most such catastrophes, but not because the rule is valid based on soil mechanics, but because most such excavations are fairly shallow, most footings are conservatively loaded, and most such excavations are backfilled or otherwise adequately retained without undue delay.

Other Shallow Foundation Problems

This section addresses two fairly common problems often associated with shallow footings, but that do not fall in the lateral movement or load categories: vegetative intrusion and hydrostatic uplift.

We have all seen woody and herbaceous plants growing from cracks in ostensibly solid materials. We have also seen the damage that plants can do to structures if their growth continues unabated. Herbaceous plants do not pose nearly the threat that woody plants do simply because of the compressive capacity of their roots and stems. Woody plants are also more likely to be perennials; the root material continues to grow in diameter from season to season. The distension created by expanding roots is considerable, but hardly mysterious. Numerous unattended trees growing out of walls precipitate partial collapses of those walls. We will not go on at length about this aspect of the problem because we all know that trees growing out of walls or roofs are not a good thing. Furthermore, it does not require a course in building pathology to detect trees growing out of façades.

Roots, particularly tree roots, can extend into cracks as small as 10 micrometers, and they can develop lateral pressures of several hundred pounds per square inch. Although an individual root can develop substantial radial pressure per unit area, the area over which a single root exerts the pressure is quite small. When, however, a mat of hundreds of fine roots work into and then expand in a small crack, the collective lateral force can produce several hundred pounds of force. This magnitude of force is demonstrably capable of fracturing mortar bonds or even the stone or brick unit itself.

There is no secret to dealing with vegetation growing out of roofs and walls, other than to kill the plant by, first, pruning, followed by the topical application of an herbicide to larger woody specimens or, preferably, to young dry foliage of small and herbaceous specimens. Although the decaying roots of a deceased plant are potentially inviting to rot, termites, and other critters, dead roots in the wall are less risky than trying to pull a developed root system out by the remaining stem.

Far more insidious than the root of a plant visibly protruding from a building are the roots of plants seemingly growing outside the perimeter in soil adjacent to the building. But their under-foundation roots can grow to dimensions capable of lifting entire structures. Tree roots can fracture pavements, invade and fracture sewers and catch basins, and collapse walls. Limbs can fall and damage buildings, and trunks can split with much the same effect. W.H. Ransom reports a devious behavior of deciduous species, namely that they will occasionally desiccate cohesive soils below footings, causing the soil to consolidate and the footing to settle or collapse into the void.* In short, trees and plants generally can invade and damage buildings. That is not news, and the remediation is not exotic.

When the plants are running vines and creepers, the issue becomes altogether more complicated. The discussion of that complexity is reserved for Chapter 4, but a mention here is warranted relative to foundations and structures; that is, generally, the damage associated with vines and creepers may be greater below-grade than above. The analysis and, ultimately, the decision to remove or retain clinging vegetation almost always centers on the damage and/or aesthetic properties above-grade, and that is the discussion we will leave for Chapter 4.

To repeat, water accumulated against the exterior surface of the foundation wall will exert a lateral pressure to the wall. That pressure is cumulative with depth, as explained in the preceding section. Once water is encountered, ordinarily that water will continue to permeate the soil below that point. There are exceptions to this rule, particularly at the geological formation scale. At the scale of most buildings, however, the rule is reasonably valid: once water is encountered it continues with increased depth.

*Building Failures: Diagnosis and Avoidance (London: E. and F.N. Spon, 1987).

It is possible, even fairly common, for rock layers to be more or less permeable, and to contain significantly different quantities of water per unit of volume. Layers of rock that are especially porous and permeable may act effectively as water conduits, called *aquifers*. Because the rock layers typically are not absolutely parallel to the erosion plane, the rock will intersect the surface at a shallow angle. This exposure can be and usually is quite large in area because of the angle of intersection. When the rock layer that intersects the surface at the shallow angle is also an aquifer, the exposed surface acts as a large water collection area called a *recharge area* or *zone*. The water enters the aquifer recharge area and flows down the shallow incline of the rock pipe, sometimes to distances of hundreds of miles. Much of the well water in Texas is drawn from a deep aquifer that is recharged in the Rocky Mountains.

If the water table, either permanently or temporarily, is above the top of the finished basement floor, the water will accumulate under the basement floor and exert an upward force on the slab in an effort to "float" the building. The pressure under the slab is approximately equal to the hydrostatic head adjacent to the foundation. If the water table is two feet above the floor slab, the upward pressure under the slab is approximately 125 pounds per square foot. This by itself is sufficient pressure to lift the slab off the soil, break it, and send tons of water into the basement. There are, however, some mitigating factors.

To begin with, two feet of hydrostatic pressure outside a residence is extraordinary. Two feet of hydrostatic pressure outside the foundation wall of a Chicago office tower is not. Taller, heavier buildings are often founded close to or below water tables. In such buildings, water intrusion is less of an issue than in residential construction, for a number of pragmatic reasons. The foundations are more likely to have been designed specifically for the site, with considerable time and attention devoted by geotechnicians and structural engineers to accommodating the water. In contrast, residential construction is more likely to have been built based on generic conditions and without the services of specialized engineers. Larger buildings, whether hydrostatic pressure is a major design consideration or not, have sufficient budgets for foundations, and the incremental cost of accommodating ground water is financially critical to the project. Adding into the residential project budget sums sufficient to accommodate significant ground water problems can or should result in cancellation of the project. The point is, residences are more often inadequately designed to accommodate excessive ground water, and home owners often find they have inherited long-standing water problems, with which owners of tall buildings that have foundations far deeper into the water table don't have to deal.

Other mitigating reasons follow from much the same premises as described regarding foundations: More expensively constructed buildings typically have sophisticated sumps, ejection pumps, slabs designs, perimeter waterstops, slab edge gaskets, exterior water proofing, perimeter drains, underslab drains, and crews to mop up the water. When the building is larger, meaning more expensive, two factors typically follow: a greater likelihood that potential hydrostatic pressures will be identified; management of such pressures will be dealt with in the design of the building.

If the building has an earthen floor or other highly permeable surface at the basement level, there is little accumulated hydrostatic pressure because the water will run freely into the basement. This will result in other problems of moisture accumulation in the basement—which we will address in subsequent sections—but there is less likelihood of structure damage. In order for pressure to accumulate under the slab of a basement or against the wall of a retaining structure, the water must be static, meaning at or near rest. If water is in motion, it must be moving in the direction of lower pressure. Water cannot simultaneously be moving to lower pressure and accumulating static pressure. The two phenomena are mutually exclusive.

Intervention in hydrostatic pressure mechanisms is difficult to undertake inexpensively for the same reasons that most foundation-related problems are tedious and expensive (see Table 3.4). The source or sources of the problem are large in scale and scope, and they are not readily accessible. Any approach that tries to deal with the problem from within the perimeter of the building is, by definition, *not* dealing with the source of the problem, which is water accumulating outside the building. Water percolating through a basement wall or up through a basement floor slab cannot be defeated by treating the surfaces of the wall and slab with impermeable coatings. The owner may gain only momentary relief if the coating is tightly bonded to the surface.

When coatings and films, including cementitious parging, are used on interior surface to defeat water originating at the perimeter and under the floor, the basic mechanism is unaltered. The pressure is still present, which means that, as the water migrates through the substrate, the bond between that substrate and the coating is in direct tension. The bond is saturated and begins to deteriorate from its own deterioration mechanism; the substrate continues to deteriorate; the coating itself begins to deteriorate. Applying of interior coatings to hydrostatic pressure deterioration mechanisms is widely done, but that does not make it effective. A strong and developed industrial front supports this approach, as the wet basement problem represents a huge potential market. Industry has, indeed, developed entire families of impermeable coatings largely to meet requirements to line tanks and other holding vessels such as pools and reservoirs.

The obvious distinction between a pool and a basement is that, in a pool, the water pressure is pressing the membrane against the substrate; in a basement, the water pressure is pushing the membrane away from the

Table 3.4 INTERVENTION MATRIX: HYDROSTATIC PRESSURE

Necessary and Sufficient Conditions	Intervention Approaches					
	Abstention	Mitigation	Reconstitution	Substitution	Circumvention	Acceleration
General Mechanism	Accept the consequences of the hydrostatic pressure.					Fill basement with concrete and delete space as a volume below the water table.
1 Static water pressure above finished surface of interior volume		Reduce amount of water accumulating adjacent to the structure.	Pump water away as it accumulates against structure.		Flood the lower level with water, thus providing back pressure and precluding intrusion by exterior water.	
2 Pressure-resistive floor structure		Add dead load to surface of pressure slab.	If fractured, demolish existing slab and replace in kind.	Demolish pressure slab and replace with stiffer/stronger slab.		Remove slab and allow water into basement.
3 Absence of pressure-relief device		Punch holes through slab and drain off water from sump inside the building.		Remove perimeter soil and replace with porous backfill connected to outlet.	Remove slab; replace with underslab drainage; replace slab.	

substrate. If the exclusive manifestation of the hydrostatic pressure problem is vapor migration through a reinforced concrete slab or wall, for example, and if the concrete can be dried and remain dry long enough for the adhered membrane to adequately cure, then the efficacy is positive. If the substrate is friable, if the water is migrating in liquid form, if the path of migration is through cracks of optical dimension, and/ or if the membrane is applied to a wet substrate, then interior surface treatment is not recommended.

Another fairly common and ineffective approach is to install dehumidifiers to mitigate the problem. Certainly, enough dehumidification run at a sufficient rate will draw water out of the ambient air, and produce a sensation of dryness to an occupant. What is absent from the solution is any mention of a reduction of moisture in the materials or reduction in hydrostatic pressure. In fact, the pressure under the slab is relatively unaffected by dehumidification.

Ironically, the migration of water through the various materials is actually exacerbated by dehumidification. Analogous to heat flowing from hot to cold, moisture flows from wet to dry. If the air inside a room is dry, it acts as a sponge to any moisture in the vicinity of that block of air. If the surfaces of the walls and/or slab are moist, the dehumidified air will absorb the moisture. As those surfaces, in turn, become drier, they act as blotters of moisture contained in the adjacent material and so forth. The chain reaction of moisture migration initiated by a well-intended effort to dry a damp basement necessarily assures that *more* water will move through the walls or floors than was the case in the damp condition.

Chapter 4 extensively covers evaporation and condensation, but for our purposes here, we point out that dehumidifiers work by using two sets of coils, as in a refrigerator or an air conditioner, except both coils are in the same space. In a refrigerator, the cold coil is inside the box and the warm coil is in the room. In an air conditioner, the cold coil is inside the room, and the warm coil is outside the building altogether. In a dehumidifier, the cold coil is on the back side of the unit; the warm coil is on the front side. A fan pulls air from the back and pushes it out the front into the room. Thus, the relatively moist air passes over the cold coil first, and water condenses on the surface of the coil. The water runs down the coil, drips off, and, typically, is collected in a bucket. The now drier, but cooler, air is pushed over the warm coil, where it recollects the heat lost to the cold coil, so that it reenters the room at or near the same temperature at which it entered the back of the machine originally.

Dehumidifiers are effective, inexpensive to purchase, and relatively cheap to operate. About the only operational catch is that they either must be emptied by hand or be connected directly to a drain line that dumps into a sanitary waste container. Older models often overflowed if not emptied, and that will still occasionally occur if the float valve sticks or fails to turn off the machine. If drying of air is required, dehumidifiers are good tools. But in a damp basement, for the reasons cited, simply drying air does not ordinarily solve the problem. This is not to say that sensible dampness is not reduced; the room or basement will *feel* drier to the occupant. The reduction of relative humidity of the air, however, may not, probably will not, positively affect hydrostatic pressure; and it certainly will exacerbate the deterioration of the intervening materials.

The approaches indicated in Table 3.4 range from the mundanely obvious to the marginally amusing. This is typical of most of the matrices we have examined, and that we will examine in this book. Again, the point is to demonstrate that more than one option is always available, and that an absurd option in one set of circumstances may be the key to a satisfactory outcome in another situation.

The mitigation of water outside the structure is probably the definitive hydrostatic pressure palliative, absent the cost. This has the altogether desirable characteristic of addressing the source of the problem. To reach the source, however, requires that the soil immediately adjacent to the foundation wall be removed. This is normally accomplished by earth-moving equipment such as a backhoe. But, the earth itself may be part of the problem because of its composition, so the excavated soil must be removed from the site and disposed of properly. Once the foundation wall is exposed, particularly if the wall is relatively old, it may require repairs such as repointing and reparging. Then some combination of layers of protective materials may be added, such as waterproof or dampproof films and/or a patented drainage layer. There are a number of such products on the market. One such product is applied in sheets and is adhered to the surface, although the adhesive need work only until the backfill is placed against it. The product has three layers: the contact layer, which is placed against the wall, a drainage layer, and an outer surface against which the backfill is placed. The outer layer is permeable to water but the pore size is small enough to exclude soil fines. The drainage layer is constructed something like an egg crate: the two outer surfaces are maintained a set dimension apart, but water can pass vertically without obstruction.

Obviously, once the water enters the drainage layer, it passes quickly to the base, and cannot exert any lateral pressure to the inner surface. When the water reaches the bottom of the drainage layer it must be effectively transported away from the building; otherwise, the water floods the drainage layer and there is no net benefit.

As a matter of common practice, the excavation is then lined with filter fabric, followed by a layer of granular material in the bottom of the excavation. At this point, there may be a porous pipe perimeter drain installed, provided that such a drain will connect to an adequate discharge point. Then the balance of the excavation is backfilled with granular material. The filter fabric lining the sides of the excavation is now between the granular backfill and the soil, with the loose top ends exposed. Those loose ends are lapped over the top of the backfill to complete the wrapping of the backfill in filter fabric.

Adequately discharging the collected perimeter water can be the most expensive part of the fix. Because of the invert elevation of the perimeter drain, simply connecting the line to a storm water system may require a cistern and a lift pump. If there is no available storm water system, the water may have to go into a detention basin of dry well. If either of these is used, it has to be deep enough to allow interception at the invert elevation and have enough additional depth and capacity so as to not backflood the drain lines. A less likely option is to connect the drain directly to a convenient water course. But more and more water districts are requiring on-site water disposal, and are not allowing storm water into streams and other surface water features. Even if it is allowed, the designer must be aware of the risk of backflood.

The top of the wrapped backfill is usually a few inches to a foot below finished grade. One option for detailing those last few inches is to continue with gravel to grade, thus exposing a strip of gravel at the surface. Another approach is to install a layer of compacted, relatively impermeable clay above the drainage layer. The clay is installed with an upper surface slope to direct surface water outboard, away from the building altogether.

A technically closely related approach to perimeter drainage is underslab drainage. All the principles and physics are the same, only the specific location of the porous pipe drains changes. As a retrofitted intervention, underslab drains are extremely rare. As initial equipment in new buildings, however, they are not uncommon in areas of high water tables. The installation sequence and materials are not unlike the foundation wall installation already described. As a retrofit, removing the existing slab is an added complication. Inasmuch as such slabs are often in basements, the cost of physically breaking them up and pulling them out is no small feat.

The general approach of mitigating the water at the outside surface of the wall does not preclude that water will still enter the soil, nor does it significantly depress a permanent water table problem. What it can and does do is reduce the hydrostatic pressure at the wall by diverting water away before pressure can develop. Because all the materials of such a system are below-grade in thermally stable conditions, and away from potentially harmful ultraviolet light, the durability and reliability of properly designed and installed perimeter drainage systems are very high. The degree of mitigation is often so high as to obviate the problem altogether. With all these virtues, why is this not the exclusive remedy for hydrostatic pressure?

As mentioned at the outset, this is an expensive operation. Institutions with small operating budgets typically try almost any other approach before resorting to what they term a drastic measure, when what they mean is a drastically expensive measure.

Before leaving this topic, two specific approaches require at least cursory coverage: piercing the slab and backflooding. Piercing the slab and/or walls to relieve the pressure will work. The principle is the same as for installing through-wall drains in earth-retaining walls. When through-wall drains are part of the design, a few additional pieces are required in addition to the drains themselves. If the drain is installed into a clayey soil on the earth end, the prospect exists that water migration to the drain will be significantly delayed, such that instead of draining the whole area, the area affected is reduced to a small hemispheric zone adjacent to the drain port. In a new installation, this is preempted by installing granular material behind the water to act as conduit to transport the water to the drains. Retrofit drains through slabs or walls, even if they do connect to porous backfill, may clog from accumulated fines migrating through the granular materials.

Backflooding is an option that ordinarily falls into an inappropriate category; but in some circumstances, this may be a viable option. If the source of the hydrostatic pressure is not water in soil, but a flooding river, the biggest threat may not be the water entering the building, but the mud that comes with it. If the interiors of the building will suffer more from the mud than the water, or the cost of clean-up of mud is much higher than the damages from the water, then backflooding may be an option.

The way it works is fairly simple. As the flood waters begin to rise outside the building, clean water is deliberately pumped into the interior of the building. In order for backflooding to work, the water inside must be maintained at a slightly higher level than the flood water, so there is always a slight positive pressure from inside to outside. This will mean that clean water is always pumping outward through any openings, and precluding any muddy water from entering the building. For this to happen there must be a large, convenient, and continuous supply of clean water throughout the flood; and the opening through which the clean water will be pushing must be small, such as a slit or a crack, not an open window or door.

Moving Water

Several times in the discussions of ground water, we have mentioned particles of soil moving in water: on the topic of heaving, we mentioned fines being pumped into the ballast; later, we mentioned mud included in flood water and fines clogging drain ports. The transport of fine particles of soil is evident in every muddy river, every erosion track. So far, the relevance of soil-suspended water to building pathology has been treated as auxiliary to other deterioration mechanisms. On occasion, however, moving water *is* the deterioration mechanism.

Water in motion has velocity and mass, and so has the capacity to move detached particles. The size and mass of the particle transported depends on a limited number of variables: the quantity or velocity of the water stream compared to the size and density of the particles. Although a small stream may be moving swiftly, it may not possess the momentum to initiate acceleration in a relatively massive particle at rest. A large stream moving slowly may have the same lack of effect. The greater the momentum of the water stream, the greater the potential for moving soil particles. The smaller the particle, the greater the likelihood of being transported. Momentum is a function of velocity and mass. In moving water, the variable of mass translates simply to the quantity of water expressed in cubic feet per minute or kilograms per second or some other measure that carries with it the notion of flow rate. The velocity of the water is expressed in a distance unit per time unit as feet per second. Velocity is sometimes implied in expressions of pressure. When water under pressure is released, the water is projected from the orifice or opening at a velocity that is related to the pressure. Water released from a spray nozzle at 400 psi will be moving at a lower velocity than the same quantity of water released under 800 psi. Pressure is a measure of force, so as long as the mass of the water is known, the accelerating effect of the pressure on the water, hence its velocity upon release, can be calculated.

Obviously, soil particles such as those of clay or silt are more easily transported than particles of gravel. In fact, particles of clay, once dislodged from the native clay mass, are so small they remain suspended in still water for some period of time. Such small particles are subject to transportation in moving water of extremely low velocity and low volume. The vulnerability of clays and silts to transportation in water is so great that anytime the necessary and sufficient conditions for such movement are present, the presence of the mechanism must be presumed.

For water to be in motion, it must meet four requirements: there must be a water supply or source; there must be a path for the water to traverse; there must be a repository for the water to exit or release; and there must be a motive force to accelerate the water into motion. Flowing or moving water is not, however, sufficient to establish a deterioration mechanism relative to a building, but it is necessary.

The supply of water is typically via precipitation, but there are others. Investigations involving moving water normally do not begin *because* we identify a source of water; they normally begin because of the manifest damage resulting from the moving water, from which we attempt to find the source. While we must always suspect precipitation as the primary source of moving water, we must consider too the complete list of potential sources, which is not short: leaking or broken mains or sewers, roof drains and leaders, industrial waste, laundry operations, ejector and sump pumps, condensate returns, dishwashers, even underground streams. Once there is a source, there must be a path or potential path through which the water can flow, and there must be a sink or repository to accept the outflow of the moving water.

The motive force to initiate and sustain the movement is normally gravity, but not always. Water can be pushed or pulled through channels by pressure differentials. For example, water escaping from a water main may not immediately incur a channel through which to flow. As the pressure of accumulated water increases, a channel may occur by fracture. Water may build up and climb to a level above the original rupture before encountering a potential path; the main, in effect, pumps water up to the escape path. There are many ways to transport water to a point of release into or through a potential path.

Once moving within channel, however small the volume, however low the velocity, there is the potential for moving the finest soil particles, beginning with clay and silt. If the path is, in fact, through or adjacent to a face or cleft containing these materials, then that channel is altered as the particles are removed, generally resulting in an enlargement of the channel. If the flow of the water is restricted by the opening dimension of the channel, as material is removed from the channel, the open channel dimension increases, and the capacity of the channel to transport fluid increases. If the supply is available and the capacity is increased, then the volume of moving water increases. As the volume increases, the momentum increases, and the size of suspendable and transportable particles increases. The water stream begins to erode the channel.

The moving water is carrying soil at increasing rates, but still no deterioration mechanism exists relative to a building unless the building is, in fact, constructed of the soil that is being removed. In order for there to be building pathology, there must be a building, and that building must deteriorate as a consequence of the environment in which it is located. The deterioration mechanism is a function of the environment *and* of the properties of the building. The part of the building most at risk from moving water is the foundation, in particular shallow foundations and footings. When the moving water is proximate to the footing, there is the risk that the fines and particles being transported are from below the footing itself. The unresolved extension of lost soil is reduction of bearing and support. A washout below a footing can precipitate a settlement of the footing into the eroded void, bringing with it the wall above. Erosion below a supported slab may result in a partial collapse of the slab.

In-ground tanks, whether containing water or other fluid can develop large voids adjacent to leaks because of long-term transportation of fines in the vicinity, resulting in a major void. The pressure in the tank or cistern ruptures the tank wall at the point of lost support, and the mechanism suddenly accelerates.

Intervention in deterioration mechanisms involving moving water proffers all the frustrations of other below-grade problems, namely inaccessibility and cost (see Table 3.5). It also has the added complication of identifying the source of the water. Moving water problems often are not even identifiable as problems until the damage is manifest and of a magnitude as to virtually preclude minor mitigation. Doing nothing is rarely an option because of the catastrophic potential of the consequences of an unchecked mechanism. Given that something must be done, the general intervention classes that do not address the source all carry grim concerns about the time gained versus the costs incurred.

Underground water is always a dangerous condition; migrating underground water is even more so. This pushes the diagnostician to locate the source of the water and eliminate it. Because the number of sources is great, and the access to the locus of activity is restricted, this is a particularly difficult analysis. Given the difficulties of observation and testing, trial and error is not unreasonable logically, but the costs of each trial can be exorbitant and the costs of error can be even higher.

In terms of remedies, the entire class of options directed at stopping the water at the point at which it is manifesting the problem are fraught with the same prospects as those associated with hydrostatic pressure. If the water is coming under the footing and running across the basement floor to the point that soil is being lost below the footing, dry packing the hole under the footing will not solve the problem. The water will momentarily pile up on the far side of the footing until it finds another point of vulnerability; subsequently, the mechanism reactivates somewhere else. Now there are two locations with washed-out footings. Moreover, there is always the risk that the second problem will go undetected for longer and with greater consequent damage than the first. Sometimes it is better to deal with the devil that you know.

No one argues that identifying the water source is the primary objective. It is reasonable and has the prospect of a definitive intervention. The problem with source identification is, of course, cost and time. Unless the source is self-evident, such as a broken water main flooding an entire neighborhood, some sleuthing will be required. One objective of the investigation pattern is to avoid excavating as much as possible. Investigation by excavation is analogous to taking biopsies blindfolded. The theory is sound, but the execution is costly, and the outcomes are uncertain at best.

Table 3.5 INTERVENTION MATRIX: MOVING WATER

Necessary and Sufficient Conditions	Intervention Approaches					
	Abstention	Mitigation	Reconstitution	Substitution	Circumvention	Acceleration
General Mechanism	Accept consequences of moving water.					Demolish footing or other structure threatened by moving water.
1 Water in motion		Reduce water momentum.				
2 Transportable soil			Replace transported soil with same material.	Replace transported soil with alternated material; e.g., granular material.	Consolidate or stabilize soil.	
3 Collapsible footing			Repair footing as it collapses.	Replace collapsing footing with different material.	Replace footing with different support system; e.g., deep system.	

The first step in the investigation for an unknown source of moving water is to develop an investigation plan. Avoid wandering around hoping for a message from the divinity. Help is more likely to come from following a map. Next, discuss this plan with the person paying for the job. Explain what it is you are looking for and why that is important. Explain what you will do when you find the source. Show the client the intervention matrix for moving water. Explain how much the remedies may cost if all goes well, and how much if all does not go well. Explain the cost of executing the search. Explain the risk of not being able to identify the source. Above all, explain the risks and costs of doing nothing. Ask the client to approve the plan in writing. If he or she is unwilling to approve any reasonable and prudent plan, seek other employment.

The tactical investigation plan for identifying the source of moving water begins with a list of any and all possible sources of water. This list need not be reasonable nor realistic. The only requirement is that it be comprehensive. Next, organize the list into the order of probabilities, with the statistically most likely candidate at the top. Beside each potential source make a sublist of techniques that will prove or disprove the potential source as the actual source. Next to each of the techniques enter the cost of the technique or test. Next to the cost of each definitive test enter any risks or hazards associated with the test itself. For example, we could excavate the perimeter of the building in search of a fractured perimeter drain, for which the risks would include fracturing more perimeter drain, collapsing the excavation on an excavator, obstructing traffic with spoil, and so forth.

Next, examine the list for tests or techniques that are attached to more than one possible source; highlight those lines. Now look for the cheapest, most risk-free test possible on the entire list that has the greatest prospect of identifying a likely source. If the list is thorough and the test costs are reasonable, there are two inexpensive, low-risk, noninvasive investigations that will be at or near the top of your testing checklist.

Possibly, the first item on the list is that you send a sample of the water to a laboratory and have it analyzed. In addition to inorganic analysis, have an organic analysis done as well. If the water has colliforms, sewer water is a likely source. If the water contains high levels of chlorine, a water supply line is a likely source. If the water contains high levels of hydrocarbon distillates, the water is likely to be runoff from streets. If the water contains any abnormal marker mineral or chemical, that marker not only can help narrow the search, but can confirm possible sources once they are identified. By comparing samples against the candidate source, you can confirm or eliminate a candidate.

The other relatively inexpensive and noninvasive investigation is to conduct a literature search of documents and records, looking for any and all water sources. Find terrain maps that indicate surface drainage. Design and construction documents will show supply lines, roof drainage, sanitary sewers, perimeter drains, and site drainage. Water department maps can provide the locations of catch basins, storm sewers, sanitary sewers, utilities of adjacent properties. Older site maps from the local historical society can provide locations of old stream beds, abandoned reservoirs, and other water features. Old Sanborn maps may show prior construction of the site, suggesting old privies and cisterns. Geologic surveys will provide pitch and slope information for ground-water paths and aquifers. The point is, a wealth of information is to be found in documents and archives, which can save tens of thousands of dollars of excavation.

Given that invasive techniques are recommended, begin with the cheapest, high-efficacy techniques. If the building is surrounded by open land, renting a backhoe and hiring a soils engineer for a day can yield 8 or 10 strategically placed test pits, each of which will be 15 or more feet deep. If any of these pits intercepts a water source, it will fill with water. This water can be tested for a match with your base sample, and the rates of filling can be compared to the problem site for flow correlation. If they do not fill with water, then the elevations of the bottoms of the test pits should be correlated with the elevation of the point of entry or with damage from the moving water in the building. If it is reasonable to conclude that the source is lower than the bottoms of the pits, selective borings may be the next step.

Each increment in the testing plan must progress the investigation by adding specific information leading to an identification to eliminate potential candidates. Investigations in building pathology differ from medical diagnostics in that baseline and background information are not normally collected by testing. Test data must progress the investigation and do so economically and directly.

DEEP FOUNDATIONS

The term deep foundations includes a class of foundation techniques that are deep relative to shallow foundations; depth per se is only a loose way of classifying foundations. Deep foundations include driven piles and caissons, which may be shorter in actual length than the piers of some shallow foundations. That said, our job is not to reclassify the inventory, but to understand its failings.

There are very few progressive, potentially fatal deep foundation deterioration mechanisms, but there are many types and variations, and certainly their design and installation are complex. The fact is, once in place, not much will unsettle a well-placed deep foundation. If there is a general problem, it is at or near the surface, not at the depth of the deep foundation itself.

The general nature of deep foundations is that they do not settle. One of the basic parameters of deep foundation design, which distinguishes deep foundations from shallow foundations, is that they are designed to resist the design load *without* subsidence. The basic nature of shallow footings is that they compress soil, which plastically deforms over time. This is not to discount the fact that many shallow footings do not bear on plastically deformable soil. Shallow footings can and do often bear on rock or consolidated granular material. Where the substrate is purely elastic, the issues of differential settlement are rarely a concern. The design premises of deep foundations depend on elastic deformation of the bearing materials, such that long-term deformations are rarely an issue.

This is accomplished in two fundamental ways: end-bearing and side friction. End-bearing deep foundations are analogous to table legs bearing on a floor. The ends of the piles or caissons go through the unreliable material until they reach rigid material such as rock. The number of piles or caissons employed, and their respective bearing areas, are calculated and designed to support the load without long-term consequences such as differential settlement. The conditions at the tips of end-bearing caissons do change. The rock will continue to weather from chemical dissolution and mechanical stress, but the rate of deterioration of bedrock at depth is measured in increments of hundreds of thousands of years. We have not been building deep foundations long enough for the first generation of them to be threatened by bedrock material deterioration.

There are instances where deep foundations do not bear on rock, but on granular materials, generally consolidated gravels or sands. These materials may not have the ultimate strength of most rock, but they are elastic, and do not display the long-term deformation properties of cohesive soils unless the granular materials are liquified or frozen while saturated. Both freezing and liquifaction are rather far-fetched eventualities when the bottom of the caisson is a dozen or more feet below grade, the soils are drained, and the material is consolidated.

The other basic approach to deep foundation support is to wedge a shaft into the ground so far and so deep that the displaced soil firmly clamps the shaft so that it will not move even at full design load. The technique for driving these members is not particularly sophisticated. Anyone who has ever hammered a spike into the ground knows all the fundamentals of the process. A shaft, which may be the trunk of a tree or a precast concrete pole, is held vertically by a portable derrick. A weight is hoisted further up the same derrick and allowed to fall or is power-driven onto the upper end of the shaft. The blow drives the shaft some inches or, occasionally, some feet into the soil. The hammering process is repeated until the shaft refuses to move despite repeated blows by the weight. The point of refusal is reasonably predictable based on the soil density and frictional properties, the diameter of the shaft and its frictional properties, and the level of consolidation of the soil. The mass of the weight and the height from which it is dropped are calibrated against the properties of the soil and the pile such that the pile will refuse to move at an impact by the weight that corresponds to the design load of the building. If the pile refuses to move under a sufficiently high level of impact, theoretically, it will not move under the static equivalent load of the building.

The correlation between pile refusal and load capacity is largely empirical. Although there are formulas for determining blow counts and resistance that at first glance appear to be mathematically precise in their configuration, further examination reveals that they are more processes of doing what worked than scientific deductions. In the text all the basic variables of pile resistance were listed except the length of the pile. In the design phase, the soils mechanic estimates a pile length based on the soil characteristics discovered from boring samples. The piles delivered to the site are several feet longer than the calculation. If the pile refuses to advance at the calculated depth, the excess can be removed. As is often the case, however, the pile continues to advance at the calculated depth; there is some reserve length.

If a pile does not hit refusal at it full length, all is not lost—although the soils mechanic will have to make some rather rapid decisions. He or she can add piles to the pile group, increase the length of other piles within the group, or increase the diameter of the remaining piles in the group. Deciding which to do is based in part on the blow count log, a record of the number of inches a pile advances per blow, or, in denser situations, the number of blows to advance a certain distance. If the blow count at full pile length is 10 per foot, the soils mechanic may simply sink a second pile to confirm the condition. If the first pile was advancing three feet on the last blow, he or she may declare that pile "lost" and recalculate the entire group.

Misadventures in pile driving can be a bit bizarre at times. For example, one of the risks of driving any pile is that it will drift off vertical, which can happen for a number of reasons, the most common of which is an encounter with a boulder. In the case of steel piles, there is the added potential that the drift is not announced back at the driving end. Everything seems to be going well until the tip reemerges from the ground, having executed a U-turn below-grade. Sometimes the pile simply snaps. Occasionally, the pile is driven into an underground cavern or void and suddenly disappears.

The long-term changes in the ground that can affect a friction pile are a bit greater than end-bearing piles, but not by much. Soil conditions can change at depths that can affect soil density and water content, in particular. If significant quantities of surcharge are removed from the area above the pile group, theoretically, soil density at depth could change as the soils expand. While these are possibilities, the circumstances in which piles are driven to begin with rarely lend themselves to these sorts of alterations.

In short, deep foundations are very durable and stable at least relative to the ground into which they are placed. If deep foundations are flawed in their design, such flaws normally are manifest during installation. If those flaws survive installation, they are likely to become manifest soon after service load is applied. Deep foundations are generally located in stable environments with few hostile and proximate agents. Issues such as moving water, which can be devastating to shallow foundations, will rarely affect

FIGURE 3.6 Years of water moving with velocity within the core of the wall has resulted in a major collapse exposing the core of the major potential failure.

deep foundations. Heaving is virtually impossible, and sliding is no less improbable.

Although quite stable, there is at least one general condition that can degrade deep foundations and timber piles, in particular, and it is related to the ground table. Most deep foundations are founded below the permanent water table; most buildings are not. This means that somewhere between the bottom of the caissons and grade, the foundation has to make that transition from wet to dry. In terms of the long-term durability of the foundation, it is not of major significance where that transition occurs as long as it does not fluctuate very much. Materials submerged in water may deteriorate, dry materials may deteriorate, but rarely is either of those conditions, if they are static, as aggressive as a fluctuating condition, which is alternately wet and dry.

If the deep foundation contains metals, most likely ferrous metal in the form of steel reinforcement, then alternating wet and dry cycles, particularly in saline environments, will exacerbate corrosion. The connection between alternating moisture condition and corrosion is not a direct connection, but related to differential hygroscopic expansion, which was discussed earlier in the chapter. Steel that is impermeable is not hygroscopically responsive; however, the concrete that typically encases the steel is. Not only is the concrete differentially responsive compared to the steel, but the constituent materials that comprise concrete, namely the sand, cement, and aggregate, have varying hygroscopic properties. The property differences indicate there are dimensional differentials as the moisture content of the materials change. This results in internal stresses within some of the materials and along the boundary planes of others, which can result in cracks. In addition to the hygroscopic differential, there are also moisture gradients within the individual materials as a result of the fluctuating moisture conditions, which also result in internal shear stresses, which can result in cracks.

In addition to the constituent materials that comprise concrete, there is ferrous hydroxide, which pacifies the steel against corrosion. Ferrous hydroxide typically forms on the surface of the steel during the hydration phase of concrete formation. This tightly bonded, impermeable compound, once formed, tends to protect the steel from further corrosion. Ferrous hydroxide has its own hygroscopic properties that subject it to the same set of stresses as other components of the concrete. In addition to its other properties, ferrous hydroxide is relatively brittle, and when it cracks, there is an unprotected avenue of approach to the base metal, and the corrosion of the steel is initiated.

If the material is timber, alternating wet and dry cycles will exacerbate fungal rot. The necessary and sufficient conditions for timber rot were discussed earlier. In summary, they require the timber itself, moisture content at or near fiber saturation, oxygen, moderate temperature, and the fungus. In submerged conditions, the rotting of the timber is inhibited to a considerable extent by the significant reduction in the available oxygen. If the timber protrudes above the water line, water can saturate the wood fibers horizontally, and progressively wick water vertically to a zone where oxygen is plentiful. Because lateral absorption is slow, rot is relatively inhibited. If, however, the surface of the wood is alternately above and below water, the rot can operate cumulatively on the recently wetted fibers. The fibers in the tidal zone disintegrate, which exposes the end grain of the timber above the tidal mark, which absorbs the water easily, and rot progresses upward in the piling. As long as the piles remain completely below the water table, conditions are relatively stable. If the water table drops, the upper end of the piles will be exposed to increased oxygen levels, with absorbable water below. In the case of all steel piles, this will accelerate corrosion directly. In the case of reinforced concrete, the casing cracks that provides avenues of corrosion attack of the reinforcement. Also in the case of concrete piles, salt absorption will exacerbate concrete deterioration. In the case of timber piles, exposure of the caps of the piles will accelerate fungal rot of the exposed pile ends.

In all of these cases, the risk is that the upper ends of the piles will deteriorate and be crushed under the weight of the building above. The crushing of the pile ends means that the building will subside, which, in turn, can generate differential stresses and large-scale cracking in the superstructure.

Intervention in pile-cap deterioration is fairly limited because the change in water table levels that result in the exposure of the pile ends to air are rarely reversible. Intervention almost invariably means accepting the change in water level as a given and operating from that premise. The most common general approach is to accept the amount of subsidence to date and to stabilize the building at its current elevation. This is normally accomplished by some form of encapsulation.

One method of encapsulation is to inject grout into the region of the deterioration. Theoretically, the grout will wrap around the deteriorated area, starve it of the moisture and/or oxygen required for further deterioration, and extend a jacket of grout down the pile to engage undeteriorated material. To increase the prospects of encapsulation, the pile ends may be excavated, exposed, and backfilled with concrete. This will provide positive assurance of what the injection technique presumes based on logic.

At Boathouse Row in Philadelphia, there are two boathouses, which are excellent examples of this technique. Both are on timber piles, the caps of which became exposed to air as the result of a drop in the water table. Although both are on a river bank, the water table is not at the river level, but slightly above the surface of the river. This is typical of rivers and the banks of rivers. The rivers are in part charged by ground water moving down to and into the river. By definition, there must be a head of water behind this movement. The greater the head, the steeper the gradient of the water table as it moves away from the edge of the river. In the case of the east bank of the lower Schuylkill, the head was diminished when East River Drive (now Kelly Drive) was paved. Instead of water seeping into the bank soil and then into the river, the water runs on the surface into the river. The charge is more rapid and the permanent water table is depressed below the pile caps.

SUPERSTRUCTURES

The superstructure of buildings includes those structural forms and elements above the foundations or grade. Most of what we see of a building is the superstructure. In this text, the term is somewhat more restricted to the deformation-resistant elements of the building. For reasons of categorization, the closure systems are not included here, but will be addressed in subsequent chapters.

The following sections go through the various components of the superstructure beginning with compression elements and proceeding through tension elements and elements in bending, concluding with a section devoted to the issues associated with connections. We want to generalize these categories enough to also include minor elements and portions of elements that may not rise to the level of a structural member, but that can be and often are the loci of many deterioration mechanisms and building pathogenesis.

The analysis and assessment of the current capacity and of the diminishment of that capacity over time is the essence of structural pathology. We are particularly interested in that class of structural elements that have endured, but that can and will fail in the future due to the progressive nature of one or more deterioration mechanisms. For this class of structural elements that did not fail early, there are only three possible explanations:

- The element survived because the critical load was never achieved, but it can or will fail now or in the future because of an increase in the service load.
- The element has survived to date, but because of gradual diminishment of capacity, even the current service load will exceed critical proportions now or in the future
- Both of the preceding or some proportional combination of both are simultaneously true.

There is a common twofold presumption that, one, inadequately designed members fail, or failed, upon application of service load, and, two, its corollary, that members that endure the initial application of service load are adequately designed; both are dangerously flawed, particularly the latter, which is called by some "the test of time." There is no such thing as the test of time except to the extent that the phrase implies durability. Unfortunately, those who use the phrase in the context of aging buildings usually believe that a building that has endured for some considerable amount of time will continue to endure for the indefinite future at whatever service load is currently imputed to its design. There is a gloss of validity to the logic even if the ending is tragic, that being that grossly inadequately designed or constructed buildings or portions of buildings often fail during construction or shortly thereafter. The logic extends to imply that the absence of such an early failure is indicative of adequate design and construction—and of continued safe use if that use is not discernibly different from the historic use. This, simply, is not true. For example, take an elevated porch, built in 1822, suspended by hanger rods, that has endured without recorded failure for 175 years: the presumptions are that it was designed and constructed remotely close to modern definitions of safe standards, and that the materials have not deteriorated in the intervening years. Both are without logical or factual basis. People went onto the porch intermittently in "large numbers" during those many years and as recently as six months ago. Is the porch, therefore, safe; and, if so, for what loading? Do these sort of experiential load tests mean anything?

To answer these questions, begin with the realities of design in 1822: there were no codes; there was very little of what we think of as engineering. Structural design, such as it was, relied on mimicry or patterns to achieve stability and strength. To illustrate, we use the Carpenters' Company of the City and County of Philadelphia, founded in 1724 and still active today, a guild company of master builders who promoted the use of reliable construction patterns. The early pattern books are still in the Carpenters' Company library. Using those patterns books, in conjunction with datable construction in Philadelphia that followed those patterns, we can analyze the structural capacity of the various elements. In general, individual early nineteenth-century members were reasonably well designed as regards compression, torsion, and bending—meaning that from trial and error, the patterns that emerged incorporated what we consider today enough material to satisfy modern safety criteria from these sources of stress. On the other hand, the members are notably underdesigned for shear and only marginally designed for tension. Connections as a whole were weak, and there was no apparent awareness of buckling.

If the construction digressed from the patterns, anything was possible in terms of outcomes because there was no mathematical modeling of the behavior of materials or structural members. Care may have been exercised to ensure that construction met some standard of full load testing, but there is no record of such an experimental approach to design until the mid-eighteenth century, notably with Captain Eades at the St. Louis Bridge. And French engineers certainly used basic calculation techniques in Suez, but miscalculated in Panama. Railroad engineers began using algebraic devices for designing bridges in the mid-nineteenth century.*

Materials science, as we know it, did not exist, and there were no public inspectors or permits. To summarize, for a hanging porch constructed in 1822, we have no idea what the design criteria were, although we suspect there were none. There were no codes or other external guidelines for design other than the aforementioned pattern books. There were little, if any, widely available analytical tools even if the designer were to seek them. What we do know is that materials and systems degrade over time, so whatever the original capacity was, which we do not know, that capacity has degraded to some extent or other. So let us reform the earlier question: Does this structure deserve more or less scrutiny than a comparably configured contemporary structure? If public safety has any meaning at all, then the older structure should be more suspect and, therefore, requires more direct analysis than a modern counterpart. The more indeterminate a structure is, the greater the prerequisite to investigate to confirm the condition.

Despite the compelling logic of investigating and confirming capacity, a contingent of practitioners still cling to the belief that a given structure is safe simply because it has not killed anyone to date. To those who agree, consider this caution: the law of torts conflicts with this notion of preservation in an extreme and potentially costly manner.

Property owners have varying degrees of obligation or duty to occupants of their property, depending on the status of the occupant. Occupants may, after all, be trespassers, who may be further classified as undiscovered trespassers. The other broad classes of occupants are licensees and invitees. The degree of obligation on the part of the property owner is generally less to trespassers than to licensees and less to licensees than to invitees, who are also numerically the most likely group to be on or in the property. An invitee is a person who enters the property at the express or implied invitation of the owner; he or she is a member of the general public entering the property when it is, in fact, held open to the public. Invitees as a class include employees and delivery persons, customers, and anyone accompanying them, as well as garbage collectors, meter readers, health inspectors, and census takers, but not normally police or firepersons, who are usually classified as licensees.

The duty of an owner to an invitee differs slightly depending upon whether the risk is associated with something the owner is doing on the

---

*For an excellent and entertaining read on the subject of preanalytical building and construction see *Building Construction before Mechanization*, by John Fitchen (Cambridge: MIT Press, 1989).

land, an activity, or a risk associated with an encounter with the property itself. In the case of aging buildings, the primary concern is the latter. It is entry into the building itself that poses the risk. An owner's duty to an invitee is to keep the property reasonably safe, to stabilize any known unsafe condition, to warn of any known uncorrected hazard, and *to make reasonable inspections* to discover dangerous conditions.

Compression Elements

Structural elements in compression at the scale of the building as a whole consist primarily of columns and walls. Minor elements in compression are wind braces, knee braces, and window jambs. It is important to distinguish elements designed to withstand compression, those that are in compression only from self-weight, and those that are placed in compression because of deformations of adjacent elements.

Prototypical compression elements are subject to two classic forms of failure: crushing and buckling. How a compression element fails is a matter of which of the two classic modes wins the race to collapse. If the element is relatively short and squat, it will fail by being crushed, which is essentially a material-level failure magnified to the level of a structural element. Buckling will win the race if the element is relatively slender. (Characterizing an element as slender is done based on whether or not the element will fail in buckling before it fails by being crushed. It is a sort of self-fulfilling definition.) The two forms of compression failure can be analyzed based on the properties of the material and the geometry of the element. The maximum or critical load (P_{CR}) for a particular specimen will be the lower of the two values computed for crushing or buckling: $P_{CR} = \sigma_{ULT}A$, which means that the column will crush at load equal to the ultimate stress times the area, or $P_{CR} = \pi EI / (kl)^2$, which means the column will buckle at a load equal to the product of its stiffness (E) and geometry (I), divided by the square of its effective length.

At any point in the life of a given compression element at which the actual load exceeds one or the other of the two potential failure values, the element will either crush or buckle. If either value occurred because the element was severely underdesigned, there is a higher than average probability that the element would have failed upon entry into service. That this is not necessarily so, however, is an important factor to remember when assessing aging elements. Underdesigned elements do, in fact, survive initial entry into service for a limited and not unusual number of reasons.

One reason inadequately designed elements survive initiation is related to the definition of adequacy. An adequately designed structural element is one that will reliably carry the design load, plus a reasonable additional amount beyond that design load. That additional amount beyond the design load is a sort of extra capacity held in reserve for a wide variety of reasons, ranging from our suspicion that we can "cheat" by sneaking additional load into the building to our concern that manufacturers err and deliver members with lower capacity than specified. Experience (which is the result of what we get when we did not get what was specified) teaches us that the added cost of a reasonable reserve in structural capacity is so low as to qualify as a good investment against the unlikely and unpredictable. The generally accepted term for this deliberately constructed reserve capacity is the *factor of safety*.

We, as designers, attach factors of safety at various stages of the design process. Officially, we attach different factors of safety to different materials based on their relative reliability. The factor of safety for steel is approximately 50 percent; the factor of safety for wood is typically 200 percent; the factor of safety associated with soil may be as high as 1000 percent. The correlation of the factor of safety and a scientifically based risk analysis are practically nonexistent. Some are historic accidents, and some have obscure rationales, but the habit has generally served us well, particularly considering the relative inexpense.

Unofficially, we roll into the design of structural elements a variety of fudges and round-offs, which collectively put more capacity into the element than is otherwise proscribed by engineering logic, including an official factor of safety. One common and not insignificant source of unofficial increase in capacity is simply that manufactured elements are made in discreet increments of moment of inertia and section modulus. Suppose, for example, the calculations demand a beam with a moment of inertia (I) of 1370 in.^{4}; but the closest economical standard shape to that number has an I value of 1550 in.^{4} There are beams with I_x closer in value to 1370 in.^{4} than 1550 in.^{4} which is the I_x of a W24 × 68. The trick lies in the use of the word economical to describe the selection. A W14 × 120 has an I_x value of 1380 in.^{4}, which is satisfactory, but at the expense of 52 pounds per linear foot. Unless the dimension is critical, it is far more beneficial and less costly to go with the deeper beam. Despite our most precise calculations and economical design efforts, the stiffness of the member just increased by 13 percent with a corresponding increase in section modulus, weight per foot, and shear capacity, all because of the increments of manufacture combined with the extreme reluctance of engineers to round down and select a lighter section, even if the variation is seemingly small.

There are other sources of unofficial round-off and round-up that are more likely to occur when the critical design criteria involve shear or buckling capacity. The logic of this so-called fudge factor has to do with how elements fail. In general, given a forced choice, engineers will select bending as the preferred mode of structural failure—about which more will be said later in this chapter. The least preferred mode is a shear failure, followed closely by buckling. Shear failures, especially, and buckling failures, only to a slightly lesser extent, occur with little warning. The deformations preceding catastrophic failure in shear are minuscule; there is little energy release prior to fracture. The fracture, when it occurs, is total, meaning that it progresses completely through the member; and fracture progression is virtually instantaneous. Buckling failures are only slightly less frightening. In the final analysis, crushing failures in compression elements are, at the crystalline level, shear failures multiplied to the elemental level. For these reasons, when there is discretion, particularly, in the design of compression members, engineers hedge in one direction, away from the failure envelope.

The wary and vulnerable public gets involved in the design criteria of buildings through codes and regulations. Designers and building owners often lambaste construction codes as being overly conservative and expensive, until there is an echo of the Regency Hyatt collapse, and they are silenced again, if temporarily. Building codes *are* rather conservative. To actually squeeze human bodies to a density of 100 pounds per square foot requires that only the taller, denser members of the species be allowed on the floor, at which point they must crowd together tighter than on the E train in New York City at 5:05 PM, *and* they must at that point start bouncing up and down in time to, say, a polka. As fantastical as that prospect is, remember that the operative code requirement for the elevated walks at the Regency Hyatt was 100 pounds per square foot. When they fell, and more than 180 died as a consequence, the live load at the time of the failure was under 30 pounds per square foot.

When something as tragic as the Regency Hyatt failure becomes news, the public outcry is often reflected in modifications to the relevant building codes. The proscription by code that building "of public assembly" be capable of carrying 100 pounds per square foot is another way of including a safety factor to construction. True, such loads are arguably unrealistic, but what other enforceable means does the public have to protect itself from careless or unscrupulous designers and owners?

Once in service, an element will fail in compression because the service load increases to a critical level, the capacity of the element degrades to a critical value less than the service load, or some combination of both. Buildings constructed before codes were promulgated may have been designed to standards comparable to current ones, but without analysis there is no way of deducing what they were. To assume that the operative design standard of a precode design is comparable to a modern standard because the building has endured for 175 years is wantonly illogical.

With the sources of reserve capacity now defined, it is possible to begin to understand how grossly inadequate designs are constructed and how they endure. First, because our current live load standards are inherently conservative, it is possible, even probable, that a place of public assembly has never actually incurred a live load close to what is now required by code. It is, therefore, also possible that the implicit design load capacity of a precode place of assembly comes closer to the actual load than a modern building, but at the expense of that factor of safety associated with designing to higher service loads than current practice requires. Third, it is possible, even probable, that precode buildings were designed to lower standards than modern equivalent construction. It is, therefore, possible that a hanging porch built in 1822 was built to no standard or to a lower standard than would be code-conforming today; but at the same time, the actual load may never have been equal even to the lower design standard.

If the old hanging porch was built, absent articulated standards, to an implicit standard of 40 pounds per square foot, it is still probable that in the entire 175 intervening years the actual load never exceeded 30 pounds per square foot. The fact that the service load never reached the ultimate strength of the construction is very different from assuming it is safe for 100 pounds per square foot because it has not fallen down.

The factor of safety imposed on materials such as timber may be 100 percent greater than the ultimate strength of the material. The consequence of such factors when combined with code-required live loads is that a hanging porch properly designed by modern standards has the distinct prospect of carrying 200 pounds per square foot before collapsing. This is the product of the factor of safety of 100 percent or more of the materials, times the code-required design load. This is the case in a situation that we acknowledge as unlikely to see an imposed load greater than 30 pounds per square foot in its entire service life.

In the precode porch, however, there is no assurance or reasonable prospect that materials were addressed in terms of ultimate strength or allowable stress. The concepts were not invented 175 years ago. The timber and metal in an 1822 hanging porch may have had reserve capacity in the materials criteria or not; no information exists one way or another. It is even mathematically possible that the materials were undersized even for what we consider ultimate strength, but because the loading history was so low, such a gross deficiency was never exposed.

The marvel of very old construction is not that it has endured but that we have not categorically condemned it all as presumptively dangerous. To the contrary, most codes allow precode construction to remain in service not because it is has passed the test of time, but because of the economic impact to society. On top of all this is the inescapable fact that materials deteriorate over time. The progression is inevitable, and barring specific and active intervention, the result is always negative. The only valid test associated with the progression of time is how long it will take before a particular structural member will fail, not whether or not it will ever fail.

In summary, it is possible, even probable, that our hypothetical 1822 hanging porch was not designed to any recognized or recognizable standard, and was built with materials of unknown capacity and that are to some degree less competent today than they were 175 years ago. Although the current code may allow the continued use of the construction, nothing about the survival of the structure should offer us any comfort or confidence that it is safe.

Columns

Humans have been shaping columns to prop up structures for a long time. If a pole was not the first fabricated structural element, it was a close second. Columns are simple structures, and have been in the building inventory for so long, it seems we have an intuitive sense of their proper proportions. When a column is properly sized, there are only a limited number of eventualities that can crush or buckle it: the load increases to a critical level, the capacity of the column decreases to a critical level, or some combination of both. This reforms our definition of deterioration mechanisms for structural members. The mechanism was defined as a progressive degradation of the material as a consequences of a set of material properties operating in a hostile environment. The structural equivalent to material properties is the member characteristics and the capacity of the member resulting from those member properties. The structural environment is the load placed on the element in compression, in this case.

The basic member characteristics that determine the structural capacity of a column are the variables that define the critical bearing capacity of the column. For short columns, the variables are area and ultimate strength in compression. For long columns the variables are the modulus of elasticity, the moment of inertia, and the unsupported length of the column which takes into account the fixity of the end conditions. Other variables affect the performance of compression members, but these are the fundamental ones that will provide us with a procedure even if we later elect to add to the sophistication of the analysis. As we progress through our analysis of structural elements, we will reuse this procedure many times. It is as follows:

1. Find the evaluative formulae for critical components of the structural capacity of the element.
2. Array the variables that contribute to that capacity.
3. Inspect the member for signs of diminishment of one or more of the independent variables.
4. Evaluate the loss of value and recalculate the remaining capacity.
5. Compare that remaining capacity to the design requirement.

As we will discover, the origins of the diminishment of capacity of a structural element will very often, if not generally, lie in a material-based deterioration mechanism. We will also discover that there are certain sources of diminishment that are more related to changes in geometry. In many cases, there is a relationship between the material and the geometry of the member. For example, timber columns typically are square or rectangular. Steel columns are generally wide flange sections, where the nominal depth and nominal flange width are equal, and they are often hollow pipes or tubes. Stone or brick masonry columns are typically solid cylindrical sections. Reinforced concrete columns are round or square solids. The point is that the material and the geometry are typically interconnected due to the economy of producing the section.

There is also a strong correlation between the combination of materials and geometries and the nature of the deterioration mechanism. Corrosion of unprotected columns will evenly affect the entire exposed perimeter of a steel column. The rotting of a timber column will typically proceed from the core outward, beginning at end grain cuts such as the top and/or bottom of a post. Freeze-thaw degradation attacks masonry columns more frequently on the windward exposure. All of these examples also result in a net reduction of cross-sectional area.

The specific variables related to the crushing capacity, hence to potential crushing failure, are the cross-sectional area of the material and the ultimate strength of the material. The variables related to resistance to buckling are the modulus of elasticity of the material, the moment of inertia of the member, the length of the member, and the end conditions of the member that affect fixity. Any variation in area will necessarily affect the modulus of elasticity, but the inverse is not necessarily the case; there are ways to alter modulus of elasticity without affecting area.

The loss of cross-section area in a compression member directly and linearly reduces the load-carrying capacity of short elements. For slender compression elements, the loss of area is complicated by the location and geometry of the damage. We can construct a typology of deterioration based on the configuration of the damaged area, from which we discover that there are significant differences in the consequences of that damage, depending on where the damage occurs and the pattern of the damage. For compression members, damage to the extreme edges is disproportionately threatening if the capacity of the member is driven by its relative slenderness. There is a disproportionately lower overall threat to a compression member from the loss of the center, or core, of the member.

The investigation of eight patterns of degraded cross-sections are shown in Figure 3.7. The area of the damage in all cases is 25 percent of the original area, which for purposes of the typology is a nominal 6 × 8 rectangular section. If the member is timber, the actual dimensions are 5.5 inches

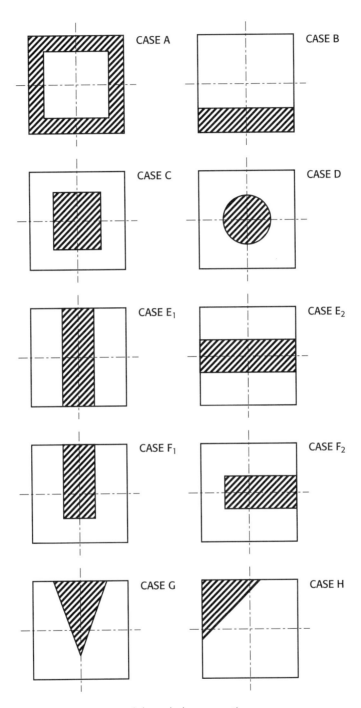

FIGURE 3.7 Patterns of degraded cross-section.

by 7.5 inches; area (A_O) = 41.25 in.^{2} The area of the damaged material is, therefore, (A_{LOSS}) 10.3125 in.^{2}, leaving 30.9375 in.^{2} in sound material (A_{NET}). The question underlying the comparison is whether or to what degree geometry affects the change in moment of inertia, which affects the critical load of the member relative to buckling. The eight geometric configurations are as follows:

A. Deterioration of the entire perimeter of the section of uniform thickness. For A_{LOSS} = 10.3125 in.^{2}, the depth of deterioration (t_{LOSS}) = 0.424 inch.

B. Deterioration of uniform depth across the entire width of the section aggregated, one side only. The depth of deterioration (d_{LOSS}) is 1.875 inch.

C. Deterioration of rectangular pattern proportional in dimension to the original shape symmetrically located about both axes at the geometric center of the section. The dimensions of the deteriorated area are 2.32 inch × 4.32 inch.

D. Similar to C except the pattern of the deterioration is circular rather than rectangular. The radius of the deteriorated area is 1.812 inch.

E. Deterioration of a band of material located symmetrically about one axis, extending from edge to edge of the section. When the band is in the long dimension, the width of the deteriorated area is 1.375 inch; when it is along the short axis, 1.875 inch.

F. Deterioration of a band that extends from one extreme edge and two-thirds of the way across the section. The resulting remaining sound material is a squared U-shape. If the slot of deterioration is along the long axis, the width of the area is 2.065 inch; if it is along the short axis, the width is 2.5 inch.

G. Deterioration of a triangular-shaped area, symmetrical about the long axis, with the base of the triangle at one extreme edge and the height of the triangle extending to the two-thirds point of the section. The width of the base of the triangle is 4.125 inch.

H. Deterioration of one corner of the section, resulting in triangle of deterioration, with sides extending along two intersecting extreme edges of the section, connected by a third side of the deterioration lying at a 45°-angle to the orthogonal axes.

Each of these eight general cases of deterioration is reasonably recognizable in practice as approximations of common degradation patterns; and each, with two variations of E and F, is analyzed in terms of moment of inertia and section modulus. The normal and expected loss of capacity is equal to the loss of area, that is, 25 percent, and the expected remaining capacity based on the loss of area is 75 percent of the original capacity of the element. If the section displays a remaining capacity compared to the original capacity of moment of inertia greater than 75 percent, then the loss of moment of inertia is disproportionately low. If the remaining capacity is less than 75 percent of the original value, then that loss of moment of inertia is disproportionately high compared to the loss of area.

Table 3.6 displays both the respective moment of inertia comparisons and the section modulus responses. For compression members, section modulus is not relevant; for beams and girders, however, section modulus is an indicator of bending capacity. These members will be addressed in subsequent sections. The shaded values indicate those significantly less than the representative value of 75 percent, which we can establish as the normative value for a 25 percent loss of cross-sectional area. As such, these are the conditions that represent a particular threat to a diminished section. None of the disproportionately low values is especially surprising, as the losses in cases A and B are concentrated at the extreme edges of the section because the diminishment of I is a function of the cube of the depth of the section.

Case E will not surprise structural engineers, but it may surprise architects and preservationists not familiar with the effects of moment of inertia and the manner in which it is computed. The message of case E is that two compression members lying side by side, or a single member that has been severed into two independent members, are significantly less

Table 3.6 TYPOLOGY OF DAMAGE PATTERNS

	I_X	$\%I_X$	I_Y	$\%I_Y$	S_X	$\%S_X$	S_Y	$\%S_Y$
Original	193.3		104.0		51.6		37.8	
Case A	114.1	59	55.8	54	34.3	66	24.0	63
Case B	81.6	42	78.0	75	29.0	56	28.4	75
Case C	177.7	92	99.5	96	47.4	92	36.2	96
Case D	184.8	96	95.5	92	49.3	96	34.7	92
Case E_1	145.0	75	11.0	11	38.7	75	4.0	11
Case E_2	27.8	14	78.0	75	7.4	14	28.4	75
Case F_1	149.7	77	34.7	96	35.9	70	36.5	96
Case F_2	171.9	97	82.9	80	50.1	97	27.8	74
Case G	119.4	62	99.5	93	26.6	52	36.2	96
Case H	106.9	55	73.3	71	23.6	46	24.8	66

competent than a single member of the same gross dimensions. In Case E, the member is effectively split into two members. Even if 25 percent of the area were not lost, the diminishment of I is 75 percent (residual value is 25 percent).

At the other end of the scale are the relative effects of the loss of material from the core of the member. Cases C and D show that, from the standpoint of buckling, compression members are relatively undiminished as a result of loss of the core, as from tunnel rot. This is not to minimize the significance of tunnel rot, especially since it is difficult to assess and monitor, but in terms of its absolute damage to a compression member, loss of the core is not a great threat.

Another consideration in assessing the loss of section of a compression member is the location of the damage relative to the length of the member. The loss of cross-section as it relates to the crushing capacity of the member, the longitudinal location of the loss, is not significant. Any loss of area linearly diminishes bearing capacity regardless of its configuration or location. If the member displays more than one location of area reduction, the bearing capacity is diminished by the value of the single lowest cross-sectional area.

If the capacity of the compression element is driven by its slenderness, the longitudinal location of the diminishment and the length over which the diminishment is manifest is very important. Imagine a rectangular column with a slice cut into one side to a depth of 25 percent of its cross-section. In the typology of damage patterns, we determined that the reduction in I resulted in a residual moment of inertia of approximately 42 percent of the original value *at that location*. This does not necessarily mean that the capacity of the column is reduced to 42 percent of its original capacity. If the loss is near one end of the member, the diminishment of the capacity of the column is closer to its diminishment of bearing capacity, that is, 25 percent. If the same loss occurs near the midlength of the column, the loss of capacity is much greater, because the vulnerability of the member to buckling increases as I decreases toward the middle of the column length. If the diminishment is not only near midlength, but extends to much of the middle third of the member, the capacity of the column will, indeed, approach the theoretical value found in the typology, namely 42 percent of its original capacity, if in fact, the value of I in that axis governs the capacity.

It is important to note that the diminished value of I, however, dramatic it may be, may not necessarily result in a corresponding diminishment in the overall capacity of the column. One reason is the longitudinal location of the damage. Another reason is found by inspecting the figures in Table 3.6 closely. Notice that in cases B, G, and H that I_x is reduced by significant amounts, enough to arouse alarm, and that alarm is represented by the shaded blocks. Despite the major percentage reductions in I_x it will still be the value of I_Y, which is the lower absolute value. If both axes are equally unsupported, then the lower value of I will govern. In these three cases, despite the alarming *relative* drop in I_X, it is the lower of the two values of the moment of inertia that will determine the buckling capacity of the column.

In brief, the loss of cross-section area is a primary measure of diminishment of bearing capacity of a compression element, and that loss may be compounded by the configuration and location of the damage insofar as it may also diminish the moment of inertia of the member. These are not, however, the only ways compression members degrade over time. Another materially related change can occur relative to the modulus of elasticity. This is a somewhat exotic alteration, but it does occur. Timber is notable for it gradual increase in brittleness and loss of elasticity. As it becomes stiffer with time, it also is more inclined to brittle fracture with lower moduli of rupture. Stone and masonry, generally, undergo modest alterations of physical properties over time, which, at least theoretically, can shift the member into a state of risk. The alterations in modulus of elasticity that are prone to contribute to significant changes in buckling capacity or internal stress are more likely to occur during the intervention period than during the deterioration period. These untoward interventions are discussed later in this section.

Two potential sources of pathogenesis are related to the unsupported length of the member. The buckling capacity of a compression element is, in addition to the modulus of elasticity and the moment of inertia, dependent upon the length of the member. More specifically, it is the unsupported length that is critical. The difference is that a column braced against lateral movement at its midlength, for example, cannot move that point out of plane. This means that the braced point is fixed in space against buckling. The effect to the column is that the curve that forms if and when the column buckles changes from a simple single curve to a compound double curve resembling a shallow S.

The two columns may be of the same absolute length, but one can bend once from one end to the other, while the second must bend twice. The double curvature is as if two columns were stacked atop one another: one buckles to one side, the other to the other; together they display the same shallow S pattern as a single column brace at the middle in double curvature. In other words, bracing a column at its midlength has the effect of reducing the unsupported length by half. Inasmuch as the critical load capacity of a slender column is inversely proportional to the square of the length, reducing the effective length of a simply supported column by half may increase the critical load by four times. The reason is that such a reduction in length may change a long column into a short column, thus crushing capacity, not buckling, becomes the determinant.

It is important to remember that buckling can occur on any axis. In orthogonal sections, this is normally expressed as the X and Y axes with their respective moments of inertia. If the stiffer of the two axes is the one that is braced, column capacity is unaffected because neither the effective length nor the moment of inertia of the weaker axis has been increased. It is for this reason that sections with significantly different values of moment of inertia in the two axes are often braced in the weak axis and unbraced in the strong axis. The products of the respective resulting values relative to critical load capacity are more similar, one from a higher initial moment of inertia, the other from a reduced effective length.

Although buckling in the X and Y axes should be the primary concerns of both column designers and pathological analysts, they do not account completely for all the ways a compression element can or will buckle. A compression member may buckle in rotation about the third axis, which is to say the Z axis, which runs down the center of the member. The concept may be foreign as it applies to a building, but we are familiar with the phenomenon. Consider as an example, a plastic shower curtain rod cover, a cylindrical plastic tube with a slit along its length. It can be easily twisted because of a lack of torsional stiffness, a characteristic of open sections, meaning structural sections that despite any folds or turns are essentially plates. In this case, the slit tube is torsionally no stiffer than a flat plate rolled into a circular shape, and no stiffer than a flat plate of the same area and flattened dimension. If the plate is folded into a Z-shape, the torsional properties are unchanged. If instead of twisting the slit tube, we compress it, there is a distinct possibility that the slit tube would again rotate, displaying similar slippage along the slit. Eventually, the slit tube would fail, in a sort of spiraling collapse, not distinctly buckling about either the X or Y axis, but seemingly both simultaneously. This is a torsional buckling failure.

If the slit is taped tightly closed, thereby fused as a continuous material, the section is now a true tube or pipe, which, as a class, are closed sections. The increased resistance to twisting is immediately and profoundly evident. Closed sections, whether circular or rectangular, are preferred when the determinative external force is torque, as the resistance of closed sections to torsional buckling is similarly and beneficially affected. From the standpoint of building pathology, therefore, the concern with closed sections is that they degrade so as to convert to open sections. The resistance to torsion and to torsional buckling will be profoundly reduced.

The actual amount of force that a brace may have to exert laterally on a column in order to maintain its alignment and, therefore, the braced effective length, is remarkably low. The stress on a column brace is rarely a determining design factor. By the time the dimensions are determined for fabrication purposes, rarely has the member increased in its capacity because of the lateral force required of it to perform as a structurally adequate brace. This is not to say, however, that the force exerted by the column brace is zero, or that it absence is not potentially critical. The loss or deterioration of a structurally necessary column brace can, in fact, pose a serious threat to a structure. With an effective midlength a column may be capable of carrying 100 tons. If the brace fails, that capacity may drop to 25 tons. Because the force in the brace at the time of its failure is small, there may be little discernible deformation in the brace prior to its failure. In fact, bracing failure is far less likely to manifest as a fracture in the brace than as a failure in the connection of the brace to the compression element.

Conceptually related to the subject of bracing or the absence of bracing is the topic of combined loading on compression elements. The concept is related only in that bracing restrains lateral movement, whereas combined loading induces lateral movement. Beyond that sort of inverted relationship, the consequences are mathematically and structurally quite divergent. The situation involving combined loading begins with a member in axial compression. The member is then subjected to some additional force, which is *combined* with the axial load of the compression member. There are three common sources of such loading: lateral loads applied to the face of the member, axial loads applied eccentrically to the axial center of the compression element, and external rotational moments applied typically to the ends of the member. Such moments may occur anywhere along the length of a compression element and result in either bending or torsion.

The mathematics of combined loading are addressed under the deceptively simply title of "stability problems." Structural stability or, more accurately, structural instability is typically the end stage of progressive deterioration, as well as a set of common structural problems. It is, in fact, impossible to design columns such that there are no eccentric loads or external moments. The fundamental problem with combined loads is the inverse of the benefit of bracing—that is, that a small external force can have profound consequences on compression members.

Similar to a brace that can significantly increase effective length without itself absorbing much load, an external lateral force or moment can affect profound deformation of a compression member without itself being of significant magnitude. The difference in axial load alone between an adequate and undeformed member in compression and a buckled member can amount to many tons. The same member, however, can fail utterly with a modest application of external lateral load or moment. The accelerating effect of such external loads means that the unexpected and, therefore, uncompensated application of such externalities can be particularly hazardous to the stability of the compression member.

This hazard is especially worrisome because such slight loads can so easily become attached to compression members. The addition of loads, which axially are not a problem, can, if applied eccentrically, induce moments that are perilous. Floor loads framing into the column may change. The axial change may be nominal, but an increase in the eccentricity of the loading at the framing point may be devastating. A well-intentioned consolidation of the surface material of a stone column that is not uniformly applied can induce asymmetrical response in the thermal or hygroscopic movement of the column, which, in turn, can result in eccentric response and moments. Any of these seemingly innocuous additions or alterations can result in a hazardous combined loading situation for a compression element.

Another method of altering the effective length, in addition to bracing, is through various methods of fixing the ends of the column. Fixity, as a term, is somewhat obscure, but as a concept is quite familiar. What prevents a diving board from simply falling into the pool is that one end is fixed. Likewise, one end of a flag pole is fixed, and the other end is free to move laterally in any axis. A table leg is fixed at one end, as is a flag pole, but the lower end is set in the carpet or on the floor and, therefore, is not altogether free to move laterally; but it is not fixed either: it can theoretically rotate relative to the floor, but lateral movement is restricted. This is the definition of a *pinned end*. There are a total of three possible end conditions: fixed, free, and pinned; and we are familiar with each from common experience.

The complication and confusion associated with end conditions stems from two factors. First, the ends of a column may have different end conditions in different axes; second, columns may have different combinations of conditions at their respective ends. If we start with the latter complication, there are a limited number of possible end condition combinations (see Table 3.7). Each of the combinations of end conditions affects the capacity of a given column to support load. From the standpoint of building pathology, this means that if an assumption of structural stability is predicated on a particular end condition, then it is critical to the assessment of the deterioration of that element to determine the degree of change in the end connections.

For the other complication, namely, the variation of the end conditions as they are present on different axes, it is possible to fix a column in one axis at the same time that it is pinned in the other axis. As with bracing, it is of little structural significance to fix a column on its strong axis if the weak axis will continue to govern the overall capacity of the member.

If the end conditions are reliant on fixity, then deterioration of the fabric at those connections may be critical to the stability of the element.

Table 3.7	END CONDITIONS		
	Upper End	**Lower End**	**Effective Length (kL)**
a.	Pinned	Pinned	1.00 L
b.	Pinned	Fixed	0.75 L
c.	Fixed	Fixed	0.50 L
d.	Fixed	Free	2.00 L
e.	Fixed	Fixed, but free to slide laterally	1.00 L

Suppose, for example, that the member is deteriorated from tunnel rot at the precise location that the bolts pass through the column to fix the end against rotation. The loss of cross-sectional area may not have jeopardized the member nearly as much as the loss of fixity in the end connection. The loss of area is linear, meaning that a loss of 25 percent of the fabric at or near the end of the column reduces the crushing capacity of the column by 25 percent. If that very material is, however, essential for bearing of the bolts to develop a fixed connection, then the loss of the material may result in the release of the connection.

The release of a fixed connection can mean one of two things. First, the connection fails utterly, and the member disengages from its attachment point altogether. As catastrophic as this may be, it is an unlikely event. For column connections, failure frequently means that the fixity of the connection is lost, and that the column end can rotate at its connection. In other words, the fixed connection has converted to a pinned connection through the failure of the connectors. The significance of such a conversion may be as catastrophic as if there were a complete disengagement. If the column end conditions were both fixed, and one of them converted to a pinned condition, the buckling capacity of the column would be effectively reduced by 55 percent.

For a column with both ends fixed, the effective length = 0.5 L. This means that the critical load, (P_{CR}), is computed as follows:

$$P_{CR} = \pi EI / (kL)^2.$$

For kL = 0.5L, $(kL)^2 = 0.25L^2$, and $P_{CR} = 4\pi EI / L^2$. When one of the fixed ends releases, the effective length increases to 0.75L and $(kL)^2 = 0.5625L^2$. This means that $P_{CR} = 1.78\pi EI / L^2$. The relative capacity of the two conditions is $1.78/4 = 44.5$ percent of the original condition.

Though this book is primarily concerned with the deterioration of materials and connections, we need to be aware that if the external load increases to a point of critical proportion, the column may fail. Arguably this is not the focus of building pathology; it is, nonetheless, a common enough condition that practitioners assess any change in external loading over the course of the history of the building. Such changes are rarely of such magnitude as to account for a failure unilaterally; but when combined with normal degradation, the combination can precipitate a catastrophic failure.

By addressing several factors that can and do influence the capacity of a compression element to carry load, along with the diagnostic significance of measuring alterations and degradation of those factors and variables, we do not imply that these variables, though significant, are the only factors by which to measure the capacity of a column or the degradation that will reduce the capacity of a compression member.

We now turn our attention to the intervention in degradation mechanisms. We begin by emphasizing that analysis of and intervention in the condition of structural members should be left to licensed professional engineers—not conservators, not preservationists, not architects. Only engineers have the education and experience to understand all the variables that affect the stability and performance of structural elements.

With that admonition the question arises, why discuss the subject in this context at all? The reason is twofold: First, it is those in allied professions who will most likely retain the services of the structural engineer, and they will benefit from a basic understanding of what structural engineers do, in order to evaluate and hire their consultant. Second, they need enough information to know which questions to ask and, then, to be able to understand the answers. And let us not forget, as in all professions, not all structural engineers are equally competent, and their clients need enough education to make a discerning choice.

The basic approach to intervention at the element and member levels is not unlike that to intervention at the materials level, which is to array the intervention approaches and the structural variables into an options matrix (see Table 3.8). We begin the exercise with an analysis of the consequences of doing nothing. This option takes on added significance at the member and systems levels. Because the mechanisms associated with structural deformation are the result of external loads, or moments, that may either self-arrest or be of sufficiently slow progression, the degraded condition may be acceptable without intervention.

As with the options matrices of materials-level deteriorate mechanisms, the options listed in Table 3.8 are not intended to be definitive or exhaustive; they are intended to motivate thought and action. Particularly fruitful for structural members is the category of circumvention,

Table 3.8 INTERVENTION MATRIX: COMPRESSION MEMBERS

Necessary and Sufficient Conditions	Intervention Approaches					
	Abstention	Mitigation	Reconstitution	Substitution	Circumvention	Acceleration
General Mechanism	Accept the condition of the element without intervention.					Demolish member or building in its entirety.
1 Diminishment of area		Address source of loss of area as a materials-level problem.	Replace lost area or member as a whole in kind.	Replace lost area or member as a whole with alternative material.	Shore or duplicate member.	
2 Alteration of moment of inertia		Ditto	Ditto	Ditto	Add material to member.	
3 Alteration of effective length due to bracing			Replace brace in kind and type.	Replace brace with alternate material and/or type.	Add bracing other than as originally designed or intended.	
4 Alteration of effective length due to end conditions			Repair connections to original condition and operation.	Replace connections with alternate type but same operation.	Redesign type and/or material; install added or alternative connection types.	
5 Application of combined load or moment		Reduce external loads/moments.			Reinforce member for a level of accommodation of added eccentricity.	
6 Increase in external load		Arrest or reduce load; redistribute loads.	Remove loading; return to original loading.			

specifically, reinforcing procedures. Where there was one column, maybe there are now two. Where there was a timber post, maybe there is now a timber post with a pair of steel plates lagged to each face. The prediction of performance, as well as the anticipation of untoward complications for reinforcing devices at the structural level, are relatively simple. For the reasons explained in the sections on thermal and hygroscopic compatibility, it is necessary to review the differential movement potential of the subject element and the reinforcing members or elements. Obviously, these concerns are much greater when the member at issue is exposed on the exterior of a structure rather than on the interior where environmental conditions are more stable.

Walls

It is tempting to think of walls as really wide columns. It is, in fact, not unusual for wall design to be taught as a succession of one-foot-wide columns placed in contact with each other. For uniformly distributed loads lying along the top of a wall, such an approach, while conservative, is not unwarranted. There are differences, however, between columns and walls. The mathematics to explain those differences are beyond the scope of this book, but suffice to say they are never more manifest than in end-stage collapse. Imagine a series of in-contact columns arrayed in a plane as though they were a wall. Opposite the columns and parallel to the alignment of the columns is a true wall. Both are in a state of catastrophic collapse. We will look at two cases: one where the ends of the wall turn and continue at right angles; the other where the wall does not return at its ends. In other words, in the first case, the ends are braced; in the second, the ends are unbraced.

If each of the columns in the line of columns is loaded to the expected critical load for buckling, some will, indeed, buckle. A few will actually buckle prior to the design value; others will endure added load. The differences that account for which will buckle and in what order is idiosyncratic to each column. The range of critical loads for ostensibly identical columns will be narrow, but the specific values will vary, for reasons buried in the peculiarities of asymmetry of material distribution, indiscernible exposure variation, or any one of dozens of flaws embedded in the fabrication and construction of the members.

The consequence of the variation from column to column means that, while an average value of the critical load for the typical column is predictable, the specific buckling load of an individual column is predictable only within the range. The precise order and sequence in which the columns will fail is, therefore, not predictable. Unless the loads are somehow linked, or the columns are somehow connected, the order of collapse will be random; there will be no pattern or order.

Walls, in contrast, are far more predictable in terms of the pattern and order of collapse. To begin with, walls will fail at a more predictable load than columns in series. As described above, the idiosyncrasies of columns will necessarily result in a dispersion of critical loads; the same idiosyncrasies of construction in a wall are distributed throughout the wall in a sort of averaging process. In a wall, the same flaw that might have inclined a column to buckle prematurely is likely to be mitigated by a corresponding flaw, which would otherwise result in extended capacity. In columns, one would fail early; the other, late. In a comparable wall, the two values are averaged, meaning that, within limits, one value is mitigated by the proximity and magnitude of a counterposed value. Whereas ten columns might fail at values ranging from 80 tons to 120 tons, a wall equal in dimension to the same 10 columns is more likely to fail at 1,000 tons plus or minus 5 tons. The major qualification to such a claim is that the 1,000 tons are evenly distributed at the top of the wall.

When the wall does fail, it will do so not at random locations, but in a distinctly predictable pattern based on the restraints at the boundaries. In the two cases at hand, one is unbounded and the other is braced at its vertical edges. In the unbounded case, the wall will collapse first at it ends; each end will flop over like a dog's ear. The failure will progress from the ends toward the center until the entire wall has toppled. If the vertical ends are restrained, the failure will begin at the top center as a bulge, which will dissipate laterally and vertically, shaped somewhat like half of a cone. The bulge will eventually split, as a crack will form at the outside of the bulge, beginning at the top of the wall and running downward. This fracture will release two newly formed, unrestrained edges, which will continue the failure longitudinally along the wall. After the first major collapse of the restrained wall, there will be a zone of missing wall, shaped something like a blunt-pointed V. From there the wall will continue to unravel in the direction of the restrained vertical edges.

If as a third case we restrain the top horizontal edge, as well as the two vertical edges, the failure will commence as a bulge in the lower center of the wall, shaped somewhat like a pot belly in the wall. The first fracture may be vertical, horizontal, diagonal, or some combination of an orthogonally oriented pair of cracks originating at the point of maximum bulge. The first major catastrophic collapse will be a slumping of the wall above the bulge such that the resulting missing zone will be larger than the preceding case, more U-shaped than V-shaped, and generally larger than the failure of the wall with the unrestrained horizontal edge. The common denominator among all of these wall collapse cases is that the sequence and pattern are more predictable than a comparable assemblage of individual columns; and within these scenarios is a major advantage associated with the intervention in failing walls that is often unavailable with failing columns.

Because the initial deformation and subsequent collapse scenario of a failing wall are so predictable and characteristic, we can intervene at a relatively early point, and with relatively more reliable results than with individually failing columns. In short, walls are more forgiving than columns, if for no other reason than that their leading indicators are clearer and generally slower to develop.

All of the basic approaches available for intervening in column failures are also available walls. When it comes to specific applications, however, there is a wider range of options than for columns. Because a symmetrical, free-standing column can theoretically buckle at either of two axes, and in either of two directions in each axis, the bracing of incipient column failure must be prophylactic. Column bracing must tie back to a rigid structure, one column at a time, or reinforce the column from base to top, one column at a time. The reinforcement of walls can take advantage of the already existing tendency of the wall to act as a plate, and, in limited instances, to reinforce that aspect of the structure of the wall without actually tying the wall to its top, base, restraints, or a rigid independent structure. The device typically employed to accomplish this minor feat is called a *strongback*.

The strongback is a form of splint. It can be applied to columns, but not as effectively as with walls. The strongback may run vertically or horizontally. A pair of strongbacks may splint an entire wall, whereas it would ordinarily take a minimum of two to splint a single column. A fairly common application is to run channels with toes against the wall, vertically, in the zone between windows, and bolt through the wall into the floor plates. The strongbacks brace the wall vertically and tie it back to the floor much as a star bolt at the same time.

Strongbacks can also be used as temporary bracing until the permanent solution is installed. In practice, we have on a few occasions run the strongbacks vertically or horizontally outside the building and across the window openings, then run a rod or cable all the way across the building to a strongback on the opposite side, and cinch them together, using the window openings as the access ports to the strongbacks on the outsides of the building. Of course, the application of the principle is not restricted to the exterior. Stainless steel can be needled through a structure on the inside surfaces, which has the advantage of protecting the surfaces from weather, but at the price of having to find a path for them through the structure of the building.

The point of this digression is less to advocate strongbacks as the solution to all wall problems and more to point out the distinguishing nature of walls as panels or plates in compression, as opposed to sticks. The lateral and vertical distribution of deformations generally means that walls are relatively more deformable in more predictable ways and with greater warning than columns.

Exterior Walls

The previous discussion points out the major distinction between the prototypical generic wall and the specific nature of exterior walls; that is, that generic walls are perfect plates. Exterior walls are far more inclined to be pierced with holes, better known as windows and doors. Openings in exterior walls are little more than large cracks with straight edges. Structurally, there is almost nothing good to say about windows or doors.

When windows are tightly spaced, the narrow segments separating the windows effectively convert the wall plate into the series of columns we lambasted in the previous sections. The area of material that carried the building above the openings is now concentrated into substantially less material; and even though these segments may visually resemble the wall, they are, if fact, thin flat columns. The reentrant corners of the windows provide stress concentration, from which cracks easily propagate and collectively tend to further isolate the segments from the field of the wall.

For all these reasons, a wall problem is more likely to manifest at these segments than in other places; but we emphasize that the phenomenon is not unique to the segments of walls between windows. The phenomenon is called *delamination*, specifically *structural delamination*. For a wall to delaminate, it must be constructed of layers, a condition typically satisfied with a wythe of brick or of stone masonry backed with a layer of block, tile, or more brick. The bond between the face material and the backup is vulnerable to fracture, more so in the segments between.

Wall delamination is not inherent to masonry construction. It is quite possible to construct a brick or stone wall that is not subject to cleaving between layers of masonry. Any wall can be laid in such a way that there is virtually no way of telling whether there are distinguishable layers through the use of masonry headers. This means that, intermittently, a piece of the same material that comprises the face material is turned 90 degrees and bridges to the next inboard layer. The process is repeated between the second and third layers in a pattern that mimics, but is offset from the first set of headers. Normally, the first pattern can be repeated without impediment between the third and fourth layers, if there are that many. The greater the frequency of headers, the tighter the bond between layers. In brick masonry many patterns are possible because bricks are rectangular objects with three different dimensions.

For any rectangular section, $I = bd^3 / 12$. When the depth of the wall splits into two evenly divided sections:

$$I = 2 \times b(d/2)^3 / 12 = 2\, b(d^3/8)/12 = bd^3/4(12)$$

which is 25 percent of the original value. If the original pier were operating at any more than 25 percent of its critical buckling load before debonding

The Brick

The brick has three distinctive faces, each one rectangular. The long dimension of each face may be horizontal or vertical, so there are a total of six possible orientations of a single brick. Each orientation has a specific name. The long, narrow face, seen typically in running bond, is called the *stretcher face* or, simply, stretcher. That same face turned vertically is called a *soldier*. The short face, or the end laid horizontally, is a *header*; laid vertically, a *rowlock*. The long, broad face laid horizontally is a *shiner*; set vertically, a *sailor*. The term header is specifically the name of the face of the brick that is seen when the brick is turned in-plane from a stretcher. Because the headers are commonly used as the tie between face and backup, the term has become synonymous with virtually any brick used as a tie between layers. It is, however, altogether possible to use a rowlock as a tie between layers where the field is composed of shiners. Still, probably they would be referred oxymoronically as a rowlock header.

occurred, the debonded pier would immediately be subject to buckling. This means that the resistance to buckling is similarly reduced. Once the outer face is an independent panel with vertical edges free and horizontal edges semifixed laterally, the form of the delamination will begin at the free edges, buckling outboard, further separating from the backup layer. Sometimes the window frame will move with the outer face; other times it is anchored to the backup, and slips out of the exterior layer. If the window is held by the backup and stays with it as the face layer buckles outboard, the now-open collar joint is quickly and easily exposed, and water flows into the void and saturates the lower wall. At this point, the deterioration mechanism will accelerate rapidly, to delamination of the lower reaches of the wall, with accelerating bulging of the wall as a whole, leading to collapse.

A small-scale example of delamination and subsequent lateral displacement of a wall can occur at the shelf angle of veneer walls. This is increasingly evident among buildings constructed in the 1920s and 1930s with shelf angles, which have aged to the point at which this problem is manifest. Either the veneer was constructed as a cavity wall or as a solid wall; regardless, the bond or ties immediately above the shelf angle have disengaged. As material from above sloughed off the sides of the now-separate layers, it accumulated at the bottom of the cavity or within the now-divided layers. This material acts as a wedge when it becomes saturated and freezes, and the lateral pressure pushes against the back of the face material. If the bond between the bottom course of face material and the shelf angle fractures, the face material can ride off the angle. If the bond holds, the face material bulges outboard in a band approximately eight inches above the shelf angle, and the wall rolls off the shelf angle.

A special case or class of exterior wall problems belong to the portion of the wall that projects above the roof, namely the parapet. This portion of the wall is subject to the harshest exposure, yet it is often constructed with little accommodation to the changed environment between the lower reach of the wall and the parapet. The parapet by its nature is exposed to the open environment on both vertical faces. It gains no mitigating benefit from heat emanating through the wall from the interior when the parapet is saturated and/or freezing. Moreover, the coping is a constant source of vertical water intrusion. Water can and often does flood the base of the parapet. The winds riding up and over the edge of the parapet can increase velocities by more than double their average velocity. It is not remotely surprising that parapets present problems. It is only surprising that, like their close neighbors, roofs, they manage to hang on as long as they do.

Delamination is certainly a potential deterioration mechanism of a parapet. A mitigating circumstance is that, by virtue of its being at the top of the building, the parapet is not subject to significant vertical load. It is, however, subject to significant lateral loading from wind. Inasmuch as it is also a vertical cantilever, the end condition multiplier (k) is 2.0. As long as the parapet is bonded solidly, the moment of inertia is ample to accommodate the lateral loads, both as a compression element unbraced horizontally at it upper edge and as a member in bending as a vertical cantilever. When, however, the layers of the parapet delaminate, the moment of inertia degrade grades to 25 percent for a two-layer bifurcation to 12 percent for a three-layer trifurcation. As the value of I is degraded to a fraction of its original value, so is the value of the section modulus. The resistance to bending is reduced, and the proportionate is vulnerable to the failure mode described for a wall, restrained at the vertical edges and released at the top horizontal edge.

Another phenomenon peculiar to parapets is unrelated to delamination. It is called *parapet ride-off* or *parapet jacking*. The origin of the problem is traceable to the fact that the parapet is subject to significantly greater temperature swings—that is, harsher conditions—than the wall below, in conjunction with the fact that the parapet is usually constructed of the same material and of the same detailing as the wall below. Even if the lower wall were to be exposed to the unrelieved conditions, the parapet would still suffer more than the wall below.

The temperature swing impacting the parapet can easily reach 120° F, meaning that the absolute temperature of a parapet can range from –10° F to 110° F within a single year. Even if it takes two years for the parapet to achieve these values, the point is, it will occur, along with many other intervening swings of lesser magnitude. The mitigating effects of interior climate reasonably assure that no other building wall will undergo as dramatic a swing in temperature. At the same time that the parapet is subjected to vast temperature swings, it is also subject to major moisture content changes and, therefore, major dimensional changes due to hygroscopic expansion and contraction. Sometimes, the temperature and hygroscopic changes will cycle together to effectively amplify the dimensional shift of the parapet.

The magnitude of the dimensional shift of the parapet by itself is not significant. What is significant is the dimensional shift of the parapet relative to the shift of the wall below under the same conditions. The dimensional shift in the lower wall is less than that of the parapet for a number of reasons, the leading one being the mollifying and mitigating effects of the interior on the exterior wall. As discussed earlier in this chapter, differential movements between materials or, in this case, portions of the wall, generate internal shear stresses.

As the parapet increases in dimension relative to the lower wall, the lower wall, in effect, acts as a restraint on the movement of the parapet. The point of maximum restraint is at the horizontal line roughly equivalent to the roof line on the back side of parapet. Thus, while the lower portion of the parapet is undergoing the same temperature and hygroscopic change as at its top, the lower portion is restrained in that movement more than the top. The top expands more than the bottom.

The differential in dimensional change generates internal stresses, which are greatest at the extreme ends of the parapet because this is where there will be the greatest absolute dimensional change. If the magnitude of these stresses exceeds the tensile capacity of the masonry, a crack will form, typically running from the bottom of the parapet near to corner, diagonally upward toward the top of the parapet and away from the corner. As the face of the crack itself weathers, debris accumulates in the crack so that when the parapet cools, it cannot contract back completely to its original location. As long as the crack is open, airborne debris can deposit there, further adding to the relatively rigid material located within the crack and preventing its return to complete closure.

This rigid material also absorbs water, which freezes; the increase in dimension of the freezing water pushes the corner of the parapet further. The capacity cycles of parapet expansion, contraction, debris accumulation, saturation, freezing, and thawing converge to gradually push the corner of the parapet off the building, as though it were being gradually jacked away.

The basic approaches to prevent parapet jacking constitute one of the great lessons of architectural design, because the forces that will eventually fracture and ultimately destroy a parapet are similar in origin and effect to many forces found in building movement. The architect or engineer has a limited set of options associated with the accommodation of the internal stresses derivative of building movement. He or she can ignore the stresses altogether with the obvious ultimate result. Two alternative approaches accommodate building movements: resist or release.

The resistance approach entails recognizing that the forces exist and where. Once the locations, directions, and magnitudes of the forces are known, reinforcing is designed sufficient to resist the forces without otherwise damaging the construction. In the case of a parapet, this usually means wrapping the corner with enough steel embedded in the masonry to simply overwhelm the dimensional shift of thermal and hygroscopic elongation. The longitudinal steel may have to be supplemented with rods doweled into the lower portion of the wall, to distribute the shear stresses vertically as well as to resist the forces horizontally.

There are problems with the resistance approach. The reinforcement does not eliminate the forces; it simply provides enough material at critical locations to absorb the accompanying strains. Reinforcement also has the secondary effect of redistributing the stresses; at times, the redistributed

FIGURE 3.8   The lateral restraint offered by the roof, relative to the wall, failed, and the coping has moved four inches (10 cm) laterally out of plane.

stresses are to zones and materials that are under-reinforced for the consequential increase in stress. Instead of cracking at the corner, the parapet cracks some 10 or 12 feet from the corner. Likewise, there are problems associated with burying materials inside masonry. In principle, metals corrode, a subject about which much will be said in a subsequent chapter. Aside from the loss of the metal, the by-products of corrosion typically are of greater volume than the original metal. They expand inside the masonry, thereby exerting lateral stress on the parapet materials. If the lateral pressure exceeds the tensile capacity of the material, it will crack and possibly spall.

In the release approach, the forces are recognized and defeated by eliminating the opposition to them, hence the stress in the materials. The basic device for accomplishing this feat is a joint, which is a lingo term for an intentional crack. When we design the crack into the building, we call it a joint. If we neglect to crack the building ourselves, nature will do it for us, but our cracks are straighter and, for no other discernible reason, preferable to jagged, natural cracks. Releases can even be amplified into design features such as slots, which are wide cracks disguised as art.

Releases, or joints, are effective, but they, too, come with a price. Joints fragment the wall or parapet and structurally weaken the construction by breaking up the horizontal integrity into a series of shorter segments, each operating independently. The joint is also a potential avenue of entry into the structure for water and vapor. This can be countered with ample applications of sealant at the joints, but the liberal use of sealant to defeat water intrusion is a dead giveaway that the designer did not understand water intrusion. The subject of leaks is another topic we will spend much time discussing in a subsequent chapter.

Interior Walls and Partitions

Obviously, not all walls are exterior. There are many walls inside the structure, subdivided as load-bearing or nonload-bearing. Actually, unless an interior wall is very carefully detailed to not absorb load, it becomes load-bearing regardless of the intentions of the designer, for a reason that has nothing to do with the sizes of the studs or the thickness of the block, but with the flexibility of the structure and the rigidity with which nonload-bearing walls are inserted into buildings.

Whether intended to carry load or not, interior walls and partitions are scarcely distinguishable. The studs in a bearing partition may be slightly larger or closer together; or the block may be eight inches thick instead of six inches thick if it is load-bearing. Initial shrinkage of the load-bearing and nonload-bearing construction are very similar, but the load-bearing construction will also be reduced in dimension because it is under compressive stress. If the nonload-bearing partitions are built tight to the undersides of the construction above, there is every prospect that the building will settle onto the nonload-bearing walls and partitions, which means they are now load-bearing, regardless of the original intent.

When any interior partition is compressed, whether it is load-bearing or nonload-bearing, the same potential exists for stability and crushing problems as for their exterior counterparts. The laws of physics are valid indoors as well as outdoors. Because nonload-bearing partitions and walls are, in fact, constructed very similarly to their load-bearing cousins, the capacity of nonload-bearing elements is often quite significant. The distribution of loads, however inadvertent that may be, into nonload-bearing partitions normally has greater consequences on the joists and beams that are carrying the partitions than on the partitions themselves. Where we have observed the phenomenon, the redistribution was virtually insignificant, except for the deflections in the floors. In one case, in a late Victorian townhouse, the hall partition parallel to the party walls had inadvertently become a bearing wall along the centerline of the house. Three stories of such walls were aligned, and they carried three floors of load onto the ground-level joists, which spanned party wall to party wall without support in the basement because the partitions above were not supposed to be load-bearing. The joists on the ground level deflected approximately an inch and a half, a measure that was replicated in all of the floors above because at each of those levels the joists were headed off at an adjacent stairwell and source of the redistribution. Despite the deflection, we saw no identifiable problem until we inspected this same partition at the top floor below the attic. The attic structure spanned from party wall to party wall and did not bear at all on the interior partition, but because the interior partition was stacked on three stories of displaced partitions, it had pulled away from the ceiling the full inch and a half that the ground floor joist had deflected.

During the course of the life of a building, compression cracks may occur in the walls, but they are frequently confused with temperature cracks, simply because cracks are cracks, and are not distinguishable as to origin. Compression cracks, however, may be cosmetically treated more successfully than true temperature cracks, which are more likely to return regardless of treatment. Interior compression cracks tend to not be of great concern, and for good reason: they are seldom consequential—that is, until the wall or partition is altered. Alteration is more significant in this case than removal. If a nonload-bearing partition that has secondarily been placed in compression due to settlement or shrinkage is removed altogether, the overall structure is not diminished because the nonload-bearing partition was never intended to support the structure; therefore, its absence cannot be a determinant to the stability of the same structure. If, on the other hand, a nonload-bearing partition that has inadvertently been put into compression is reduced in area, such as by the insertion of a doorway, the resulting diminishment of cross-sectional area may prove disastrous to the partition as a whole. The result is seldom threatening to the building, but it may come as a major unanticipated expense to the homeowner.

Critical alterations are not all intentional alterations. The partition may be diminished by water damage, termites, ants, or a number of other reasons. The manifestation may take the form of a compression failure simply because of the loss of area and/or moment of inertia.

Tension Elements

Pure tension structural elements are relatively rare in buildings, except as they occur in trusses, for two reasons, one primary, the other secondary. The primary reason that tension elements are rare is, simply, that the direction of gravity is down, not up. This is not meant to sound flippant; it is the reason. The general objective of members is to keep their greater structures up, which means interposing rigid construction between the earth and the desired structural elevation. Tension elements placed perpendicular to the earth's surface are, by definition, pulling up, and, therefore, must eventually be connected to a compression element in order to transfer load back to grade. This indirect way of accomplishing a basic structural tenet is generally inefficient and costly.

The secondary reason tension elements are rare in buildings has to do with the way tension members fail—utterly and with little if any warning. Most tension elements found in buildings are not primary, but secondary. Secondary elements are those included to brace the primary elements from wind or side sway, or that brace columns to reduce their effective length. Consider curtain wall hangers, which are designed as tension elements: knee braces, which brace the hangers laterally, may be thrown into compression or tension. Sway braces in roofs and shear walls absorb wind loads in tension only if they are rods; in tension and compression, if they are resistant to buckling. Sag rods are tension members, as are lateral bracing rods for deep, long-span trusses.

This section examines the peculiarities of both primary and secondary tension members. Tension elements acting as primary structural elements though rare, when they do occur, are particularly important for the simple reason that if and when they fail, the consequences are profound. Other than in trusses, with which we will deal below, primary tension elements are seldom redundant, or the redundancy is very limited. Consequently, in the event of a failure, there is little if any capacity or time for the loads to redistribute. The general configuration of primary tension elements is such that there is not a compression element below; therefore, the failure of the tension element will probably result in collapse, typically falling some distance before coming to rest. The obvious presumption is that the tension element is vertical and suspended above-grade. This, however, is the stereotypical condition; it does not describe all primary tension elements. Long-span arches and portals often have tension ties connecting the contact points of the structure with the ground. The ties are often below-grade, buried below the floor slab. The tension element itself may be a cable or a bundle of reinforcing rods embedded in concrete. The descent means acceleration and subsequent impact with an unyielding surface.

Incipient tensile failures are scarcely identifiable, not because the material does not elongate prior to fracture, but because the amount of elongation at ultimate strength is normally limited to a very short, specific segment of the member. Steel in tension, for example, will theoretically elongate approximately 20 percent prior to fracture. (See ASTM standard A 36/A 36M – 96. §8 and Table 3.) This assumes that the entire length over which the elongation is observed is of uniform area and stress at the time of fracture. In fact, this condition exists only in laboratories with very short specimens.

Although the elongation at ultimate strength may be 20 percent, the elongation at yield stress is only one-eighth of 1 percent (0.124 percent). For a hanger 15 feet long, the elongation at yield is less than one-quarter of an inch (0.223 inch). This means that over 99 percent of all elongation occurs at stress levels above yield and prior to rupture.

Suppose, for example, that the hanger is three-fourths inch in diameter (ϕ) and is 15 feet long. If design capacity is set at 2/3 F_Y = 24,000 psi, the design load for the hanger is $2F_Y/3 \times$ Area = 24,000 × 0.44179 = 10,600# ±. If F_{ULT} = 72,000 psi, the critical cross-sectional area is 0.147 in.^{2} which is the equivalent of one-third of the original area. If that critical reduction in area were to occur in a graduated manner inside a 4-inch-deep timber member attached to the bottom of the hanger, the total elongation of the rod above the critically reduced area would be slightly over one-eighth inch and the elongation within the timber would be less than one-fourth inch at the moment of fracture. Keep in mind this example is taken at full design load of 24,000 psi.

In practice, the reduction of diameter or area of a critical tension member must be as much as two-thirds of the original area, a loss likely to be noticeable if it were to occur at a location clearly visible. It does not require a masters degree in structural engineering to recognize that something is wrong when a member is corroded to only one-third of its original area.

A major problem with tension elements in failure is that the critical reductions in the cross-sectional area are likely to occur where visibility is obscured, such as by rods or hangers imbedded into ceilings, floors, or other structural members. The critical reduction in area may occur over a very short length or segment of the element, such as that portion that is obscured. The elongation associated with ultimate stress—which, if it occurred over the entire length would be blatantly obvious—in fact, occurs over a very short segment, and is not particularly noticeable. The exposed portion, which is to say most of the overall length, may appear to be in excellent condition, whereas the nominal length that may be seriously diminished will go undetected. By itself, this characteristic of tension elements is not debilitating, but when combined with the consequences of the failure of a primary tension element, the conclusion is that the mere presence of a primary tension element in a building is a matter of immediate and automatic concern to a responsible steward or professional. No primary tension member can be presumed to be secure until it has been examined from one end to the other and found to be safe. The margin for error is so narrow and the consequences so profound that anything less than total inspection is careless.

There are, of course, situations when visual observation may not be easy, meaning that some removal of covers or soffits may be necessary. Even if these soffits and covers are attached to national shrines, the practitioner must do his or her duty. To help in this effort, techniques such as X-rays, resonance devices, low-velocity radar, magnometers, and other nondestructive techniques are available. Owners may object to the cost of some of these techniques, but they are positively cheap compared to defending a wrongful death claim.

The modest amount of redundancy that primary tension elements may contain usually is not because another member duplicates the primary element, but because the intersecting member is continuous, and can, therefore, engage in some redistribution if the primary member fails. If, for example, the hanger rods were threaded through continuous beams, and if those same beams were carried by other hangers distributed along the length of the beam, then in the event of a failure it is possible that one of the rods in the beam may be able to span between two other surviving hangers. Even if a beam also were to fail, that failure would be in bending, which, as is discussed below, is a preferred mode of failure.

In nonredundant tension elements, while redistribution is possible, the likelihood is far greater that the failure of one primary member will trigger a sequential failure of adjacent members as loads are redistributed. A common term for this sequential failure is *unzipping*, to reflect that one failure triggers another, which triggers another, until the entire structure has opened. One of the more dramatic and tragic examples of unzipping was the Regency Hyatt collapse (July 17, 1981 in Kansas City, Missouri), which was triggered by an initial failure of a single connection. The structure acting somewhat like a slab distributed the unsupported weight to adjacent hangers, which in rapid succession failed, and the walk-way fell.

Although, as stated repeatedly, primary structural tension elements are somewhat rare, secondary elements in tension are quite common. Determining the difference between primary and secondary members is done using a conceptually simple, but, at times, technically indeterminate, principle: If the removal of the element *will* result in collapse or significant deformation, then that member is a primary element; If the removal of the element *may* result in collapse or significant deformation, then that member is a secondary element. Though there are severe limits to the application of this test, as illustrated below in the discussion of trusses, the idea is reasonably sound and useful for classification purposes.

The broadest reasonable way to describe the function of secondary members is that they typically brace primary members or systems. The class also includes elements that are sufficiently small or that carry a sufficiently small area or load such that their loss cannot jeopardize the structure. Examples of the former are knee braces, sag rods, and wind bracing. Examples of the latter are hangers, framing around openings, and lintels. Some of these, particularly the bracing elements, will or can experience tension.

The problems associated with secondary members are rarely a function of the inherent nature of the forces and more the result of the peculiarities of the situation, such as abuse by humans. For example, because of the contingent nature of the function of many secondary members, they are often removed or damaged in the course of an alteration or even in the course of repairing a primary member. Obviously, any scenario that can adversely affect a primary member can degrade a secondary member; and there are no unique conditions that affect secondary members, such as hangers or sag rods, which are in a constant state of tension, or lintels, which are in a state of bending (see following section). Often, the secondary members that are alternately in compression or tension are braces, and the problems associated with braces are usually the result of their absence, not their presence. From the standpoint of building pathology, it is improbable, indeed, for the building to fail as a whole because of the local degradation of a secondary member, much less a brace.

Braces are members that, as a class, fix connections against rotation, typically, by triangulation. If two primary members are joined at right angles, there are a limited number of ways to ensure that they remain at right angles. They can be fixed as to location and angle by external supports, such as buttresses or abutments; they can be joined so that the joint is frozen or fixed, and rotation about that joint is effectively eliminated; or the two members can be triangulated by the addition of a third member.

Braces can be used in all three applications. In the first instance, braces take the form of props and are usually temporary in nature; for example, the shores used to prop up a sagging porch or to secure an excavation. As the examples in the previous paragraph suggest, when such props become permanent fixtures, such as buttresses or abutments, we tend to affix other names to them that suggest they are something more or different from shoring or props. They are not different; they are merely larger and decidedly more permanent.

In the second instance, fixed joints, the fundamental problem is that the forces present at the connectors of the joint must be the equivalent in strength to the members they join. Fixed connections may be stronger than the joined members, and while that may allow the designer to sleep better, there is no real point in designing the joint beyond the capacity of the joined members. Fixing joints is not particularly difficult if the material being joined is steel; full penetration welds will normally do the trick. Fixing timber joints is not as simple.

One general technique for designing a fixed joint is to enlarge the depth of the joint; that is, the connector increases in dimension in the direction of the crotch formed by the intersection of the two members. As the connector migrates, it joins with the primary members farther and farther away from the center of the joint. Analysis demonstrates that the farther the connector migrates away from the crotch, the lower the stress in the connector material. This is because the connector is operating farther away from the centroid of the joint, and the mechanical advantage of the connector increases with the distance. In fact, it is possible that at some point, the edge of the connector farthest away from the center of the connection will actually separate and become a separate item, connected diagonally to each of the primary members. At this point, the joint is still as fixed as it was with a frozen joint, but the fixity is accomplished with a separate member, which is a brace.

Having disengaged from the corner joint, the brace may continue to migrate along the respective lengths of the two primary members and connect at almost any location and at almost any angle. If the angle is very shallow, respective to either of the primary members, the mechanical advantage of the brace and the relative fixity of the joint both decrease. The generally accepted angle of optimum economy is 45° relative to each of the orthogonal primary members. If the brace migrates so far away from the joint at issue that it engages the ends of the two primary elements, even if the resulting relationship to the members is not 45°, the members are at this point fully triangulated. This is the third of the available options.

The significant difference between fixity using a brace and full triangulation is that fixed joints always impose bending into the member as the joint is worked. Consequently, as external effort is applied to the members so as to open or close the connection, if that same joint is fixed either at the connection or by a brace, there will necessarily be moments imparted to both primary members. By itself, such an added moment may not be consequential; it is, typically, referred to as a secondary moment. As with secondary members, it *may* be significant, whereas primary moments are the reason to have a primary member, and are by definition significant.

The advantage of using braces to fix joints is that the connection technology is simple, and the bracing members themselves are generally simple and economical. The disadvantages are the imposition of secondary moments and of diagonal members into the construction, which may have to frame or work around them. The advantage of full triangulation is that it does not impart secondary bending; but it generally means taking the longest path between the ends of the members. The location along one of the great diagonals implies that the member itself may be long, and that it will necessarily bisect what may otherwise have been a transparent wall or opening.

As mentioned above, the deterioration of secondary members is rarely a source of building-threatening proportions precisely because the members are secondary. It is also unlikely that such members will deteriorate without the same mechanism attacking the primary members directly. More likely is that the secondary members were never installed or were compromised by the building occupants. When the secondary members are intended to restrict deformation, and they are compromised, obviously the deformation may occur. Although they may or may not lead directly to instability or collapse, even limited deformations can result in cracks, which, in turn, admit water and pollutants, which then degrade the fabric. The point is, the deteriorations or even the absence of secondary members are rarely of pathological proportions, but they can pose a long-term indirect threat to the building.

Spanning Members

The problem of getting from one side of a void to the other is as old a problem as exists in structures and structural engineering; and it is as fundamental as gravity and as essential as breathing. Imagine a world without spanning between two supports: there would be no tables, diving boards, Gothic cathedrals, airplanes, rails for trains, concrete slabs for roads, tall buildings . . . the list is endless.

The spanning problem is solved in many ways, but only one of which is fundamental physically. The essence of spanning is the diversion of a force perpendicular to its line of action to at least a pair of reaction points that are offset from that line of action. Spanning solves the fundamental problem that a force that exists at a location will, if unimpeded or intercepted, result in an undesirable consequence. A train will wind up in the river because the force of gravity acting on the train will accelerate it toward the center of the earth if and only if a force acts in the opposite direction of gravity, in this case, and of a sufficient magnitude as to result in the static position of the train relative to the surface of the river. The major caveat of all spanning problems is that the requisite equal and opposite reaction does not occur directly below the train, because there is no reasonable way to match the crossing of the train with a moving vertical support immediately below the location of the train.

The weight of the train must be opposed by a force or a set of forces equal to the weight of the train, the sum of which act in the opposite direction of the train, but that cannot act in the same line as the force of the train. All this is intuitively obvious; but there is a catch: these two or more reactions not only must balance the vertical force of the train, they must do so in such a way as to be among themselves balanced against rotation. Imagine the 100 ton train balanced by two abutments; the train is in the exact middle of the bridge. We do not need a degree in engineering to know that each abutment will carry 50 tons, but how do we know that other than by an innate sense of symmetry?

We know that a train in the exact center of a bridge supported by two abutments is supported by reactions equal to one-half of the weight of the train, because the train is not spinning. The fact that the train continues on without spinning off into space is confirmation that when the train was precisely in the center of the bridge, the reactions at the ends were identical. If the reactions had been otherwise, such as 60 tons and 40 tons at the exact moment that the train was in the middle, the abutment with the higher reaction would have risen from the banks of the river and begun arcing over the river with the train now rolling rapidly downhill toward the opposite shore. If the train escaped, perhaps the abutment would reverse its arc and crash back where it started. If the train were too slow, the abutment would continue until it inverted itself on the opposite shore. Neither scenario has happened, nor will happen because there is no imbalance of the forces of the reactions.

Assuming that all the force is concentrated in a hypothetical point, when a 100 ton train launches onto a bridge, the near abutment carries all 100 tons. As the train traverses the bridge, less load is carried by the near abutment and more is carried by the far abutment. The sum of the two reactions is always equal to 100 tons, which we know because, like the spinning fantasy, the bridge does not launch into the sky nor fall into the river. The whole business of how much of the weight of the train is carried by each abutment is self-correcting, and that correction is proportional to where the train is relative to the ends. The logical nature of loads and reactions to loads is apparent; at times, the arithmetic confuses us, not the logic.

Regardless where the train is on the bridge, we know that the reactions at the two abutments will add up to 100 tons. We know that the two reactions will be equal only when the train is dead center on the bridge; and we know that when the train is closer to one abutment than the other, that abutment will carry proportionately more load. We know that each of the abutment reactions is trying to curl up the bridge and fold the bridge over on itself, and it would do so but for the stiffness of the bridge. The science of bending is all about determining the capacity of the bridge to resist these simple forces.

Although there are several specific ways to accommodate all the forces at work in the typical spanning problem, as noted, there is really only one fundamental physical problem: the structure must in some fashion or other absorb internally the work that is occurring externally; that is, to carry a load that is at a distance from its supports. This section investigates a variety of basic approaches to the problem of spanning, along with the typical deterioration mechanisms associated with each approach.

Rafters, Joists, Girders, and Other Beams

The general category of beams includes many that we associate with floors and other similar horizontal construction, and they are what we will concentrate on in this section. To begin, we point out that the definition of a beam is not its orientation. Although many familiar applications of beams are, indeed, horizontal, the orientation in bending is not limited to that direction. Vertical beams are what resist wind load in curtain walls; the principles are the same.

A beam is a structural member (1) operating primarily in bending, which is (2) continuous in its construction, and (3) supports the external load or loads between two or more reaction points. What is excluded from the definition of the beam concept is as important as what is included. Included are beams that also may be compressed or tensioned, but for which the dominant structural consideration is bending; and beams that may exhibit both positive and negative moments. Excluded by virtue of number 2 above are trusses (see the following section); and by virtue of number 3 so are cantilevers (see the section following trusses). Beams may have unjointed or segmented webs. The webs need not be solid; they may be pierced, but they are essentially a single piece, and integral with the flanges, which is why trusses are distinguishable. Cantilevers are excluded because the problems associated with them are somewhat different than for beams and because they are distinguishable based on their support arrangements.

To understand bending, it is key to avoid being confused by the numbers part of the problem. Visualize a simple beam supported on each end with a single point load near, but not exactly, at the center of the span. We know that the reactions are not equal and that the one closer to the load is the greater of the two reactions. Imagine the beam upside down: the reactions are now hanging from the ends, and the point load is supporting the beam and the two reactions from a thread connected to a magic skyhook. The reactions are now free to swing or rotate or do whatever it is that reactions do when they are turned loose. The beam is free to rock or roll or do whatever upside down beams do. What will the beam and the reactions do? The beam will curve downward from the skyhook to the two hanging reactions; that it is all. The values of the reactions are self-adjusting depending on precisely where the point load skyhook is located. They will always be in balance so that the beam remains perfectly balanced.

The moment generated outside the fabric by the left-hand reaction hanging down at a distance from the skyhook is the same in value, but opposite in direction, to the moment generated by the right-hand reaction, which is of a different value hanging at some other distance. The product of the left-hand reaction and its distance from the skyhook equals the product of the right-hand reaction and its distance away from the skyhook. That the beam is balanced and not spinning is proof that these two moments are equal. The fabric of the beam must be sufficient in capacity to resist the efforts of both reactions to fold the beam in two.

If we could enlarge the face of the beam, particularly near the point load, and if we could somehow actually see the stresses within the beam itself, we would see that the sum of all the stresses within the beam were equal to the forces and moments outside the beam. In the hanging mobile image, the bottom of the beam is pushing from the hanger point outward, at the same time the top portion of the beam is pulling toward the hanger. Thus, in this case, the bottom is in compression and the top is in tension. This necessarily means that there must be a place within the cross-section of the beam where these forces are neutralized. If at that location we take the sum of the products of the compression area and tension area, we find that the internal resistance to being folded in two is identical to the effort of the reactions to fold the beam in two. How can it be otherwise?

Now imagine the beam turned right side up. All that changes are the directions of the loads and reactions; their values stay the same. The top surface of the beam is in compression and the bottom is in tension, but the values of the stresses stay the same. The beam is still balanced. All the numbers are the same; even the deflections are the same, except that the curvature is reversed. The important point here is that the moments inside the beam are a reaction to moments generated from the loads, and those moments inside are the result of tension and compression on opposite faces of the beam.

If the shape of the beam is such that most of the material contributing to the resistance to the external moment is concentrated at the extreme edge, such as with a wide flange beam, then most of the tensile force and concomitant compression force will be at the extreme edges. If the shape of the beam is evenly rectangular, such as a common timber beam, then the force stresses are distributed across the face of the cross-section in proportion to the distance from the neutral axis. The farther from the neutral axis, the greater the stress. One of the major considerations in designing a beam is to reasonably ensure that the stresses that do occur at the extremities of the cross-section—or, as they are called, the extreme fibers of the section—are below an acceptable design level. In designing a beam to resist bending pay attention to protecting the extreme fibers, as well as the entire cross-section, from stresses above the designated design level.

In the deterioration phase of the life cycle of a beam the opposite is the case. We are particularly concerned with fabric deterioration, which will result in the aggravation of the stress levels of the extreme fibers. Knowing that the extreme fibers of the tension and compression faces are disproportionately vulnerable in a bending situation means that any deterioration mechanism that results in the diminishment of the material at those extreme fibers is of particular danger and, therefore, interest.

The measure of the distribution of material at the extreme fibers of a given cross-section is called the *section modulus*. This value is determined by first computing the modulus of inertia (I). I is divided by the distance from the centroid of the section to the farthest extreme fiber. The resulting value is the section modulus (S). Earlier in this chapter when we discussed the effects on buckling capacity as a result of diminishment of the moment of inertia, we computed the reduced section moduli at the same time (see "Columns," previously).

FIGURE 3.9 The apparent sweep in the ridge is almost four inches below dead level. This sort of deformation is common among buildings with simple, inclined, paired rafters.

From that analysis we learned that the diminishment of area at the extreme edges of the direction of investigation resulted in significant reduction in moment of inertia (I). Now note that significant reductions in moment of inertia correspond closely with significant reductions in section modulus. This is because both bending and buckling are disproportionately reliant on the distribution of material at a distance from the neutral axis. For moment of inertia, that degree of disproportionality is the cube of the depth of the member; for section modulus, the square of the depth. Any mechanism that disturbs the area at the extreme edges of beams or columns poses a serious threat.

With our imaginary ability to see stress, upon further investigation we will notice that the magnitude of the tensile and compressive stresses in the extreme fibers is not uniform over the entire length of the beam. If, for simply supported beams, the external source of the moment is the force of the reaction times the distance of that force away from the reaction, then as we come closer to the reaction point, that distance decreases and the force of the rotating moment decreases. These stresses in the extreme fibers do not, however, explain all of the behaviors of the beam; they explain only the accommodation of the external moments.

The other major component of beam behavior has to do with the rather simple and obvious balancing of the translation forces, as opposed to the rotating forces or moments. Externally, the translation forces are the various loads and the corresponding reactions. We know that the loads and the reactions are necessarily equal in magnitude and opposite in their respective directions of action. Imagine a situation in which the external load is positioned directly over the left-hand support. It follows that the reaction directly under the load is equal to the load and opposite in direction to the load. There is no bending in the beam because the reaction is not any distance from the load, so there cannot be an external moment; no external moment means no internal moment.

Now move the external load a small dimension to the right of that left-hand support. The left reaction is still very close to the total value of the load, with a tiny fraction now carried all the way over at the right-hand support. The external moments are very small in value because the distance from the left-hand reaction to the load is very small; and while the distance from the right-hand support is great, the right reaction is very low. Either way, the products, which is to say the moments, are small.

The load is now poised a short distance away from the support, and is pushing down, let us say. The reaction is, therefore, pushing up. These two forces are acting like the leading edges on a huge paper cutter trying to slice the beam off at or very near its support. If we draw an imaginary line starting at the external load and cutting right through the beam, and if we do the same thing directly at the face of the support, we have a salami slice of material. Again using our imaginary ability to see stress, if we now examine the slice of material, we will find that the fibers of the material are aligned on a slope, or angle, pointing up toward what was the support side and down toward what was the line of force of the external load.

Zoom in closer and we see that the fibers are stretched along this short diagonal. Stretching means elongation; elongation means tension; sufficient tension results in fracture perpendicular to the axis of maximum elongation, which is at right angles to the lines of tension. These fractures are the result of a phenomenon generally called shear; but upon close examination, we can see that it is equally explicable as a tensile fracture (recall the earlier discussion on Mohr's Circle and transformation theory).

Certainly there is no harm in using the term or the concept of shear, except that it may lead to the conclusion that forces and stresses are of different types—that shear is inherently different from tension or compression. They are all the same; force is force; stress is stress. Only our analytical mythology changes, not reality. If viewing the beam problem as consisting of two components helps to analyze the problem, there is no harm as long we remember that in the end the beam and is constituent materials respond only to primary forces and stresses.

Back at the salami slice, the fibers are elongated along a diagonal from upper left to lower right. The resulting cracks, therefore, run from lower left to upper right, because they will form perpendicular to lines of maximum elongation. If the external load migrates to the right another fraction of an inch, more of the total reaction is carried by the right-hand support. Moment begins to increase because the distance from the left-hand support is increasing to the left of the load, and the magnitude of the action at the right-hand support is increasing even as the distance is decreasing. The paper-cutter action of the load and the left-hand reaction is still there, but the magnitude of the left-hand reaction is decreasing; thus, the shear stresses are diminishing on the left side, but they are now measurable and increasing on the right side. The magnitude of the shearing force is the same all the way from the reaction point to the point of external load, but the value of the shear goes down as the point of loads slides away from the left because more of the load is being carried by the right support.

Despite the semantic distinctions between shear and bending stresses, it is appropriate to recognize that these are the generally accepted terms, and at least for analytical purposes, the primary sets of coincident forces occurring in a member in bending; each set must be accommodated. As mentioned, any deterioration mechanism that jeopardizes material at the extreme fibers, particularly in the zones of maximum bending stress, is a severe threat. Similarly, any deterioration mechanism that jeopardizes the material integrity of the cross-sectional area is also a threat to the member, particularly when that diminishment occurs at locations of high shear stress.

Of the two types of failures, in bending and in shear, the bending failure is much preferred. Bending failures indicate that the extreme fibers are in failure. The bottom fibers are snapping in tension, the top fibers are crushing in compression, or some combination of the two. As the beam loses capacity because of failure of the extreme fibers, it will sag. The sag, or deflection, which may be gross and may occur fairly rapidly, will be accompanied by a significant release of energy stored in the highly stressed fibers as they fail. The consequence of this release of energy is that failures in bending do not take place in silence. Bending failures are loud, dramatic, and frightening. People instinctively run away; and when the failure is, indeed, one of bending, there is often sufficient time for them to do so.

Shear failures are altogether different and much more frightening. The prelude to fracture is barely detectable, whereas bending failures are preceded by gross deformations. Shear failures are through-body, complete and virtually instantaneous. By the time an occupant is aware that there is a problem at all, the member has already completely severed and the building is accelerating toward the center of the earth. For this reason, the design criteria for shear are more conservative than for bending, and appropriately so.

In a manner analogous to the analysis conducted earlier for buckling, we can abstract the factors affecting bending and shear. The response of a member in bending to the bending stresses is a function of the maximum moment (M) present externally, hence internally, in the beam. Maximum moment is a function of the magnitude and locations of the loads, the length of the beam, and the fixity of the ends of the beam. At this point in this text, none of these factors should come as a surprise. The computation of the value of the maximum stress is at times a tedious process of accounting, but it is not normally a conceptually difficult issue. We will, nonetheless, leave the discussion of the combinations of loading, end conditions, and other variables to others.

Given a knowledge of the location and magnitude of M, we can compute the requisite section modulus (S), which will satisfy the condition that the member sustain the moment without the extreme fiber reaching an unacceptable level of stress. This is accomplished by multiplying M by the allowable stress we will accept at the extreme fiber (f_b). The quotient of M/f_b is the value of the minimum section modulus (S_{min}) that will satisfy the condition. The second general condition is that the area of the cross-section (A) be sufficient to accept the maximum value of shear (V) such that the shear stress is less than an acceptable level (v). For reasons having to do with the inherent risks associated with shear failure, we typically multiply the total shear by 50 percent. The two criteria expressed in mathematical terms become:

$$S_{min} \geq M/f_b \text{ and } v \leq 3V/2A$$

As mentioned, the critical factor associated with the former is reduction of extreme fiber area, which reduces S; the critical factor associated with the latter is anything that results in the reduction of A. By inspecting the two formulae, it is also apparent that either situation can be impaired by a negative reduction in the allowable values for extreme fiber stress in bending or the allowable shear stress. Inasmuch as material properties can change over time, these factors are not trivial.

Different materials respond differently to bending problems, and, depending on the material properties, one material may be more vulnerable to shear problems than another; one more vulnerable to bending than another. For example, timber, which is effectively a laminated material, is relatively vulnerable to shear stresses. The fibers of trees tend to align along the long axis of the trunk. The cells form bundles of fairly stiff fibers in that axis, particularly the late wood. When the timber member is in bending, the extreme fibers are aligned with the dominant direction of the principle forces of tension and compression; and timber members perform reasonably well in bending.

In shear, as determined earlier, the direction of the principle stress is diagonal relative to the axis of the member; and, in the case of timber beams, the strong axis of the material fibers. This diagonal force can be analyzed as the vector product of two component forces, the vertical shear and the horizontal shear. It is very important to emphasize that vertical and horizontal shear forces are mathematical abstractions, not physical realities. They *derive* mathematically from the reality of the diagonal force in the member; they are not the point of origin that results in the diagonal force. If this seems adamantly worded, it is to balance the fact that this subject is often described in abstract terms, using obtuse rationalizations, causing students to lose the thread of reality and flounder in the minutiae of the arithmetic.

Accepting for the moment the abstractness of vertical and horizontal shear, the timber fibers are oriented in such a way as to be particularly resistant to forces trying to cut the fibers in two, but weaker against forces trying to slip one fiber along another. In short, the timber displays a different response depending on the orientation and direction of the external force. As a consequence, timber tends to split horizontally at the locations of maximum shear, usually at or near the ends of simply supported beams.

Because this concept is so important and so frequently misunderstood, let us restate what is happening when a timber beam splits horizontally at or near its extreme end. The split or crack is the result of the relative material weakness of timber parallel to the long axis of the fibers, as compared to the transverse resistance to the same force. The external force at issue is the shear force caused by the reaction force offset from the line of action of the load. The result of these two forces is a line of tension operating on the diagonal relative to the plane of the beam itself. The horizontal split or crack occurs not because the force is oriented parallel to the crack plane, but because the material is disproportionately weak in that plane, compared to other planes, which are, in fact, stressed to the same degree as the plane in which the material failed.

Another factor affecting the performance of a spanning member is the cross-section geometry which, of course, directly influences the section modulus and area of the member. It is an irony of engineering that the more efficiently a member is configured, the more vulnerable is that same member to diminishment at the critical aspect of that very same economical design. A steel wide-flange beam, for example, is a marvel of technological sophistication and design. The typical modern wide-flange profile is a marvel of economy, especially in bending. The material is aggregated toward the extremities, so that every pound of material is placed at its maximum advantage, leaving sufficient material in the web of the beam to accommodate shear and crippling in virtually any reasonable bending situation. At lengths where shear or crippling is likely to occur at full bending capacity, the beam is usually not efficient in bending, and, therefore, is unlikely to be selected by the designer (see the following sidebar).

For all of the advantages and efficiency of the design of the modern wide-flange beam, however, its disadvantage is that deterioration mechanisms, which have the effect of degrading critical sections of the beam, can severely diminish the capacity of the beam with nominal reductions in area. In the case of the steel wide-flange beam, material lost from the lower exposed surface of the bottom flange, generally, is far more damaging than a comparable amount of material lost from the web. In the case of a rectangular cross-section, the loss of 25 percent of the material from the extreme bottom portion of the beam will reduce the bending capacity of the beam by approximately 45 percent. Loss of the same amount of material oriented vertically will reduce the bending capacity of the beam by an amount in direct proportion to the loss of total area, that is, 25 percent.

The method of manufacture of a given member may be associated indirectly with material properties and, therefore, with the geometry of a member. A timber beam is typically cut in a rectangular profile primarily because of the method used in the manufacture of timber beams, which is to use vertically cutting blades, efficient slicers of the material. The tool used is effective at producing rectangles. The material properties of timber are not coincidentally consistent with the typically manufactured profile. Timber is relatively competent in tension and incompetent in shear. If we manufacture a timber beam in a profile resembling a wide-flange beam, we are maximizing the tensile advantage of the material, but also exposing the critical vulnerability of the material in shear. Such a profile may have an application where the span is long but the loads are light, such that deflection is the driving determinant. In typical load-bearing applications, shear will dominate timber beam design; and exacerbating the inherent material weakness is inefficient. In other words, we can make wide-flange timber beams, but they will have very limited application. Rectangular timber profiles have the consequence of putting relatively greater amounts of material in the web portion of the beam, thereby, providing a more generally useful beam.

Shear Versus Bending

If a W21 × 62 steel (A36) beam is protected on its top flange by a supported slab above, and if the remainder of the section is exposed to a corrosive environment, what is the critical depth of corrosion before the beam fails, and how will it fail?

Based on allowable stresses for bending and shear, the length at which both allowable bending stress and allowable shear stress are equilibrated is 13.4 feet. This is surprisingly short, which means that this specific section, and probably rolled sections generally intended as beams, are rolled with relatively thick webs precisely so that when area is lost from all exposed surfaces, a beam that is economically sized for bending will not fail in shear even in aggressively corrosive environments.

The maximum allowable shear for a beam is typically $3V/2A$, which for a W21 × 62 means that $V_{allow} = 75.9$ kips given $v_{allow} < 0.4F_y$, and $M_{allow} = S_x F_b = 3{,}048$ inch kips. Because $M/V = L/4$, then, for the allowable values of M and V, $L/4 = 13.4$ feet.

If L is greater than 13.4 feet, then the beam will be driven in its design by the bending capacity of the beam, such that at a more economical length of 30 feet, for example, the actual shear stress at full bending capacity is only 30 percent of the allowable shear stress. If at such a length, 30 feet, cross-section area is lost from all surfaces except the top surface of the upper flange, the beam will degrade approximately 30 percent for each $\frac{1}{16}$ of an inch of corrosion penetration. At that rate, the extreme fiber will reach yield at a corrosion depth of approximately $\frac{3}{32}$ of an inch; but even at that depth, the actual shear stress will still be less than 60 percent of allowable shear stress.

At the point at which the extreme fiber stress reaches yield stress, though a common indicator of incipient failure, it is, in fact, necessary for at least one plastic hinge to form in order for the beam to actually collapse. This means that the beam must continue to degrade to the point the plastic modulus (Z_x) is diminished to the point that $F_y Z_x$ is less than M. In order for this to happen, the corrosion depth must be approximately $\frac{7}{64}$ of an inch deep. Even at the point when a plastic hinge does form and, assuming simple supports, the beam folds in two, the shear stress is still less than 70 percent of the allowable level.

A few other variables affect beam performance and can be disproportionately significant in end-stage building pathology. We will discuss three of them: warping, end conditions, and combined loading.

If the bottom flange of a member is in tension, which it generally is, the length of the bottom fibers are elongated as they are stretched. The path followed by the extreme bottom fibers is not the shortest distance between the two support points. As the beam is deflected, it acts as if it were a large leaf spring, which arguably is exactly what it is. Energy is stored in the deformed, elastic shape of the deflected beam. If the vertical alignment of the beam is disturbed, meaning that the bottom beam is rotated slightly to one side, there is a powerful tendency for the stretched fibers to continue moving laterally and upward. This tendency is a likely response because the stored energy in the deflected curvature is reduced if the bottom fibers can straighten out—which is to say that they resume their original length. They can reduce their length by assuming a horizontal position relative to the supports.

As the bottom of the beam flips sideways and upward, the top surface, which is in compression, is effectively doing the same thing as it rotates onto its side. For a brief instant, the potential energy stored in the leaf spring-cum-beam is released, and the beam reestablishes its original undeformed dimension and shape, albeit on its side. As always, though, glory is fleeting. The load acting under the influence of gravity begins to accelerate with reduced resistance from a horizontal member of substantially lower capacity than when its long axis was vertical. The sideways beam is now carrying the same load as it did in a more efficient orientation, and it is doing so with significantly less success. This phenomenon of twisting or flipping and rotating is called *warping failure*, and it is a pronounced risk with deeper elements.

Warping is defeated by bracing the extreme lower flange or portion of the beam against lateral sway. This is accomplished in long-span trusses and other deep members with sway bars (actually, they are nonsway bars) or ties attached to the bottom flanges, which connect one bottom flange to the same point on adjacent members. It is possible that all the so-connected members will warp simultaneously, so we typically tie the continuous sway bar to the end wall or some other rigid structure.

In the case of common timber floor joists, the bracing of the bottom edges of the timber joists is accomplished with diagonal bridging. Bridging typically consists of short lengths of wood wedged into the gap between two joists such that the end of the bridging extends from the top on one joist diagonally to the bottom edge of an adjacent joist. This is repeated in the opposite hand such that the two pieces of bridging form an X between the two adjacent joists. The bridging acts as a diagonal brace for the bottom edge and quite effectively restricts warping. The bridging pattern is repeated in a line between all of the joists, and the lines are replicated at the third and/or the quarter points of the joists, depending on their depth, length, and load.

When the bridging is continuous in this manner, it serves a second function in addition to restraining the joists from warping. If a load is imparted to one joist, that joist will, naturally, deflect in response. As the joist position descends, the bridging connected to it must also move vertically downward, but because the bridging pieces are connected to parallel joists on either side, any movement is transmitted to the adjacent joists, and they are effectively dragged along with the first joist. As these joists are deflected, they continue to engage more joists through the bridging. The consequence of having bridging between the joists is that loads on any one joist are distributed through the bridging to the entire array of joists. The combination of joists, bridging, and the flooring itself behaves as an integrated plate, rather than as an assemblage of individually operating joists. This is particularly advantageous when heavy point loads or rolling loads are placed on a floor.

Note that this distributional benefit of bridging is not present when the bottoms of the beams or joists are braced against warping by continuous lines of sway bars or strips. These continuous strips or bars may restrain the bottom edges of the joist or beam from moving laterally, but because they are not triangulated, they cannot redistribute loads to adjacent joists, hence to the floor as a whole.

The loss of bridging and other antiwarping items is quite common, but not so much because of a predictable, environmentally derived deterioration mechanism. The primary threat to bridging and sway bars are people, for whom these items are impediments to some other objective. Because bridging is highly replicative, the loss of a single piece of bridging is barely measurable; and because bridging provides a secondary structural function, the loss of bridging altogether is significant if and only if the joist or beam warps out of plane. For these reasons, it is not unusual in the course of rewiring, remodeling, or other alterations for bridging to be damaged or removed.

A more benign loss of bridging, particularly diagonal bridging, occurs when the floor is subjected to heavy rolling or other cyclic loading. As described above, when a specific joist is heavily loaded, the bridging attached to the top of the deflected joist reacts upwardly toward the load, which means they are in compression. They push downward and outward to the bottoms of the adjacent joists. As described earlier, this has the effect of dispersing the force of the point load throughout the floor. At the same time that the bridging attached to the top of the joist is put into compression, the bridging attached to the bottom of that same joist is put into tension. If the connection of the bridging in tension is sound, these elements will transmit movement and redistribute forces as well as the pieces in compression. The reality of bridging connections, however, is that they are toe-nailed and not particularly effective in tension. As a consequence, if the floor is placed into high levels of cyclic tensile loading, the connections at the bottoms of the bridging tend to back out or fail.

Bridging failure is not a particularly serious condition. Because it is quite visible, it often looks far worse than it is, especially if the number of damaged items are limited and the damage is dispersed. It is more serious when there is a concentration of failed items in a relatively small area. In this cases, not only is the loss of the bridging itself more problematic, but such a concentration of lost bridging may also indicate a past traumatic event that fractured the items in a single event. If that happened, there may be more damage than just the loss of bridging, warranting a closer inspection for joist fractures, in particular shear fractures.

In the previous discussion of compression members, we emphasized the importance of end conditions relative to the effective length of a column. There is an analogous importance associated with end conditions of beams. A beam is supported when both ends are free to rotate vertically on the support point. When the ends are fixed, the ends are not free to rotate, and the net effect is that the beam that was previously bending in one direction is now bending in two directions. Looked at in elevation, the deformation of a beam with fixed ends is similar in character to a column in buckling with fixed ends. The curves, while seeming similar in shape, are quite different mathematically.

A beam may be fixed at only one end and simply supported on the other. If it is fixed at one end and not supported at all on the other, it is a special class of beam called a cantilever. The point here is not to analyze all the possible end conditions for beams, but to indicate that end-conditions may vary, and that those variations are quite significant in a determination of beam capacity and performance. From the standpoint of building pathology, end conditions are especially significant because the deterioration of a connection at the end of the beam is often difficult to detect or to correct if it is detected.

Connection or end-condition deterioration can have a profound effect on the performance of a beam. For example, a beam designed with fixed ends under conditions of uniformly distributed load will experience both positive and negative moment. The maximum absolute moment for a beam with fixed ends occurs at the supports, and the value of that moment for a uniformly distributed load is WL/12, where W is the total value of the load (load per foot times length) and L is the total length. For a simply supported beam, the maximum absolute moment occurs at midspan; its value is WL/8. This means that, for the same clear span, a W24 × 62 with fixed ends can carry the same load per foot as a simply supported W24 × 84, so why pay for the added weight when, structurally, they are equivalents?

The design-based reason is that fixed ends are not free; they are, in fact, both monetarily and structurally more expensive than simple supports. The design implications of fixed ends transmit through the structure and particularly the supports, thereby increasing the demands in other parts of the structure which must absorb the moments generated by the fixed ends.

The pathological reason is in some ways more compelling, if not as popular. Fixed-end conditions often are barely discernible from simple supports once they are fabricated and installed. If the item providing the fixity is a knee brace (see the previous section on bracing and fixed connections), then the detail providing the fixity is relatively apparent; but if the material providing the fixity is a welded joint between a beam and a column, the identity of the fixed condition is scarcely distinguishable from the simple connection. If the fixed connection deteriorates, but is misidentified as a simple connection, the consequence is that the connection may release, whereupon the beam will convert to a simply supported condition. Even if the loading is sufficiently below the design load, a connection failure is not necessarily catastrophic. The risk is, however, that the W24 × 62 which was competent to perform with fixed ends is abruptly called upon to perform the task of a W24 × 84. The results can hardly be flattering.

The fixed/simple connection interchange can work the other direction as well. What was designed and fabricated as a simple connection can become frozen or fixed. This is typically the result of corrosion of metal members and connections, but is more a problem in bearings than in hinged-beam-to-column connections or similar conditions. (The problem associated with bearings will be addressed later in the discussion on combined loading in beams.) Another instance of pinned joints becoming frozen occurs in trusses (which will be discussed in a subsequent section dedicated to trusses and truss-related problems).

When a single spanning element is supported at more than two points, it is a continuous beam. At a minimum, at the interior support points, the beam is in negative bending. The ends of such a beam may also be fixed, but it is the portion of the beam running continuously over one of the interior supports that is the object of our attention. A continuous beam may extend over three, four, or more supports, limited only by its overall length as that relates to the feasibility of fabrication and installation. The negative bending at interior support has much the same effect as a fixed end for single span beam. The difference is primarily a function of how the rotation of the beam at the support is resisted.

If the beam is a single span with a fixed end, the negative moment at that fixed end is resisted by the vertical support. The moment is resisted by the fixed connection, which, in turn, is resisted by the column or wall. In the case of a continuous beam, the negative moment at the interior support is resisted by an equal and opposite moment of the same beam on the other side of the interior support. In other words, the beam provides its own stiffness and resistance to bending, This is rather intuitive and only gets confusing the more we try to explain it in engineering terms. Continuous beams may also be fixed to the vertical support, which means that the moment is distributed both into the adjacent bay of the beam *and* into the vertical support.

When a continuous beam is loaded in alternate bays or spans, the span that is loaded will deflect, resulting in negative moment at the support. The continuity of the beam means that the moment at the interior support is resisted by the adjacent span; however, there is no counterbalancing load in that span. The adjacent span, or backspan, as it is often called, will rotate upward, restrained only by the support at the far end of the backspan, which ties the far end down to restrain it from flipping up. Absent any downward load, the backspan may be in negative bending over its entire length. It follows then that, at the support point, the beam will pass over the support at an angle or slope that is not going to be absolutely horizontal.

If the beam is too long to be made of a single piece, but the designer wants or needs the moment to be transferred across a connection, he or she can design a moment splice, which, as the name implies, is a connection in the plane of the member through which moment is transferred without forming a hinge. A moment splice in a steel beam is identifiable by a plate on the top surface, which spans the break in the beam and effectively replicates the cross-sectional area of the flange. The splice plate must then be attached to both sides of the splice with welds or bolts sufficient to resist the total force of the top flange in tension. A similar detail is feasible in timber connectors, but the length of engagement or development length on either side of the splice is considerable, given the relative weakness of connectors in timber. In addition to the tension splice plate, the webs of the two pieces must be spliced in order to transmit shear through the splice material and onto the adjacent member. The bottom flanges must also be engaged; however, because they are in compression, this relies as much on alignment as on a physical connection.

The reason for going into this much detail about continuous beams and moment splices is that, once installed, they are not always apparent. When they deteriorate, usually through corrosion, the deteriorating items are often embedded or hidden from view. This is particularly true of steel connections at steel columns. It is altogether possible to detail a simple seated connection and a fully fixed, continuous connection using a seat such that they are barely distinguishable once installed. The critical and distinguishing elements of a moment connection are the discontinuity in the member itself, the tension splice, and the shear splice, all of which can be obscured. The first indication of a major problem will be the otherwise inexplicable deflection of the primary spanning member (see following section dedicated to connections and connection-related problems).

In the case of a continuous beam, the top flange of the beam is in tension as it crosses over the interior support. It is also not unusual for the top surface of the beam to be obscured or even embedded in the slab or floor construction above. The deterioration of the top flange is a serious matter even when it is in compression. When the top flange is obscured *and* in tension, the potential seriousness is compounded. The deterioration of timber-beam top surfaces were analyzed in a number of examples in the section on column deterioration. Note in particular case G, as it may occur in the top of a beam. The structural member in that instance is reduced by 48 percent. In the case of rolled-steel sections, the entire top flange may be embedded in concrete, obscured, and deteriorating.

In a manner analogous to the combined loading of columns, beams can be subjected to axial loads, especially compression, in addition to bending. Simply supported beams are typically restrained from lateral movement and allowed to slide on the other. If it is allowed to rotate but not slide, the connection is called a *pinned connection*. If it is allowed to slip,

Composite Construction

In the last 20 to 25 years, we have seen the increased popularity of composite construction of steel wide-flange shapes, welded studs, and concrete. The concrete slab may be formed such that the top flange is embedded in the slab or is flush against the slab, with the studs welded to the top of the beam. Frequently, corrugated material deck is used to form the slab and to weld the studs through the deck onto the beam in the field. The studs grip the concrete and provide shear transfer to the concrete, so that the steel and concrete act as an integrated structural unit. The structural advantages of composite construction are considerable. Steel depths can be reduced significantly without compromising stiffness; spans can be increased without increasing beam depth or weight.

Most such construction occurs under interior floors, but rarely, at roof decks due to the likely presence of lightweight concrete. As a floor construction, the hostility of the environment is significantly mitigated by the protective envelope of the building as a whole. As the years progress and the buildings with composite construction are abandoned, and where there is a prolonged roof failure, the exposure of the studs within the concrete to a corrosive environment will become more of a problem. Probably the loss of a single stud will not devastate a beam, but the local loss of a cluster of studs may trigger an unzipping of the composite boundary and the consequent conversion of the composite beam to a simply supported beam in a fraction of a second.

it is called a *free* or *sliding connection*. The reason for this combination is that it allows the beam to adjust in length as the result of thermal change or, possibly, hygroscopic changes or shifts in foundation conditions. If the sliding bearing corrodes or fouls with debris, it may freeze. The restricted travel in the bearing will arrest movement, and if the restraint is in the extension of the beam, it will be compressed. The combined stress is not essentially different from the combined loading of a column, but the external conditions are different, as are the corresponding manifestations of the combined loading.

The combined loading of a column normally accelerates buckling, in large part because the compressive force is unrestrained. Consequently, as the column buckles, the load travels with it all the way to complete collapse. In a beam in combined loading, the compressive force is normally strain-limited; therefore, as long as the strain of the compressive stress is absorbed either by elastic reduction in length, lateral buckling, increased deflection, or a combination of deformations, which add up to the same reduction in length, the compressive force is mitigated. In other words, whereas combined loading is a severe threat to columns, it is a somewhat more limited threat to beams. It is precisely for this reason, along with warping buckling, that the bottom flanges of beams are braced. Typically, the top flange is already braced by virtue of the manner in which the loads are imposed, but where the beam is not laterally braced, combined loading can precipitate buckling in a manner as catastrophic as can occur in columns under similar circumstances.

For all the concern about collapse and failure of beams, deflection is very often the determining design factor. This is particularly true when the span is long or the moment of inertia is low or the modulus of elasticity of the material comprising the beam is low. Timber beams are often controlled by deflection if shear is not already driving the design. Steel members over 30 feet, and almost all aluminum spanning members, are deflection-driven. Concrete beams are often deflection problems long after the strength issue is resolved.

The typical criterion for acceptable deflections of beams is expressed in terms of some fraction of the length of the span, such as L/360. This means that for a span of 30 feet, with a deflection (δ) of L/360, $\delta = 1.0$ inch. Some state building codes effectively legislate such criteria. Massachusetts, for example, is quite articulate about various applications of deflection criteria, ranging from L/120 in some circumstances to L/240 and L/360 in others.

The origins of deflection criteria are traceable to the nineteenth century, specifically with regard to plastered ceilings. Trial and error determined that if live load deflections were limited to L/360, the plaster on the ceiling below did not crack. We now have the tools to verify such a conclusion based on mathematics and analysis of the material properties of plaster; indeed, L/360 is a valid criterion for plaster ceilings. That, however, is about as far as the science goes; the rest is largely habit supported by very little mathematical rationale.

Though there may be a relative absence of science, there is an abundance of logic. Spandrel beams at the exterior façade, which also carry masonry, need to be quite stiff or the masonry will crack. Roof beams may be somewhat more flexible because the anticipated deflecting load, snow, is intermittent and temporary. The point is, the deflection criteria typically associated with beams have little to do with the performance of the beam and more to do with the tolerance of the deflection by attached and adjacent materials. L/360 has no particular connection with beam design and everything to do with ceiling design.

Regardless of the origins of the criteria, a sudden and unanticipated increase in deflection can shatter a precious ceiling, cause doors to jamb, and fracture plate-glass windows. The fact that the beam may still be in place and has not physically separated from the structure or folded in two does not mean that it has not failed. Recall that the functional definition of the structural system is to minimize deformation, not to avoid collapse.

Fortunately for those of us in the diagnostics business, the factors that affect beam section modulus also adversely affect moment of inertia, which simplifies at least that part of the analysis. Section modulus is determined by dividing the moment of inertia by the distance from the neutral axis to farthest extreme fiber. When the cross-section is asymmetrical, there may be two different values for the distance from the neutral axis to an extreme fiber. When the section is in bending, the fiber that is farthest away will have the higher level of stress, and that fiber, therefore, will reach the allowable stress of yield stress or fracture or any other designated level of stress before any other fiber in the section. The use of the section modulus as a design device tends to protect that extreme fiber to the virtual exclusion of other design considerations.

As diagnosticians and pathologists, we are not required to critique the design side of the house, but we are required to understand how the designer operates. The better we understand how a member was designed, the better equipped we are to predict performance and vulnerability. In the case of bending and spanning members, generally, the design process tends to guard the profile from the development of a localized stress approaching the yield stress or some other designated stress level. Having established the "to-be-avoided" boundary, the designer then backs off a sufficient distance from that level and assigns that value as the working level (see previous discussion of factors of safety). Nothing is wrong with this approach; it is rational, replicable, and fairly easy to teach and practice. As pathologists, we may take some comfort in this design approach for members in bending because it is, indeed, rather conservative and, generally, leaves a considerable level of reserve capacity in the member.

As a rule, when members are driven in their design by deflection considerations, probably the strength consideration has already been satisfied, but for the usual reasons listed above, deflection remains excessive. The primary device for decreasing deflection is to increase moment of inertia. By definition, any increases in moment of inertia result in increases in section modulus, hence in the bending capacity of the member relative to strength. The more that deflection drives the design, the greater the reserve capacity of the beam relative to strength. The pathologic conclusion is that gross deflection alone is not an indicator of failure except to the extent the member was intended to resist deflection. By no means does this imply that a grossly deflected beam is *not* a problem and can be ignored. It points out that a grossly deflected beam is not, by definition, a problem and requires further investigation.

A few clues help to distinguish between gross deflection and failure. As mentioned above, there are circumstantial considerations: long span, flexible material, low moment of inertia relative to strength. There are also specific indicators. If a beam is grossly deflected *and* in a state of failure, there must be fibers identifiably in a state of fracture. The two locations of greatest probability for simply supported beams are the extreme bottom fibers at midspan and the fibers at midheight near the support points. If there is no perceptible fiber damage, the next step is to investigate and analyze the beam based on an operating assumption of deflection-driven initial design.

The approaches to intervention in beam related pathology are arrayed in Table 3.9 in a manner similar to that given for compression members. The variables that influence bending capacity are arranged in the column to the left. The variables are very similar to those for compression member intervention, which should, at this point, not come as a surprise. Material properties and material geometry consistently drive structural deterioration mechanisms. The basic principles are relatively few and fairly simple. It is the permutations and combinations of the applications that are tedious and confusing. In plain language, building pathology diagnostics is the application of simple principles to explain the complexities of specific cases.

We reemphasize that the matrix is intended to stimulate creative thinking and options development; it should not be regarded as a definitive source of answers to all bending problems. Also, it is important to remember that material issues need not necessarily be addressed with material-based solutions. Options development benefits by constant reference to the objectives of the intervention.

A number of the options here refer to a general consideration of reduction of external load; this is viable if and only if the load is physically removable. Even if sufficient load can be removed to relieve the immediate prospect of failure, the question immediately arises whether such a

Table 3.9 INTERVENTION MATRIX: SPANNING MEMBER

Intervention Approaches

Necessary and Sufficient Conditions	Abstention	Mitigation	Reconstitution	Substitution	Circumvention	Acceleration
General Mechanism	Accept the condition of the element without intervention.					Demolish member or building in its entirety.
1 Diminishment of area		Address source of loss of area as a materials-level problem.	Replace lost area or member as a whole, in kind.	Replace lost area or member as a whole with alternative material.	Shore or duplicate member.	
2 Reduction of section modulus		Ditto	Ditto	Ditto	Add material to member.	
3 Lateral instability due to warping buckling		Reduce external loads/moments	Repair any damaged bridging or sway rods.		Install bridging or sway rods.	
4 Alteration of effective length due to end conditions			Repair connections to original condition and operation.	Replace connections with alternate type, but same operation.	Redesign type and/or material; install added or alternative connection types.	
5 Application of compressive load or external moment		Reduce external loads/moments.			Reinforce member for a level of accommodation of added eccentricity.	
6 Increase in external load		Arrest or reduce load; redistribute loads.	Remove loading; return to original loading.		Duplicate member; shore or reinforce member.	

reduction is permanently or only temporarily acceptable. Even if load-permanent restrictions are tolerable, can they be enforced over time? In general, restricting load is a valuable tool for gaining time during which we can conduct an investigation; this option is less valuable for long-term mitigation. Posting of occupant limits is, nonetheless, an important and necessary device, but it is difficult to police. If a structure is in a state of sufficient peril as to impel an owner to restrict loads, then excluding occupancy until the investigation is complete is a more prudent approach.

Another recurring theme among the intervention options is the general category of reinforcement of the member. As a category, reinforcement includes two general subcategories, depending on the intervention objectives and the aspect of the beam being reinforced. The first subcategory deals with reinforcement of the member itself; the other, with load relief.

Reinforcement of the member itself can be organized around the member properties of material or geometry. Material reinforcement requires an actual alteration of the material, such as epoxy injection (substitution) or the addition of material to some surface or another (circumvention). The addition of material is a particularly appealing tactic for dealing with beam-related problems simply because we can apply the additional material in a wide variety of ways with considerable effect and at relatively low cost.

If we want to increase bending capacity, we can access and address the bottom flange at midspan. If the member is timber, for example, we can add a steel plate to the bottom; this can significantly increase bending and deflection performance. Or we can add plates to the sides to increase shear capacity. In fact, a vertical steel plate has been and, occasionally, still is sandwiched between two pieces of timber to form a flitch beam. Steel is used to reinforce concrete. Concrete decks and slabs are bonded with studs to steel wide-flanges. The practice of using one material to reinforce another is both common and economical. Furthermore, while the arithmetic may be more tedious, the process of calculating and designing such composite members is conceptually no more complicated than the calculations described in the following sidebar for a steel plate added to a timber beam.

When two materials are bonded compositely, and the moduli of elasticity are different, which generally they are, there is necessarily a transfer of force from one to the other. The stress along the boundary plane is often the source of considerable problems. With laminated materials, the problem may arise from yet a third material, such as an adhesive that is inferior to the two primary materials. With bolts or lag screws that attach a steel plate to a timber beam, the problem may simply be that bolt size and spacing are inadequate for the shears involved.

Shoring beams is the second general method of reinforcement, in which no material is added to or otherwise affixed to the beam itself. The

Calculating Composite Members

The addition of a steel plate of even nominal dimension can radically alter the performance of a timber beam. For example, a timber beam finished to a true 4 inch wide by 10 inch deep has the following properties:

$$A_w = 40 \text{ in.}^2 \qquad E_w = 1 \times 10^6 \text{ psi}$$
$$I_w = 333.3 \text{ in.}^4 \qquad S_w = 66.7 \text{ in.}^3$$

Add a steel plate, 4 inches × ½ inch, with $E_s = 29 \times 10^6$ psi, to the bottom surface of the timber beam. Compute the revised values for moment of inertia and section modulus. This type of problem is quite common. When a member is fabricated of more than one material, which act integrally, it is classified as a composite member. Given the disparities in material properties, the question is always how much work each material is contributing to the combined effort. The trick to solving the entire class of multiple material problems is based on what is called *compatible deformations*. When two materials are bonded and undergo the same deformation, their respective resistance to being deformed is a function of their areas and their moduli of elasticity (AE). Another way of describing the concept of compatible deformations is that composite materials perform in proportion to their respective stiffnesses.

In this particular formulation of the basic problem, the solution is most easily derived by transforming the properties of one element into those of the other. The transformed area is superimposed over the centroid of the original, and the geometric properties of the transformed section are computed. If the steel is "converted" to timber, the transformed area (A_t) of the new material is equivalent in stiffness to the 2 in.² (4 × 0.5) of steel. This is determined by multiplying the actual area by the modulus of elasticity of the steel, divided by the modulus of elasticity of the timber: $A_t = A_s \times E_s/E_w = 58$ in.² The next step is to determine the neutral axis of the transformed section. Taking moments from the top surface where the values of y are the dimensions from the top surface to the respective neutral axes:

$$y_t = A_w y_w + A_t y_s / [A_w + A_t] = 8.11 \text{ in.}$$

this means that the neutral axis of the transformed section is 8.11 inches down from the top surface. From this determination we can compute the revised moment of inertia (I_t) where the values of h are the respective distances from the neutral axes of the segments from the neutral axis of the transformed section:

$$I_t = I_w + I_s + A_w h_w^2 + A_t h_t^2 = 985.8 \text{ in}^4 = 2.96 \, I_w$$

(continues)

Calculating Composite Members (*continued*)

The increase in moment of inertia is almost triple that of the original moment of inertia. The increase in section modulus is not quite as dramatic, but substantial nonetheless:

$$S_t = I_t \, / \, c = 121.6 \text{ in}^3 = 1.82 \, S_w$$

Attention must be paid to the capacity of the attachment detail of the steel plate to the timber beam to ensure the composite behavior, but for the cost of approximately $5 per foot the increase in bending capacity for strength is nearly doubled and the resistance to deflection is tripled.

shores are normally perpendicular to the beam, which, for most beams, means that the shores are vertical. The term shoring is not rigorously defined but refers generally to a class of items that brace other members and are normally in compression. Shoring is the term used to describe members that brace the restraints holding back the earth in an excavation; it is also used to refer to bracing beams, typically from below. In their capacity as bracing for beams, shoring comprises little more than added support points installed after the fact. It is also fair to say that the term implies that the members are temporary, although there are certainly many examples of temporary shores in place for prolonged periods of time.

As added support points, the contribution of shoring to beam reinforcement is directly related to where along the beam length the shoring is located. If it is near the original support point, then it will relieve the support point itself, which is often the very end of the beam. The force in the shore will or can approximate the force in the support itself; and shoring is an effective device for moving the support point slightly inboard of the original support. If the end of the beam is corroded or rotted, this slight shift may have the effect of engaging sound beam material.

When the shoring is closer to midspan of the beam, its function is one of two types: load relief or beam protection. The difference is actually only a matter of how much force is induced into the shore when it is installed. Load relief means that at the time of installation, the shore is raised, typically by screw mechanisms or with driven wedges, such that it not only engages the bottom surface of the beam but actually lifts the beam. Even if the lifting dimension is a minor fraction of an inch, given the rather stiff nature of materials, such a minor movement may represent the transfer of many tons of load from the beam into the shoring. The positive benefit of load relief is that, instead of being carried by a damaged or weakened beam, the load passes directly through the beam and into the shoring. The negative consequence may be that as a result of the beam now being a continuous member supported on more than two points, the beam immediately above the shoring point is now in negative bending. Although very few symmetrical sections will be damaged by such a reversal, it can damage finished surfaces above which may have conformed to the original deflected shape. Asymmetrical shapes such as concrete and steel wide-flange composite floors may be rather seriously compromised, hence will require analysis prior to the installation of shoring after the concrete has hardened. It is also possible in some situations to rotate the ends of pocketed beams back into the now-packed pockets so that thrust in thrown laterally into the support walls. This will, at a minimum, compress the beam, which compounds the stress state of the beam or blows out the wall at the ends of the beams or joists.

Shores installed as items of deflection control differ from load relief items primarily in the degree to which they are tightened below the beam. Whereas load relief items are tightened to the point that they actually lift and, therefore, transfer load, deflection control items are placed and tightened only until they are snug to the underside of the existing condition. If, subsequent to installation, the beam, for whatever reason, deflects or subsides from its elevation at the time of shoring, the beam will compress the shoring and effect a transfer of load into the shoring. The idea behind deflection control is to provide assurance and protection to the beam without necessarily incurring the potential negative consequences of load reversal in the beam. If there is no impetus for additional deflection, there is no load on the shoring.

The reasons a beam may continue to deflect—or, in lay terms, to sag—vary from the blatantly obvious to the deceptively subtle. The obvious cause of increased deflection is an increase in the external load. If this is the concern leading to a decision to install shoring, then shoring is a proper and logical response. If the reason for increased deflection is the continuing degradation of the beam itself, shoring may arrest deflection, but with the prospect that the added deterioration will be disguised. If, for example, shoring is added at midspan to reinforce a weakness in bending, and load is, indeed, imparted to the shoring, the shoring becomes a support point, and the beam is continuous over the shoring. The beam is now in negative bending over the new support, and, more importantly, there is a major point of shear in the beam at a location where previously there was little or no internal shear stress. The risk is that what began as a bending problem, is converted to a shear problem with all the attendant undesirable attributes of shear failure.

If the increased deflection is a prospective anticipation of creep in the beam, then the shoring may be installed to arrest deflection, and it well may provide such deflection protection. But because no good deed goes unpunished, as the shoring actually absorbs what would otherwise have resulted in added deflection of the beam, there is an imposition of force into the shoring. Though rather obvious, this truism is often dismissed or discounted. If a beam deflects, there is a reason: added force or reduced resistance. If that movement is arrested or reduced from whatever it might have been otherwise, it can only be because of the application of an external reactive force. In this case, that reactive force is from the shoring. The point is, regardless of the motive behind the installation of the shoring, if the result is that movement in the beam is averted or arrested, the shoring must be in a state of stress, and that state of stress is, typically, compression.

From this discussion it may already be apparent that the shoring installation sequence may also spell the difference between load relief and deflection control. Deflection control can be accomplished in one of two ways. In the first, the beam is lifted with jacks (which, of course, potentially reverses the bending), the shore is inserted to fit the snugly, and the jacks are released. The beam will necessarily impart load to the shoring. The second approach is to place a shore—which itself has a built-in jacking mechanism below the desired point—and raise the beam in order to impart load. Either approach is valid structurally, and choosing between them is very much a matter of cost and facility. If the shore is intended to be temporary, then the cost of the lifting device is recoverable; but the shore in permanent built-in jacking mechanisms can be costly. One advantage of built-in mechanisms is that they can be adjusted at a later time with no appreciable added cost.

If the shoring is intended as deflection control only, the same basic options are available, except that the existing beam cannot be lifted either prior to or coincidentally with the installation of the shoring. Although it may be obvious here, it is persistently missed in the field, with the result that shoring is either too loose or too tight for its intended function.

Trusses

Trusses are beams with a good public relations agent. When truss design is taught to architects and occasionally to engineers as well, it is often with an air of sublime reverence: trusses are pure; trusses are refined. The fact is, there are many trusses in this world, but very few of them conform to the theoretic truss, with fully articulated pinned joints, and whose members change size and shape as the forces change. These trusses are rarely subject to lateral loads, warping, or unbraced lengths.

In real life, truss joints are bolted or welded; most of the pieces are the same size; and they are constantly being racked, twisted, and buckled. The few ideal trusses, those with articulated, pinned connections and stress-differentiated member sizes are, in the main, over a hundred years old, and the pins have frozen in the connections due to corrosion. Modern trusses for modest spans of under 30 feet are likely to be preengineered items called, simply, steel joists or bar joists. These trusses are characterized by continuous top and bottom chords composed of paired steel angles that sandwich a steel rod bent into a zig-zag and placed between the paired angles on the bottom and at the top. The bent bar or rod performs the function of the web members.

A similar item is available in timber. The preengineered timber joist is part of standard use for residential-scale roof construction. Placed at close centers of 24 to 36 inches, these items can economically span 30 feet or more. They are fabricated from stock 2 inch material and connected with gang-nail plates, which are made of sheet metal. The sheet metal is pierced in a pattern of sharp V shapes. The tongue of the V is folded perpendicularly to the plate, to form teeth. The teeth are the nails that can connect two or more pieces of wood together in the same way as a gusset plate.

Deeper and heavier preengineered steel trusses suitable for spans of up to 60 feet differ from bar joists primarily in the web items. In the long span joist, the web is likely to consist of angle iron welded at each end to a gusset plate, which is still joined to and sandwiched between paired angles that comprise the top and bottom chords. In both the bar joist and the long span joist, the zig-zag pattern of the web members is typically found in diagonals only, not in vertical members. The economy and prevalence of preengineered trusses is so great that for modest spans, individually designed trusses are rare.

For spans or loads beyond the range of preengineered trusses, custom-designed trusses still conform to the manufacturing dictates of economical fabrication. Paired angles as primary chords, diagonal angles as web members, welded connections, and gussets sandwiched between the chord members are all standard practice even for large, customized trusses.

The essential difference between the theoretical truss of the classroom and the practical truss of the industry is at the joint. According to truss theory, the joints are pinned, which means that any member connected to that joint is free to rotate about the joint. Such rotations can and do occur, but not because the members begin spinning around the joint or any such behavior. When the truss is loaded, as a rigid plastic body, it will deform; that is, the joint location will move relative to its original location. In fact, all the joints will move, however slightly. All the members will change ever so slightly in length as well, some longer and some shorter. The changes in joint locations and member lengths indicates that the original geometry is also changed, which in turn means that the angle of intersection of the members relative to the joint also is changed. A change in these angles of intersection have the same effect as the same members rotating about the joint, some in one direction, some in the other.

If the members are restrained from making these very small rotational adjustments, they can only accommodate the relocation of the joint if, in addition to elongating or shortening, they also bend slightly to adjust to an altered angle of intersection at the joint. This bending is contrary to truss theory; and if the magnitude of the bending is great enough, it can jeopardize the integrity of the truss as an effective spanning member.

When mathematical techniques were developed to enable the analysis of pure trusses, no concomitant techniques were available for calculating the magnitude of these bending stresses if the joints were frozen. Rather than risk the disadvantages of fixed joints, truss designers in the mid-nineteenth century tended to detail truss joints in a somewhat pure interpretation of the pinned joint. In the 1840s and 1850s, many of the nation's preeminent structural engineers worked for railroads. Wendel Bolman worked for the B&O Railroad as did other engineers with trusses named for them, and they designed railroads and railroad structures to include trusses. They designed them in an openly competitive way, naming their new designs after themselves. In fact, their names are known to us today primarily because of this self-aggrandizing behavior.

Bolman left the B&O to set up a truss manufacturing company in Baltimore, Maryland. His company manufactured preengineered trusses, a few of which survive. One is still in service after more than 110 years in Williamsport, Maryland. That bridge has pinned connections, with as many as seven individual members joined. Unfortunately, the pins have corroded to such a degree that the members are locked onto the pins and, therefore, must make any small angular adjustments by bending. The bends occur close to the end of the respective members as they engage the pin. As the load dismounts the bridge, the load is released, and the bending reverses. For the most part, the stresses generated by this bending near the joint remain in the elastic range; however, in few instances, the stresses in the member exceed the elastic limit. The loading and unloading of the members over such a wide stress range, often in cold weather, means that they are prime candidates for fatigue failure.

For most trusses, this type of localized bending induced by rigid joints is not a problem because the member was designed not only with the capacity for the axial load but for the added secondary stress of the local bending as well. The added material is inexpensive compared to the costs of detailing and fabricating pinned joints, which may lock once they are put into service. Secondary bending stresses are acceptable and tolerable in truss design, not because they conform to accepted truss analysis, but because their elimination is diseconomical. Whether the truss is timber, steel, or a combination of the two, it is probable that the chords are continuous, not segmented at each node or joint, and that the web members are rigidly connected to those chords.

Taken to its extreme, meaning an extreme extension of the notion of the rigid truss joint, we can rely on the rigid joint and the bending capacity of the members to absorb deformation without resorting to diagonal web members. Such a truss looks like a conventional solid web beam with exceptionally large holes cut into the web. If the holes are cut to a smaller diameter, we are back to the proposition that a truss is a beam with holes cut into it.

As trusses age, the array of age-related maladies are much more a function of the materials involved than the fact that the materials are arrayed in the form of a truss. As mentioned above, metals, whether or not they are connectors of members, corrode; timbers rot. As the materials deteriorate, the truss is threatened similar to the material loss discussed in previous sections. On a structural level, independent of specific material problems, trusses are vulnerable to the same issues as beams, generally. As such, trusses are especially vulnerable to lateral stability issues, owing to their relative depth to internal stiffness.

The objective of a truss is to achieve depth, thereby increasing the efficiency of the flange material without incurring proportional weight in the web. The result is that trusses are thoroughbreds; they are fleet and lean, but relatively fragile relative to sudden changes in direction or bumps. Trusses do not absorb impact well, nor do they fare well changes in support conditions.

A classic failure mechanism for a truss is load reversal. Whereas a wide-flange beam is theoretically as capable in one direction as another, a truss is designed to be loaded in a specific manner, generally vertically and downward. If, for whatever reason, the load is reversed, as in a wind uplift condition, they may become unbalanced, and collapse utterly. If the individual compression members buckle, as happened at the Hartford Coliseum, the redistribution may trigger a rapid succession of collateral failures, resulting in catastrophic collapse.

In older trusses, it was not unusual for the designated compression members to be effectively wedged into position, relying on the force of the load to hold them against their respective bearing points. If the trusses deflect, the geometry of the truss might change, such that the distance along the compression element might increase, and the bearing points pull away from the ends of the compression element. The element becomes loose and potentially disengages and slips out of position. The source of the deflection might not be load-induced; it may be the result of creep or long-term plastic deformation or slippage at the connections. At any rate, the loss of compression in the compression member, absent a positive connection to the bearing point, may result in a dislodgment of the compression member.

Approaches to intervention in truss problems either may be related to material deterioration mechanisms or globally related, and are often a combination of both. A case in Baltimore will serve to illustrate the point. The trusses, 25 full-length items, 100 feet long, were constructed in 1858 and designed by the aforementioned engineers of the B&O Railroad. The trusses span two adjoining sheds, which were designed as railroad car repair facilities; the bottom chords are approximately 40 feet above track level. The configuration of the trusses is that each truss is symmetrical and gabled, meaning that the top chords are inclined and join the bottom chord at the exterior walls. Each truss is divided into six panels; the diagonals are in compression—they slope from the upper joints down and away from the centerline of the truss; and the vertical elements, which are rods, are in tension. All members other than the vertical tension elements are timber.

What distinguishes these trusses other than their length and age, is the design of the fittings—that is, the seats at the ends of the members where they are joined—the splices or couplings at the bottom chords, and the elaborate boots into which the top chords and the bottom chords join, and join the walls. The cast iron boots are attached to the members by two bolts, which pass first through the vertical wall of the fitting, next through the outer lower chord member, then through the paired, inclined top chord members, next through the far side bottom chord, and, finally, through the far side vertical wall of the fitting or boot. The vertical walls are connected to each other by a horizontal bottom to the boot (the fitting is like a shoebox into which the chords are nested and through which the bolts pass). The top chords are in compression, and are, therefore, pushing into the boot; they are not particularly problematic. But the bottom chord, which is in tension and trying to pull out of the boot, is much more difficult to restrain. The two bolts acting alone were perceived by the designers, and correctly so, as inadequate for the task. Therefore, the designers added fins, or cleats, on the inside faces of the vertical walls of the boot, and similar lugs upstanding from the bottom, which fit into slots cut into the bottom chord. When the bottom chord pulled away from the boot, the shoulders of the slots pressed against the cleats, which were integral to the boot. (The bolts were more for erection purposes than actual restraints).

In theory, the entire assembly works quite well. The configuration is complete; the major members are slightly undersized, specifically the extreme ends of the top chords, as well as the bottom chords generally, but not dangerously; the diagonals and the vertical elements are reasonably sized; but the connections at the ends were and are completely inadequate. As brilliant as these engineers were, they did not, nor could not, reasonably evaluate the shear stresses at the end connection, because, at that time, the concept was scarcely understood academically, much less at the applications level. As the cleats of cast iron boot pressed against the shoulders of the slots cut into the bottom, the effect was to shear the captured wood off, such that the bottom chord pulled out of the boot, leaving behind the shear blocks of wood still in contact with the cast iron cleats. As the bottom chord slipped out of the boot, the erection bolts moved with it until the bolts came into contact with the edge of the hole in the vertical wall of the cast iron boot, whereupon the bolt was put into bearing against the boot wall in the direction of travel of the bottom chord, and against the back surfaces of the hole through the bottom chord, in the opposite direction. Once the cleat device failed, the only items resisting the pullout of the bottom chord were the two erection bolts.

When the problem was discovered, more than half of the trusses' end connections had sheared the cleated connections and were totally reliant upon the erection bolts. In those cases, the bolts themselves were highly deformed into U-shaped rods, such that the heads and nuts were rotated relative to their original orientation by 15 or more degrees. The slippage of the bottom chords relative to the boot connectors meant, among other things, that the length of an individual bottom chord had increased by the amount of the slippage; and the sloping top chords thrust against the boot and moved the tops of the walls outboard by a dimension equal to the slippage of the bottom chords. Consequently, the slope of the top chords decreased, and the relative distances of the top chords to the bottom chords decreased as the trusses assumed a geometry compatible with the elongation of bottom chords. The vertical members intended to be in tension were in fact placed in compression; and because the vertical rods pass through the bottom chords in anticipation of pulling on the joint, the rods simply punched out the bottom of the connection and were disengaged from the bottom surface of the bottom chords in several locations. The distance between connector locations for the diagonal members increased, and several of them slipped out of their respective upper seats, coming to rest against the vertical rods.

Intervention was required; the trusses were in a precarious state of incipient, catastrophic collapse. For two winters, the staff of the B&O Railroad Museum operated under standing protocol that, in the event of a forecast of snow, much less actual snowfall, they were to turn the heat to full capacity to warm the uninsulated roof to minimize accumulation. On more than one occasion, the protocol was invoked and worked, but probably only because there were no major snowstorms for those two years.

The general approach to intervention was to reinforce the system, to leave as much original material as practical in place, and to accept the distorted configuration as found. The cast iron boot was circumvented by applying gussets to the near and far sides of the intersection of the sloping top chord and the bottom chord. Those gussets were then connected by a third plate, which spanned below the bottom chord and connected to the two gussets. Then a pair of wire rope cables spanning the length of the truss and connected by turnbuckles at each end of each cable to the underslung plates were spaced vertically by blocks at two of the interior panel points. The cables were tightened only to "snug point," absent any appreciable slack.

Additional timber was sistered onto the top chords; the end connections provided additional capacity in bearing, as well as resisting the disengagement of the bottom chord. The deformed condition was maintained; therefore, no untoward load reversals occurred and the trusses conformed to code.

Cantilevers

Cantilevers comprise a special class of beams. They are characterized by the fact that they have an extension in bending, unsupported at the far end. They can be developed in one of two general manners: The fixity at the reaction point can be accomplished by locking the connection at that point, or by continuing the same member over the reaction point and creating a backspan to the cantilever.

The fixed connection approach used, for example, with poles buried into the ground that form a vertical cantilever fixed at its end by embedment into the soil. Backspan applications are implemented in such common situations as the extension of rafter ends to support eaves. What distinguishes cantilevers from other bending members is the relative magnitude of the deflection at the unsupported end, and the requisite fixity at the reaction point.

If a common beam is supported by fixed-end connections, the loss of that fixity means only that the beam will convert to a simply supported condition at that end. This does not necessarily mean the beam will collapse. To be sure, collapse is a possibility if fixity is lost, but it is not a physical requirement. But with a cantilever, the loss of fixity at the primary reaction point does mean the cantilever will fail. In the case a locked joint, the joint will fail followed closely by the collapse of the cantilever. In the case of a continuous member with a backspan, the loss of fixity necessarily means that the member is no longer continuous; and while the backspan may remain in place, the cantilever will fail.

Deterioration mechanisms peculiar to cantilever diminishment attack the fibers of the material at the fixed end, specifically those that diminish the capacity of the fibers in tension. In the stereotypical cantilever, the fibers in maximum tension are those near the fixed end and on the top surface, inasmuch as the cantilever is typically in negative bending relative to similarly situated simply supported beams.

The primary reason for devoting a section to cantilevers has more to do with their deflection characteristics than their actual deterioration. The deflection of a cantilever is as much as 16 times that of a simply supported beam of comparable length, beam properties, and loading. This relative flexibility implies, among other things, that changes in the properties or capacity of a cantilever are manifest in significant changes in the deflection of the extreme end of the cantilever. Changes in the deflection of the cantilever include those in the backspan, where there is a backspan. For example, the reduction of load to a backspan can have an effect on an associated cantilever of increasing the deflection.

Increases in deflection in and of themselves do not damage or reflect damage to a member, although they do signify change. As building pathologists, we are always attuned to changes in the homeostasis of any building condition. Changes in deflection indicate that there is an unaccounted-for change somewhere in the system; and until the source of the change is attributed, the change cannot be dismissed as trivial. It is common, however, for changes in cantilever deflection to result in an alteration to the cantilever support condition itself. Many curtain walls, for example are carried off the main frame of the steel structure by short cantilevers. If the properties of the cantilevers at one level change, the resulting increase in deflection will bring the curtain wall to bear on the panel below. This has the effect of converting the panel below into a compression element, and converts the cantilever into a propped beam. Because of the disproportionate deflection associated with a cantilever, it is relatively easy for such a redistribution to occur when cantilevers are stacked, as illustrated in the example of curtain wall panels.

When cantilevers are constructed of concrete or any other material for which creep can be an issue, there is the distinct prospect that the cantilever will continue to deflect over time. For structures such as cantilevered retaining walls, the absolute deflection is not normally a problem; however, if the structure is restrained by, for example, return walls or abutments, stress will develop at the intersection with the restraint. The deterioration mechanisms associated with cantilevers are identical to those of any other beam. Any loss of section modulus, change of moment of inertia, reduction of area, increase in load, or, in some cases, merely the location of the load, have amplified significance relative to cantilevers because of the consequential acceleration of deflection.

Arches and Catenaries

People are fascinated with arches, partly because they are graceful and seemingly effortless, partly because they are nature's own structural form, and partly because of the mystery of building a spanning element out of small pieces. People also generally believe that arches are very stable and durable, which stems from images of Roman arcaded viaducts surviving after 2,000 years of low maintenance.

An arch may be an interesting structure, but it is not particularly mysterious. We begin our study of the arch with a simple gable, two inclined members leaning against one another to form a triangle. As load is added to the two inclined members, they distribute some of that force axially and some of it in bending to reaction points. The portion distributed axially is absorbed at the reaction point both by a vertical and a horizontal component. The horizontal component is the thrust that, if the sliding force is not restrained, will cause the lower end of the inclined member to slide outward, and the gable to collapse. The gable is a successful structural form as long as the thrust is restrained, and as long as the bending capacity of the inclined members is not exceeded.

Imagine now that we increase the number of elements from two inclined members to four in the form of a gambrel (a mansard in the United Kingdom). Each of two symmetrically inclined members supports two other inclined members, which are at a shallower pitch. From the gambrel, we can begin to see both the strengths and the vulnerabilities of the segmented arch.

First, there is still the issue of thrust, as was the case with the gable. The lower inclined members of the gambrel intersect the base at a steeper angle; therefore, the vector of the force is such that the horizontal component is reduced, but there is still some thrust at the base. A far more difficult problem exists at the joint of the upper and lower inclined members. The two upper members act as a gable, and apply thrust against the upper ends of the lower members, forcing the respective joints outward laterally. When this occurs with the gable attached directly to a base, there is nothing to exert any lateral restraint. The lower elements rotate outward; the upper elements collapse vertically, followed closely by the lower members having rotated outward enough to accommodate the inversion of the upper gable, resulting in reverse rotation and collapse inward. This geometric instability is called a *mechanism*, and a gambrel is the prototypical mechanism. In the case of symmetrical mechanisms, there is at least theoretical stability; but the problem with gambrels and other symmetrical mechanisms is that, if there is the least imbalance or idiosyncrasy in loading, the mechanism is destabilized, and collapses.

Some of the methods available to stabilize a mechanism are familiar from the previous discussion of beams and trusses, namely triangulation and fixity. The joints of the gambrel could be locked, at which point the segmented span would be fused into a single crippled beam. The individual segments could be triangulated with additional members, which would have the effect of converting the gambrel into a truss. Both methods will work as a general approach, and both are, in fact, used to stabilize gambrels. However, to begin to understand segmented arches we must recognize that neither of these methods of stabilization advances the argument regarding arch stability because neither is applicable to the typical segmented arch.

The segmented arch is composed of a large number of individual pieces, which are not fused and for which internal triangulation is not applicable. Certainly, fixity and triangulation are valid, but their implementation is self-defeating if the objective is to preserve the open character of an arch and the segmentation of the pieces. If the joints between segments cannot be fixed, and connecting segments by triangulation is not permitted, then the approach to stabilization must be outside the perimeter of the arch.

Return to the gambrel for the moment and the specific sequence of mechanism formation: If the reaction points are fixed against sliding, in every sequence of collapse it is first necessary for at least one joint to move *outside* or *above* the original perimeter of the gambrel. If all the joints are precluded from ever moving outside or above the original perimeter, the mechanism cannot develop, and the gambrel remains stable. In other words, although the geometry of the gambrel, hence the segmented arch, is unstable, it is a geometric requirement that at least one joint move effectively outward in order for the collapse mechanism to develop. If every single joint is somehow or other precluded from that initial movement outward, the collapse sequence cannot develop, and the gambrel and the arch remain aloft.

Suppose that a great deal of heavy material were piled on top of the gambrel. The stresses in the individual segments would obviously increase. The load of the overburden would push the individual blocks downward; and at least in the one block that is top dead center, this downward force would tend to shear the block from its in location in the arch. To push past the adjacent blocks on either side, one or both of those blocks must move

FIGURE 3.10A This 1906 commercial building was constructed using unreinforced, "flat arch" floors flush framing into encased steel girders and beams.

FIGURE 3.10B Looking up at the remaining edge of the collapsed "flat arch" floor structure along the intersection with the wall. Note that the wood floor is still spanning the collapsed area.

laterally; and because they are slightly angled relative to the blocks adjacent to them, any lateral move must push one of the blocks slightly *upward*, which is precisely the opposite direction of the force against both the displaced block and the blocks adjacent to it. As long as the overburden is great enough and evenly distributed over the arch, any upward movement is precluded, and all the blocks remain in position.

The sum of all the forces in a given arch is such that there is lateral thrust; and this lateral thrust must be resisted, either within the geometry of the arch itself or from the exterior. The arch can either be tied or braced. Tying the arch means that a tension member is placed horizontally across the span of the arch. For aesthetic reasons, such ties are often placed at the spring points of the arch, although the structurally most advantageous point for tying the arch is slightly above the spring point. That said, the structural inefficiency is not great if it is placed at the spring point, and almost every culture that has adopted the arched form has also adopted a tie location below the structurally optimal elevation.

By bracing the arch, the thrust is resisted by a rigid element or form beyond the lateral boundaries of the arch. Typically, such counterforce is provided by an adjacent arch in a series of arches or an arcade. Obviously, this sort of "passing off" the problem has to reach a termination in the last arch of the series, and that end arch must be braced or tied to preclude the destabilization of the entire set of arches. Roman engineers rather cleverly used intervening hills as the external braces for their long arcaded viaducts. The arcades dissipate into the land, creating the illusion that the arcades are unbraced, when in fact the mass of the rigid hills are essential elements to the stability of the overall form.

This short course in the structure of segmented arches is the foundation for our discussion of arch pathology. We begin by listing all the structural requirements touched on in the preceding discussion:

- The arch must be weighted by the overburden.
- The overburden must be evenly distributed.
- The thrust of the arch must be balanced by an equal and opposite force.
- The segments must be shaped to resist shearing through or out of the arch.
- The loss or diminishment of any one or more of these requirements can and, if sufficiently diminished, will result in the collapse of the arch.

If the overburden is removed by excavation or erosion, the arch can rupture, normally above the spring points, approximately one-third to half of the vertical dimension from the spring point to the top dead center. If the load is uneven or substantially reduced eccentrically, the arch will rupture at a similar location, modified to some degree by the exact location of the eccentricity or surcharge loss.

If the lateral restraint is lost, the arch will spread at its base. If such an opening of the arch is symmetrical about the center of the arch, the joints on either side of the keystone will open at the bottom of the joint. This will change the angular relationship of the shoulders of that stone to the two adjacent blocks in such a way that the keystone will drop; and if the spread is great enough, the keystone can even drop out of the arch, at which point the adjacent blocks collapse toward each other, and the entire arch usually collapses. This sort of dislocation can occur at points or blocks other than at the keystone, although the top center of the arch is the most vulnerable to this sort of movement.

If the blocks are held in place in part by mortar between the blocks, and the mortar disintegrates for any number of reasons, such as moving water, the upper block will move to close the gap between adjacent blocks. This changes the total circumferential distance of the arch, thereby reducing the radius of the arch. As the blocks compress against one another, they also migrate toward the geometric center of the revised and smaller form. It is, after all, possible to dry-set the blocks without mortar and to achieve a stable arch, so the loss of mortar between blocks could mean only that the blocks convert from an arch, the circumference of which is the sum of the blocks and the mortar joints, to a smaller arch with the circumference of the sum of the blocks alone without the mortar joints. The theory, though intriguing, is flawed. Even if there were an arch with such very well-behaved mortar, the reduction of the radius means that the diameter is proportionately reduced, which also means that the distance between the base support points is reduced. Even if the arch shrinks from the loss of its joints, the spacing of the supports will not be so accommodating. The arch is not nearly so likely to slip off its supports, as the spring points will be held in their location by the impertinent supports, resulting in effectively the same net consequence as if the arch had spread. The difference is that, when the arch spreads, the distance between supports increases and the arch fails to accommodate the increase. In the latter case, the arch decreases its dimension, and the supports fail to accommodate the change. In both cases, the manifestation of the alteration is that the blocks rotate relative to one another as the arch attempts to flatten out. If the magnitude of the flattening is great enough, one or more blocks will dislodge, with the others following close behind.

The apparent stability of segmented arches can be attributed more to association than structure. When arches or arcades are made of relatively inert materials, such as masonry, particularly dry-set masonry, and the climate is relatively dry, such as in Spain, the external lateral thrust is absorbed by rigid land forms such as valley walls, and the intermediated arches remain unbroken, the form can and will endure for thousands of years. In other words, the Roman viaducts in Spain, while quite splendid, are not examples of the general rule of arched construction, even for the Romans.

The segmented arches we are more likely to encounter are far less glorious than those of the great Roman viaducts. They tend to be round headed windows and doors spanning 3 feet, not 40. The lesson is, however, that scale is rarely a determinant in arch pathology. Because all the forces are scaled to the span, the prospects of thrust imbalance or load asymmetry remain the same.

In modest openings such as windows in solid walls, above every flat-headed window formed by a lintel, there is an implicit arch or round-headed window. In other words, deciding whether to insert a round-head or a flat-head over the window is a structural toss-up. The structural equality of the two forms cannot extend to any scale, but at the scale of a residential window opening, the differences are insignificant. One of the consequences of this structural equivalency becomes apparent when a lintel deflects or sags in an opening. When the lintel deflects, the material immediately above the lintel follows what is supporting it, namely the lintel. In an extreme case, maybe the lintel rots out altogether and collapses into the opening. Interestingly, the amount of material that will follow the lintel is that which is circumscribed by an arch over the same opening. In other words, the wall will lose material until the forces within the wall restabilize in another form, and that form will resemble an arch.

Such losses look more triangular than curved because the construction of the wall in rigid blocks will corbel to form the structural arch even though the lost material looks more like a wedge than an arch. We can take some comfort in this phenomenon when a hole is punched into a wall. Instinct tells us that the wall will collapse into the hole; in fact, the wall will arch over the hole and remain erect. If the requirements of overburden, distribution, thrust balance, and shear distribution are met, the opening will self-arch.

The way an arch works is simple and elegant. At the top dead center of the arch is a keystone. For that one stone to move down, it must displace laterally the two stones on either side. If those two stones are restrained from moving laterally, the keystone cannot move vertically. It's that simple. If the stones on the lower portion of the arch cannot move laterally, the stones above them cannot move vertically.

The ways arches fail structurally are all traceable to the imbalance in the weight of the elements trying to move straight down under the influence of gravity, their having to wedge lower elements laterally to do so, and the restraint of those lower elements from moving laterally. There are two general approaches to the overall lateral restraint or thrust of arches: tie the arch together or push it together.

The tying approach is often accomplished with a cable or rod spanning the opening of the arch. The rod will be in tension inasmuch as the respective bases of the arch will try to move apart from the thrust. The ideal vertical location for the tie is approximately one-third of the distance up from the bases or spring points. That elevation is approximately the resolution point of all the lateral thrust generated by the forces acting downward. Designers have seldom been pleased with the appearance of a horizontal rod at the third point of the arch, so they locate it in line with the spring points because it looks better. Unfortunately, this puts the restraint of the thrust out of line with the cumulative outward thrust. As a consequence of these noncoincident forces, rotation occurs at the bases: the arch bulges out and upward at the third points; the upper elements move down, and arch collapses. To keep the third points from blowing out requires enough counterweight, particularly at the third points, to restrain the movement resulting from the tie at spring points. If the restraining force is from external compression, the same concern applies. There must be enough counterweight to counterbalance the tendency to blow out the third points.

The geometric form of most arches is semicircular, and this is part of the reason that many arches are no more stable than they are. The ideal form of an arch can be demonstrated simply by hanging a cable. The form of the resulting curve is called a *catenary*. It is a fourth-order smooth rational curve often confused with the curve formed when a beam deflects under load. While both curves are fourth-order equations, they are not the same; nor should either of these shapes be confused with parabolas, which are second-order curves. The catenary is much closer to the ideal curve for an arch than a semicircle. Although designers rarely have seen the catenary as an aesthetically preferable arch form, there is no shortage of them as tension forms. The curve formed by the cables of every suspension bridge is a catenary. Antoni Gaudi, the famous designer of the cathedral in Barcelona, applied this simple form of the catenary as an arch form. He was able to arrive at the shapes by hanging weights from wires and inverting those shapes into compression forms.

Connections

Throughout this chapter we have referred to connections and conditions affecting connections. This section addresses the issue of connections directly. Without explicitly restating the point in the previous discussions of connection-related problems, we can form a typology of connections based on function. From the typology, we can deduce the necessary and sufficient conditions for the success or failure of the connection using the criteria established by the function.

For our purposes here, we will use the terms connection and joint interchangeably. The basic issue of joinery is the same: two separate pieces are brought into proximity to each other with the expectation that they will remain in the same relative positions. In addition to relative proximity, joints may be required to transfer stress or to release stress; the joint may join similar or disparate materials; the connection may invoke still other pieces or not. Clearly, not only is there a fairly broad category of joinery details, but at least an equal number of joinery problems.

In the broadest terms, there are three categories of connection types: stress transference connections, stress release connections, and connections of accommodation. The first two are self-descriptive; the third is enigmatic. Connections of accommodation are connections that are the result of one of three common conditions. The first condition results from a requirement that the material at issue change direction, and that such a geometric modification cannot be accomplished with a single piece. An example of this type of accommodation is a picture frame. The second type of accommodation is the result of dimensional limitations, which are the consequence of how the material is manufactured. An example of this type of joint is lap seam between two rolls of sheet goods. The third type is simply the contiguous location of two disparate materials, where nothing more is required than that the pieces maintain their relative locations. All connections of accommodation are required only to maintain the pieces in proximity to each other.

Stress transference and release connections also must maintain proximity, but they also must either fix the two pieces structurally or release the two pieces to avoid selected stresses from being transferred. In fact, most release connections will release the materials in one or two axes, but fix the pieces in the remaining axis or axes. Even though connections of accommodation may not have fixity as a requirement, they also often develop some degree of fixity simply as a result of the construction technique used to affect the joinery.

Fixed Connections

We begin our detailed examination with fixed connections. A fully fixed connection means that the connection is capable of transferring all the stress developed in one member through the connection interface and into the second member. Such connections are of two general types. Fully fixed connections may be accomplished by bridging the discontinuity between the two members by elements that affect the transfer of stress through the added elements and into the second material. The second general method is to fuse the two base pieces into a single element, either by literally fusing the materials or by inserting another material between the two base pieces, thereby acting as a continuous bond across the face of the discontinuity.

One of the more common examples of the fully continuous connection is a full-penetration weld between two steel members. Welds may be inadequately designed—that is, the weld is too small for the task—but, otherwise, welds are as strong as the material joined by the weld because the weld and the joined pieces are, in fact, a single piece. The weld metal and the two pieces are melted, not glued together. There is no boundary line between the pieces or between the pieces and the weld metal; they are completely indistinguishable.

It is possible to achieve nearly the same results without using any weld metal, but simply by fusing the two pieces with a torch. The reason fusing pieces is not done more is that it is very difficult to fuse the pieces and maintain the precise thicknesses of the materials being welded. The molten portions of the pieces tend to flow away from the joint site, which leaves the joint partially cavitated—that is to say there is an indentation in the material at the site of the weld. By itself, a cavitated joint is not necessarily a failure, in that the joint may have considerable structural capacity. The reduced cross-sectional dimension may mean, however, that the locally effective section modulus and moment of inertia are less than the base material. Thus, the bending capacity may be reduced, and there may be a geometric stress raiser in the throat of the cavitation. In short, fusing is an effective means of joining two pieces of material, but the resulting joint may not be structurally equal to the base material even though the material properties involved are the same. Welding accomplishes the fusing process at the same time that it adds enough material to prevent cavitation, and avoids the production of points-of-stress concentration.

The concept of structural continuity across the physical discontinuity of the joint is eliminated by welding or fusing, in that the plane of discontinuity is obliterated or, more accurately, blended to form a single, homogeneous element. The fixity of the connection is self-evident, inasmuch as there is solid material where once there were distinctly separate pieces. The joint is as strong as the base material because the joint material is one and the same as the base material, and with the same or greater dimensional characteristics.

Although metals are frequently used in welding and fusing, they are by no means the only materials that can be used for these processes. Bituminous-based products are often thermally reversible, hence fusible. Some themoset plastics are fusible and, therefore, weldable. Concrete is typically viewed as a fused joint or monolithic material, although the continuity of concrete joints is often more a matter of appearance than fact.

Once the concept of joint fusion becomes apparent, it is only a small step to understand the adhered joint. The adhesive is a material sandwiched between the two base pieces; it bonds to each of the base pieces, hence one to the other. With both bonded and adhered connections there are several sources of pathological deterioration. The adhesive or bonding agent is itself a material, and, therefore, is subject to some form of deterioration mechanism, because, as stated repeatedly, all materials are subject to some form of deterioration mechanism. Water-soluble glues fail in the presence of moisture; polystyrene can crack when cooled. The bond between the adhesive and the base material may fail in the manner described for discontinuities of composite materials.

One form of adhered joint is a *brazed connection*. Brazing is the bonding of two pieces of metal using a second metal, namely an alloy of copper and zinc as the adhesive. Solder is a form of adhesive for other base metal combinations, and comes in two varieties: hard solder which is an alloy of silver and brass, and the far more common soft solder, which is an alloy of lead and tin.

Soldered joints are excellent examples of the typical problems that can beset the general class of adhesives. The bond between the solder and the base metal is dependent on many variables present at the time of production that affect the long-term performance of the joint. If the surfaces of the base materials are contaminated, the bond will be weakened in whole or in part. Even a small local flaw provides the ground from which cracks may progress. The sources of such surface defects can be the coating on the base material or a layer of corrosion product on the base material. Over time, the base material will thermally expand and contract, thereby stressing the bond between the base and solder metal. As the bond fractures, other deterioration mechanisms may ensue in the interstitial space of the crack, such as chemical or galvanic corrosion of either the base metal or the solder metal or both. (The consequence of the corrosion of the fracture face will be discussed in greater detail in the subsequent section addressing corrosion.) It is sufficient at this point to say that the accumulation of corrosion product on the exposed corrosion surface causes the fracture to progress. This general process of bond fracture and subsequent fracture-surface deterioration is a common mechanism, and can afflict any adhered connection.

Solder can also crack through the body of the solder joint itself. The motive force is typically provided by the thermal contraction of the base material. At the same time that the contraction of the base metal is providing the eternal stress, the brittleness of the solder metal increases. Although a flawless solder can and usually does withstand the stress resulting from conditions such as these, a flawless solder joint or anything else flawless is rare. The initial flaw in a solder joint can be an entrapped air bubble or, more often, a trapped piece of flux. (Flux is a generic term for materials added to solder and weld metal to reduce the melting point and to increase the wetness of the fluid metal so that it coats the base metal and runs into fine cracks and discontinuities.) The flux in supposed to vaporize, or at least float, to the top of the molten puddle; but instead, small fragments are caught in the body of the solder and are frozen inside the joint. As we are by now well aware, when stressed, the initial flaw produces a stress concentration, which effectively amplifies the average stress and from which a crack will propagate.

The general adhesive issues are either related to bonding or to the properties of the adhesive itself. The propensity of solder to crack, especially in colder conditions is the peculiar vulnerability of solder. Another bonding material may have a different vulnerability, but every adhesive has a failure mode. For some, the failure will occur at the bonding plane; others will fail through the body of the adhesive; still others will "fail" through the base material.

In the last clause, above, we enclosed failure in quotation marks because failure through the base material is not a failure of the adhesive nor, arguably, a failure at all. Technology has in some instances, notably wood joinery, produced adhesives that are stronger and last longer than the base materials. In such cases, if a fracture occurs, or when such a fracture occurs, the adhesive stays with the base material, and it is the base material that fractures. This may, at first glance, seem to be a good thing, to have adhesive so strong as to outperform the base material, but there a couple caveats worth noting.

First, for reasons associated with long-term maintenance and repair, it is often desirable that the joining material—in this case, the adhesive—be the sacrificial material. because if there is a failure, it may be easier and cheaper to replace the adhesive than the base material. This is certainly the case with stone or brick masonry, where the adhesive, the mortar, is deliberately retarded in its properties specifically so it will yield to movement in the brick or stone. When the mortar weathers sufficiently, we repoint the wall; and while repointing is not cheap, it is cheaper than replacing the brick or stone.

Second, the relative strength of the adhesive may be a bit illusory: it may be the adhesive that adversely affected the properties of the adjacent base material. The surface of the base material that is infused with the adhesive may effectively be consolidated, whereas the next micrometer of base material may be detrimentally altered by the adhesive, or, more likely, the activators or vehicles in the adhesive. Such considerations are applicable for consolidants as well as adhesives. The activity for which this issue is particularly relevant is the joining of thermoset plastics. Certain hydrocarbon adhesives actually dissolve the surface of the base material as it fuses the pieces together. The resulting joint may be stronger than the base material, but that may in part be due to the possibility that the adhesive or the vehicles in the adhesive degraded the substrate.

The implication of this discussion is that the two pieces of base material are the same. In fact, continuous joints are not limited in this manner.

True, welding, fusing, brazing, and soldering are generally restricted to specific materials, and the two base pieces are usually of the same material. The task of bringing a single base material to the point of heat to enable fusing or adhering with a second metal is difficult enough. The welding of aluminum, for example, is so delicate an operation that it is only feasible with a special welding machine called a heliarch. The heliarch machine is a tube within a tube. The inner tube is a flexible pipe through which passes an aluminum wire—the weld metal. Surrounding the weld metal tube is a second tube through which argon gas, which is virtually inert chemically, is pumped. (Originally, the inert gas was helium, hence the name heliarch, but argon has long since replaced helium.) The argon gas is blown over the point of weldment as a high-voltage electrical current bridges the gap between the protruding aluminum wire and the base pieces. Heat from the arc fuses the two base pieces and the weldment wire into a homogeneous solid unit. If not for the argon, the entire ensemble would burst into flames, as the aluminum, reacting with atmospheric oxygen, would burn, producing aluminum oxide gas. This rather insidious consequence is the nightmare of aircraft pilots, and the reason aluminum is welded primarily in the shop and not in the field.

Despite the difficulties of fusing or adhering just two pieces of metal together, more generally, at least theoretically, it is possible to adhere any two pieces of base material together using a third material as the adhesive. The problem of satisfactory bonding properties is, however, now multiplied. If all we wanted to do was to adhere one piece of linoleum to another, for example, there are numerous adhesives that could do the job quite nicely; but once we decide that linoleum works better when adhered to wood flooring, the list of suitable adhesives dwindles rapidly to a paltry few, particularly if we throw into the specification that this same adhesive must resist water without deteriorating. Finding anything that will bond with materials such as glass or plastic laminate is difficult enough, but to then bond with metal window frames or wood countertops is tedious and difficult to accomplish economically and durably.

Determining which adhesives to use in a given situation is a constant challenge for any designer, but more so in the current environment of technological discovery. Few fields are changing more rapidly—or hold greater promise—than the general area of adhesives. The first unfortunate consequence of this rapid development is that almost as soon as a combination is learned and deemed reliable, it is replaced and no longer available, which leads to the second unfortunate consequence. In periods of rapid technological advance, the designer is particularly vulnerable to and reliant upon manufacturers' representatives for reliable information and data.

The subject of emerging technologies, discussed in detail in the last chapter of this book, is no small matter in a field that sees the entry of thousands of new products or variations on existing products every year. There is no reasonable way anyone can have experiential knowledge about every product or procedure. Therefore, one of the primary objectives of this book is to consistently apply and refer to basic principles of behavior and logic to arrive at assessments and predictions, and to accept nothing on faith alone. If an adhesive is promised to last for 50 years, we want to see the tests. We ask logically, how can a product be said to last 50 years that has only been in existence for five months? And assuming if it can last 50 years, can it be replaced at that time? If it fails prematurely, can it be removed? If it is warranteed, will there be anyone to back the warranty for the intervening 50 years? Remember this: your client deserves nothing less from you than to demand answers and proof. Your reputation and practice are on the line every time you accept an unsupported assertion.

Having addressed continuity by homogeneous methods, and continuity by intermediating agents or adhesives, we now examine how we can achieve structural continuity using discreet secondary elements or pieces without fusing or bonding any of the pieces together. In a fully fixed connection, each axis in which stress can or will occur will be made continuous across the plane of discontinuity. There are numerous devices for this process, but they accomplish the objective of stress transfer in a more limited number of ways. The connector itself is in a state of rotational stress, lateral stress, or axial stress. The connectors in a state of rotational stress are torqued; those in lateral stress are in a state of shear or bending; those in axial stress are either in tension or compression. The analysis yields to following the forces in the primary element, slowly tracking those forces across the discontinuity through the connectors, and resolving them in the second basic piece. No forces can be left unaccounted at any time during the traverse.

If the connection is a moment splice between two steel beams, the forces at the end of one beam include the bending moment and the vertical shear. Each set of forces is addressed separately. If the member is a wide-flange beam, the majority of the bending stresses are absorbed in the flanges. The bending component across the joint, therefore, is implemented by joining a piece of steel to one side of the joint, spanning the joint, and attaching the same piece to the opposite member. The splice plate on the top flange is typically in tension, and must be sized to the match the area of the top flange.

The purpose of the attachment detail may be either to weld the splice plate to the first side or to bolt the plate to the first side. Because the splice plate is not in the same plane as the top flange, the weld or the bolts will themselves be in single-sided shear, and must be sized according to their shear capacity for a force equal to the tensile force in the beam flange. This means that the actual cross-sectional area of the connectors may be greater than the area of the flange or the splice plate. To develop this much attachment area, the splice plate may have to run along the length of the top flange for a seemingly long distance. Numerous bolts may be necessary, but that is the price of moment splices.

The bottom flange must also transfer compression across the plane of discontinuity. If the bottom flanges can be brought into contact and alignment, added material across the discontinuity is redundant. The only reason to provide a compression plate is to ensure transfer where contact is less than certain, where a modest amount of fabrication tolerance is possible, or where subsequent misalignment is possible. If the compression splice plate is expected to transfer the same force as the tension plate, then the sizes and connectors will be comparable.

Once we have accounted for the bending forces, we can turn our attention to the shear forces, which are concentrated in the web of the steel beam. The same basic technique of using a spanner plate is typically used for shear connectors. Here, too, the plate can be bolted or welded to the first piece, and is usually bolted to the second piece because the final connection will occur in the field where bolted connections are easier to effect than welds. Field welding is certainly possible, and there are occasions when it is preferred, but there is usually a hefty cost premium associated with the process. With the bending splice, the plates were axially stressed and the connectors were in shear. In the case of the web connectors, the plates are in a state of shear and must be sized accordingly.

If the connection is steel to steel, but is a column instead of a beam, the same basic process is applicable. The forces for which there must be accountability are, primarily, compression and, secondarily, buckling. As mentioned, technically, compression requires little more than alignment of the upper and lower shafts, and contact between the bearing surfaces. In fact, plates are added to the flanges and web of the column primarily to assure alignment. If the splice location is an acceptable place for a hinged connection, little more is required. If, however, the splice point must be continuous, the splice will take on the characteristics of a moment splice; the plates will approximate the flange sizes, and the connectors must resist shear that is equivalent to the moment exerted by a lateral buckling load.

If the joined members are pieces of timber, and the connectors are spikes or nails, the same principles apply: there will be a spanning plate, sometimes of wood, sometimes of sheet metal, and the spikes will be driven into and through the timber members. The peculiarities of driven versus threaded fasteners comprise an important distinction. Driven fasteners rely on much the same principles as driven, side-friction piles. The fastener pierces the fabric; and in the course of doing so, displaces the fabric later-ally. The displaced material, in its effort to rebound to its original position, clamps the fastener and resists pullout because of the friction developed along the sides of the shank of the fastener. The pullout force is not gener-ally an issue unless the fastener is placed in tension, but indirectly it does affect the capacity of a fastener even when the shank is in shear.

The resistance to pullout is a function of the following factors: surface area of the fastener, coefficient of friction between the fastener surface and the base material, and clamping force of the base material. If the shank is larger in diameter, or the length of the shank is longer, or both, the surface area increases, which increases pullout capacity. If the surface of the shank is ribbed, pleated, or otherwise roughened, the coefficient of friction in-creases. If the modulus of elasticity of the base material is higher, the clamping force per unit of displacement is greater. The designer of the con-nection must constantly juggle the options presented by these variables, plus, obviously, the total number of fasteners in an array, to satisfy the stress requirements of the connection.

As building pathologists we must work backward from the design to the terminal condition. In most such connections, the pathology is related to material-level deterioration mechanisms. The long-term success of the connection is decidedly more related to the deterioration of the connector than to the connection as a whole. If the latter were the case, the connec-tion probably would have failed upon or prior to entry into service. There are obvious and catastrophic exceptions to that statement. The failure of the Regency Hyatt elevated walk is a prime example of a flawed detail that survived initial entry into service only to fail under load. Most connection failures, however, are related more to the peculiarities of the fasteners than to the structural capacity of the connection as a whole. Bolts corrode; nails do, too, unless the wood around them rots first.

Two other connectors or connections require additional commentary here: the special connector is the rivet, and the special connection is the friction connection. Though structural rivets are no longer being installed, many are still in the inventory. Furthermore, future building pathologists will have to restore some of the largest steel structures in history, and they are riveted together; to name just a few: the Empire State Building, the Chrysler Building, and the Golden Gate Bridge. So even though they are no longer part of the designer's toolkit, rivets will continue to be very much a part of the building pathologist's kit of problems.

Rivets were developed at a time when threaded fasteners were well un-derstood and relatively inexpensive. The decision to use rivets instead of bolts was largely theoretical, not practical. The theory had two prongs: one, a presumed advantage of the rivet; the other, a disadvantage of the bolt. Both have since proven to be broadly inconsequential; nevertheless, they pose compelling arguments. The presumed advantage of the rivet was that it could clamp the base pieces together more tightly than the bolts available at the time, so clamping was an advantage. The disadvantage of the bolt was that, to develop the full strength of the bolt in tension, it had to be torqued to such a high degree as to risk shearing the shank in torsion before developing the same shank in tension. As with many theoretical positions, there is some truth to both propositions.

Rivets can develop substantial shank tension, accomplished by employing—to our advantage—the property that most materials (certainly including metals such as mild steel) shrink as they cool. If a mushroom-shaped rivet were heated to a cherry red, inserted into a hole through two pieces of base material, and its stem end hammered into a matching cap, the cooling shank would contract, thereby effectively clamping the two pieces tightly together. Indeed, this is exactly how rivets behave. Not only do the shanks contract, but they can contract so much that the stress developed within the shank exceeds the ultimate strength of the steel; hence the shank fractures. This is why rivets must undergo 100 percent testing after installation. That may sound onerous, but the test is cheap and crude: it entails whacking the button of the rivet with a sledge; if the button pops off, the rivet was no good. This is a reasonably accurate test and, when combined with the redundancy built into riveted structures, results in reliable buildings.

If it survives initiation, the rivet squeezes the two base pieces together with such a force that even if ultimate strength is not exceeded, the plastic limit of the steel probably is. If the stress in the shank of the rivet does not exceed the plastic limit, the sledge hammer test is supposed to expose the malefactors by dislodging any "loose" rivets, which will rattle in the hole when whacked by the sledge. This is probably possible only if the rivet is so under-hammered initially that it did not develop the shank in tension at all, in which case, the clamping action of the rivet is so great that the faces of the two joined pieces will be unable to slide, relative to one another, because of the friction developed along the joint plane. This is partly true; rivets do develop significant clamping force; the shanks are developed into the plastic range. One downside is that rivets do not completely fill the holes into which they are inserted. It stands to reason that if they shrink longitudinally, they also shrink in diameter as well. The shank is not, therefore, bearing on the shoulder of the hole. When a shearing force is applied laterally to the rivet, the thinly forged seam between the button and the base metal cracks. The base piece slides relative to the shank, so that it is now bearing against one wall of the hole; but there is now a path for moisture accumulation under the lip of the button and along the wall of the shank. This also means that the shank that is already in a state of plastic deformation longitudinally is now also being sheared by the primary forces of the base elements working through the connection.

As we will describe in detail in the chapter on corrosion, this is an ideal corrosion cell. Moreover, it occurs in conjunction with an element that is in an extreme state of stress. The significance of stress relative to corrosion has not been widely reported, but there is evidence that high levels of stress will accelerate corrosion. If true, then the shank of a rivet is a prime location for corrosion. Based on personal observations of bridges in particular, the evidence supports the hypothesis. On various occasions when inspecting aged rivets, we used the same postinstallation techniques to test 75- to 100-year-old rivets. We found that a substantial number of rivet buttons were held in place by the accumulated paint; the shanks had corroded enough to snap the shank, usually right at the intersection with the button.

As long as the rivet is working according to the theory, however, there is considerable clamping action and a semblance of a friction connection. True friction connections probably were only partially achieved with rivet technology. The rivets clamped the pieces together, but the grade of steel available at that time for the production of the rivets made it unlikely that the pieces were clamped firmly enough to develop a friction connection worthy of the name. The development of high-strength alloys for bolts, conforming to ASTM standard A325 and better, also occurred at the same time that it was determined that torquing of the bolt did not, in fact, affect the tensile capacity of the bolt significantly. These two developments gave designers the luxury of creating a true friction connection; but it is still possible to fail to achieve the desired results due to improper contact area preparation.

Friction connections are difficult to achieve reliably in other materials; and, generally, even attempting to achieve such a connection in any other material than steel is not advisable. Stated in these terms, open disagreement is unlikely. Even vague familiarity with connections is probably ample to dissuade a practitioner from attempting to develop a friction connection in timber. That said, however, it is a common enough practice to over-torque connectors, though certainly not by all trades or professions. Auto mechanics, for example, amateur and professional, would rarely if ever consider torquing a cylinder head down until the bolts reached refusal any more than a cabinetmaker would drive a screw to refusal. The former knows that the shanks may snap, and the latter knows the threads may strip. We submit that over-torquing and overdriving connectors to be one of the leading causes of premature connection failure, but unfortunately, there is no theory or text that can prevent humans from believing that tighter is better. It is not.

One additional type of connector deserves mention here, namely expansion anchors and related devices. This group comprises dozens of variations but with a common denominator: one end of the fastener achieves fixity by being embedded rigidly into the body of one of the joined pieces. The most common embedment materials are concrete and masonry. The installation process typically entails drilling a hole into the concrete or the masonry; the anchor is inserted into the hole and turned such that the particular device expands in the hole of the concrete or masonry. This may be accomplished by inserting a shield into the hole first, by building the expansion shield into the bolt shank, or by aligning the second piece with the base material and inserting the bolt after both pieces are in place.

The variation in the sequence and in the types, sizes, and shapes of anchors is amazing. There is even an entire family of expansion anchors that precludes the hole-drilling part of the sequences by shooting the anchor into the base material with a blank .22-caliber round. Actually, these shot-in devices, known by the brand name Ramsets, are more related to single-stroke nails than to expansion anchors, but the consequences are related. The problem with expansion anchors and, to some extent, with driven spike anchors as well, is that they exert lateral stress at the point of insertion. To achieve the level of pullout required to perform as connectors, the magnitude of the lateral pressure may exceed the capacity of the base material. This is particularly true if the base material is brittle or friable.

The expansive pressure or displacement pressure may not immediately fracture the base, but it can and will initiate radiating microcracks from the locus of stress. These microcracks can progress, coalesce, and lead to long-term spalling, or they can be avenues of moisture infiltration. The point is that any expansion or displacement device that is inserted into brittle materials has the potential of damaging the material at the time and location of installation.

When the method of insert is to fire live ammunition at the building, these cautions are amplified. Shot anchors are always dangerous, and should be avoided when the target is within three inches of a corner or when there is the prospect of hitting a reinforcing bar, resulting in a ricochet. The rules for shot anchors are as follows:

- Never shoot anchors into block or structural clay tile.
- Do not shoot anything you are not prepared to replace.
- Avoid shooting into anything if it is possible to go through the material.
- Do not shoot at historic fabric.

There are only three essential resources in this business: time, money, and safety. They are not interchangeable.

The chemical grout anchor is not new, but it is enjoying increased popularity for the reasons cited above regarding expansion and displacement anchors. This anchor started out 30 years ago as the epoxy anchor, and has since grown to include a broader class of grout agents, notably styrenes. To use these anchors, the hole is drilled into the base material. Then instead of inserting the expansion shield or expansion anchor, a vial of resin or chemical grout is placed in the hole. Next, the bolt is placed into the hole, after which it is struck hard enough to rupture the vial and disperse the resin into the void between the shaft of the bolt and the wall of the hole.

Two advantages of chemical grout anchors is that they firmly grip the wall of the pilot hole and they do not exert lateral pressure on the perimeter of the hole; but much remains unknown about anchors. Though the epoxy resin anchors have been in the inventory for a number of years, as have epoxy resins generally, and we have a reasonable understanding of how the epoxy resin reacts with various substrates, the reaction and weathering properties of other chemical grouts are not so well known. We do not, for example, know how long a chemical grout will hold up under continuous stress, much less cyclic loading. True, that there is good reason to believe that the natural deterioration of these chemicals in a benign environment will take a very long time, we do not otherwise know how these chemicals may react when loaded in a more aggressive environment: we do not know how well they perform on impact, nor how brittle they may become when very cold. Probably, the answers to these questions will only surface over time and, unfortunately, as the result of failures, because there is no monied interest funding the costly research to track down and document the weaknesses of these and thousands of other new products.

Released Connections

In contrast with joints, which are designed to achieve structural continuity in one or more axes, is another large class of connections specifically designed to *not* transfer stress. In fact, these joints are designed to do just the opposite; they are intended to ensure that stress is not transferred across the discontinuity in one or more axes of either translation or rotation. Many such joints may transfer stress in some axes, but what distinguishes these joints is the requirement that they release one or more axes, not incidentally, but by design. Releases can occur at any scale, but among the more readily identifiable are those that occur at the structural level. After exploring large-scale releases, we will examine some of the more exotic and smaller-scale released connections.

Among the most familiar of all release joints are the paving joints in a sidewalk, of which there are two types, both of which we all have seen thousands of times without giving them much thought. They illustrate the fundamentals of releases very nicely. The first has a dark asphaltic strip sticking up through the joint; the other is the groove cored into the concrete when it was set.

The former is an unconditional release; the latter is a conditional release. The trade term for the asphalt-impregnated strip is *premolded filler*, and it is used not only in paving but along the edges of slabs that abut walls and other similar intersections. Premolded filler is used commonly, but not exclusively, with concrete and masonry. The reason for its popularity is its cost, which is only slightly higher than dirt. In fact, good-quality soil (at a certain price point, dirt becomes soil) is probably more expensive. Premolded filler is compressible and relatively durable. At the line formed by the premolded filler, the paving is completely and totally separated. The two pieces of paving on either side of the premolded filler are completely independent. If one cracks, the crack will not bridge the premolded filler. If one side settles or subsides, the opposite side will remain unaffected. The two sides may settle for the same underlying reason, but neither one will be dragged down by the other. The two segments of paving may shrink, swell, twist, crack, bulge, settle, or anything else because the two pieces are unrelated except by proximity and possibly "date of birth." If the filler were removed, the immediate consequence to either piece would be nil. If the premolded filler had never been installed, the short-term consequences to the paving would have been indistinguishable.

We inject premolded filler into slabs at regular intervals for two basic reasons. First, the premolded filler allows the contractor to pour long strips of paving without interruption; second, because, as we will see, narrow gaps between large masses can be the source of significant problems. During concrete pouring in particular, the strips of premolded filler act as small barriers against which to interrupt a pour either to change trucks at the end the day or to limit the length of a continuous pour.

Our immediate concern at this point in the discussion has to do with the consequences of not breaking the paving into independent segments. If we pour concrete without the isolating effect of the premolded filler, any subsequent movement of one segment will be transmitted through the entire slab. If a void develops under a portion of the paving and the consequential subsidence fractures a portion of the paving, it is possible, even probable, that cracks radiating away from the fracture site will traverse several meters through otherwise undisturbed and relatively unstressed paving.

The presence of the premolded filler means that each segment is contiguous to the adjacent paving, but relatively immune to the problems that may beset a neighboring segment. Though the segments are located approximately one-half inch apart, otherwise they are of no consequence to each other. If the filler were not installed—or, more practically speaking, it were removed shortly after curing—the negative consequence would be that material far less compressible than premolded filler would lodge in the gap. That material would, in time, absorb water, for example, and freeze. The expansion of the freezing water would exert a lateral pressure to the edge of the paving. If the paving should expand for whatever reason, and the material in the gap were rigid, as opposed to the compliant premolded filler, the rigid material would not compress, and the result would be an increase of compressive stress in the paving. Every so often, the premolded filler in the paving serves as an all-purpose protection device, preventing whatever misfortune may befall one piece of paving from contaminating its neighbors. At the same time, it benignly fills the void and prevents the gap itself from becoming the source of a problem. The behavior of the premolded filler is not conditioned on any particular event or set of events; its benefit is unconditional.

The grooves in a sidewalk are altogether different in concept and in operation. The generic term for these sorts of grooves and slots is *control joints*. What they control more than anything else is the configuration of cracks. To much of the world, cracks are signs that something is wrong; the absence of cracks means all is well. Designers are aware that a piece of four-foot-wide, concrete paving more than about eight feet long will crack. When it does, the path of the crack will be the result of thousands of variables beyond the control of anyone, meaning the crack is likely to wander anywhere, and it will be jagged and unsightly. To direct the paving to crack in an orderly manner, the designer precracks the concrete with grooves, so that the rest of the crack will propagate downward through the slab emanating from the bottom of the slot. Consequently, the paving cracks, the crack is disguised by the groove, and the casual observer sees unblemished paving.

The conditional part of the situation derives from the fact that the crack does not have to occur. It is highly likely to occur, but not necessarily. If it does, the release is there. The reason that paving is predisposed to crack is that during the initial curing concrete shrinks. As it shrinks, it is restrained by the substrate below and on which the paving is bearing. The friction between the paving and the substrate is sufficient to exert a restraining force on the paving as it attempts to slide across the substrate surface as it contracts. The restraint by the substrate results in tension in the paving, which at some rather predictable level will fracture the slab into segments, approximately equal in length to where we place the grooves or control joints.

The other reasons for shrinkage were discussed at considerable length in previous sections on thermal and hygroscopic expansion and contraction. The common denominator among all these is that the source of the stress that is relieved by the control joint is internal to the material of the paving, whereas the source of the stresses that are unconditionally relieved by the premolded filler are movements external to the fabric of the paving.

The general name applied to a joint such as in paving with premolded filler is an *isolation joint*. As the name describes, the function of the joint is to isolate the movement of one of the elements from that of the other. In sidewalks, complete isolation is achieved by inserting premolded filler completely through the thickness of the slab. Another familiar example is where columns penetrate a slab on-grade. It is common to see at those locations a three-fourths-inch-wide gap filled with an elastomeric compound in the plane of the slab surrounding the base of the column. The most common pattern is a diamond, relative to the orthogonal axes of the column. The purpose of this joint is to completely disengage any movement in the slab from the column, and vice versa. As we explained in the section on shallow foundations, settlement is a constant issue. In this case, there are relative settlements of the slab and the column footing. If the footing subsides relative to the slab, it will drag the slab down and crack the slab. If the slab moves down relative to the footing, the slab will hang up on the footing and crack. Either way, the slab loses. Of course, there is always the prospect that the slab will shrink, usually during initial curing. If the slab is poured tight to the column, particularly if it is a wide-flange section, the concrete will lock onto the column. As the slab attempts to shrink away from the column, the column will restrain the slab, and the slab will crack. To preclude these cracks, the slab must be released from the column both vertically and horizontally in both horizontal axes. The isolation joint must, therefore, be triaxial in its isolation of the column from the slab.

The use of synthetic elastomeric compounds for such joints is largely a cosmetic decision. Premolded filler would work just as well; but whereas we are accustomed to and tolerant of asphalt filler as paving, we are less tolerant of the same material when it is inside and where we can see it. The range of compounds used for this relatively undemanding situation is staggering. The main requirement of the substance we choose to use in slab isolation joints or, for that matter, any isolation joint, is that is remain elastic and not harden over time. It may also be desirable that it not shrink and crack, and perhaps that it even hold paint, but the essential requirement is that it remain soft.

When the isolation joint is not horizontal, resting on-grade, but vertical or horizontal above-grade, the situation becomes more complicated. Imagine a major addition to an existing building. The addition can either be fixed to the structure of the original construction or it can stand free. If the addition is rigidly connected, the foundation must be able to absorb the additional load, which is usually not the problem. The common problem is precisely the opposite. Even though the addition may be adding weight to the existing foundation, it is the perimeter of the addition that settles relative to the original construction. If the addition is, in effect, an open-sided box backed up tightly against the original building, and if the box is attached to the original wall, then the weight at the old building line is increased by the tributary area of the floors and roof, relative to the edge attached to the older building.

Further assume that the older building has been bearing on the soil along the attachment wall as an outside exterior wall (read: relatively heavy) for a protracted period of time, and the soil below this segment of original wall is consolidated and compacted. Along comes an addition that has no exterior wall at the plane of attachment (read: relatively light), but three other exterior walls and respective floor and roof loads bearing on virgin soil (read unconsolidated and uncompacted). The soil under the three new walls consolidates at almost the precise moment that the builder's warranty expires, but the attachment edge remains perched on the old wall without movement. The net result is that the addition appears to be rotating away from the old house. A crack opens at the roof resulting in a wedge-shaped gap down the walls at the line of contact between the old and the new. Water begins to enter the premises, prompting calls to the builder, followed by calls to the architect, followed by calls to the attorneys.

A wise designer builds the addition as a completely independent building, located very, very close to an existing building. In fact, the distance between the two buildings may only be an inch or so. Where the buildings are likely to be shaken about, this dimension is critical, because if they shake enough to bang against one another, what may have been thought a passing tremor becomes a major case of building battery. In calmer locations, the dimension is inclined to be modest indeed. Once in service, the addition may well subside, but will do so more uniformly and predictably. It is necessary to allow this largely vertical movement to occur without ripping the addition from the old house. The new must be able to slide down the face of the old without disruption and without providing an avenue for water penetration.

The joint that allows this movement at the roof plane is different from the joint that allows the same movement but in a different orientation in the wall. The movement at the roof is perpendicular to the joint, whereas the movement at the wall is parallel to the joint. Both joints are typically constructed of strips of sheet metal attached to either side of the joint, folded to allow one strip to slip relative to the other, while closing the gap from air infiltration and water passage. Carefully tracking building isolation joints around and through the adjoining buildings is one of the most tedious and least rewarding tasks in all of architecture. The designer will never be praised for a job well done because perfection is the basic standard from which deviation can occur in only one direction.

Conditional releases, such as the grooves in the sidewalk, also occur in other circumstances. For example, a long free-standing brick garden wall attached to a relatively rigid footing can elongate and/or shrink. Inasmuch as the footing is a restraint, the upper portion of the wall may be placed in a state of tension, and may, therefore, subsequently crack to relieve the stress. Even though this is a strain-limited type of problem, the wall is, nonetheless, cracked; and as we are now aware, the crack will wander in an impolite manner. We, therefore, precrack the wall with a control joint. For reasons that will be discussed extensively in Chapter 4, the designer may wish to exclude or preclude water across the depth of this prearranged crack, and, therefore, fills the crack with a substance. Again the substance of choice is normally similar to, if not the same as, the material used in the column isolation joint, though probably with greater emphasis on color. The one added requirement is referred to simply as the *sag properties*, which are rarely quantified but mean that the substance will not droop in a pattern suggesting relaxation; and, of course, the substance cannot flow out of the joint. These properties are related to viscosity, which are addressed in a subsequent chapter on fluids.

Another set of materials, collectively referred to as *gaskets*, can often be substituted for—and generally are preferred to—elastomerics because they are more durable and less prone to physical damage and vandalism. These materials are formed or extruded neoprene, nylon, or other inert, tough synthetic elastic materials. Their shapes and sizes are designed for use in very specific conditions. As general fillers they are quite successful, and are also often reversible, meaning that they can be stripped out when desired. Along those lines are a vast array of tapes and strips made of air-entrained materials formed into tapes. They may have adhesive backing, which allows them to be adhered to one side of the open joint during construction. Some have mild expansive properties so that they can be installed in a gap, then slowly expanded to completely fill the void. A family of foams is available that can be installed from a compressed air canister. The foam expands and solidifies into a substance somewhat similar to Styrofoam.

If it seems that the list of joint filling materials is endless, that is because it is, and it continues to grow at a seemingly exponential rate. There are, in fact, so many that no single person can master the properties and performance of all of them; no one should try.

To segue to the pathology of releases, we first examine the basic design criteria for successful joint design. The first requirement of good joint design is to recognize the need for a joint and, given that the joint exists, what it must do. If the joint is to release one body from another, that release must occur across the *entire* contact plane between the two bodies. If there is a gap between the two bodies we must query whether that void must be filled or can remain unfilled. If it must be filled, we must ask whether it can be filled with an elastic solid, as opposed to an elastomeric material. All these questions are partially answered in the subsequent chapters on closure pathology, but some explanation is pertinent here.

Joints, whether conditional or unconditional, are often misapplied, both in terms of reflexive misapplication, where they are unnecessary and inappropriate, and in terms of failing to recognize the need for a joint, much less how to make one. The manufacturers of the materials often used in the fabrication of joints often over-recommend the ease to design types of joints, at the same time that they are mute about the tough ones, which cause premature aging of designers. The craft of appropriate and properly fabricated moving joints is a measure of the technical skill of any designer.

Recall the sections on expansion and contraction. If the material at issue will expand or contract, then the question is whether it is restrained in any direction by another material. If it is restrained in its potential movement, is it sufficiently elastic to absorb the deformation now or as it ages? If it is or will become brittle, then control joints are recommended to guide the direction and character of the crack. Control joints do not prevent cracks, they only allow the designer to select the direction and visibility of the crack. We can crack these materials, or nature will do it for us. Some of us actually prefer that nature do the job.

Given that the control joint (read: crack) will occur, does it have to be filled? If it can remain open, it will almost always perform better than if it is filled. Fillers often harden, notably oil-based caulks and putty. Elastomerics also embrittle, and tend to crack themselves unless vandals pick them out of the joint first. Many ooze and seep into the adjacent materials. Despite all the claims made by filler manufacturers, a joint that can remain open will out-perform any filled joints. The risk to leaving the joint open is that debris will accumulate in the joints, which, over time, can choke the joint with what becomes rigid material, thereby defeating the purpose of the open joint.

If the joint must be a gap and must be filled, look first to gaskets and tapes. If cosmetics is an issue, consider setting the gasket back from the surface of the joint. If vandalism is an issue, then recessed joints are particularly appropriate. If the joint must move in a single direction, then bellows and other folded-sheet devices are effective. Any material used in a joint should be as durable as possible, even at some added expense for the material itself or the detail to house the material, because regardless of the good intentions of the owner, joint-filler maintenance is very far down on the list of priorities.

With all that said, the path to deterioration is relatively apparent. Release joints fail, first and foremost because they cease to release; and the leading causes of such failures are that the filler material hardens or the joint becomes choked with rigid debris. There are certainly other sources of release joint failure, and we will discuss some of them, but those are the primary causes. If the joint is a conditional release, then the prospects of failure are far less than for those of an unconditional release, because conditional releases are less likely to be filled, and because their failure is less problematic.

Other common sources of release joint failure include the failure of the designer to adequately predict the amount of travel required in the joint, or the failure of the substrate to perform according to expectations (for example, the concrete was improperly cured or mixed too rich to begin with; or unforeseeable settlement occurred because of a below-grade condition). In general, the joint either locks, or circumstances require it to travel outside its tolerance. In either case, either compression or tension develops in the connection, and that force traverses the plane of discontinuity. If the resulting stress on either side of the joint is sufficient, then the adjacent materials may fracture or deform.

The third general type or class of connections is defined more by what they do not do than what they do do. Simple alignment connections may transfer stress, but that is not a defining function. Simple alignment connections may release the adjacent materials from stress, but that is not a defining function either. Simple alignment connections occur at locations where there is a change of direction of the material, or where the material simply plays out and another piece is spliced to the first piece. The common picture frame is an example of the former; the beveled splice joint of a continuous crown mold is an example of the latter.

Simple alignment connections are uncomplicated breaks in the continuity of an element to accommodate whatever geometric limit the piece may have. Little more is expected of the joint than that it reconstitute the continuity of the original profile or element.

We dwell on simple alignment connections not just to be comprehensive in our survey, but because they often generate problems despite their lowly expectations. While there may be no particular structural function associated with simple alignment connections, they can and often do transfer stress or precipitate stress at the joint interface. Because the simple connection is for alignment purposes and, therefore, primarily, cosmetic, the tendency is to minimize the visibility of the joint. This is typically accomplished by joining the two pieces together with sufficient force to squeeze the line made by the joint into nonexistence. Such joinery is the close cousin of the torque-it-'til-the-threads-strip school of screws and bolts. Close simple alignment connections are accomplished by finely fitting the opposing faces, not by squeezing the faces together.

If the simple connection is compressed, force is imparted across the plane of discontinuity, and the joint is, by functional definition, a partially fixed joint. If, as the result of thermal or hygroscopic expansion, the pieces push toward the joint even more, the compressive stress will increase further and may exceed the ultimate strength of the adjacent materials. In short, a simple connection with no particular requirements was converted to a potential hazard because the joiner tried too hard to make it disappear.

Simple alignment connections can also fail subsequent to fabrication because the joint is altered in some fashion or other that causes its conversion to a fixed or partially fixed connection. The two tools that contribute to the demise of the simple connection are paint and finishing nails. Take as an example a picture rail. The rail is a piece of milled lumber, attached to the wall at intervals with nails into or through the plaster or drywall. The rail is milled in 16-foot lengths, and the room is 18 feet by 22 feet. The rail will be spliced at no fewer than four locations, plus four corner joints if it completely wraps the perimeter of the room. The four splice joints will be highly visible so the finish carpenter uses his or her very best miter saw and bevel-cuts the pieces to precise plane and fit. He or she snugs the joints without compressing them, then sands the face of the joint, and pauses momentarily to admire his or her good work before leaving the room.

The next day, the painter arrives, and he or she applies a coat of primer, which consists of a relatively thin medium and a low concentration of pigment. The next day the painter comes back with a finish coat of heavier vehicle and high concentration of pigment, and happens to apply the paint in the opposite direction of the bevel splice. The consequence is that the splice floods with paint, which completely closes and seals the joint. The very next day the ambient air humidity and temperature rise so much that, over the course of a few days, the internal temperature and moisture content of the wood rail also rise, causing the wood to expand. Instead of further closing a slight gap at the splice, however, the bevel rides over the bottom segment and outward away from the wall. A well-intentioned maintenance worker passes by and notices the out-of-plane rail, so he or she dutifully takes a 6d finishing nail and drives it through the bevel joint to bring everything back into alignment and to keep all the pieces where they are supposed to be.

That's not the end of the story. The following week, the weather becomes cooler and drier than when the rail was originally installed. The rail, too, also gradually dries and cools, and, therefore, shrinks. The rail pulls away from the splice, but is restrained by the nail. The likelihood of the finishing nail effectively restraining a shrinking piece of wood is negligible, so naturally the outboard segment effectively pulls the nail out of its backside and through the tongue of the bevel splice. One week later, the ambient temperature and humidity rise, and the rail elongates once again. This time, instead of compressing against the rigid paint, the pieces try to compress the nail, which is trying to go back into the hole from which it came; but whereas it came out the back-side, riding on its smooth shoulder, it is now trying to reenter the hole with its blunt head. It hangs up on the sides of the hole and pushes the outboard segment further off the wall.

We can exacerbate the situation at any time by repainting the entire assembly, followed by the installation of still larger nails, until the splice is damaged beyond repair and/or the rail disengages from the wall. This very common tale illustrates in gory detail the plight of moving joints. Even when the original designer and installer recognize the need for play in the joint, there is no way to assure that subsequent action will not defeat the objectives and operation of release joints, as well as simple alignment connections.

Intervention in failed-joint conditions can be every bit as challenging as their original and proper design. In fact, it is often during the "repair" that many joints are defeated or compromised. Intervention in failed joints requires first and foremost that the professional recognize the original purpose of the joint and how it was designed to operate. The intervention is almost always an exercise in reconstitution, unless the original design was in some way flawed and the joint could never have worked properly. In those cases, the approach may become an exercise in substitution. It is unlikely that the approach will be one of mitigation or of circumvention, because less than complete travel in the joint is precisely that, less than complete. Bridging the joint with some sort of circumventive detail does not address the issue that the joint may be frozen and will continue to apply stress across the plane of discontinuity in an untoward manner.

4 Vertical Closure Systems

4.1 Functional Definition

The function of the vertical closure system is to provide a selective barrier between the interior and the exterior of the building. The vertical closure system not only provides a selective barrier, it delineates the inside of a building from the outside the building. The vertical closure system, when combined with the horizontal closure system, which is discussed in the following chapter, is synonymous with what we refer to as the *envelope*. The vertical closure system is differentiated from the horizontal closure system not only by orientation but by function as well. Though both systems define volume, the dominant function of the vertical closure system is an exercise of separation. As explained later, the dominant function of the horizontal closure system is something more specific.

This chapter addresses the function of the vertical closure system in exhaustive detail. As in Chapter 3, we begin with a description of the components and their assigned functions, move on to material-level problems and the deterioration mechanisms associated with those materials, then progress to systems-level issues associated with the components and assemblies that comprise the major elements of the system as a whole.

Note: The material in this chapter makes reference to principles and applications of mechanisms discussed in detail in Chapter 3. Those not reading the book sequentially, but rather as a topic-specific resource, may need to backtrack to make the most of this discussion.

4.2 Components and Description

The dominant component of the vertical closure system is the wall. For the purposes of this book, the generic term *wall* is used to describe all of the vertical components of the vertical closure system that are *not* openings. The term *field*, more restricted than wall, is used to refer to the opaque, solid, nonoperating portions of the wall, which are above-grade and provide self-support to the system. Below the field, and below-grade, is what we will call the *foundation wall*. The voids or openings are of two general types, windows and doors, although there are many specialized types of openings, such as louvers, vents, and mechanical sleeves.

Though the field and foundation wall provide a coincidental structural function for the building, this chapter concentrates on the nonstructural aspects of the field, foundation wall, and upper wall, generally. As stated at the beginning of this chapter, the vertical closure system provides a selective barrier; however, selecting what to exclude and to what degree are major complicating factors associated with the system. Among the components that comprise the vertical closure system are portions, such as the field, that are highly exclusive, whereas, other portions are operable and, therefore, variably exclusive.

Choosing where to begin our investigation of this highly complex subject is a bit like unraveling a large ball of twine. There is no single end on which we can reliably pull, rather dozens, if not hundreds, of ends, all of which either snag on other segments or play out to little apparent progress. We, therefore, begin the investigation at *a* point, as opposed to *the* point. We then proceed in a direction—namely, materials pathology—followed by an extended discussion of systems pathology.

FIGURE 4.1 This random ashlar "serpentine" wall was subsequently stuccoed. That is now failing and reexposing the highly deteriorated stone. The origins of the deterioration lie in the expansive nature of the captive mineral olivine.

4.3 Materials Pathology

We begin by assuming that materials issues are dominated by the materials associated with the construction of the field, and that the field is constructed of porous, brittle material. Neither of these assumptions is more than loosely defensible, but we will pull on this string for a while anyway.

To penetrate some of the more remote aspects of field material issues, we delve into a specific deterioration mechanism: freeze-thaw cycling. This mechanism will be used as a platform for discovering and exploring the basic principles associated with, and underlying, the more general phenomenon.

4.3.1 Freeze-Thaw Cycling

We introduced the term freeze-thaw cycling in Chapter 2, and defined the necessary and sufficient conditions for the phenomenon. Here we will discuss the details of this deterioration mechanism, and use it as an introduction to several aspects of materials physiology, namely thermal transmission, temperature and moisture gradients, and the peculiarities of water as it approaches freezing and eventually freezes. An understanding of these subjects is essential before proceeding, as these topics offer insight to other thermodynamic and hygrodynamic issues. Recall from Chapter 2 that freeze-thaw activity is a deterioration mechanism that afflicts porous, absorptive, brittle solids. To review, the necessary and sufficient conditions for the origination and progression of freeze-thaw cycling are:

- Porous, permeable, absorptive materials
- Surficial moisture saturation
- Relatively low modulus of rupture of the affected material
- Temperature declining through freezing point

The completion of the cycle is that the material subsequently thaws, resaturates, and refreezes; but the minimum essential elements that account for the deterioration mechanism begin with these four necessary and sufficient conditions. (Note that this is not a complete list, rather an essential one.)

The sequence of events that lead to the freeze-thaw mechanism is as follows:

1. Water, generally in the form of rain, runs down the vertical surface of an exposed porous, permeable, absorptive surface such as concrete, brick, or stone masonry.
2. The water is absorbed by and into both the crystalline solids and the capillaries, which erupt at the surface of the material.

3. The water freezes, which results in expansion of the water and subsequently the ice.
4. The expanding water and ice exert pressure inside the fabric; if the pressure exceeds the modulus of rupture, a crack is initiated.
5. The water subsequently melts, and the process is repeated, but henceforth, the newly initiated cracks absorb water as well.
6. Pressure inside the nascent crack extends the crack, and the cycle begins again.

At this simplified level, the deterioration mechanism is relatively clear and uncomplicated: freezing water results in cracked brick. Several interesting details about the process are not as simple. We know the thermal resistance property of the brick; we know the temperatures of the extreme outside and inside surfaces, hence the temperature gradient, which is a sort of temperature profile, of the brick. We can, therefore, identify the location within the material where the temperature will be low enough for water to freeze; this is the depth to which the water within a saturated brick will freeze. But that is too simple: a brick saturated to some depth is not a thermally homogeneous material; it is thermally two materials, saturated brick and unsaturated brick. Each has a different thermal transmission resistance, and the location of the line where freezing occurs must be analyzed as a two-layer situation, not one. Moreover, as the ice forms, a third layer of thermally distinct material emerges, namely brick, which is both saturated and frozen. This means that as the ice forms in the brick, the line of potentially freezing water is shifting. The computation of the depth of penetration of frost within a saturated brick is a multivariable, dynamic problem. As the phenomenon occurs, the variables change, which of course alters the phenomenon itself.

THERMAL TRANSMISSION

Heat is not easy to understand, for it is colorless, weightless, and odorless, and it lacks mass; but it is powerful in its effects. Heat is tangible, however, in that we can directly and physically sense the presence or, in some cases, the relative absence of heat. It is also intangible, in that we cannot separate heat from the object that is heated—or only abstractly. The fact that heat is only abstractly separable from the heated object is the cause of our difficulty in confidently grasping the reality and palpability of heat. Partly for these reasons, teachers have traditionally resorted to invoking metaphors and analogies to convey the concept of heat.

The model we will use to describe heat goes directly to the issue of how heat works. We will employ the molecular model and take some liberties with our understanding of nuclear physics. Let us begin with crystalline solids. Imagine a cube made of roughly spherical solid particles connected orthogonally to six other similar spheres, one in each direction of all three principle axes. Now imagine that the connectors are springs capable of absorbing compression or tension. The connectors can, therefore, be elongated or shortened depending upon the relative positions of the spheres; but all the spheres will be maintained in the grid unless there is some form of externally induced distortion. If we accept for the moment that the equilibrium position of all the spheres is on a rectilinear, triaxial, orthogonal grid, then every sphere is at right angles to every other sphere. Now imagine that we can reach into the cube, somewhere near the middle, and grasp one of the spheres and pull or push on it, moving it off the grid. As we pull or push the sphere, the springs connecting that sphere to the others will either elongate or shorten depending on the direction that the selected sphere is moved. As the sphere is moved, the force required to accomplish the movement will increase; the magnitude of the translation is related to the force required to effect the translation. The greater the movement, the greater the force.

We have already noted that the relocation of one sphere will result in elongation or abbreviation of all six of the connectors attached to it; therefore, all six surrounding spheres will also move, although not necessarily to the same extent. The reason they all move is because they are all attached with flexible connectors to the one sphere being moved. The reason they will not move as much is that each of the six surrounding spheres is itself restrained to some extent by the other five spheres that are attached to each of them. In other words, the movement of a single sphere within the three-dimensional matrix of spheres will cause every other sphere in the matrix to move to some extent, however minor.

Having dislocated a selected sphere and, therefore, having stretched or compressed every spring connector in the cube, now we let go abruptly. When the sphere springs back to its original position, it behaves like a plucked string: it does not suddenly stop when it reaches the point of origination; the sphere travels past the point of origin and returns in an oscillatory manner, only gradually coming to rest.

The reason the oscillating sphere comes to rest eventually is because of *damping.* Damping is the term given to the collection of forces that contribute to the resistance of any object in motion through vibration or oscillation. For example, the forces that damp harmonic motion for larger objects are wind resistance, friction in a bearing or hinge, internal stiffness of the string or spring, or in the case of the crystalline solid, the countervailing flexible connections to all the other spheres in the crystal. The effect of damping is to dissipate the energy manifest in the oscillating spring or string. At the same time that the selected sphere is quieting down, all the other spheres in the crystal are being damped as well. Gradually, all the spheres return to their original positions, and everything is still.

The length of time required for the entire system to return to relative stillness is a function of several variables relevant to our discussion. First, the oscillation period is subject to the magnitude of the original displacement, that is, the distance we moved the sphere from its position of rest before releasing it. Second, the period is subject to the collective and individual stiffness of the connectors to both the displaced sphere, as well as all the surrounding spheres. The more flexible the connectors, the longer the respective oscillatory action of each sphere. Third, the period is a function of the mass of the displaced sphere and the masses of the other spheres in the system. The greater the mass, the longer it will take to quiet an excited system.

Now imagine that more than one sphere is displaced and released. The sum of the stored energy will be dissipated in a complex interaction of the respective oscillating spheres. Some oscillations will amplify the motion of adjacent spheres in a manner similar to the way harmonically tuned strings can effectively reinforce each other. If one string is plucked, if it is harmonically tuned to a second string, the vibration of the first will set the second string in motion through the force of the compressed and uncompressed air waves generated by the sound of the first string. But consider that it is also true that disharmonic strings will damp each other when both are in motion. A similar phenomenon occurs in crystalline solids when multiple spheres are simultaneously excited.

Suppose the spheres in the matrix are of radically different masses. Some spheres may be massive, others less so. If a less massive sphere is excited, it will have a less pronounced effect on an adjacent massive sphere than it might if its neighbor were less massive. To further tax the imagination, suppose that, instead of being attached to six other spheres in a rectilinear, triaxial, orthogonal grid, the sphere is attached to other spheres in a grid that is hexagonal and, therefore, has six axes of rotation. Then, instead of being connected to six other spheres, each individual sphere will be connected to 12 other spheres not 6. The *packing arrangement* is significant. A packing of similarly sized spheres is densest if the arrangement is based on a hexagonal grid rather than an orthogonal grid. Denser packing means that the distance between centers of adjacent spheres is less, which, in turn, means that the energy required to sustain and maintain the arrangement is less. In other words, if a set of equally sized spheres can be packed either orthogonally or hexagonally, the hexagonal pattern will result in a denser package, requiring less energy to build. In a Newtonian world that seeks to minimize the energy state for any given condition, hexagonal dense packs are preferred.

The inorganic, crystalline solids that comprise the bulk of what we know as the solid world, according to the atomic theory, are composed of complex arrangements of variously sized, roughly spherical particles called atoms. The composition and stability of the atoms located within a given crystal are dependent upon the electrical balance among all the associated elements and a compatibility of arrangement that allows that particular set of atoms to maintain a stable order arrangement. The combination of these two requirements means that not just any combination of elements can form a crystalline solid. Certain combinations in electrical balance may be impossibly sized for packing purposes, for example. Another combination may be beautifully composed for packing purposes, but be electrically unbalanced.

Once the electrical charges and the size issues have been satisfied and the compound has been formed, the specific type of connectors further define the properties of the substance. The bonds are the connections between the atoms, and they come in different flavors depending primarily on the manner in which electrons in the outer shell of the respective atoms are shared or distributed. For covalent bonds, the electrons orbit the nuclei of the respective atoms simultaneously. This bond is both extremely strong and extremely flexible; it is common in organic compounds. If the electrons gravitate predominantly to the electron-"absorbing" elements and away from the electron "donor" elements, the bond is an ionic bond. It is the characteristic bond of salts that are defined as combinations of metals and nonmetals, such as sodium chloride. Ionic bonds are very strong and relatively flexible, despite what we will discover to be a vulnerability to water. Metallic bonds are characterized by a group of particles sharing the electrons throughout the crystal. In metals, electrons freely flow from one area to another within the solid. This, as explained in Chapter 5, is an essential characteristic of electrical conductivity and of the corrosion mechanism. Metallic bonds are strong, but generally not as strong as covalent or ionic bonds, and they are relatively stiff. (Two other types of bonds, namely hydrogen and Van der Waals bonds, are weak bonds and of less significance to deterioration mechanisms.)

Bond strength and flexibility are relevant to the capability of a material to store heat before the excitement level of a given particle fractures the bond, and the solid loses the capacity to maintain the regular geometry of the crystal. The regular geometric pattern of the crystal is also referred to as the *long-range order* of the crystal. When the long-range order of the crystal is destroyed, the particles, which may be single atoms or clusters of atoms, are free to roll about one another because the crystalline bonds are gone. This is the definition of a liquid. The essential difference between solids and liquids is that the excitement level of the solid is low enough for the bonds between particles to maintain a long-range, repeated, closely packed pattern based on electrical and geometric compatibility. When the particles within the solid are sufficiently excited as to break the crystalline bonds, the substance liquefies, the substance melts.

The amount of external excitement that a given substance can absorb before the crystalline order is disrupted is a function of the mass of the particles, the spacing between particles, and the bonds maintaining those particles in position. This characteristic is designated as the *heat of liquefaction*, or *melting point*, and it is a specific and experimentally reliable point for each crystalline solid measured in terms of degrees of molecular excitation. The level of excitation is defined as the temperature of the material, so the melting point will occur for a given material at a predictable level of response or crystalline temperature.

Though the temperature at which two different solids will liquefy may be the same, they may require different degrees of external stimulation to reach the same degree of temperature. If there is a single concept associated with heat that is difficult to internalize, this is it. The molecular structure of a compound, as discussed, is at least in part defined by the size, spacing, mass, and packing of the particles, and the bond strength and flexibility between the particles. The crystalline bonds will break when the oscillation of the particles exceeds the capacity of the bonds to maintain the long-range order of the crystal. The amount of work required to get the particles to move far enough apart to fracture the bonds is altogether different between any two compounds, even though melting temperatures may be quite similar.

It is also a measurable property of all the crystalline characteristics we have discussed that the rate at which the spheres become excited varies among substances. In other words, a small amount of heat will generate a rapid increase in molecular excitation, which translates into a rise in temperature. Other substances require a greater amount of stimulation to achieve the same level of molecular excitation, hence a similar rise in temperature. This is similar to the differences in stimulation that may result in liquefaction, but there is a difference between what amounts to the overall resistance or relative ease of stimulating the solid, and the rate at which such stimulation occurs. It follows, then, that materials that require higher levels of stimulation to achieve a given temperature will release that stored energy more slowly than the materials that respond rapidly. If a given material requires a great deal of stimulation without risking destruction of the crystalline bonds, then such a material can be used to store that excitation and release it back over a prolonged period. The material property that measures its propensity to hold energy is called the *specific heat*, about which we learn more very shortly.

The amplitude of the oscillation of the crystalline matrix is primarily a function of the duration and magnitude of the heat source. As heat is supplied to a surface, that energy is absorbed into the fabric and transformed into the oscillation of the molecular particles. If the heat source is removed, the oscillation of those same particles damps down; it dissipates. The rate of dissipation is a function of the specific heat of the material. Even as the heat is being supplied to the surface, and the surficial particles begin to oscillate, those vibrations are transmitted further into the interior of the fabric. Thus, even as the heat source is still active at the surface, the heat transferred to the fabric of the solid is simultaneously being dissipated throughout the interior of the solid. If the heat source is maintained at the surface, and at a constant level for a period of time, the temperature of the surface of the material will rise to at least the same level as that of the source. As the heat continues in undiminished supply, the energy is dissipated throughout the solid in a fairly even distribution. If the material at issue is completely surrounded by the heat source, the temperature will rise at the perimeter; simultaneously, the heat will dissipate from all the heated surfaces toward the interior. In this particular case, that interior is the core of the object. If the source of heat is from conduction, this heating process will continue until the entire object is at the same temperature as the surrounding heat source. This is a condition of *thermal equilibrium*. But for as long as any part of the object is cooler than the ambient surrounding temperature, the heat source continues to supply heat to the object.

A more common situation is that there is a heat source on one side of a panel of material, which divides two distinguishable volumes, the inside and outside of a building, for example. If the heat source is on the outside of the building, we can see some similarity to the completely surrounded object in the previous example—ignoring for the moment that the bottom of the box is in contact with the ground. The outside heat source begins heating the material that separates it from the inside. The surface temperature begins to rise; the heat dissipates through the panel. Gradually, the temperature of the entire thickness of the panel is rising, which means that the panel becomes the heat source for the air inside the box or, in this instance, inside the building. The air temperature begins to rise, and the outside heat source continues to supply energy to both the panel and the interior air volume until a state of thermal equilibrium is reached between the inside and the outside.

If we turn on an air conditioner, the inside air becomes the heat source for the cooling unit. The effect is that, as the temperature of the inside of the panel rises and heats the inside air, that amount of heat is pumped outside the box. The heat continues to heat the outside surface of the panel to at least the temperature of the outside heat source. At the same time that the inside surface is trying to heat the inside air, that air is being maintained at a specified temperature by the "magic" of air conditioning. Consequently, the temperature of the inside surface is at or near the temperature of the inside air. As long as the outside heat source does not run out of heat, which is unlikely, and the air conditioner continues to operate effectively, the outside of the panel will stay at a steady temperature equal to the outside heat source; and the inside of the panel will remain at a steady temperature equal to the inside air temperature. This is a state of *dynamic equilibrium*, in contrast to *static* thermal equilibrium, discussed earlier.

The typical methods of heat transfer are organized into three categories: *convection*, *conduction*, and *radiation*. Each is relevant to building systems and to building pathology. Convection is the transfer of heat via a heated gas or a heated fluid. We have dealt almost exclusively with the heat-related properties and issues of heated solids, but the peculiarities of heated fluids and gases are highly relevant to the overall topic of building materials and systems.

The general model of particle oscillation is as valid to fluids and gases as it was to solids. Heating the solid to a sufficiently high temperature will, in fact, result in the liquefaction of that solid because the oscillation of the individual molecular particles is enough to fracture the crystalline bonds and destroy the long-range order of the crystal. Incidentally, we also know that solids typically expand as they are heated, so the liquid density of a given material is lower than the same material in its liquid state. The reverse is also typically true—that is, as materials solidify from a liquid into a solid state, the solidified material sinks within the liquid portion. For example, solidified iron sinks to the bottom of the molten ferrous puddle.

Once the material is in a liquid state and it continues to be heated, the individual molecular particles continue to increase their oscillating amplitude, and the liquid continues to expand. The expansion of the heated liquid changes the density of the fluid as it is heated; the denser and cooler liquid sinks at the same time that the warmer and less dense fluid "floats" to the top. The rising warmer fluid and descending cooler fluid move up or down as the case may be, each physically displacing other fluid either above or below it. This displacement process typically occurs in rising or descending columns of fluid within the fluid mass. The rising warmer columns are balanced in volume by descending cooler columns. If the heat source is at the bottom center of the pool, for example, the rising column will occur vertically above the heat source. The displaced relatively cooler fluids typically flow downward at the perimeter of the pool, replacing the rising volume from the center. This fluid movement continues as long as the heat source persists and there is any fluid in the system cooler than the heat source. One result of this fluid movement and displacement process is the continuous cyclical motion of the rising and descending volumes, which we call *currents*. The fluid currents in this example are roughly the shape of a mushroom, rising at the stem, spreading outward at the cap, and descending at the perimeter of the mushroom.

In a fluid state, even though the long-range order of the crystalline solid is absent, the molecular particles continue to display an affinity. In some instances, within the roiling and rolling about of the fluid particles there is a semblance or order to certain molecular groupings. These ordered molecular-level arrangements, such as they are, are defined as the *short-range order* of the molecule. A fluid can be analogized to a bag of marbles rolling around each other inside a bag: the integrity of the individual marbles is maintained. (Incidentally, if we put the same marbles into a box and shake them, they may arrange themselves into a hexagonal dense-pack, long-range order.)

Not all substances have a short-range order, in which case the atoms essentially float at random in the liquid state; and not all substances set up a regular long-range order when they solidify. Substances that do not have a long-range order are called *amorphous solids*. Amorphous solids never make a distinct shift from being solid to being liquid, and vice versa. When a liquefied amorphous solid cools, like other substances, it condenses. In the case of crystalline solids, this densification will result in some increase in the thickness of the fluid, but that is followed by an abrupt and complete shift to a solid phase. Amorphous materials also thicken as they cool, but there is no abrupt shift in state. They simply continue to thicken until the properties of fluids are indistinguishable, and the properties of solids appear more pertinent. The term used to define fluid thickness is *viscosity*. High viscosity means that the fluid is thick; low viscosity describes what we generally think of as thin.

Going in the other direction, as amorphous solids become warmer, they soften, as opposed to melting. The viscosity decreases as the amorphous solid is heated, and the softening is fairly even as the temperature increases. It is arguable that amorphous solids are not solids at all, but simply very, very viscous liquids. Substances that are amorphous solids are significant to us because they are quite common in, and important, to building construction. Most asphalt-based materials are amorphous solids, as are many hydrocarbon polymers. Many members of the poorly defined group of substances called plastics are, in fact, amorphous solids. Some rocks, such as basalt, are amorphous solids. Probably the most surprising member of the class is glass. As brittle as it can be, glass never freezes; it only gets thicker and thicker.

Once a substance is in a fluid state, it may be heated to the point at which the attraction of various particles is no longer strong enough to maintain even the fluid condition, and the molecular particles segregate into a gaseous state. In many instances, the amount of energy required to convert a substance from a fluid to a gas is substantial; in other cases, it would appear that not so much energy is required, though this is largely a function of our frame of reference. The physical state we assume as "natural" for a given substance is less natural than familiar. Whether a substance is in a solid, liquid, or gaseous state is neither an issue of nature or custom, merely a matter of heat.

Gases by their nature are quite transient; they have little affinity to cohere to other particles, and their mass is extremely low compared to fluids and solids of the same basic composition. Heating a gas will produce dramatic changes in the gas density; and just as with fluids, the density differential will cause less dense gases to be displaced vertically by denser gases. As with fluids, heated gases will rise in columns, concentrated at the source of heat. If the gas is encased in a closed container, the gases that necessarily replace the vacated volume of the warmer gas produce currents in roughly circular patterns. The flow of gases that result from heating a column of gas is called convection, which is a major transfer device for heat in the built environment. For example, suppose that the heat source is an outside window or wall panel. The surface is warmed by the sun. The heat is transferred through the solid panel and begins to warm the inside surface of the wall. The air, a mixture of gases, which is immediately proximate to the inside surface is heated from the surface and begins to rise. When the warmer, lighter air reaches the ceiling of the room, it begins to spread out across the surface of the ceiling and gradually fills the room, with warmer, lighter air collecting at the top. The void left by the rising air column is filled by less dense air, which is effectively sucked into the volume vacated by the rising air. The effect of this mechanism is that the warmth of the wall is initially transferred to the air and, subsequently, to the surface of the ceiling, which is heated by the warm gases of the air. This transference of heat by currents of air or other gases is convection, or convective transfer.

If the wall were constructed of two layers of identical material, the heat would traverse the total thickness in virtually the same manner as if the wall were of a single layer of material of the same total thickness. At the plane of discontinuity between the two layers, however, there must be a transfer of heat. This is different from the intracrystalline transfer described as typical in solids because there is no crystalline bond between the two layers. The transfer by direct contact of one material with another is called *conduction*. The plane of discontinuity, however fine it may be, cannot be as efficient at transferring heat as the undisturbed crystalline solid. The difference is sufficiently small so that we may discount it at the scale of something the size of a building, but there is a difference between conduction across a plane of discontinuity and within a crystalline solid.

If the dimension of the plane of discontinuity becomes measurable, if the two layers of material are not in direct contact and the void between is filled with a gas, then the first layer could heat the air, which could convectively heat the second layer. If the void were a pure vacuum, however, then convection would not be possible, but heat would still traverse the void. This third method of transfer is called *radiation*, and it is the mother of all transfers, because it is responsible for getting the heat of the sun to the surface of the earth through the void of space. There are a variety of exotic ways to explain radiation, but they are beyond the scope of this book, and are not required to appreciate the simple fact that heat can traverse vacuums and voids without relying on conduction or convection. In fact, in our example of the warm wall and the rising air column, there is necessarily a certain amount of radiant heat transfer occurring simultaneously between the warm wall and the rest of the interior of the room. Any surface will radiate heat at all times; the net transfer from one surface to another occurs because the rate of radiation from the warmer surface is greater than the radiation from the relatively cooler surface. It is important to remember that radiation is occurring from all surfaces all the time. Even if a surface can be described as cold, some heat is still contained in the material as long as its temperature is greater than absolute zero; and as long as there is any heat in a material, it will radiate heat. To be sure, if a very cold item is proximate to a very warm item, the net exchange of radiation will result in the warming of the cold object and the cooling of the warm object. But this is because of the disparity in the rates of radiation, not because there is a one direction transfer of heat from the warmer object to the cooler object.

If, on the other hand, a cold object has a very high capacity to absorb additional energy in the form of heat, regardless of the method of transfer, that object is analogous to a thermal storage container. But when that concept is applied to inert heat containers, we do not call them thermal storage containers, we call them *heat sinks*, to describe that we can "sink" heat into them. When heat is convected or conducted, the temperature of the "receiving" body cannot exceed that of the "sending" body; when the heat is radiated, the temperature of the "receiving" body can rise above its immediate environment. If the wall panel is exposed to brilliant sunlight, hence strong radiant heat, it can rise in temperature even on a cold day. It is the air that is cold, and it is highly probable, therefore, that the wall not only will reradiate some of the gained heat but will conduct some of the absorbed heat internally, and convect some of that heat to the surrounding air. We have all felt the intensity of heat from pavement to be far hotter than the surrounding air. That heat is the accumulated *radiant solar gain*, which cannot dissipate as rapidly as it is collected.

Through Solid, Homogeneous, Single Layers

We know from this discussion that a heat source applied to an extreme surface of a solid will heat that surface and initiate a dispersion of that heat through the body of the solid. We know that if the opposite face of the solid is a sink, any heat transferred through the solid will dissipate into the sink. Barring the effects of accumulated radiant heat at the outer surface, the temperature at the heated surface will hold at the ambient level of the warm side; and the temperature of the other side will hold at the ambient level of the sink. For example, if the outside air is a steady 90° F (32.2° C), then the surface temperature of the wall will stabilize at 90° F. If the ambient temperature of the inside is 70° F (21.1° C), then the inside surface temperature will stabilize at 70° F.

The difference in temperature of the two surfaces of the same material is 20° F (11.1° C). The common notation for temperature is T. We will designate the outside temperature as T_o and the inside temperature as T_i. The difference in these two temperatures is $(T_o\text{-}T_i)$, which we will designate as ΔT. In this case, $\Delta T = 20°$ F. The heat passing through any given location within the section is the same. Heat passes from the warm side to the cool side; no more is added along the way, and none is lost.

If the material is homogeneous, the change in temperature is directly related to where in the cross-section we take the temperature. The temperature at the centerline is one-half of ΔT because the midpoint of the material is one-half of the way through the material, and the temperature is evenly distributed across the section. This is a very important point, so we'll restate it differently: The temperature of any point within a homogeneous solid is proportional to the distance from the surface times the temperature differential. If the total thickness is defined as W, and distance from the surface is X, then the temperature at X, namely T_x, is X/W × ΔT. If W=12 inches and X = 3 inches away from the warm side, then the drop in temperature at X is $\frac{3}{12}$ or one-quarter of ΔT, so $T_x = T_o - (\frac{1}{4}\Delta T)$.

We can use the section of the material as a convenient graphic device for depicting this phenomenon. If we set each face as a vertical temperature scale, we can place a tick mark at the outside face of the material at the 90° mark and another tick mark on the inside face at the 70° mark. We can connect the two marks with a straight line, and every location through the thickness of the material can be read directly from the vertical scale corresponding to the location of the horizontal scale of the material's thickness. One-quarter of the way through will be 85° F; halfway, 80° F; three-quarters, 75° F. Every intermediate temperature on this straight line connects the temperatures of the two extreme surfaces. The temperature drop across the section is proportional to the percentage of the thickness traversed. This is true no matter how thick the material is. If the material is four inches thick, the two surfaces are still 90° and 70°, respectively. The thickness of the material does not change the temperature of the two surfaces; and as long as the material is homogeneous and continuous, all the intermediate temperatures fall on the straight line connecting those two surface values.

If this is true, and it is, what is the thermal difference between a four-inch-thick wall and a four-foot-thick wall? The difference is not in the respective temperatures; it is in the quantity of heat required per hour to sustain the steady state temperatures. If heat traverses the section, and the wall acts as a thermal sieve through which the heat passes, it stands to reason that the thicker the wall, the greater the resistance to the passage of heat. If the material is homogeneous, it also follows logically that the resistance offered to the passage of heat is directly proportional to the amount of material. The rate of heat passage required to maintain the outside surface at 90°, given the 70° sink on the inside, is twice as great for a two-inch (50.4 cm) panel as for a four-inch (100.8 cm) panel. The difference in the thickness—or, more specifically, in the thermal resistance—of the panel is proportional to the amount of material, providing its resistance is continuous and homogeneous. This says that the issue is the *rate* of heat transfer, not the temperature at any given point.

To emphasize the significance of this proposition, we can turn the problem around and put the warm side inside the building. Now $T_i = 70°$ F (21.1° C) and $T_o = 20°$ (–6.7° C). It is winter, and we are trying to keep the place warm. We have a furnace that is supplying heat at a sufficient rate to maintain the inside temperature at the desired level. We know that if we turn off the furnace, or the furnace fails for a protracted period of time, the entire house will cool down to the outside temperature; it is only a question of how long that might take. In the meanwhile, we can maintain the ambient inside temperature as long as the furnace keeps cranking out the requisite amount of heat. That amount of heat, Q, must be supplied at a regular rate, because we know that at some rate or other it is leaking through the wall to the outside, which is, for all intents and purposes an infinite heat sink with an ambient temperature of 20° F.

If the rate of heating required to heat the house satisfactorily is 2Q/hour for a four-inch wall, how much heat per hour is required for a two-inch-thick wall of the same material? The second option is exactly half the thickness, and, therefore, offers half the resistance of the first wall. Because it is less resistive, heat will flow through at a faster rate if we are to sustain the inside temperature at the proscribed level. In fact, because the resistance is half, we must supply heat at double the original rate, or 4Q/hour. The impact of this requirement will be absorbed by the furnace, which now must burn twice as much fuel in the same period of time to compensate for the thermally thinner cross-section.

This is all well and good as long as the furnace has the reserve capacity to increase the thermal output of the burner. Suppose, however, that the furnace was already performing at full capacity when the wall was four inches thick: as long as the wall offered that same resistance and the outside temperature remained at 20° F, the furnace could keep up with demand. Now suppose that, for whatever reason, the thermal resistance of the wall drops to the equivalent of two inches of the original wall material— remember, the furnace is already at full capacity. Obviously, the furnace will not keep up with demand, and the steady-state condition will adjust for the available rate of heat supply at the reduced rate.

The ΔT of the first wall was 50° F (70° F – 20° F), the wall thickness was four inches, and the rate of the heating supply was 2Q/hour. The rate of heating supply remains the same, namely 2Q/hour, but the resistance, R, of the wall is now one-half the original resistance. What is the revised ambient inside temperature that will bring the second situation into thermal equilibrium? The rate of heat loss is a function of ΔT and R, namely $\Delta T/R$. If, in the first case, $2Q = \Delta T_1/R_1$, and the rate of heating supply in the second case is the same, then $2Q = \Delta T_2/R_2$. We know that $R_2 = R_1/2$; therefore, ΔT_2 must be one-half of ΔT_1 or 25° F (50°/2), which means that the ambient inside temperature is only 45° F ($T_o + \Delta T_2$ = revised T_i). It follows that the furnace must be designed not only with the expected outside temperature in mind, but with the thermal resistance of the envelope in mind as well.

The point is, we can predict the internal temperature of a given material if we know the surrounding ambient temperature conditions. We also know that the temperature drop or temperature gradient across a given cross-section is linear from one surface to another, and we know that the increase in thermal resistance to heat flow is in direct proportion to thickness of the material.

Through Multiple Layers

We know from the previous section that the thermal resistance of a panel composed of a single, homogeneous material is a function of the thickness of the panel. If the panel thickness is doubled, the rate of thermal transmission will be cut in half. Any change in the thickness of the panel will result in an inversely proportional change in the transmission rate. But what if the added thickness is not the same material, but a thermally distinct material instead?

Any additional material will increase the thermal resistance. Of course, the addition will also increase the overall thickness of the panel. If, however, instead of simply adding thickness, the second material substitutes for the original material, the total thickness will be the same as the original panel, meaning that the overall thermal resistance of the revised panel will depend on the inherent resistive value of the second material compared to the original material. If, coincidentally, the second material is thermally identical to the original material, and if the total thickness is the same, then the thermal resistance will be unchanged. If the inherent thermal resistance characteristic of the second material is less resistive than the original material, then the thermal conductivity will increase. If the second material is inherently less conductive—that is, more resistive—than the original, the overall resistance of the panel will increase and the rate of thermal conduction will decrease.

Now we will see how easy it is to determine the actual temperature drop in each layer when the layers are of different thermal properties. The total thermal resistance of a panel is the sum of the resistances of the individual layers. If one layer has a thermal resistance (R value) of 40 and the other has an R value of 20, then the overall R value of the wall is 60. The larger the R value, the greater the thermal resistance. The thermal resistance, or R value, has two components: the thickness of the material and the characteristic thermal resistance per unit of thickness, which is called the U value ($R = \text{thickness} \times U$). The decrease in temperature across any set of thermally varying materials is proportional to the R value of that layer (R_A) compared to total R value (ΣR) as the temperature drop in that layer (ΔT_A) is to total temperature drop (ΔT).

Table 4.1 gives some exemplary values of a hypothetical two-layer panel. The panel consists of two materials with a combined thickness of 12 inches and a combined R value of 56. The layers are arranged such that A is on the outside and B is on the inside. The outside ambient temperature is 20° F, the inside, 70° F. The total temperature differential (ΔT) is, therefore, 50° F. The temperature change that occurs between the extreme outside face of A and the interface of layer A and layer B is proportional to R_A as a fraction of ΣR (32/56 = 57.1%), which means that $\Delta T_A = R_A / \Delta R$ (ΔT) = 0.571 × 50° F = 28.55° F. The actual temperature at the interface is $T_A + \Delta T_A$ = 20° F + 28.55° F = 48.55° F.

Next suppose that the layers are more reflective of values and numbers of layers in an actual assembly. Table 4.2 represents the thicknesses and thermal resistance values of a typical brick cavity wall with a block backup,

Table 4.1	HYPOTHETICAL THERMAL RESISTANCE VALUES			
Layer	U Value	Thickness	R Value	Percent of ΣR
A	8	4	32	57.1
B	3	8	24	42.9
Total		12	56	

Table 4.2	THICKNESS AND THERMAL RESISTANCE VALUES FOR A BRICK CAVITY WALL				
Layer	Material	U Value	Thickness	R Value	Percent of ΣR
A	Brick	1.20	3.63	4.35	9.81
B	Cavity (air)	3.50	2.38	8.31	18.75
C	Block	1.60	5.63	9.00	20.30
D	Batt Insulation	6.00	3.63	21.75	49.06
E	Drywall	1.30	0.63	0.81	1.83
F	Plaster	0.90	0.13	0.11	0.25
Total			16.00	44.34	100.00

a nominal 2-inch cavity; nominal 4-inch studs with fiberglass batt insulation filling the stud space, gypsum drywall, and a plaster skim coat on the inside face. The unit values for the thermal resistance properties of the individual materials are for illustrative purposes only and should not be used in actual calculations.

If the outside ambient temperature is 17° F and the inside ambient temperature is 72° F, then ΔT is 55° F. Table 4.3 displays the temperature at each of the interfaces through the thickness of the wall.

Table 4.3 illustrates several valuable lessons; probably the most important is that these sorts of computations are not that complicated. The only element that will be different in these tables and similar ones produced by design professionals for the design calculations of actual buildings are the U

Table 4.3	INTERFACE TEMPERATURES				
Interface	Material	ΔT Overall	Percent of ΣR	ΔT between Faces	T° F at Interface
O/A	Outside to Brick		0.00	0.00	17.00
A/B	Brick/Cavity (air)	55.00	9.81	5.40	22.40
B/C	Cavity (air)/Block	55.00	18.75	10.31	32.71
C/D	Block/Batt Insulation	55.00	20.30	11.17	43.87
D/E	Batt/Drywall	55.00	49.06	26.98	70.86
E/F	Drywall/Plaster	55.00	1.83	1.01	71.86
F/I	Plaster/Inside	55.00	0.25	0.14	72.00

values. This is not rocket science; in fact, it is little more than basic book-keeping. This should not be interpreted as a denigration of the engineering skill required of design professionals, rather to point out that a great deal of what we do is keep track of many pieces of data. Therefore, the format is probably as important as the formula. Another valuable lesson of this table is that even if there were 50 layers, our ability to predict performance would be no more complicated, only more voluminous. The number of layers does not change the concept, only the number of rows of calculations.

The specific information provided by Table 4.3 is that when the outside temperature is, indeed, 17°F, then the entire thickness of the brick is below freezing. In fact, the majority of the cavity is below freezing as well. On the other hand, even at a temperature as low as 15°F, the entire thickness of the backup block is above freezing. There are, however, a few qualifications to this, and a major one has to do with the air in the cavity. The value provided is a compromise between what the true value might be if the air in the cavity were completely trapped, and if the air in the cavity could circulate. This single value is a crude recognition that the U value of an air column in a constructed cavity can be quite variable.

Normally speaking, still air is an excellent insulator. In fact, trapped, still air performs so well as an insulator that most insulation is an elaborate effort to trap air. In the case of batt insulation, it is not the glass fibers that provide the insulative effect of the blanket, but the insulative value of the air caught in the maze of glass fibers. The reason that cavity air is relatively less resistive is that it is more fugitive than the air ensnared in the fiber glass batt. This same principle affects the performance of our clothing and bedding in cold weather: fabric and clothing construction that can sustain trapped air pockets has a greater insulative effect than the same material that has been squashed or flattened. The warmth of a down comforter is the result of the air, not the down. The benefit of the down is that it will return to its lofted position after being compressed, thereby reestablishing the air volume. It is also lightweight and relatively nonconductive even as a material, inasmuch as the shaft of the feather is filled with air. The warmth of a wool sweater comes as much from the loft of the knit as from the wool fiber. Likewise, the principle of layering clothing has to do with capturing air between layers, not with the layers themselves. The same is true of the "clothing" on a building; the air provides insulation as long as the air is not moving around.

The stillness, or trapped property, of air is largely relative. The air in fiber-glass batt insulation is not truly trapped; it can migrate. The impediment to free flow is the labyrinthine nature of the maze of the construction. The accumulation of interlaced fibers restricts and impedes air flow to such an extent that it is effectively still. The air in a cavity that theoretically may be as insulating as the air in a batt blanket migrates depending on other factors of construction; thus, assigning a fixed thermal resistance value to the air in the cavity of a cavity wall is not reliable without knowing much more about the construction and detailing of the wall.

Another relevant detail related to Table 4.3 is the determination of the air volume immediately proximate to the inside and the outside surfaces. This air is relatively still—or at least sufficiently so that it may be considered part of the envelope. If this air is still enough, it will—or can—provide thermal resistance between the inside volume and the outside ambient condition. The obvious problem is determining the degree of stillness to evaluate this contribution. Various manuals on the subject of thermal resistance provide guidance as to when and to what degree to include this variable in the calculation of total thermal resistance, so we will not go beyond the qualitative identification of the phenomenon here. There are, however, circumstances when the inclusion of this factor is significant,

With all that said, Table 4.3 does not calculate the total amount of heat lost through the wall. We have calculated the rate of loss through a given profile of the wall for a given area, depending on the unit system being employed, not the total heat (Q) lost through the entire wall in an hour. To do that, we need to know the area of the wall for which this particular section is applicable, and the respective areas of other wall profiles as well.

TEMPERATURE AND MOISTURE GRADIENTS

If, as was suggested, we use the vertical section through the field of the wall as both a horizontal graphic scale of the wall profile and as a vertical scale of the temperature, we can plot the inside and outside temperatures at those surfaces. Recall that when the material is a single layer of thermally homogeneous material, we can draw a straight line from the point of the inside surface representing the inside temperature to the point on the outside surface representing the outside temperature. All the temperatures between the two extreme surfaces lie on this line, and the values can be read directly from the vertical temperature scale.

If the outside temperature is below freezing, the sloped line from the outside surface will rise as it proceeds laterally toward the inside surface. At some dimension away from the outside, the temperature will pass through 32° F, a point—or, more accurately, a line—of considerable importance to us, and to which we will refer as the *frost line*. The sloping line of the temperature track across the dimension of the wall is the temperature gradient of the wall. In Chapter 3 we discussed the concept and application of a temperature gradient relative to the effect varying temperatures will have on the internal stress of materials. The presence of a temperature gradient in the context of vertical closure systems is essentially the same, with a slightly different take. Here we are interested specifically in the consequences and locations of potentially freezing water in the materials; earlier, we were interested

FIGURE 4.2 This is an early twentieth-century attempt to reinforce cement stucco. The exposed wires were embedded in the stucco and attached to the stone masonry with wood inserts and bent nails.

in the internal stress generated by differential expansion (see Fig. 4.2 for an early attempt to counteract internal stress). We are not looking at two different phenomena; we are looking at different aspects of the same phenomenon.

If we can construct a graphic representation of the temperature at any location on a single-layer panel, based on the notion that thermally homogeneous materials exhibit a linear temperature change across the thickness of the material, then we can do the same with multiple layers of materials because we can identify and plot the temperature at each interface. If our wall section has five layers, and each layer is accurately drawn to scale laterally, we can construct on the vertical scale all the temperatures at the various interfaces, and connect the adjacent points to form a graph of all the immediate temperatures across the entire cross-section. Having connected all the dots, we have a sequence of line segments that may have different slopes, but they are always positive or negative depending on the consistent, relative positions of inside and the outside temperatures and the steady thermal state of the entire system.

If, as is often the case, the outside temperature fluctuates, then the series of linear line segments may not be so predictable. Consider the very common situation that occurs with the passage of a warm front. The starting point is the same example as we described earlier: 20° F outside and 70° F inside. We will revert to a single-layer panel again, which means that the temperature gradient is a straight line connecting the outside lower temperature to the inside higher temperature. Over the course of a prolonged period of time, the temperature of the outside surface rises until it is also 70° F; it holds at that temperature for an indefinite period. Given long enough, we know that the temperature gradient will reestablish along a straight horizontal line at 70° F all the way through the wall.

If the outside temperature were to increase rather rapidly—particularly if the source of such a rise were the radiant heat of the sun against the outside wall surface (see the following section for more details)—then the outside surface temperature may rise rather rapidly. If the rise is rapid enough to warm the extreme outside surface faster than the heat can dissipate to the interior, there is the prospect that, at least for the moment, the exterior surface of the panel will be warmer than the proximate material toward the interior. Remember that the extreme inside surface is at 70° F, so the inside surface is warmer than the interior of the core of the panel as well. This means that the shape of the temperature gradient is higher at the two extreme surfaces than at some location in the interior.

The precise shape and properties of this resulting temperature gradient may appear to be much more complicated than it is, so we will demystify it. By now we are familiar with the graphic representation of a section of a wall or panel with the temperatures of the extreme faces on the vertical scale. We know that the temperature gradient in a single-layer system is a straight line between the two surface temperatures. We know also that with enough time the wall will rise in temperature until it is exactly 70° F all the way through. The amount of heat required to accomplish this task is a function of the amount of material to be heated (mass), the amount of the temperature rise (ΔT), and the specific heat of the material (c): $Q = mc\Delta T$. The immediate problem is to determine ΔT, because while the extreme outside surface will rise 50° F, the inside surface will not change temperature at all. The trick to the solution is to look at the average ΔT, which is (70° F – 20° F)/2 = 25° F. The total amount of heat required to bring the wall to a uniform temperature of 70° F is 25° F, times the mass of the wall, times the characteristic specific heat for the material at issue, which we will arbitrarily say is 12 inches thick and made of masonry with a specific heat of 0.5 and a density of 120 pcf. The heat required per square foot of wall surface area to raise the temperature from an outside temperature of 20° F to a through-wall temperature of 70° F is (25° F × 120 pcf × 1.0 foot × 0.5) = 1500 BTU (1.58×10^6 J). What we do not yet know is how long that it might take, the source or sources of heat that may provide the energy, or the profile of the temperature gradient as it changes from the original condition to the final condition.

Thermal Inertia and Thermal Lag

Daily experience teaches that substances get warm and that they cool. Water does not start boiling the instant the burner is turned on. The delay is a measure of the time required to transfer the heat into the material, and then to "fill" the material with enough energy to see a rise in the temperature of the material. The next section deals specifically with quantifying the delay time; here, we accept for the moment that there is some delay, which we will define as the *thermal lag* but which we will refer to by an abbreviation, M (for minutes). Specifically, we will accept that the wall will get warm, and that it requires M amount of time to reestablish the equilibrium point of 70° F.

There are two general scenarios for the warming of the wall: slow and rapid (see Fig. 4.3). In the slow scenario, the heat applied to the outside face itself is applied at such a low rate that the temperature gradient remains a straight line throughout the process of warming. The outside temperature begins to creep up the vertical scale on the outside surface of the wall, but the incremental changes are so small and so slow that the wall is effectively always in a steady state of thermal equilibrium. Geometrically, it is as though we put a pushpin into the inside temperature point and pivot the temperature gradient about that point until it becomes horizontal at precisely the 70° F level; then it stops. This is not an impossible scenario and may come close to the actual situation, in that the delay, or *lag time*, of the material is very short.

Short thermal response times may be the result of a combination of factors. The amount of heat supplied may simply be extremely high, but that alone probably will not result in what we have described above as a straight-line thermal response. Other factors more relevant are physically thin sections, low specific heat values, and, as we will discuss in a subsequent section, high conductivity properties. If the material is thin with low specific heat and high conductivity properties, it will change temperature rapidly. The term we apply to this combination of values is *thermal inertia*.

The idea behind the concept of thermal inertia is analogous to the notion of physical inertia and acceleration. If heat is analogous to force, then thermal inertia conveys the notion that application of heat may initiate a change, but the rate and duration of that change will depend on a set of variables, just as the acceleration of a physical mass is dependent on a set of variables, namely the mass of the object, the magnitude of the force, frictional resistance to the force, and so forth. The slow scenario may be a result of extremely high thermal inertia, extremely low increases in Q, or both. Slow scenarios are not exclusively the result of thermal inertia properties, nor are rapid scenarios exclusively the result of low thermal inertia properties.

The second scenario for thermal adjustment is the rapid, or fast, scenario. The heat is applied to the outside surface in sufficient quantity such that the extreme fibers heat almost immediately. This is credible depending on the definition of extreme outside fiber. If the definition is a substantial amount or depth of material, then we are less likely to accept the premise, because it is difficult to imagine a substantial amount of material heating instantly. We are far too familiar with the delays involved in heating to accept the proposition. If, however, the definition of extreme outer

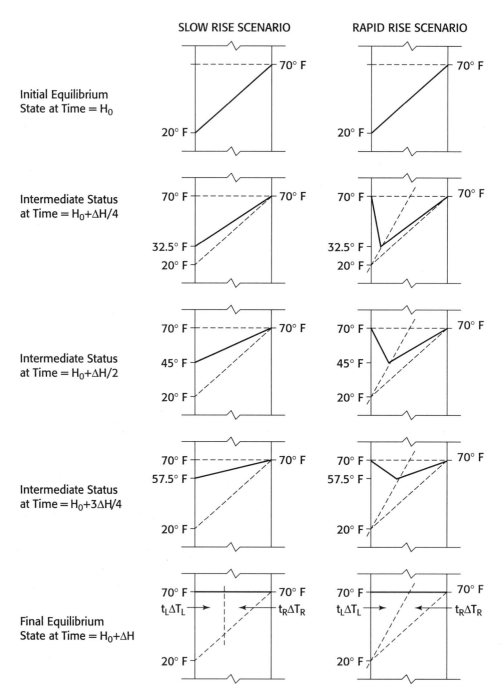

FIGURE 4.3 Graphic portrayal of thermal scenarios for slow versus rapid temperature responses.

fiber is one molecule in depth, then the premise is not only more credible, but the inverse—that is, that the molecules are not immediately affected—is less credible. As discussed, the result of the application of heat to a substance is to increase molecular oscillation velocity and amplitude. It follows then that the most immediate effect of the application of heat will be to increase the activity of the molecules most proximate to the heat source.

To continue our example, the wall is receiving sufficient heat from both sides so that the wall or panel is being heated toward the middle. This means that the extreme outside surface will reach the 70° F temperature quickly, if not instantly. The core of the section is actually at a lower temperature than either of the extreme surfaces, and although we intuitively know that the core temperature will begin to rise, the trajectory and shape of the temperature gradient of that rise is less than obvious.

The gradual warming of the interior of a chilled wall is qualitatively familiar; and perhaps surprisingly the quantitative appreciation of the phenomenon is not that complex either, and it is, once understood, informative. The hardest part, as with most endeavors, is taking the first step. In the wall-warming problem, the first step is to recognize that regardless of the relative temperatures of the two sides of the wall, heat is *always* being supplied to both sides simultaneously. What accounts for the differential is the *rates* of supply. While it is convenient to think of the issue as one of heat on one side and no heat on the other, this is simply inaccurate. The complete absence of any heat on the cold side means that it is an *infinite* heat sink, the temperature of which must, therefore, approach absolute zero. If, in fact, the temperature of the cold side is higher than absolute zero, then there must be some heat present in the air of even the coldest of conditions short of less than 400° F below zero. If the temperature on the outside of the wall reaches absolute zero—other than in controlled laboratory conditions—we all have a problem far greater than anything this book can address.

In our example, the outside temperature is initially 20° F. To us that is cold, and we may, therefore, be inclined to think of that as an absence of heat. Even at 20° F, however, the molecules and atoms are oscillating, just not as rapidly as those on the opposite side of the wall, which are oscillating at a temperature of 70° F. We tend to think of the difference between 20° F and 70° F as a large spread, but in the overall scheme of materials, a difference of only 50° F (27.8° C) is minuscule. It is only because we are particularly sensitive to biologically threatening conditions that we attach so much significance to such shifts in temperature. Compared to absolute zero, 20° F is actually fairly warm, and for our purposes here, it means that there is some heat being supplied to the cold side.

The notion that heat flows from hot to cold is valid, but with the qualification that such clear directionality of heat flow is a resulting net value, as opposed to a unidirectional absolute value. In other words, the flow of heat is the steady-state condition of our example, and could be a single vector of $\Delta T = 50°$ F or the sum of two vectors operating in opposite directions, one from outside to inside of 20° F and the other of 70° F from inside to outside. Reality is closer to the latter than the former. This field of physics is called statistical mechanics; and, while the details of it are beyond our concern here, the reality of heat dissipation and dispersion is more a summing of trends than the simple addition of vectors.

From now on, the physics and the mathematics get easier. If in our on-going example, we started with a wall supplied with differing rates of heating, the amount of heat being supplied to the outside increases such that at the reestablished equilibrium state, both sides hold at 70° F, then the rate of heating must be equal. The amount of heat reaching the interior of the wall from each side is the same per hour as long as the properties of the two surfaces are the same, which for now we will accept as being equal. If we accept the fast-heating scenario as the more probable of the options, then at any time during the readjustment period, the amount of heat supplied from each side is the same.

If we connect the three points of the example on the diagram of the wall section with the scaled temperatures on the opposite faces, 20° F and 70° F on the outside face and 70° F alone on the inside face, and if we connect the three points with straight lines, the resulting figure is a triangle. We will label the points of the triangle with capital letters: A at the outside temperature of 70° F, B at the inside with a temperature of 70° F, and C at the outside temperature of 20° F. In the fast scenario, the temperature at the extreme outside surface immediately climbs to 70° F, which means a second after heat is applied to the outside face, there is a new triangular diagram. Points A and B are the same, 70° F at the outside and inside faces, respectively. The original point C, however, cannot be the same. The condition that allowed C to be at its original location was a heat supply level that was enough only to sustain an outside temperature of 20° F, and that condition has changed. The new point, which formerly coincided with but has now migrated away from C, we will call point O. We do not yet know exactly where O is, but we know this much: O must be the interior of the extreme outside face, which is the AC line, because the extreme outside face is now at 70° F. O must be above the diagonal line connecting B and C because every temperature below that original equilibrium line means that there is a net reduction in the energy in the wall section. We know that we are adding energy, so O must be above the BC line. O is, therefore, at an indeterminate location to the inside of AC and above BC.

If we arbitrarily and temporarily place a point O on the diagram to conform to these two conditions, we can form a new triangle, ABO. The straight line from A to O is the temperature gradient from 70° F at the outside face and the location inside the wall at O. The straight line from B to O is the temperature gradient from 70° F at the inside face and the location O. It is that simple. These two line segments are the temperature gradient of the changing temperatures inside the wall; and the temperature at O is, by definition, the lowest temperature in the wall *at that particular moment.* In other words, the new triangle ABO is a snapshot of the changes occurring inside the wall on the way from the initial equilibrium state to the finial equilibrium state. The point O is an actual low temperature point along a sequence of changing temperatures, which are the trajectory of low points as the entire wall gradually warms to a uniform temperature of 70° F through the entire wall.

We do not know at this point how long the entire warming process will take, but it will eventually stabilize, and we can assign that period of time as H. Because the rates of heat being supplied, and presumably absorbed, by the wall is the same over time H, then half of the energy absorbed by the wall will be absorbed in half the period; 75 percent will be absorbed in three-fourths H. The area between the original triangle ABC and the subsequent triangle ABO is proportional to the energy or heat absorbed by the wall to arrive at the temperature gradient AOB. If the rate of heat supply is the same from both sides of the wall, the contribution to the heat absorbed, represented by this area, is evenly divided between the outside heat source and the inside heat source. The area of the difference in the toward triangles is, therefore, evenly divisible into two equal areas.

If we connect O to B by a straight line, the area between the two temperature gradients is now divided into two smaller triangles, ACO and BCO. The sum of these two areas is proportional to the energy absorbed during the warming process, but not every, or any, location of O will satisfy the condition that these two smaller triangles have equal areas. Only a specific set of locations of O will result in equal contributions of heat from the inside and from the outside faces. To solve for that set, let us begin with a simple test location, namely half ΔT, which is to say that the low temperature on the interior of the wall, O, is by definition at 45° F. We do not know how long it will take for the interior core temperature to reach 45° F, but we know that it will cross that line. We also know that it must be to the outside of the geometric centerline of the wall, because it was at the temperature of 45° F before we started the warming process. It must, therefore, be warmer than when it started because we have added heat throughout period H. O is, therefore, on the horizontal line of $T = 45°$ F, inside the outside face and outside of the centerline.

The precise location of O is such that when O is connected to B by a straight line, the areas of the two resulting smaller triangles ACO and BCO are equal. The proof of where these conditions are met is actually quite easy. The area of our original triangle, ABC is defined as $A_\Delta = [\Delta T_O \times t]/2$, where ΔT_O is the original change in temperature from point C to point A, and t is the thickness of the wall. At $\Delta T_O/2$, the area of any triangle with a point lying anywhere along that line, which is to say point O at any location along the line T = 45° F, necessarily has an area of $[\Delta T_O/2 \times t]/2$. This area is exactly one half of the original area, which means that the area outside this triangle, ABO, and the original triangle, ABC, is also exactly $A_\Delta/2$.

We now want to find the horizontal location of O such that ACO = BCO = $[A_\Delta/2]/2$. This means that the area of ACO = BCO = $[[\Delta T_O/2 \times t]/2]/2 = [\Delta T_O/2 \times t/2]/2$. Because the base of triangle ACO is necessarily the length of line AC, then the area of $[\Delta T_O/2 \times t/2]/2$ is more clearly expressed as $[\Delta T_O \times t/4]/2$. This says that the temperature of 45° F will occur at some time during the course of the warming trend; and when that event occurs, the location of the low point of 45° F will occur at a lateral equal to one-fourth of the wall thickness away from the outside face.

If we repeat the algebra for other locations, and plot the locations of O, we will define the trajectory of the low-point migration from B to the midpoint of the horizontal line between A and B, which is to say at the center-

FIGURE 4.4 Normally attributed to the expansion of the reinforcement due to corrosion, the stress sufficient to fracture and spall the concrete is exacerbated by thermal expansion and moisture expansion differentials.

line of the wall and at the 70° F line. This is the low-point migration line. (Adventuresome readers may at this point want to try their hand at figuring out the low-point migration trajectory when the final outside temperature is not the same as the inside, or when it is higher than the inside wall. The clue to the answers lies in the assumption that the rates of contribution from the respective sides of the wall are much more important than the eventual steady-state equilibrium position of the temperature gradient.)

Conductivity and Heating Rate

The remaining unknown in our example is H, the length of time required for the equilibrium process to occur, or, in other words, for the interior of the wall to come up to the temperature of the surrounding atmosphere. The answer to this question can be approached from a variety of starting positions, but we will attempt to stay with the graphic model of the wall section and the respective temperature scales of our continuing example. Heat must enter into the wall and raise each square foot of the wall from an outside temperature of 20° F to an outside temperature of 70° F which, co-incidentally, is the inside surface temperature as well. The average cubic inch of the wall section, therefore, must increase in temperature by an amount of $\Delta T/2$, or 25° F. If the thickness of the wall is 9 inches and the density of the material is 120 pcf, which is typical of masonry, then the weight per square foot of wall is 90 psf. If the specific heat of the masonry is 0.6 BTU/pound/Δ° F, then the total number of BTUs required to raise one square-foot wall is Q = mass × specific heat capacity × change in temperature = 1350 BTU (1.42×10^6 J).

The second part of the computation is to determine how long it will take to pump 1350 BTUs into and through a masonry wall that is nine inches thick. Obviously, this depends on the supply of heat at the two surfaces, but we said early on that the two rates were equal, and the same as that required to sustain the wall in its original state of equilibrium, namely 70° F inside and 20° F outside. The net flow of energy in that situation depended upon the same resistance (R), which was discussed earlier. The higher the resistance to heat flow, the longer it will take for the heat to dissipate and disperse within the masonry.

Looking again at the diagram of the wall section and the temperature scales, at the outset, the net heat transfer was from inside to outside, and the temperature differential was $\Delta T = 50$° F; the thickness was and still is 9 inches; and the area is one square foot. If we set the rate at some number of BTU (Q) per hour; to sustain the equilibrium of this arrangement, we can solve for the Q/hour according to the logic that the rate of the passage of energy through the wall is directly proportional to the temperature differential and inversely proportional to the resistance to that passage per unit of thickness and the thickness of the wall. In this case, we know $\Delta T = 50°$ F and the thickness is 9 inches. The only unfamiliar value is the actual U value of masonry which is 21.2 BTU/ sq. ft./hour /° F. Multiplying all the variables together, the Q/hour = 1 sq. ft. × 50° F/21.2 × 9 inches = 2.822 BTU/hr (2975 J/hour).

At that rate of heat supplied to both faces, the time required to supply 1350 BTUs is 1350 BTUs/2 × 2.822 = 239.2 hours, or almost ten days, without interruption to bring the entire wall up to a temperature of an even 70° F. This may seem surprising, but it is by no means shocking. The reason it may seem longer than experience suggests is that the actual and likely rates of heat gain on the outside surface are generally much greater than the low level of 2.8 BTU per hour. With that said, however, when north walls, in particular, go through seasonal temperature changes, these rates of temperature adjustment are not at all surprising. The north side of a solid masonry wall will rise in temperature more as a response to ambient air temperature and less as the result of solar gain. If the ambient air temperature during the day climbs to 70° F only to fall to 45° F during the evening, the time to reach a daytime wall temperature of 70° F may take several weeks, if it ever actually reaches such a temperature.

Solar Gain and Emissivity

A wall may absorb heat much more rapidly primarily because of ambient air and the impact of direct solar radiation on the surface of the wall. Solar radiation is an extraordinarily important element in evaluating the performance of any exterior material in actual service, as opposed to laboratory trials. A couple of factors combine to make solar radiation so important, but here we will restrict the discussion to the heat generated by solar radiation. The rate of solar gain from radiation against a surface perpendicular to the incoming rays of the sun is approximately 44.6 BTU/hour/sq. ft. Compare that to the 2.8 BTU/hour we determined to be the heating rate in the previous example. There are, to be sure, several factors that will mitigate the actual solar gain on the wall, but already we are aware that this is a major source of energy.

First, the wall may not be perpendicular to the direction of the sun. If the sun strikes the wall at an oblique angle, which is usually the case, the intensity of the radiation is diminished in proportion to that deviation both vertically and horizontally. For example, in Philadelphia, a wall facing due south at high solar noon on June 21 will receive approximately 13 BTU per hour (13,700 J/hour) of solar radiation. Ironically, six months later, that same wall will actually receive more radiation, more than 38 BTU per hour (40,000 J/hour), because the sun angle is less oblique, but the lower ambient temperature may negate the added heat. Of course, the other way of looking at this is that the lower ambient temperature is mitigated by an increase in solar gain.

The same principle is operative horizontally as the sun traverses the sky during a given day. Let's say that by 10:00 A.M. in the summer, an east-facing wall has been receiving low-angle direct radiation for five hours, whereas at 10:00 A.M. in the winter that same wall has been receiving radiation for three hours or less. In addition to the day and season, solar gain is significantly affected by cloud cover and foliage interference. Moreover, pollution scatters the sun's rays. The surrounding apron of the building may reflect radiation onto the wall or simply absorb it. The external factors that affect the actual intensity and duration of solar gain are complicated and highly variable, but given the magnitude of the potential gain, even for brief periods, solar gain is an important factor in assessing the performance of a wall.

The variation of the solar gain in and of itself is a major source of differential expansion and contraction of materials (see Fig. 4.4). The subject of differential thermal stress was addressed in detail in Chapter 3, but we did not explore the sources of some of the changes in temperature. There is no way to emphasize strongly enough that solar gain is a major source of thermal differential stress and, therefore, a stress gradient-based deterioration mechanism at the crystalline level. If direct sunlight strikes the surface perpendicularly, and the specific heat of that material is as high as 0.6 for one hour, the rise in surface temperature could easily be 70° F or more. The temperature of an asphalt roof may rise 100° F or more in a single day. If that same material then reradiates that energy during the night in a cool ambient atmosphere, the change in temperature could exceed 100° F in the course of a few hours. As potent as the a single swing in temperature may be, the effects of solar radiation on materials is compounded by the fact that they occur over and over again relentlessly for years.

Another factor that significantly affects the solar absorption rate of a surface is the character of the surface. A highly reflective surface will absorb less radiant energy than a less reflective surface. Highly irregular or pocked surfaces have more actual exposure, or "coastline," than very smooth surfaces. Dark surfaces absorb more energy than light surfaces. Though distinctly different, these three characteristics have much the same effects relative to the absorption of radiant energy: they all increase the propensity of a given surface to admit or reject radiant heat energy. Collectively, they are assigned a value, which is a unitless factor, called *emissivity*. The values of emissivity range from 0.0 for a totally reflective surface to 1.0 for a totally absorptive surface.

Using numeric factors in this way is mathematically efficient. Unfortunately, it has the untoward effect of disguising the familiar. When a number is described as a unitless factor ranging from 0.0 to 1.0, that means we are dealing with a simple percentage. When the factor is 0.0, it means that none of "whatever" is effective. In the case of emissivity, the "whatever" is radiant heat, and an emissivity factor of 0.0 means, quite simply, that no radiant heat is admitted. A factor of 1.0 means that 100 percent of the incident radiant heat is absorbed. The perfect energy-absorbing body is referred to as a *black body*; it has an emissivity factor of 1.0.

The three contributing factors just mentioned are not quantitatively distinguished in the literature, but they are relevant to our appreciation of changes in materials over time. The reflectivity of a material is also related to its specular quality, and should not be confused with the color of a material. The specular quality and the coastline of the fabric are related to the extent that an increase in coastline will almost necessarily reduce the specular quality. Specular quality typically diminishes over time primarily as the consequence of surface oxidation, corrosion, or other chemically related alterations of the substance. This is especially noticeable with metals; however, many substances are subject to chemical conversion from the exposed crystalline, which may have been highly polished to an unpolished crystalline solid lying on the substrate of the original material. The growth of corrosion products into their own crystalline structures is far more likely to result in a matte finish than a shiny finish. The simple change from a polished surface to a matte surface of otherwise unchanged properties can increase the emissivity of the surface by 10 to 30 percent above the former emissivity value.

Erosion becomes possible as the surface of the fabric corrodes or is otherwise chemically altered. The erosion indicates that at the suboptical level there are indentations and microscopic cavities exposed to the surface. When highly magnified, even an apparently highly polished surface is surprisingly irregular. At 400×, the edge of a razor blade looks not unlike the north rim of the Grand Canyon (which may bring a new perspective to dragging such an object across one's skin). From the standpoint of radiant heat absorption, the irregularity of a surface virtually assures the prospect that some percentage of any surface is perpendicular to the incoming radiation, regardless of the angle of incidence. Even if the immediate angle of impact is oblique, and ordinarily the radiation may have reflected off the surface, the valleys and overhangs provide opportunities for secondary and tertiary interceptions of the initially reflected energy. The deeper the crevices and the higher the frequency of fissures and flaws, the greater the likelihood of rereflected energy being absorbed into the crystalline matrix. The weathering of most materials universally results in the irregular diminishment of crystalline continuity and integrity. The consequences of routine materials pathology is that the surface becomes more irregular with a concurrent increase in emissivity.

Of the three properties that contribute to the emissivity of a given surface, arguably the most durable and reliable is the color. The color we identify with an object is not, technically, the color of the object; it is the color of the light reflected by the object. If an object appears to be white, that is because the object is reflecting at least some fraction of the entire visible range of electromagnetic spectrum. If an object appears to be black, it is because the surface is absorbing a large fraction of the visible range, and reflecting very little of the energy in that range. It is important to note that absent more information, based on color alone, there is no way to determine what the absorptive properties of the object are outside the visible spectrum. It is entirely possible that an apparently white object is an infrared sponge. By the same token, an apparently black object may be both an infrared and an ultraviolet reflector.

The range of commonly encountered values of emissivity is relatively narrow, with the most common values between 0.4 (40 percent) and 0.6 (60 percent). Some materials have values as low as 0.3 and as high as 0.7; very few have values above or below those values. Emissivity tends to change over time, which is true of some other properties, but the evolution of emissivity values is interesting. To the extent that a property changes with time, that change is, generally, the result of deterioration. And so it is with changes in emissivity. For other material properties, the trends are unidirectional. If ultimate strength changes at all, it will invariably degrade to a lower value regardless of the specific material. In contrast, changes in emissivity tend to move toward an average or at least a more moderate value.

Metals, which may initially be highly reflective with very low emissivity values, tend to corrode. The corrosion products often are quite light, but the specular quality is substantially diminished. The finish becomes more of a matte, and the surface will be less smooth as well. This combination will tend to increase emissivity. In darker materials, such as brick, the surface may fade or bloom with efflorescence. In either case, the emissivity value of the brick is likely to decrease. Although emissivity is affected by deterioration, we cannot always predict what the value will be over time, not only because of the varying degrees of deterioration, but because of the nature of the deterioration. If the result is the dulling of a polished surface, then emissivity will increase. If the result is a lightening of a dark surface, then emissivity will probably decrease.

Another aspect of emissivity, not insignificant relative to heat loss and gain, is that while all objects absorb energy at some rate or other, they also emit radiation. The reradiation properties of an object are also measured by the emissivity of the object. Black bodies are perfect absorbers and perfect radiators. Objects that are highly reflective and have low emissivity values are slow radiators. It is also true of all solids that the rate of radiant emission is proportional to the temperature of the object. The hotter the object, the higher the rate of emission, and the shorter the wave length of the emitted radiation. A hot iron starts to glow red, indicative that the wave length and frequency of the reradiation has entered the optical range. As the iron continues to get hotter, the red glow changes to orange, then to yellow, and finally to whitish blue.

In far more common reradiation scenarios, heat in the optical range is absorbed over the entire range of visible electromagnetic energy. The temperature of the object increases, and the object begins to reradiate energy back into the atmosphere. The object, while warmer than it was originally, is by no means glowing hot, so the wave length and frequency of the reradiated energy is well below the visible range. Suppose, for example, that the object being heated is inside a window. At the visible range, the glass in the window is transparent, which means that the energy, at least in the visible spectrum, passes through the material with nominal absorption. In fact, the passage of energy within the visible spectrum through a transparent window may be over 95 percent of the incident energy. The fraction that does not pass is absorbed in the glass itself, which warms and reradiates energy from both of its potentially emitting surfaces.

Meanwhile, the object inside the window absorbs much of the energy transmitted through the glass. As it warms, the frequency of the reradiated energy rises, but the frequency, wave length, and amplitudes are all consistent with relatively low-energy radiation. It so happens that such radiation is typically not transmissible back through the same glass that admitted the original source of the retransmitted heat. The reradiated energy is reflected by the glass back into the interior of the space, to be absorbed by other materials that repeat the cycle. The cumulative effect is that energy in the form of light enters the occupied zone only to effectively become trapped because the glass acts as a one-way filter. This phenomenon is called the *greenhouse effect*, deriving its name from the ability of transparent enclosures to collect and retain energy from light even if ambient outside temperature may be considerably lower than the inside temperature; the glass, otherwise, is an ineffective insulator.

PROPERTIES OF LIQUIDS

The peculiarity of liquids—specifically, water in its liquid state—is a matter of logistics. If the deterioration mechanism is ultimately related to the proximity of another chemical, then that substance must be transported and collocated with the fabric. This can be accomplished by bringing the second substance into immediate proximity with the base material when both are solids. However, the degree of interactivity of solids in contact is negligible; therefore, the proximity issue can be addressed more efficiently if the second substance is a gas. Because of the relative dilute nature of gases, there are distinct limitations on gas-solid interactions. We will see some deterioration mechanisms where this is the case.

For an intimate and penetrating contact by the second substance in dense concentration, the liquid state is the state of choice for the second material. If, furthermore, it is desirable for the second substance to exert interstitial pressure, then it is almost essential that it at least enter the base material in a liquid state. Mechanisms exist as a result of opportunities to reduce the potential energy state of an ordered system by whatever means are available.

The nature of liquids and gases are particularly advantageous for exploiting the tendency of crystalline solids and amorphous solids to crack. Cracks not even visible to the human eye are ideal boulevards for liquids and gases to enter and perfuse a solid, to collect in pores and fissures, and to exploit the targets of opportunity presented at exposed surfaces. To further appreciate the processes of interstitial deterioration mechanisms, we will discuss the manner in which substances, particularly water, enter and exit solids.

Water Absorption

The perfusion of a liquid into and through a solid is a very complicated process. To begin, it is important to point out that not all liquids can perfuse all solids; in some cases, a liquid may pass through a solid with very little of the liquid remaining behind. The combination of variables and the language used to describe those variables are the first knots we need to untangle.

Most of what we call solids are less than solid. Many items may be largely solid, but also contain a certain amount of void. A brick, for example, may be 92 percent dense, vitrified clay; but that leaves 8 percent, which is, simply, air. When we discuss the properties and performance of brick, we are referring to the composite of the fired clay and the void. The voids are an inextricable part of the nature and performance of brick; and, therefore, there is little if any value in viewing a brick as anything other than a permanent mixture of clay and air. Virtually all brick properties are tabulated based on the inclusion of some percent of the volume being void.

The void contained in and captured by a solid is measured as a percent of the total volume, which is, in fact, void. The name given to this concept is *porosity*, and the material is described as more or less porous. The values of porosity of various materials are extraordinary, ranging from 0.0 to over 90 percent. At almost every degree of porosity are common building materials. Counterpose this with emissivity, wherein extreme values are

FIGURE 4.5   Damage is concentrated at locations of maximum water volume and velocity.

theoretically possible, but the vast majority of materials falls within a fairly narrow range. The porosity of metals is an absolute 0 percent; the porosity of a closed-cell neoprene sponge may be 85 or 90 percent. The porosity of granite is 2 to 3 percent; brick, 6 to 10 percent; expanded polystyrene, 75 percent.

By itself as a number, porosity is not terribly informative. For values greater than zero, porosity is a percentage of the volume of the material that is not solid. Beyond that there is little that can be said to move from values of porosity directly to deterioration mechanisms associated with that property in isolation. We need to know more about the voids. The voids, or pores, derive from a limited number of sources. Understanding the origins of the voids will assist our understanding of their shape, distribution, and consequences. Some voids are the result of some aspect of the production process. These voids tend to be spherical in shape, because the pores are the result of gases formed or, in some cases, gases injected into the material when it was in a plastic state. The spherical shape is a consequence of the gas pressure directed outward evenly in all directions, combined with the principle that the gas can be contained at the maximum volume to minimum perimeter inside a sphere. Thus, the encapsulation of a sphere results in the minimization of potential energy on the part of the solid portion, or matrix, of the material. Trapped gases tend to form bubbles or spherical voids. It does not really matter whether the matrix is molten magma or extruded styrene; the bubbles are spherical. Also, generally, the size, gradation, and distribution of the bubbles are or were a function of the elapsed time between formation of the bubble and the solidification of the matrix.

If the bubble forms as the result of a chemical reaction within the matrix, and the chemicals were carefully mixed and blended into the matrix prior to gas production, then we expect uniformly sized and spaced voids, particularly if not much time elapses between void production and the freezing of the matrix. This sequence is descriptive of the reaction inside plastic concrete as the result of the addition of air-entrainment agents. For reasons we will discuss in detail later in this chapter, it is of considerable value to have small, evenly dispersed bubbles inside concrete in the amount of 4 to 6 percent by volume. For the pores to occur in a predictable and reliable manner, the chemicals that produce these bubbles are carefully blended with the plastic concrete to assure that the bubbles form at the appropriate time, and evenly, throughout the body of the mix.

If the material is a closed-cell neoprene sponge, the bubbles are blown into the monomer, and homogenized as the polymer forms. If the material is breccia, the air is blown into the molten magma as it is ejected under the force and pressure of the exploding volcano. Expanded polystyrene bubbles are chemically produced similar to air entrainment. In all of these cases, the manner and timing of air introduced into the liquid or semisolid state, combined with the properties of the matrix, are such that we may reasonably expect the bubbles to be small, uniform, and isolated.

The pores formed from the molding of a brick are altogether different. Brick pores are the residual voids that the compaction process did not eliminate. We may rightly assume, therefore, that these voids are irregular, nonspherical, and potentially interconnected. The pores present in timber are the drained interiors of individual cells. The voids in limestone probably formed long after the rock was consolidated, and are likely to be the result of a combination of stress cracks combined with chemical dissolution at the crack faces.

Porosity measures alone do not distinguish the character and nature of the voids, only that they exist as a measurable percentage of the total volume. In addition to how much of the solid is void, we need to know the degree to which the voids are interconnected. If the voids are connected, then gases and liquids can fairly readily pass into and through the solid. If the voids are not connected, then the resistance to the passage of gases and liquids is essentially the same as if there were no voids at all, because whatever may pass through the material must pass through the solid part of the substance. The property that measures the rate of liquid or vapor passage through a specimen is called *permeability*. The units are calculated as volume per square of area per unit of thickness per unit of time; so one measure is one cubic inch per square inch per inch per second, and that is defined as one *perm*.

Porosity and permeability are often confused, thus the terms are often incorrectly used interchangeably. But, they are different, which means that there are four possible combinations of porosity and permeability. A given material may be porous and permeable, such as concrete masonry units, bricks, and a good many masonry materials. A substance may be neither porous nor permeable, such as metals, which by definition are perfect membranes. Liquids or gases will not pass through metals regardless of the thickness or, more accurately, the thinness of the metal. This property is precisely why we prefer metals as flashing membranes and as foil wraps. A material may be porous, but not permeable; this describes expanded polystyrene and closed-cell neoprene sponge. Despite their extreme porosity, the bubbles are completely isolated, and the solid portion of the material is virtually impermeable. This leaves permeable and nonporous as a category, which includes a somewhat exotic but nonetheless important class of materials recognized, generally, as osmotic membranes, such as organic filters, intestinal linings, and kidneys.

In contrast to metals which, as just noted, are perfect membranes, many of the materials we use to construct buildings are not so unflawed. Stone, brick, concrete, wood, stucco, and mortar are all porous, permeable—and plentiful. Most, if not all, of the specimens that fall in these broad categories will absorb and transport water. The magnitude of that propensity is a compelling component of the various deterioration mechanisms associated with the presence of water on the interior of the fabric. Some of these mechanisms depend strictly and exclusively on whether water can intrude into the material, regardless of how it actually gets into the fabric; the specific topic of our current interest, freeze-thaw deterioration, extends beyond mere water entrance.

Most of the water that gets into porous materials fills the pores, and the degree to which the pores are capable of being filled is largely dependent upon the degree to which the pores are interconnected. Porosity and permeability together do not, however, tell the entire water penetration story. The remainder of the story follows two lines, one having to do with water entry and migration within the pores and the other having to do with water entry into and through the solid portion of the fabric. We will deal with the latter issue first, the introduction of liquids and, specifically, water into what we think of as the solid part of the fabric.

Absorption is the general term we apply to this phenomenon, but it is important to point out it is also the term we use to apply to the taking up of fluids in the pores. For clarity, for the balance of this discussion we will distinguish these two classes as *pore absorption* and *solid absorption*. Solid absorption is accomplished in at least three general ways, though there may be more. The vagueness of this issue reflects that the literature regarding the absorption of liquid into the fabric of solids is apparently not a cohesive or comprehensive area of study addressed by a single discipline. The three leading sources of solid absorption are, as we shall see, also the primary causes of hygroscopic expansion. The first one to be discussed is *adsorption*. Note: The spelling is not incorrect; this is adsorption, not absorption. The second one we will discuss is *salt hydration*, followed by *hydrolysis*. These three types of solid absorption are also characteristic of the three classes of materials to which they apply.

Adsorption

Adsorption is a general phenomenon associated with any crystalline structure that exhibits or manifests an electrically charged surface, meaning most crystalline solids. The presence of the charged surface is significant in that it presents an opportunity for positive ions or positively charged molecules to congregate at that surface. It is important to note that such aggregations of negatively charged surfaces of common crystalline solids are not bonded to the crystals in the conventional sense that electron orbits are somehow affected by the proximity of the adsorbed particles. It is possible and, in fact, rather common for sodium ions to satisfy this condition; therefore, they will congregate at the surfaces of negatively charged crystalline particles. Actual occurrences of positively charged molecules are rare, but what is as common as water itself is a molecule that by its configuration and geometry is effectively charged.

The water molecule is unusual, although not unique, in that the hydrogen nuclei are eccentrically bonded to the oxygen atom. This, combined with the substantial difference in atomic mass of the oxygen nucleus, and even the combined mass of the two hydrogen nuclei, means that the shared electrons tend to orbit the molecule toward the oxygen end of the molecule. The oxygen end is negatively charged, and the hydrogen end is negatively charged. Such a molecule is designated as *polarized* because the molecule exhibits a positive and a negative pole. A polarized molecule is, therefore, a miniature electromagnet; and while the electromagnetic pull of a single water molecule is small, the cumulative effect of a few trillion of them is enormous. If two negatively charged surfaces are parallel and close together, there is a naturally occurring repulsion of those two planes; but for external reasons they may remain in their respective positions.

To illustrate, imagine pigs lining up at a confined feeding trough. If the first pig can get its snout to the trough, the one behind can provide the push to get the first fully perpendicular to the trough, at which point pig number 2 can push past the rump of pig number 1 and move to the next feeding position. Pig 3 pushes past the rumps of the first two, rotating them slightly, and finds the first available position at the trough. The process continues until the trough is fully occupied, with pigs perpendicular to the trough with their little porcine rumps pressed against the far wall. If, in their zealous engorgement, all the little pigs simultaneously execute a swift kick to the back wall, which not one of them could do alone, it would result in the rapid lateral translation of the rear wall. If the negatively charged crystalline solid is the trough, and the polarized water molecules are the pigs, then the potential exists for the water molecules to be adsorbed by the surface of the crystal and to wedge their molecular bodies between the opposing crystalline surfaces. If an adjacent negatively charged crystalline surface is the analogue of the back wall, and adsorbed water molecules are the analogues of the pigs, then the polarized water molecules can and will exert a motive force, first to gain access to the negatively charged surface, and second, to occupy the volume between adjacent surfaces. In fact, because the "butts" of the water molecules are negatively charged, they will provide added repulsion to the adjacent surface. Of course, adsorption can and probably will occur at that surface simultaneously; consequently, there may well be a double row of adsorbed molecules between the opposing crystalline surfaces. The relative rotational flexibility of the adsorbed water molecules means that there is a potential path between the counterposed negative "tails"; but more to the point here is the prospect that the double row of water molecules occupy a greater volume than was originally the case of the opposing negatively charged crystalline surfaces. If so, then the water molecules necessarily are compressed in the occupied zone and, therefore, will exert a lateral pressure against the crystalline surfaces.

Recall from our discussion of stress and strain that no good stress goes unpunished and that strain is an invariable consequence of stress applied to elastic solids. Within the elastic limits of the compound, the adsorbed water will create an internal lateral force to the opposing crystalline surfaces and will increase the dimension between them. The cumulative effect of such internal stresses will be a change in the volume of the material due to the absorption and adsorption of water. A few cautions come with the delight. First, adsorption of water in the manner described requires that the solid contain, or be composed of, negatively charged crystalline surfaces. Second, those surfaces must be accessible to the water, taking into account the progressive nature of the prying action of the polarized nature of the water molecules. Third, not all crystalline structures are cubic in geometry;

therefore, the orientation of the negatively charged surfaces will dictate the direction and magnitude of hygroscopic expansion associated with adsorption. Fourth and very important, the adsorbed water can be dislodged because it is loosely attached to the surface of the negatively charged crystal.

The method of removal or relocation of water adsorbed between crystal layers is equally as important as the adsorption of water. If an external force is applied to the crystal perpendicular to the surfaces, then the volume between the surfaces is reduced. If the water molecules were perfect columns, joined and mutually bracing, the water might act as an elastic solid and respond to the stress by elastic deformation. This might result in the increase of potential energy in the bonds between the molecules and in the bonds within the individual molecules. But water is seldom so mutually supportive or so cooperative. The dissociated liquid state of the water allows the individual molecules to roll against, to dislodge, and to displace one another. It's every pig for itself.

As the volume between the opposed surfaces decrease, the two rows are gradually squeezed into a disoriented single row of water molecules, some still adsorbed to one face, others to the opposite face. Obviously, there is no longer the same room for all the water molecules. Some of the water molecules, particularly those near the edges or exits of the crystals, are squeezed laterally. If there is nowhere for this water to go, the interstitial pressure will increase, and the confined and now compressed liquid will begin to exhibit elastic strain.

Recall, however, that the water adsorbed into and between the crystalline layers had to have come from some reservoir, so it stands to reason that it can return via the same route. While the logic is compelling, the timing may be off. Whether the interlaminar adsorption of water is reversible without consequence depends very much on the time allowed for the water to reverse its path, and the source of the force precipitating the retreat.

Also recall that the ability of the water to be adsorbed into the solid is very dependent upon the availability of the negatively charged surfaces. If the sequence of events did, indeed, entail the sequential adsorption of individual water molecules, and the force opposing its adsorption was significant, then the period over which such adsorption would occur may be long.

This sort of interlaminar adsorption is commonly seen in the taking up of water in clay. The rate at which the adsorption process occurs is so slow that the same clay may be used as a shield against water migration. Historically, compacted clay was the reservoir liner of choice; and we still use compacted clay around the bases of buildings to divert water away from the apron of the foundations. We also know, however, that prolonged, continuous exposure of compacted clay to water will result in the taking in of that water into the fabric of the clay. We know that the rate of uptake is so slow that when we want to add water to clay we resort to blending and mixing: but that is only to speed up a process that we know will occur if we have the time to wait. In the manufacture of bricks, for example, we may have neither the time nor the tolerance for nonuniform distribution. Given that we know clay will take up water, and that the rate is slow, why would we expect the reversal of the process to be any different?

When we squeeze the clay, we compress the adsorbed water, and the liquid operating under compressive stress begins to migrate. If the pressure is that of a building footing, it is unrelenting and enduring; thus, the slow process of consolidation takes place. Even if the excluded water has to disperse to the surface of the soil, it will eventually migrate, and compaction will occur. If, in our effort to consolidate the same clay at a more rapid rate, we apply a very high level of external stress, we may exceed the ultimate strength of the soil. Instead of gradual consolidation, we may precipitate a catastrophic shear failure. The keys to the rate of the accommodation of pressure release of interlaminar adsorbed water are: the migration proximity of a reservoir; the actual size of the interlaminar dimension; the force with which the adsorbed water is held in position or which must be overcome for the fluid to flow; the magnitude of the applied external force squeezing the solid; and the inherent properties of water, in this case, to fluid movement.

These concerns are precisely the variables that distinguish a famous formula called Poiseuille's Law, named for a brilliant nineteenth-century French scientist. Poiseuille's Law describes mathematically what we described above verbally. The rate of flow of a given fluid through a conduit is determined by the precise factors described in the preceding paragraph:

$$R = \pi r^4 \, \Delta p / 8 \eta L$$

Let us introduce the players:

- R is the rate of fluid flow, which is what this formula will determine once the arithmetic crank is turned.
- The radius of an assumed circular conduit is r. In our example, the conduit is not a circular pipe; it is more like a slit, so there is not a perfect fit between the classic expression of Poiseuille's Law and our example, but the principle certainly remains intact.
- Δp is the pressure differential from one end of the assumed conduit and the other end. This is also familiar to us as the pressure gradient. In our case, the external pressure is supplied by compression of the fluid already in the slit, not by a pump attached to one end of a circular pipe. Again the specific condition changed, but the principle re-

mains that the rate of flow is directly proportional to the magnitude of the external pressure.

· L is the length of the pipe, or, in our case, the distance to the release point for the liquid.
· The term η is the mathematical expression for viscosity, which, recall, is the resistance offered by a particular fluid to flow.

The only surprise is that the radius of the pipe is directly resistant to fluid flow to the *fourth* power. The radius raised to the second power is not a conceptual stretch, because πr^2 is the cross-sectional area of the pipe, and we can at least imagine that the area of the conduit may influence the magnitude of the friction along the walls of the pipe to fluid flow or some such effect. If fluid flow is resisted by the diameter of the pipe by the fourth power of the radius, then any pore with a very small dimension, regardless of geometry, will resist fluid flow disproportionate to the size of the opening. In the case of clay particles or other thin plate crystals, the relevant dimensions may be in the range of a few microns. Certainly, fluid may pass through such small orifices, but the time required to pass a given amount of fluid may be long, indeed.

A material for which this phenomenon is critical is cement. The exact structure and the cement crystal is the subject of ongoing study, and no small amount of debate. The best evidence is that the cement crystal is, in its stable form, a laminate of complex salts separated by layers of water called *gel water* (because the technical consistency of cement is more of a gel than of a conventional crystalline solid). The wafers of cement gel alternate with layers of gel water in a stable pattern of the chemistry, which is important to us only to the extent that the layers of water are an integral and essential aspect of cement formation and stability. Given this information alone, it is not surprising that the maintenance of a minimum amount of water is necessary during the curing process. As we know, premature drying of concrete will result in weakened cement because the cement crystal will be under-hydrated. The significance and performance of the gel water will become even more evident when we discuss freezing-thaw cycling below.

Hydration

When ionizable salts are exposed to water, the polarized water molecule attaches to the charged particles of the salt crystal, in a process not totally dissimilar to that of adsorption described in the preceding section. The differences are, however, more profound than the similarities. The polarized water molecule may attach to either the positively or the negatively charged saline particles. Thus, both the potentially positive and the potentially negative particles are vulnerable to the advances of the polarized water molecules. The attachment vehicle is similar to adsorption in that the there is no chemical alteration of the water, which is strictly a matter of polar attraction.

In addition to the polarity of the water by virtue of its molecular structure, inherent to water as a chemical, there is, at any given time a certain amount of molecular dissociation of the hydrogen oxide (H_2O) into free hydrogen (H^+) radicals and free hydroxyl radicals (OH^-). In a pure water, there are present a certain number of these free ions that pass in and out of the ionic state, forming and reforming as water molecules and dissociating as free radicals. When water is in the presence of a salt, these occasional free radicals are particularly potent at dissociating the saline particles from the salt crystal.

The effect of this process will be discussed in greater detail in the section on salt damage. What is important at this point is that water is extremely disruptive to, and powerfully attracted to, ionizable salts. When the water has completely dissociated the salt crystal into solution, with the water acting as a solvent, the salt is totally hydrated. Until that point, however, a certain amount of water may be captured in the crystalline structure of the salt. In fact, certain salt crystals are distinguishable based on the ratio of hydrated water molecules that are integrated into the crystalline structure.

To the extent that salts absorb water into the matrix of the crystalline structure, that material can change volume. When that happens, the material increases in volume, an example of hygroscopic expansion. Any materials that contain—much less are predominated by—salts are subject to hydration, hence to hygroscopic expansion.

Hydrolysis

Hydrolysis is the fracturing of compounds by exposure to water, specifically as a result of the activity of the acidic (H^+) or basic (OH^-) radicals, discussed above, which are always present in water. This is particularly applicable to organic compounds, of which the most notable to this book is cellulose. For many of the same reasons that water hydrates salt crystals, it also is absorbed by and ultimately dissociates organic compounds such as sugars and cellulose into fractured fragments of the original compounds. We are interested in the immediate hydrophilic character of compounds that are subject to hydrolysis. A requisite initial step in the hydrolysis process is the attachment of the water molecule to the compound; the consequence is that water is incorporated to the matrix of the compound's molecular structure. This wetting of the structure is an inherently absorptive process, the result being the volumetric increase in the structure.

The cumulative effects of these three mechanisms of hygroscopicity is an increase in the volume of the material. Water is present within the structure, which is our most pressing and immediate concern, and we now have at least the rudiments of understanding hygroscopic expansion as well. However, several problems with these explanations of hygroscopic expansion remain. Unlike thermal expansion, which is based on a single unified theory of atomic oscillation, hygroscopic expansion is not explicable in a single physical theory. Hygroscopic expansion is based on at least three general theories, each of which is subject to considerable variation. And though we recognize hygroscopic expansion as a very real and relevant material property, we do not have the mathematical tools with which to predict the behavior of a material in a range of conditions. The data associated with hygroscopicity is largely empirical and unreliable, except in a general and conceptual manner. It is frustrating to know that the phenomenon exists, but be unable to reliably predict its behavior, other than qualitatively.

Pore Absorption

This brings us back to the water located in the pores of a given specimen. While direct crystalline absorption is not insignificant, the vast majority of water absorbed into most solids ultimately resides in the pores. This section addresses the properties and phenomena that affect and explain the distribution of water in the interstices and crevices within solids which we call the pores.

We begin with a stationery droplet of water that is present at the surface of a porous and permeable object. For that droplet to move from the surface to the interior of that same solid through the pores, there must be a motive force. If the droplet is at rest on the top surface of a brick, for example, gravity alone may supply the force to pull the droplet into and ultimately through the porous solid. Again, the rate of that movement is predictable based on Poiseuille's Law (the rate of fluid flow is dependent upon the differential pressure (Δp) from one end of the conduit or pipe to the other). In this case, that pressure is the gravitational pull on the droplet. Intuitively, as well as logically, we understand that fluids flow downward; and if there is a path from the top of the brick to the bottom, we know that sooner or later the droplet will emerge on the bottom of the brick. This simple reality is loaded with nuance, which is precisely what we want to explore.

The speed with which the droplet will traverse the brick vertically is a function of the brick's thickness. All other factors being equal, it takes longer for the droplet to work its way through a thick brick than through a thin one. The time it takes to traverse the brick is also a function of the viscosity of the fluid: highly viscous fluids will take longer than thin fluids. If the viscosity of a fluid changes with temperature, which is generally the case, then the length of time it takes to traverse the brick is indirectly a function of the temperature of the fluid. Take as an example water sitting on top of a brick: both the brick and the water are 68° F, and the water passes through the brick in 10 seconds. Along comes the sun, which warms the same amount of water and the same brick to 104° F; now the water will flow through the brick in 6.5 seconds. That is a big change, probably more than expected. But if temperature were raised to just below the boiling point of water, the same amount of water would pass through the same brick in just 2.8 seconds. The frictional resistance to flow of a fluid is greatly influenced by the temperature of the fluid, even those seemingly as familiar and constant as water.

The factor emphasized earlier, and that remains a dominant factor in the determination of the flow of fluids through solids, is the size of the pore itself. Poiseuille's Law says that the rate of flow through a pipe is related to the pipe radius raised to the fourth power. Suppose there are two pore paths, each with an even diameter; each is circular from the top of the brick to the bottom. One of the pore paths is $\frac{1}{100}$ of an inch in diameter and the other is $\frac{2}{100}$ of an inch in diameter. Poiseuille's Law says that the rate of flow in the larger diameter pore is 16 times greater than the smaller pore! We might expect something like four times the rate because the area of the larger pore is four times that of the smaller pore; 16 times is remarkable.

Another way of parsing the same phenomenon is to consider the distribution of pore sizes within a given brick. It seems reasonable that the pores in a given brick are not all the same size. (The determination of pore size distribution in any material is an extremely tedious affair, typically requiring electron microscopes.) Research confirms only that pores do occur in varying sizes and concentrations over a range of diameters. Suppose in our particular brick that the size range is from $\frac{1}{100}$ of an inch to $\frac{2}{100}$ of an inch and that the distribution is exactly even if measured in increments of $\frac{1}{1,000}$ of an inch. The last increment, that is, the largest pore of $\frac{2}{100}$ of an inch in diameter accounts for slightly more than 9 percent of the pores; but in a given period of time, they transport almost a quarter of the fluid. One-third of the pores with the largest radii carry two-thirds of the fluid, and this is when the largest pore is only twice the diameter of the smallest pore. If the distribution is more exaggerated, which is generally the case with many porous solids, the top 10 percent of pores based on size may carry 80 or 90 percent of the fluid traversing the solid.

The point is, pore size distribution is an important variable in any fluid transportation problem; and, unfortunately, its determination is normally beyond standard testing techniques and budgets. There are estimating and guesstimating techniques, but pore size distribution is an area of uncertainty in the fluid flow problem. As is the case generally with hygroscopic expansion, we have a firm qualitative appreciation of the phenomenon, but it may not enable more than generalization and estimation.

Moving on, we must deal with the fact that not all fluids reside on the top, horizontal surface of a porous solid. Rain washes down the face of a building, for example, and water gets into the brick. What is the motive force pushing or pulling water into the interior of the brick in this situation? Although we are primarily interested in the lateral component of this situation, there is a vertical component as well. In any problem associated with the horizontal perfusion of water in a material or through a system, it is important to remember that gravity does not suddenly disappear simply because we are more concerned with lateral transportation of the fluid.

If for example, we are examining the interior of a wall, and find evidence of moisture penetration, we can reasonably deduce that the source of the water is from the exterior of the building; furthermore, it is probable that the point of entry, or breach, on the exterior surface is higher in elevation than that of the manifestation on the interior surface. Again, the horizontal component is important, but it is essential to keep in mind force gravity is always present and working 24 hours a day every day. Other sources of external force may be present; they may be determinative of a specific manifestation; but there is no source of exterior pressure as relentless as gravity. If a motive force is acting horizontally on a fluid through a wall, coupled with the force of gravity, the sum of these force vectors will be a trajectory of fluid movement that will be both lateral and downward. As we will see shortly, there are situations when other forces may add to the vector train, resulting in the net upward movement of fluid, but there are no violations of the general rule of gravitational movement. If the object of our attention involves mass, then gravity must be addressed. To quote a late nineteenth-century issue of *Scientific American*, "Gravity does not give a damn whether you believe in it or not."

When fluids perfuse materials laterally, two primary sources of horizontal force can affect that movement: pressure differences between the ends of the effective pore, and capillarity. Pressure differential is an important and common source of horizontal motive force, which we will deal with later in great detail relative to how walls leak. We have not as yet addressed capillarity directly so we need to lay the groundwork for a full explication by looking at the contact plane between the liquid and the solid. At the location that a fluid meets a surface, competing forces are manifest in the behavior of the fluid relative to the solid.

The apparent competing forces are, on the one hand, the tendency of the molecules of a particular fluid to stick together, and, on the other hand, the simultaneous tendency of those same molecules to stick to the solid surface. The former tendency is a measure of the cohesive property of the liquid. If a given liquid is highly cohesive, it has a strong tendency to stay together, not to run out over the surface of a surrounding solid. The property of cohesion is specific to the fluid; it has nothing to do with the particular surface. A given liquid will behave just as cohesively on a glass surface as on a steel surface. The tendency of a fluid to adhere to a surface is analogous to friction, in that adhesion is a consequence of the combination of materials—in this case, the liquid and the solid. If a given fluid has a high adhesive response to the surface, the fluid will spread over that surface with relative ease—assuming, of course, that the coincidental cohesive property of the fluid does not offer a countervailing tendency.

If a fluid is dropped though a vacuum, or even though air, the cohesive nature of the fluid will tend to minimize the surface relative to the volume. Because every molecule is attracted to every other molecule in an equal manner, if by no other means than trial and error, all the molecules will be drawn toward the geometric center of the minimum volume for that particular mass of fluid. The shape of the resulting volume is a sphere, whose surface area is the minimum surface for a given volume. The natural and inevitable shape of a droplet of fluid not in contact with a rigid surface is a sphere. When molten lead shot is poured drop by drop from the top of a shot tower, the liquid lead droplets form naturally into nearly perfect spheres as they descend through the air. The height of the tower is constructed such that by the time the lead droplets get to the bottom they will have solidified into lead pellets or shot. At the bottom of the tower is a pool of water so that the shot plunges and decelerates without squashing against the floor.

When a droplet of fluid is placed on a dead-level surface, the cohesive property of the fluid will tend to hold the droplet as close to the minimum volume of the sphere as the cohesive force will allow. The surface of the droplet will remain smooth and undistorted, except for the contact plane with the solid surface. The surface of the fluid is necessarily in tension, and the capability of the fluid to maintain a level of surface tension is another aspect of the cohesive nature of the fluid. The perfect shape of the sphere distorts by flattening out on the "side" compressed against the horizontal surface. The mass of the droplet is pulled by gravity to the lowest possible elevation while the cohesive property attempts to maintain some semblance of the sphere. Thus, the surface area is, by definition, no longer the minimum surface for the given volume, and the tension in that surface is increased.

If the droplet is quite cohesive, the distorted shape of the sphere is minimal. The contact area of the liquid against the surface is circular in pattern. The angle formed by the intersection of the horizontal surface and the fluid resting on the surface is a characteristic angle called the *meniscus*. In our example of an extremely cohesive fluid, the distortion of the spherical volume is slight, and the meniscus is large. The meniscus is measured as the angle formed by the intersection of the fluid surface and the horizontal *solid surface*, viewed from the inside of the fluid. If, as in our current example, the fluid is extremely cohesive, the meniscus may actually be an obtuse angle, meaning that it is greater than 90 degrees. If, for whatever reason the fluid flattens out against the horizontal surface, the meniscus will become an increasingly acute angle, and that degree of flattening will be reflected in the meniscus. We can, therefore, characterize fluid-flattening properties by measuring the meniscus.

The meniscus is an easy indicator of the tendency of the fluid to flatten out on the surface due to a combination of factors, not just the cohesiveness of the fluid. Before discussing those factors, however, it is important to consider the distinction between cohesion and viscosity. Cohesion is, as stated, that property inherent to any fluid that accounts for the tendency to cling to itself in a minimum volume. Viscosity is a measure of the tendency of a fluid to resist being dragged across a surface. The test of viscosity is conducted by placing a fluid between two parallel surfaces and dragging one of them across the other. The resistance of the fluid can be measured by the force required to pull the upper plate along. This means, among other things, that viscosity is a measure of the shear resistance of the fluid; hence, our inclination to refer to viscosity as the measure of fluid stiffness. It is theoretically possible to have a fluid that is highly cohesive but with a low viscosity. Somewhat more familiar is a fluid that has high viscosity, but relatively low cohesive properties, such as molasses. Initially, a droplet of molasses will rest on a flat surface with little deformation, as we expect of a highly cohesive fluid. The molasses will, however, given enough time, flatten out to a thin layer on a glass plate such that the terminal meniscus is, in fact, rather low. The apparent cohesion was actually viscosity. Eventually, the highly viscous molasses displayed its relatively low cohesive property and flattened out on the plate; it simply took a while. The point is, we are really looking at different properties. The confusion over viscosity and cohesion comes from failing to take into account the time a fluid requires to assume its naturally low meniscus value.

Temperature is a factor that dramatically affects the meniscus. As temperatures rise, the meniscus decreases; fluids spread out more. Cohesive properties are reduced as molecular activity increases; viscosity also decreases as fluids tend to lose shear capacity as temperature is increased. Another identifiable factor is adhesion, which is the inherent tendency of a fluid to attach to a surface. It is adhesion that accounts for the tendency of fluids to climb and spread on a surface, despite the countervailing cohesive tendency. Temperature tends to enhance the adhesive properties at the same time that it diminishes the cohesive properties.

All of the factors—namely, cohesion, adhesion, meniscus value, and viscosity—taken together describe the *wetting property* of the fluid. Obviously, because adhesion is on the list, we need to know not only which is the wetter, but also what is being wetted. We know that wetting is enhanced as temperature increases, and we know that wetting may be enhanced with time, but for any combination of fluid and surface, there is a predictable limit to the extent of wetting. In short, there is an unaided spreading of fluid across the surface of the solid.

If those factors are less than desired, we can enhance the wetting of a surface with what is called a *surfactant* or, simply, a wetting agent. The best-selling surfactants are among that class of chemicals called detergents. We add detergents, specifically, and soaps, generally, to water to improve its wetting properties. Another thing we can do is to agitate the fluid, such as to apply mechanical propulsion to accomplish what adhesion could not. Once the fluid is mechanically distributed, the cohesive tendency may cause it to retract into a less dispersed pattern, but the net coverage would still be greater than if the fluid had not been mechanically spread at all.

This discussion seems to be leading directly to laundry. We now understand the variables that affect the washing of clothes or any other household article. Dirt and soil on clothing is either absorbed into the matrix of the fiber or it is lodged on the surface of the fiber. If absorbed into the fiber itself, it is called a stain, and must be managed in a specialized way. If the dirt is lodged on the surface of the fiber, there a good reason it may not be easily dislodged by brushing or shaking. (The attachment of fine particles to surfaces is the subject of a subsequent section in this chapter.) If we want to remove more of the soil than can be accomplished by shaking and brushing, we have to resort to some sort of washing. When we wash something, we are using the mass of the moving fluid to break the bond with the surface. Once the bond is broken, the particle is then flushed away from the surface and dispersed in the fluid. The process is inhibited by the hydrophobic properties of the surface and the limited wetting properties of the fluid. The wetting properties are enhanced by heating the fluid and by adding surfactants to the fluid. This facilitates the penetration of the fluid into complex patterns of fabric surfaces. These fabric surfaces may have extremely high surface-to-volume ratios, particularly if the fabric is fibrous. We must also be aware of the limits of the fluid to suspend large quantities of particles, so that when we segregate the fluid from the fabric we do not simply redeposit the particles back into the fibers. This we accomplish with a second cycle of clean fluid, to rinse the residue of loosened particles and surfactants from the now fully wetted surfaces. It is for this reason that rinsing fluids are generally applied at a lower temperature and without the aid of additional surfactants. For a variety of reasons beyond the scope of this book, conditions may dictate that water cannot or should not be used as the basic wetting fluid. In those cases, other fluids are substituted for the water. These techniques are often called "dry" cleaning, an obvious misnomer in that the articles are immersed in fluids. The use of the term serves only to distinguish the method of flushing from the "wet" technique involving water.

Believe it or not, basic laundry physics helps to understand building pathology. When a particular surface is flat, we can see how the effects of wetting play out on the surface. When the surface is, however, rolled into a tube, the consequences of these same properties are manifest in interesting ways. The very same adhesive and cohesive properties will tend to attract the fluid or repel the fluid within the tube. If there is a net adhesive advantage, the meniscus will ride up the sides of the tube; therefore, the intersection of the fluid against the wall of the tube will resemble a cup turned upward inside the tube. If the net cohesive properties predominate, the fluid will drag against the tube wall, and the intersection shape will be a cup turned downward.

Where there is a net affinity between the fluid and the tube wall surface, and the adhesive properties predominate, the fluid may actually ride up inside the tube for a distance above the ambient level of the fluid outside the tube. This may seem to defy gravity, but only if gravity responded to defiance. In any situation involving fluids inside tubes, the mass of the fluid will be acted upon by gravity. This alone will maintain the level of the fluid at the identical level of the surrounding ambient external fluid level. For the level inside the tube to be either higher or lower than the level of the surrounding fluid, some other force or forces *in addition to gravity* must be acting on the fluid in the tube. The hydrophobic or hydrophilic character of the tube surface, combined with the adhesive and cohesive properties of the fluid, can account for this net additional force. If the combination of surface wetting factors are highly positive, then the meniscus will turn up inside the tube. The affinity of the fluid to the surface, and vice versa, allows the fluid to climb inside the tube. The limits of this climb are set by the mass of the fluid drawn up into the tube being pulled back down by gravity. The same principle applies to fluids that are depressed in the tube below the ambient level.

If the diameter of the tube is very small, the mass of any fluid inside the tube is low. If the adhesive properties of the combination of fluid and surface are relatively high, the net force drawing the fluid along the surface of the tube may be decidedly higher than the countervailing force of gravity on the very small mass of the fluid. In these cases, the fluid may actually rise inside the tube several millimeters. If the tube is inclined, the distance the fluid travels along the tube increases, although the absolute rise within the tube remains the same. If the direction of the tube is horizontal, the absolute distance the fluid traverses in the tube can be quite great. This is called capillarity, and it can be an important factor in the introduction of fluids into the interior of porous solids. While capillarity has distinct limits when it comes to the vertical transportation of fluids, it is a particularly important factor of horizontal transportation of fluids. Actual samples will have a wide range of capillary diameters, which takes us back to the issue of pore distribution. Recall that large pores disproportionately carry more fluid than small pores, but that small pores may be able to spontaneously transport fluid longer distances than large pores. Thus, the combination of pore sizes in a given sample comprise a critical factor in the performance of the material relative to fluid transport, generally, and the vertical component of that transport, specifically.

Internal and Interstitial Condensation

We now have established avenues for water intrusion into porous solids vertically and horizontally when water is in a liquid state. As the subsequent discussion of rising damp will make clear, water can climb vertically from the bottom of a specimen, but this is an unlikely source of fluid in a freeze-thaw scenario. More probable as a source of water is migrating vapor. For water to ultimately freeze at one atmosphere of pressure, water vapor must pass through a liquid state prior to transforming into a solid state.

Atmospheric pressure in the passage from one physical state to another is dependent not only upon temperature, but pressure as well. If pressure is sufficiently reduced, and temperature is adjusted accordingly, it is possible for water to pass directly from a solid state to a vapor state without passing through a liquid state. This process is called *sublimation*. And while it is quite difficult for ice to sublimate solely due to a drop in atmospheric pressure, it is not unusual to see the phenomenon occur with "dry ice," which is solidified carbon dioxide. At one atmosphere of pressure, it is not possible to maintain carbon dioxide in a liquid state. This same principle, that pressure as well as temperature is essential in the determination of the state of a material, is also true for the conversion from liquid to solid and from solid to liquid. If pressure increases, typically the freezing point decreases, even if only slightly. If ice is very close to the melting point, and pressure is applied to the surface, the solid ice will liquefy even at relatively low applications of pressure. On very warm ice, the pressure of an automobile tire may be enough to liquefy the surface, whereas the pressure under a skater's blade is approximately 1,000 atmospheres, so the melting point is decreased by several degrees. The ice liquefies under the skater's weight, thereby providing a lubricant below the blade. As soon as the pressure is relieved, the water refreezes. If the ice is too cold, then even the pressure of the blade will not produce liquid water and skating is not possible.

The factors that affect the condensation of water are relatively simple, so we will defer that discussion for the moment. Liquid water does, in fact, precipitate from water vapor, and that, at least in part, occurs as a result of declining temperature. If the water vapor is chilled to a sufficiently low temperature, the vapor will convert to a liquid state, a process called *condensation*. If the reduction in temperature is sufficient to result in condensation because of a proximate surface, which itself is sufficiently cool, then the condensation may be given the alternate name of *dew*, a term normally applied to meteorological conditions as opposed to condensation on the surfaces of pores deep within porous solids. The point here is that liquid water can, indeed, form within the body of a porous solid from a combination of an adequate concentration of water vapor and sufficiently cool surface temperatures.

The amount of water that will condense on the surfaces of the interstices of a porous solid is important. Like many of the phenomena discussed in this book, they may appear to be debilitating qualitatively, but if quantified, become a nullity. But condensation is anything but a nullity. The condensation "engine" can be a powerful injector of liquid water into the fabric of the material. The determining characteristics are the ongoing availability of water vapor and the continuing proximity of the condensing surface. The quantity is enough to potentially fill pores with water even when there is no immediately available bulk source of water.

Moisture Mass and Content Delay

At any given time, a material will be a certain temperature. As discussed earlier, this means that in the body of the crystalline structure of the material, by virtue of its specific heat, the material has a certain amount of stored energy in the form of molecular kinetic energy. If additional external heat is applied, there is some delay in its transmission through the body of the material and some delay in the temperature response of the material as the heat is absorbed into mass of the material. As also explained, when the heat source is eliminated, there will be a gradual release of the stored heat, and the average temperature of the solid will decline; but temperature will not drop immediately. The rate of heat loss and heat gain are functions of heat transfer and the specific heat properties of the material.

At any given time absorptive materials will have a certain moisture content. Recall that moisture content is a collective term for moisture absorbed into the crystalline structure of the solid as well as the pores. Specifically, in the case of wood, the water absorbed into the cellulose fibers of the cell walls is called the *bound water*, and the water that collects in the voids of the hollow cells is called the *free water*. These are descriptive terms for other solids as well, and we will distinguish between the two aspects of moisture content using them. Following the analogy, it is also true that when there are abrupt changes in the environment we tend to interpret them in terms of their effects on the materials. The environment, particularly the immediate contact at the surface of the material, is mitigated by the heat or, in this case, the stored moisture of the material. In other words, the material affects the environment arguably as much as the environment affects the material. As building pathologists, we might regard this interactivity as mildly interesting, but of little consequence to the building. Doing so, however, would be to overlook the considerable consequences of thermal and moisture inertia as direct ameliorative benefits to the materials. In Chapter 3, we discussed temperature gradient and moisture gradient as sources of boundary stress. One of the major sources of steep temperature and moisture gradients is abrupt environmental change.

If the material has a low specific heat, in the presence of an abrupt drop or rise in temperature is the potential for a rapid surface response, hence a steep temperature gradient at or near the surface. If the material has a low specific moisture, there is an analogous potential for a steep moisture gradient in the presence of an abrupt change in atmospheric humidity.

The more immediate consequence of moisture mass or moisture inertia involves absorption rates as well as drying rates. The effect of the capacity for specific moisture is that the material tends to level out the occasional and rapid spikes in moisture, which can drive a collections curator batty, who is chasing hydrothermographic data despite the very distinct prospect that the surface is responding far less dramatically than the monitoring equipment indicates.

As moisture inertia relates to the immediate subject of freeze-thaw cycling, the latent moisture content of the material may be indicative of the material's vulnerability at the time of the drop in temperature. Cement, for example, has an inherently high level of latent moisture, which we referred to earlier as gel pore water. The gel pore water is the major source of the relatively high specific heat of cement, and constitutes a high level of bound water in the structure of the cement colloid. This also means that even in "dry" conditions cement has a high moisture content and a relatively high vulnerability to freeze-thaw damage, whether it is saturated with free water or not.

If the material at issue has pores that tend to be extremely small in diameter, migration through the capillary network will be slow. Moreover, materials predominated by extremely small pores will surrender moisture more slowly than materials of equal porosity, but predominated by large pores. The consequence in a freeze-thaw situation is that the smaller-pored material may be caught with a high level of free water in the pores at the onset of frost conditions, whereas the larger-pored material will have already drained.

Freezing Water and Ice

The final necessary and sufficient condition for the deterioration mechanism of freeze-thaw cycling is expanding water. As widely reported in architectural and preservation literature, crystalline water expands 9 percent from the moment it turns solid until it reaches maximum volume. Water is, indeed, rather peculiar stuff. The molecular polarity alone sets the water molecule apart from other compounds. There are other polarized molecules, ammonia and methyl alcohol to name two, but neither is numerically common, much less quantitatively prevalent. In addition to being polarized, the water molecule is also geometrically eccentric. There are many consequences to these peculiarities, some of which are crucial to our experience of what we call living, certainly others, to our understanding of building deterioration.

The peculiarities of water are not limited to temperatures below freezing. In fact, the behavior of water below freezing is triggered by behavior immediately above freezing. Recall that the water molecule is geometrically asymmetric. In a simplified form, the water molecule is shaped like a Y, with the two hydrogen atoms located at the ends of the two arms, and the oxygen atom located at the intersection of the three branches. This is the short-range order of the water molecule. As long as the water is in a liquid state, the legs and branches of the respective prongs of the molecule can nest in the crotches formed by the branches. Because of this peculiarity of geometry, the liquid water molecules can compact themselves, precisely because they are not arrayed in a regular long-range order.

As the liquid water cools, it behaves in a manner consistent with other liquids, meaning that it contracts as it cools. At approximately 4° C or 39° F, the water molecules begin to arrange into a long-range order such that, loosely speaking, the heads and tails of the molecular dipoles begin to align. The arrangement at this temperature is far from rigorous, and there is still a considerable amount of fluidity, literally and figuratively. As the liquid continues to cool, the regularity of the geometry increases, and the long-range order improves in its rigor. For this to occur, more of the branches must disengage themselves from the compact nesting of the liquid state, and array themselves in a less compact head-to-tail pattern. This, in turn, means that the long-range order of the crystalline material is actually more open and less dense than the nested short-range order of the liquid. Liquid water is at its densest at the temperature of 39° F (4° C). From that temperature to freezing, even in its liquid state, water expands. This is quite different from the prototypical earthly substance.

We accept that at its freezing point a given substance converts from a liquid state to a solid state. However, it is not as simple to actually effect the conversion as it is to cool the same substance one more degree. At the precise moment of conversion from the short-range order of the liquid state to the long-range order of the solid state, or vice versa, there is a substantial energy transfer. In either case, the amount of heat either given off or absorbed when the conversion is from a liquid to a solid or from a solid to a liquid is called the *heat of fusion*. In the case of water, the heat of fusion is 144 BTU per pound of water. This is nine times more energy than is required, otherwise, to raise the temperature of a pound of water a single degree. It is, therefore, a fact that at the moment of solidification, that there will be 32° F water immediately adjacent to 32° F ice. At that precise moment, all the ice must be at 32° F because if it were possible for a grain of ice to drop another degree, it would immediately become a sink for heat in the liquid water. The momentarily colder ice properties would, therefore, reheat back to 32° F and remain there until the entire brew had given off sufficient energy to convert entirely to ice or to gain enough energy to remain entirely liquid. Heat of fusion is not unique to water by any means; it is characteristic of all conversion between liquid and solid states. Although the heat of fusion is higher for water than most other materials, that is not what is unusual. The unusual part is that, at the moment of solidification, the crystal formed by that process is actually less dense than the surrounding water, and the ice *continues* to decrease in density for at least another 18 to 20° F, depending on the literature source. This means that, at least down to a temperature of 12 to 14°, F ice continues to expand.

As previously stated, the magnitudes of the expansion in the liquid phase and in the subsequent solid phase are often reported in conservation, preservation, and architectural technology sources as being 9 percent. The consistency of this value is remarkable, considering that it is not attributed to an experimental source or reference; the statements are partial statements. What is well established is that materials change volume as a result of changes in temperature. This is so thoroughly experimentally documented that, for most solids, we can look up the coefficient of thermal expansion in a reference manual.

Peculiar though it may be for a substance to expand with decreasing temperature, it is not difficult to deal with mathematically. It means, simply, that over a certain range, the coefficient of thermal expansion is a negative value. In at least one physics text, the coefficient of thermal expansion of ice is reported (a) as a positive number and (b) as a value of 0.00093 percent/100° F, which is identical to the coefficient of thermal expansion of copper and of plaster. The problem is that we know from other sources that while this may be true at colder temperatures, there is a contrary tendency between approximately 12° F and freezing. The best information available suggests that the coefficient of thermal expansion of ice in this range is approximately –0.0033 percent/100° F, which is, by thermal expansion standards, a very high value, but not remotely close to the reported value of 9 percent over an unstated temperature range. Even more significant is that it is a negative value. This discrepancy is not explicable based on the statements within the literature; we suspect it is indicative of a technical gap, which is relatively easily closed, but requires specific and direct experimentation.

What we can conclude and use from this are some working hypotheses, namely that from approximately 39° F to as least 12° F, initially in the liquid state and subsequently in the solid state, water expands. The coefficient of thermal expansion over that range is, therefore, a negative value. The magnitude of that value is not well established, but the best evidence is that it is negatively a good deal more expansive than other materials, which are positively expansive over a comparable range.

We also know parenthetically that the bulk modulus of water is 335,000 psi. If between 39° F and 32° F, the coefficient of thermal expansion of water is as much as –0.000033 percent / °F, then the total expansion over the temperature range is 0.00023 percent. The pressure required to confine the thermally expanding fluid is 77.4 psi. As the ice forms and the liquid converts to a solid, the ice will necessarily develop a modulus of elasticity, which we will assume to be approximately 1,000,000 psi. If the continued thermal expansion between 32° F and 12° F is approximately 0.00066 percent, the confining pressure required to maintain zero expansion is 660 psi. While these values are probably less than expected and at times less than have been reported in the literature, they are much more credible and consistent with experience. An interstitial pressure of 660 psi is not an inconsequential value. It is sufficient to crack brick and certainly enough to advance an existing crack in a flawed material. However, as we will see in the subsequent section, which addresses the cracking mechanism, the actual stress manifest at the interface between the pore surface and the ice will be less than the full expansive capacity of the ice.

THE FREEZE-THAW SCENARIO

We now have all the necessary and sufficient conditions in place for freeze-thaw cycling to occur:

- A porous, permeable material
- Water in liquid form to be present at the surface of the material
- A motive force to move the water into and through the fabric of the solid
- A capillary size distribution conducive to retaining the water under stress
- A solid material sufficiently brittle to crack at internal pore stresses of approximately 660 psi

- A declining temperature that will fall through and remain below freezing for a protracted period of time.

As the preceding discussion made clear, achieving this set of conditions is not a common scenario. Environmentally, there are vast regions where these conditions may be met only rarely. In very cold climates, even though they may incur substantial precipitation and prolonged periods of freezing weather, there may be relatively few occasions when the temperature starts out above freezing and falls through the freezing point. Some climates, notably deserts, may have a substantial number of temperature conditions that meet the necessary pattern relative to temperature, but not in the presence of liquid water. Places such as New Orleans have plenty of water on the surfaces of brittle, porous materials, but it rarely freezes.

In the United States, the motherload of freeze-thaw activity occurs in the Mid-Atlantic, New England, and, to a lesser extent, the Great Lakes regions. Not only do these areas experience substantial periods of declining temperatures and precipitation, they also have large inventories of buildings built with the requisite types of materials. Of course, other areas also experience freeze-thaw cycling, but less often and for shorter periods or to lesser degrees.

Because freeze-thaw damage is cumulative, the consequences of freeze-thaw activity is a function of the *number of cycles*, not the *rate* at which the cycles occur (see Fig. 4.6). By a cycle, we mean the pattern of events beginning with a dry, or virtually dry, relatively warm material. The material is then wetted, typically by rain. The period of wetting must be sufficient in duration. Degree and rate of inundation are other factors that produce significant penetration of the fabric with water. Again, once the material is wet, the temperature must fall from above freezing to below freezing, and freezing conditions must be maintained long enough to actually freeze the water in the material. After the water in the material is frozen, it must then thaw. Thawing is a significant feature in the completion of the cycle. Once the water has thawed, the cycle is completed, and so may commence again. Drainage of the water is not mandatory, but for the next cycle to be maximally damaging, more water must be added to the system.

When freeze-thaw cycling occurs in nature, we call it weathering; when it occurs in buildings, we call it a deterioration mechanism. As we go through the cycle in detail, we will use a masonry wall as our example, but it is important to remember that the deterioration mechanism of freeze-thaw cycling is a pervasive, naturally occurring phenomenon that affects many materials at every scale and condition. The basic mechanisms that deteriorate building materials are, in the end, no different from those that erode mountains and turn rocks into soil.

FIGURE 4.6   This wall displays loss of mortar and damge to brick face largely due to freeze-thaw cycling.

We chose a common masonry wall to demonstrate the freeze-thaw deterioration mechanism because common brick and mortar are prime examples of the porous, permeable materials that are prime subjects of freeze-thaw activity. Other porous, permeable materials vulnerable to the phenomenon are stones, particularly loosely consolidated sandstones, open-matrix limestones, and conglomerates. Metamorphic and igneous rocks as a class are, generally, less subject to freeze-thaw activity because of their low porosity and permeability properties, although there are certainly enough exceptions to this generalization to lead investigators to base an assessment of vulnerability not on nomenclature, but on basic properties of porosity and permeability.

Even within the classification of brick there is wide variation. There is a fired brick rated at 10,000 psi compressive strength and a porosity value of less than 1 percent. It hardly matters that this material is technically a brick; it may as well be granite. Even if we manage to jam some water into the fabric, it will not amount to much; and with a compressive stress capacity of 10,000 psi, there is a decided prospect that the modulus of rupture is greater than 660 psi. At the other extreme are bricks with porosity values greater than 10 percent. Bricks this porous will seldom meet the standards of an FBX or FBS brick according to ASTM standards. (FB stands for face brick; the X and S stand for dimensional standards.) Both are specifically designed for exterior weathering conditions, and neither is likely to be as porous as 10 percent. That said, this may be true for new walls, but there are at least two general exceptions. One is that such standards are not and, historically, have not always been, observed. The other general exception occurs when what was originally an interior, and inferior, wall is exposed to the elements. This happens all too often in aging urban areas, when one address in a row is extracted from its block. The resulting exposure of what were two interior party walls typically results in exposing an inferior brick to the elements for which it was neither designed nor intended.

The point is, highly porous bricks are occasionally exposed to weather. These bricks also tend to be under-fired and rather friable; the skins are thin or nonexistent, and they are weak, meaning that the modulus of rupture is very low. The potentially redeeming feature is that they are so porous, and the pore size distribution is so gross, that while they appear to be prime candidates for freeze-thaw deterioration, they may be unable to maintain saturation because they drain so rapidly. In other words, if the porosity pattern is such that the water cannot actually collect in the material, then water cannot be held inside the pore long enough to freeze. And even if water is suspended in a large pore long enough to freeze, there may be enough room for expansion, in which case, confinement cannot develop. If confinement cannot develop, pressure within the pore cannot develop. As we will see, avenues of pressure release are as important as any single factor in the mitigation of frost pressure.

Another porous, permeable material of considerable interest to building pathologists is concrete, specifically, the cement paste. If there is one material that will define building pathology for the next 25 years or more, it is probably reinforced concrete. Concrete is a complex material, and reinforcement adds to its complexity. We are improving the properties of the material every year, but that is a slow process; and because the test of the value of an improvement can take years, the learning curve for concrete continues to be long and slow. Still, concrete continues to be the material of choice for many building applications, and the historic cumulative inventory is huge as well. Early installations are now approaching one hundred years of age; and with age, comes an increased value. Older concrete is deteriorating, and current of remedial procedures are tedious and expensive. Put all these factors together and in concrete we have a veritable mountain range of future work.

To reiterate cement paste is actually a colloid composed of alternating layers of gel pore water and wafers of crystalline cement, which are predominately hydrated aluminates. The gel pore water is an integral part of the cement colloid, and without sufficient gel pore water, the cement wafer disintegrates, which is why it is so important to maintain sufficient water in the cement during curing. Without it, the colloid forms improperly, and so is never really cement at all. And once the aluminate crystals are formed, it is impossible to introduce water interweaved with the crystalline structures to rehabilitate poorly cured concrete. The damage is done, and the concrete cannot recover.

Given the proper formation of the cement colloid, it is possible to move water into and through the gel pore space. Keep in mind Poiseuille's Law which tells us that the rate of flow is the function of the width of the conduit, which in this case is measurable in angstroms in width. Water can move through the gel pore space, but the volume is low and the rate is slow. Once the cement paste is cured, introducing additional water into the gel pore space is very difficult; so too is the absolute removal of fluid water from the gel pores. When concrete is subjected to freezing temperatures, the water in the gel pore space can freeze and form crystalline ice. The temperature at which gel pore water will freeze is somewhat depressed by the solution of soluble salts in the gel water, a subject we will address in detail in the subsequent section on salts.

This discussion of gel pore water is not to dismiss the fact that reinforced concrete will have grosser forms of voids and fissures, which, compared to gel pores, are hundreds of times larger and still barely reach the optical level. These sizes and distributions are very important relative to the response fluid and the concrete to freeze-thaw cycling; but at the suboptical and optical range, the vulnerability of concrete to freeze-thaw cycling is similar to any other porous, permeable, absorptive, brittle material. What distinguishes concrete and cement from other materials is the water in the gel pores.

Given a vulnerable substrate, water must then be interposed into the material. It does not matter from where the water originates, only that it is present in sufficient quantity at the time that freezing occurs. The most common source is rain washing down the exterior surface of a wall, but the source is not relevant to the consequences. Other common sources include faulty gutters and downspouts, particularly the latter, which deluge a portion of the wall. The areas affected by such identifiable bulk sources may be relatively small compared to the building as a whole, but the quantities of water from these and similar faulty drainage systems are huge. (We will discuss drainage systems in detail Chapter 5, paying close attention to the manners and modes of failure of the components of the horizontal closure system.) The resulting damage of those failures is often visited upon the vertical closure system in highly concentrated areas of extreme saturation. For buildings with direct runoff from the roof, the water at the surface of the wall may be from ground splash or because the eaves do not direct the runoff far enough away from the wall plane. When water falls, even through still air, depending upon the volume and discharge orifice, it tends to splay out and disperse as it descends. The fan resulting from this dispersion may intersect with the wall.

It is also possible that the water is introduced to the wall as water vapor, and that the liquid water is the condensate from the vapor. The source of the water may be a leaking pipe within the building, which saturates the fabric from the inside. The only thing that matters relative to freeze-thaw cycling is that the fabric is wet at the moment of freezing. The source of the fluid is relevant only to the extent that intervention is facilitated.

Once the vulnerable fabric is in contact with water, it is necessary to inject the fluid into the fabric. Water flowing over the surface of the material is not enough for freeze-thaw action to produce damage; the water must be *in* the fabric, not on the fabric. Although it is a common assumption, not totally without merit, that water at the surface of a wall will traverse the material, there must be at least an explanation for the lateral movement. Newton's Laws of motion are still valid: bodies in motion remain so in straight lines unless acted upon by a force. Water has mass and is, therefore, a body. As it descends the face of a wall, it is accelerated by the force of gravity, minus the frictional force of the fluid as it is drawn across the surface of the fabric. The lines of action of these forces are vertical; and, unless the wall surface is inclined, the lines of action are parallel to the surface of the wall. There is no absolute motive force to cause the water to move laterally.

Lateral Motive Forces

Four primary sources of motive force can account for the lateral transportation of fluid into and through a porous, permeable solid such as concrete, stone, brick, mortar, plaster: pressure differential, capillarity, crystalline adsorption, and inclination. Missing from this list is a potential force, namely dynamic impact. It will be addressed first because, arguably, it is the single greatest source of confusion and misapplied physics in the entire field of building pathology and envelope design.

As a source of moisture penetration, the dynamic impact of water is a misapplication of a concept which is often confused with pressure differential. Dynamic impact describes the action of a particle or droplet of water being hurled against the wall, typically, by wind. That water penetrates the interior of porous solids because of impact is, however, independent of the presence of wind. Wind may be the usual method of delivery, but it is by no means the only way to propel a droplet of water against a surface. A fire or garden hose or spray nozzle will accomplish the same thing. The theory is that the moment the moving particle hits the orifice of a crack or pore with sufficient momentum, it will push itself into the matrix of the fabric before the mass of the droplet has decelerated to a velocity of zero.

If a pellet is fired through the air, its momentum is measured by its mass, times its velocity. The kinetic energy of the moving droplet is the product of its mass, times the square of its velocity, divided by 2. To put this example in some context, suppose that the pellet is one one-thousandth of an ounce, and that it is moving at a speed of 40 feet per second (approximately 60 miles per hour): the momentum of the pellet is 0.0025 ft lb/second. If a pellet moving at this speed is directed vertically, the force of gravity will decelerate the pellet to zero in approximately 1.25 seconds, and the distance traveled in the time required to decelerate it to zero will be approximately 50 feet. All of which is to say that pellets this small in size, moving no faster than 60 miles per hour, are not going to knock anyone down. We are hit by bigger raindrops moving at faster speeds in a driving rainstorm.

When and if this pellet hits a surface, it will impart at least some, if not all, of its kinetic energy to the second mass, namely that of the rigid surface. Now imagine that precisely at the point of impact is a small recessed cup, which is the exact diameter of the pellet, so that when the pellet flies into the cup it is captured without any rebound or splash; therefore, all the energy of the pellet is imparted to the solid. For a small fraction of a second, the pellet will apply a force to the bottom of the cup. Now imagine that at the precise dead center of the bottom of the cup is a small tubular pore $\frac{1}{1,000}$ of an inch in diameter, and that the pellet is, in fact, water. The size of the pore is an important variable in a problem of this type, so we do not want to prejudice the example with a hyperbolic hypothetical pore size. A pore $\frac{1}{1,000}$ of an inch in diameter is equal to the thickness of one-quarter of one page of legal paper—small, but still in the optical range. In fact, by the standards of actual, naturally occurring pores in sandstone (as measured by Seymour Lewin of New York University) $\frac{1}{1,000}$ of an inch (0.025 millimeter) is huge. In his experiments, Dr. Lewin used as a standard pore diameter a pore size of $\frac{1}{100}$ (0.00028 millimeter) of our hypothetical pore. If anything, our hypothetical pore is excessively large; however, making the point with such a large pore clarifies that smaller pore diameters will be less penetrable. Another reason for selecting a larger size is to include fissures as well as pores. While pore sizes are relevant to our discussion, optically sized cracks may be more important to water intrusion than pores.

The question is, how far into the pore will the water push itself before the kinetic energy dissipates and the balance of the water, if any, runs back out of the cup and down the exterior surface of the object? A droplet of water weighing $\frac{1}{1,000}$ of an ounce is approximately a sixteenth of an inch in diameter, which is a fair-sized droplet, and approximately 60 times larger than the diameter of the pore. Using Poiseuille's Law, and solving for the length of the capillary to be filled under these conditions, at mean sea level and 70° F, the droplet will push water into the pore approximately $\frac{25}{10,000}$ of an inch. A single sheet of legal paper is the approximate thickness the water will penetrate into the surface. Over 99.99 percent of the droplet winds up running down the face of the wall. For reasons that will become clear later, the water in the pore is not likely to simply run out immediately, so it is possible that another droplet will hit the cup precisely in the same spot and, therefore, advance the water a bit further. If the surface is sprayed at a rate of 10 gallons per minute per square foot of surface, which is the equivalent of eight inches of rain an hour, then the droplets will hit a given spot at the rate of about 270 drops per second spaced at approximately two inches apart. The first drop will penetrate $\frac{25}{10,000}$ of an inch. The next drop has to overcome the same friction as the first, plus the added friction pushing the first drop further into the pore. The force of the second droplet will be able to advance the combined depth of penetration half as far as the first drop; the third, a quarter as far; the fourth, one-eighth as far, and so forth. After one-tenth of one second, the water will have penetrated approximately $\frac{50}{10,000}$ of an inch, the thickness of two sheets of legal paper. After that, the force of an added drop hitting the end of the pore will not provide enough force to drive the column of water inside the pore a measurable amount.

If this seems a bit exaggerated, think of all the automobiles riding around with cracked windshields. Now picture those cars moving along a highway at 60 miles per hour in rain, our selected velocity of choice. How many of the cracked windshields will leak? The answer is, none of them, because a droplet of water hitting squarely on the edge of the crack, which is far narrower than our hypothetical pore, does not have enough kinetic energy to drive the water through the crack. Keep in mind that this assumes that the droplet hits the end of a perfectly positioned tubular pore on a trajectory perfectly aligned with the axis of the pore. If the droplet strikes the surface at an angle, or the pore is not perpendicular to the surface, much of the energy of this scenario will be expended into the plane of the surface, not into the pore. The kinetic energy of a raindrop simply is not enough to propel it through a porous surface whose pores are small. Even if the pores are larger than the droplet, and window is wide open, not all the rain drops are going to come through because the angle of incidence is less than perpendicular to the surface.

An argument to this conclusion is that rain is not a series of drops, but a stream. This is the fire hose model. Water does not strike surfaces as discrete spheres, but as a steady, moving cylinder. Suppose we do direct a steady stream of water perpendicular to a surface such that the pressure at that surface is 100 psi. What is the rate of flow of 70° F water through a crack in a windshield $\frac{1}{1,000}$ of an inch wide by four inches long where the glass is 0.125 inches thick? The answer is two cubic inches per *century*.

If the crack is $\frac{1}{100}$ of an inch wide across the thickness of a brick and the depth of the head joint, how much water will get as far as the back edge of the brick? In a full minute of steady deluge of 1,000 psi at the point of impact, the total amount of water getting through the vertical crack is less than the amount in a large drop of water. If pressure is 1,000 psi at the orifice of a 0.25-inch diameter nozzle, but the contact area of the stream at the point of impact is five square inches, the pressure at the point of impact is 50 psi and the amount of water getting through the crack after a full minute is a drop one-sixteenth of an inch in diameter, which is the same as our ongoing hypothetical droplet of $\frac{1}{1,000}$ of an ounce in volume. Obviously, no rainstorm is ever going drive water against a surface in a solid stream at 1,000 psi, but a power-washing process might. This analysis suggests that though driven rain may not penetrate a brick very far, it is possible that pressurized sprays might. (This will become relevant to subsequent discussions about the cleaning of wall surfaces with pressurized water.)

Arguably these analyses are prejudiced because cracks and pores come in many sizes; and, as was pointed out in the discussion of Poiseuille's Law, the width of the crack is critical to the quantitative outcomes of the problems. If the crack in the brick is one-tenth of an inch in width and the height of the brick, then the potential for water penetration is huge. If a hose is held perpendicular to a such a crack, and the water pressure at the face of the surface is 1,000 psi, then the volume of transmission increases to over 15 cubic inches of fluid per minute, which is a large amount of water by any standard. If such a volume of water were delivered through a standard three-fourths-inch diameter garden hose, it would require three people weighing more than 150 pounds each just to hold the hose down, much less keep it pointed at the wall. If the hose analogy were valid, the cracks were as large as hypothesized, and everything else we have assumed were true, then it would be correct to submit that under such circumstances large amounts of water would be introduced deep into systems.

There are two general models, or metaphors, for pressure against a surface. We met one already as it related to moving fluids. We dubbed it the fire hose model. The other is the balloon model. Both have merit and application. Unfortunately, they are often interchanged and, therefore, misapplied. As with the fire hose theory of fluid dynamics, the fire hose model of moving air obstructed by a barrier is based on a metaphoric analogue of the mechanics of moving masses. If a mass is moving with velocity, its continued movement is governed by Newtonian laws of motion. That mass will, among other things, continue to move in a straight line unless acted upon by an external force. Thus, for that mass to come to a halt, it must be acted upon by a force sufficient in magnitude and duration to decelerate the mass and arrest the motion.

As with liquids, where the source of water is wind-driven rain, the stream of water is not steady. The impact of rain is a succession of discrete droplets that impact the surface at various angles, rarely perpendicular to the surface, at velocities less than 100 feet per second, and for periods measurable in terms of minutes or possibly hours. The pore size distribution is measured in thousandths, even millionths, of an inch. Under these circumstances, the dynamic force of the droplet by itself cannot account for penetration of the fabric to depths more than a few 10-thousandths of an inch in depth. The key variable that changes from the dynamic impact model to the pressure differential model is the duration of the application of the pressure. In the case of ballistically propelled droplets, the impact of the droplet is dissipated in one or two one-thousandth of a second. If pressure were sustainable for even a whole second, much less longer durations, the magnitude of the issue would change a thousandfold or more. For small pores, the time required to traverse the depth of a brick may require more time than the pressure can be sustained, but the depth of penetration will increase significantly if the pressure can be sustained. If we take a likely scenario of rain cascading across the surface of the brick, and if the pressure differential across the thickness of the brick is consistent with that which may occur in an unequalized wind pressure exerting 40 pounds per square foot, the total time required to fill the pore from outside to inside will require over 10 hours of constant pressure. That kind of wind pressure may be sustainable in typhoons, but not the ordinary change of weather. More likely are gusting winds during a storm lasting 30 minutes or an hour. Even if the pressure differential of 40 psf is sustained for only 30 minutes, the depth of water penetration in a $\frac{1}{100}$ of an inch diameter pore is $\frac{38}{100}$ of an inch. This may not seem like a great deal, but relative to freeze-thaw action it is quite significant. Compare this result with the $\frac{25}{10,000}$ of an inch penetration due to the errant-flying-droplet theory.

As the diameter of the pore size increases, the potential for penetration increases. Yet, as the pore diameter increases, so does the relative difficulty of maintaining the depth of water coverage at the supply end of the pore. This reverts to a previous point about the prospect of large pores draining more rapidly than they are filling, particularly if the pores are even slightly inclined upward. If the inclination of the pore is downward, then gravitational pull of the water inside the pore will assist in siphoning water from the exterior toward the interior. The point is that pressure differential is an effective method of lateral transportation of fluid into a porous solid. The depths of penetration are limited, but dimensions are significant in terms of deterioration by freeze-thaw cycling.

This discussion leads us logically back to the influence of capillarity on the overall transportation of fluid into fabric. Recall that capillary action alone, acting against gravity, cannot reasonably pull fluid up very far. By itself, the vertical potential of capillarity can lift a fluid in a hypothetical porous material a few inches, given continuous fluid supply and extensive amounts of time. In the tubular void of a timber cell, the capillary diameter is approximately $\frac{1}{1,000}$ of an inch. The theoretical rise of fluid due to capillarity in a tree trunk is almost 48 inches. In a pore or fissure of $\frac{1}{100}$ of an inch, the theoretic vertical rise in a capillary is less than five inches. If, however, the capillary is inclined rather than vertical, then the lateral distance, which is what we are interested in at this point, could be as great or greater than the vertical rise. If the capillary is depressed in inclination below horizontal, then the fluid could theoretically flow through the capillary as it would through an open pipe. Time, however, is required in all these situations.

Where the pore diameter is $\frac{1}{1,000}$ of an inch, and is sloping downward at a steep angle away from the face, the length of time to fill a 30-inch-long pore with water is over 20 hours. If in the course of such a descent, the pore also traverses the wall, then after two days of constant rain, the water will have descended 30 inches and traversed the dimension of the inclination. If the supply of water is interrupted or discontinued in less than 20 hours, the fluid will continue to slowly descend in the pore; but absent a continuous supply, the pressure behind the column will decline and the rate of descent will be even slower. If, for example, the supply of water stops after 10 hours, the pore will only be filled one-quarter, not half, its length. For that water to descend the remaining 75 percent of the 30-inch length could take two weeks.

The time consideration is extremely important in evaluating water and heat problems because by the time the magnitude of the issue reaches a level of significance, the source of the problem may have already ceased or reversed. In the case of water penetration, the qualitative appreciation of the problem may result in a diagnosis that water can and will traverse the wall or significantly penetrate the material. In the absence of a time computation, however, a prescription may be rendered based on the qualitative analysis alone. This may result in an intervention that costs money and brings with it a certain amount of fabric disturbance and risk of other untoward consequences. If the time factor had been included in the analysis, the magnitude of the issue may have been trivial, meaning that while the diagnosis was essentially correct, the magnitude of actual potential damage—or, in this case, depth of penetration—was trivial. If professional analysts do not execute the quantitative analysis at the same time, they run the distinct risk of over-prescribing remediation and intervention.

For freeze-thaw cycling to be damaging, a few fractions of an inch of penetration are sufficient to put the outer perimeter at risk. The time factor is the critical element, and is subject to the same restraints as the example given above relative to pressure differential; but when added to pressure differential, capillarity can symbiotically amplify the effects of pressure differential, and vice versa.

All of the factors come again to bear on crystalline absorption. There is no question that some solid crystals, particularly salt crystals, can absorb fluid. With hydration of crystalline salts, there is the countervailing effect of freezing-point depression. And where water is absorbed into the cellulose of a timber cell wall, time is required to achieve saturation. If the fabric is already near saturation, for a variety of reasons, high prolonged humidity being the prime candidate, then wetting by intermittent rain may achieve fiber saturation. If, however, the fibers are not near fiber saturation at the onset of wetting by precipitation, it is unlikely that there will be sufficient crystalline absorption to achieve saturation.

This leaves a broad area of lateral motive force, called *inclination*, which is a variation of what we have developed regarding capillaries and pressure differentials, that is, that water will flow more easily downward than upward. To this point, we have stipulated that the exterior surface is vertical, and have implied that it is also smooth and flat. But walls, particularly aged walls, are not vertical, nor are they flat or smooth. The porous, permeable materials we so often use to construct walls are, at close range, highly irregular and rough. At the microscopic level, water running down a masonry wall looks more like white water rapids than a smooth film of fluid. Each and every one of the ledges and protrusions offers a horizontal surface to incline the water back toward the surface, rather than smoothly down the face.

At the next level of magnification are the bond breaks and mortar joint projections and indentations, the protrusions of fabrication features, as well as chips and dings from damage. At the construction level is the assemblage of pieces, many of which provide planes of inclination toward the interior of the building. The protruding arch at the head of a window is a natural collector of water. Water cascades around the form of the arch, which increases in depth and velocity. If on the downward portion of the protruding arch, the now relatively deep and rapidly moving water encounters a construction, with enough momentum and energy, the water pushes through a small crevice in the masonry, such as a debonded mortar joint. Simply by virtue of the inclination of beading planes, and the joints and surfaces perpendicular to the exterior surface of the building, the pieces and assemblies in their gross form can act as diverters and planes of travel for water toward the interior of the fabric, and to the building as a whole.

Regarding lateral transportation of fluid through porous, permeable solids we reiterate here that the size of the pore determines the rate of fluid movement. Lewin found that the distribution of pore sizes was more important than the absolute value of porosity, thus it is possible to imagine a material that is 40 percent porous, but that does not have a single pore larger than a 10-millionth of an inch in diameter. For all practical purposes, such a material is impermeable even if the pores are continuous and connected. We can concoct a test that will push fluid through the test block, but would have nothing to do with reality. If the permeability of a material is such that it will pass a cubic inch of water per century, it is irrelevant that it is technically porous and permeable. On the other hand, it is very possible to find a specimen that is only be 5 percent porous, but whose pores all transport fluid exceeding well.

Relative to Poiseuille's flow, statistically significant pore sizes begin in the range between $^{12}/_{10,000}$ and $^{16}/_{100,000}$ of an inch in diameter. This range covers an order of magnitude yet is still at least an order of magnitude smaller than the examples we have been using. These sizes are not necessarily those that conduct the greatest amount of fluid; the point is that pore diameters smaller than this range do not contribute significantly to Poiseuille's flow. Larger pores can and will conduct disproportionately more fluid than small pores, and at a much faster rate; however, their capillarity properties deteriorate inversely to their diameter. The problem with pore size distribution determination is most readily solved by working backward from experimentally induced absorption rates to a hypothetic average effective pore size. (As noted earlier, a direct count and measurement of pores requires the use of electron microscopy and an accomplished pore size observer. This is not impossible nor even prohibitively expensive, but it is a bit of a tough sell to clients, and it is inherently destructive.) The use of average values does, however, cause professionals to either overprescribe or under-prescribe because numeric values are based not on the material at hand, but on a hypothetical material or the average of several similar materials.

Back to the subject at hand. Introducing water into the fabric, even expanding the water inside the material is not enough to constitute a deterioration mechanism. The material must be damaged, which in the case of freeze-thaw cycling, means that the fabric must crack. If the modulus of rupture is greater than 660 psi, and our estimation of the expansive potential of ice is correct, then the pressure inside the pore will not be enough to initiate a crack. For many reasons this may be less of an issue than it appears. As discussed in Chapter 3, the force required to advance a crack or flaw is less, potentially much less, than the force required to initiate a crack in an unflawed specimen.

The possible sources of crack initiation are many, some of which we discussed in Chapter 3. At the moment that freeze-thaw cycling is operative, the material becomes more brittle. Even if no single source of internal stress is sufficient to initiate a crack, there is an increased prospect that, at a lowered temperature, the sum of the sources will crack the fabric. Once the crack is formed, however minor, those same forces can extend the crack even at lower stress levels. The initial flaws in many brittle, porous, permeable materials are latent flaws that occurred during the fabrication process, which is especially common to materials that were cooled during manufacture, such as brick. The first cooling of a brick, which is carefully controlled to avoid a through-body fracture, may not be sufficient to avoid microcracks at the pore-size scale. In short, the chance of ever having a flawless specimen are remote, so the requirement of a threshold stress in excess of the modulus of rupture, while technically accurate, is unreasonably high. Materials will experience freeze-thaw damage at internal stress levels well below the theoretical stress fracture.

The quintessential condition for freeze-thaw cycling is freezing. Even though water expands between 39° F and 32° F in the liquid phase, As a practical matter, the water must actually freeze to produce damage. Shortly, we will describe a scenario in which expanding liquid water could crack a specimen, but the likelihood of such an occurrence is improbable in the extreme. Moreover, the fact that the water expands is not by itself sufficient for internal stress to develop; water must also be confined. In a closely fitted space, wherein the water completely fills the volume between the opposing faces of a crack or pore, water in its solid form is self-limiting in its ability to migrate. In the liquid state, water under stress can migrate. Water trapped in the same pore may expand, but as it develops pressure, it is free to leak out of the pore or crack into a relieving reservoir or bubble.

Imagine a saturated specimen being cooled from 39° F to 33° F. We know that the water inside the material will expand. If that failure is confined, there will be pressure exerted against the fabric. The water, too, will be under stress. Unlike solids, fluids can redistribute pressure anywhere in every other particle. This is the essence of hydraulics. We can pressurize a tank of fluid, pump that fluid through a pipe into a cylinder several feet away, lift a piston, and raise an elevator. If we pressurize fluid in a saturated specimen, and as a result the fluid expands, that resulting pressure will be transmitted throughout the fluid mass. If there is a single outlet for the fluid, and some of it is squeezed through that outlet, the pressure for the entire mass of fluid will be relieved if not totally, at least in part.

Pressure relief is not always practical, but, generally, fluid water under pressure can retrace the path it took to get into the fabric as an avenue of relief. If the material is part of a wall, and the mode of entry was from the absorption of rain washing down the face of the wall, then the expanding water can squeeze itself out through its original entrant pore, and exit through the face. The idea that relief from expansion in the liquid state is sufficiently easy may be difficult to imagine. Circumstances may arise that cause damage to the fabric from expanding water that does not actually freeze, although this describes a sequence of events during the freeze-thaw cycle that may qualify for a distinction.

The key determinant of whether the expansion of freezing and, subsequently, frozen water will produce damage is again related to time. If the process occurs very slowly, even the fluid inside the smallest pores can redistribute stress, hence relieve pressure. If the process occurs very rapidly, water could be effectively trapped in the pore, not because there is no escape path, but because the escape path diameter so small that even under pressure, not enough fluid could move through the capillary to relieve the pressure.

The more rapidly temperature falls through freezing, the greater the prospect of trapping water inside the pore, particularly in small capillaries, and, consequently, the greater the risk of damage to the fabric. Once the frozen fluid is trapped inside the fabric, it follows that the depth of the freeze will exacerbate the damage, for two reasons: first, ice continues to expand below freezing for at least 20° F or more, therefore the potential expansion of ice continues to increase with declines in temperature; second, the toughness of the material declines with temperature, which means that the material may be able to withstand the internal pressure of the expanding ice at 32° F, but at 25° F the brittleness of the same material will increase, and from the decrease in resistance to cracking alone, the fabric may fracture. This phenomenon is independent of the ice continuing to expand through this same temperature.

The best examples of the reduction in toughness with declining temperature do not involve the expansion of ice, because the expansion disguises the reduction in toughness. The classic examples of toughness reduction can be found in reports of nineteenth-century railroad accidents that occurred during winter. It was not unusual in that era for iron members under stress to suddenly snap. A member, which at one temperature was capable of sustaining the imposed stress, would undergo a sufficient change in material properties. As a result of the depressed temperature, a critical flaw would accelerate through the body of the material, resulting in complete separation. Similar factors are present during freeze-thaw activity, but the expansion of the ice is often credited as being the sole perpetrator when it is more accurate to say that the increasing pressure of the expanding ice *combined with* the decreasing toughness of the solid material to result in a fracture or extension of a fracture at temperatures below the freezing mark.

Another aspect of the time variable is that the depth of frost penetration—which is also to say the depth of risk to the fabric—requires time to develop. As explained, there is a thermal lag associated with heat reduction. The fact that outside air temperature may drop to 32° F does not mean that brittle fabric immediately starts to disintegrate. The fabric itself must be chilled to below freezing, and the water must lose the heat of fusion for crystallization to begin. These events require time. The speed at which the deteriorating effects of freeze-thaw action can ensue depends more on the absolute temperature drop than the relative rapidity of the drop. If temperature drops from 39° F to 30° F in an hour, we might agree that the rate of temperature decrease is high, but the deleterious effects of such a drop will not, in the end, be as great as a drop from 39° F to 21° F ($\Delta T_2 = 2\Delta T_1$) even if the time required to accomplish the total drop is three or four times as long. In terms of temperature drop, there are two factors: rate of drop and magnitude below freezing. The more rapid the temperature drop, the greater the amount of water will be trapped inside the material. The greater the temperature drop, the greater the combination of expansive stress and material toughness reduction.

Frost Progression

The classic freeze-thaw cycle begins in a rapidly moving cold front proceeding ahead of a polar air mass descending through Manitoba. In the winter, the ambient air may already seem cold to humans, but compared to one of these Arctic cold fronts, cold is strictly relative. The air ahead of the front may, in fact, be 40° F and still be warm compared to what is behind it. When cold fronts advance, they tend to move across the ground quickly, sometimes reaching speeds of 40 miles per hour. The temperature drop across the front may be from ambient to 10° F within a few hours. As a rule, the more rapid the passage, the more violent the storm. The advancing cold air is denser than the ambient, relatively warmer air, so the cold air mass tends to push the ambient air up and over the top of the cold front. The resistance and mass of the ambient air presses the advancing edge toward the ground into a wedge shape as it plows under the warmer air. The relatively warm air cools as it is lifted over the colder air; and, as it does, water precipitates out of the ambient air and begins to descend, either as liquid rain or as ice, depending on the degree of temperature change before descent can occur.

As the precipitation falls through the leading edge of the cold front, it may continue as liquid if the air temperature in the wedge is above freezing or the rain simply gets through the leading edge faster than it can freeze. In either case, it will probably be colder than where it started. In some cases, the ice descends through the leading edge and melts as it passes through what is technically a cold front. Because air tends to cool with altitude, the ambient air is warm relative to the incoming air *at the same altitude*; but as it descends, it passes through decreasingly cold air regardless of which side of the front it is located. The consequence is that the descending ice passes through a cold air mass that is absolutely warmer than the "warmer" air at a higher altitude.

Another scenario is that the precipitation starts out as water and freezes as it passes through the leading edge, striking the ground as frozen pellets called sleet. In rare instances, the conditions are such that the water is so close to freezing during descent that it strikes the ground in a supercooled state and freezes practically on contact with a surface. These so-called ice storms are beautiful to witness, but treacherous to deal with. Yet another scenario is that the ice simply passes unchanged through the leading edge, and hits the ground in form of snow. An intriguing aspect of cold-front passage is that any combination or all of these can occur during the passage of a single front, depending on only very slight shifts in the temperatures of the two air masses. The shifts of interest to us in the context of freeze-thaw cycling are those where the water strikes the surface as a liquid and remains a liquid long enough to be absorbed into the fabric of the wall.

One minor mitigating effect on the situation is that the water being absorbed is cold, hence its wetting properties are depressed. Warmer water is "wetter" than cold water, so the adhesive properties of warmer water are greater than of colder water. In the overall scheme of things, this is a small consolation. Of arguably greater significance is that the fabric is also cold, thus the voids have contracted and slightly reduced in the available volume of the voids. Again, thermal contraction is still a minor factor.

In our example, we will assume that the precipitation coming out of the sky is liquid. We will also assume that the volume is sufficient so that in a relatively short period of few hours significant fabric penetration occurs, based on available water volume. Finally, we will accept that factors such as decreased wetting and pore volume reduction are present, but minor, and that within a very short period of time, say five or six hours, the temperature will plunge to 10° F. At this point, the single most important factor to influence the degree of frost penetration will be wind direction and velocity. As noted in the section on lateral motive force, wind is a common source of differential pressure across a resistive boundary. Air movement is also a major factor in the effective cooling of a relatively warm surface.

The direction of the wind determines which aspects of the façade are potentially positively and negatively pressurized. The reason for including these potential conditions is that wind alone does not necessarily lead to the windward façade being positively pressurized nor to the leeward façade being negatively pressurized. For example, the interior of the building may be overpressurized, causing even the windward side to operate at a net negative pressure even though the wind is applying an external positive pressure to the façade. If there are no other intervening factors, and the surface of the fabric is pressurized by wind direction, then it is absolutely relevant to know the direction and velocity on the saturated faces. It is possible to reasonably predict wind direction based on meteorological conditions, but that discussion is beyond the scope of this book.

As a result of frontal passage, there will be a shift in the direction of the wind, which raises two possibilities: that all primary faces will be exposed to some positive wind pressure during the passage, and that no one façade will be exposed to full positive wind pressure for the duration of the passage. During a typical cold front, the wind direction at the outset will bring the strongest gusts and, often, the heaviest rains. This suggests that the façade usually subjected to the initial wind of passing cold fronts will experience the most freeze-thaw damage Generally that is true; but there is also the possibility that the front will move through very rapidly, the wind will blow very hard; and the rain will fall heavily but for only a very short period of time. Another factor is that at the outset of the storm the temperature is still at a relatively elevated level.

As the front passes, the wind direction changes, and the façade initially leeward is now windward; but it is catching wind from the trailing side of the front. The direction has shifted and, typically, the velocity is less than before; and the rainfall is usually reduced. All this *suggests* a less hostile environment, but a couple of other factors occurring simultaneously must be accounted for. It is more likely that the drastic plunge in temperature will occur with the trailing wind, and probably the reduced rainfall will last for a more protracted period. In other words, the original windward wall may experience net drainage before the onset of freezing temperatures, while the original leeward wall is still taking on water when the temperatures take a dive. Though it is still accurate to assume that over the course of time the initial windward façade will endure the most damage, it would be a mistake to assume that the leeward side is exempt by any measure.

The time factor is of paramount importance. The actual time and rate of the inundation with water is critical to the outcome of the damage caused by the cycle. Whether there is little rain or an intense five-minute downpour, the resulting damage cannot exceed the depth of water penetration. It is a simple equation: damage is a function of the water absorbed, not the water applied. The amount absorbed is related to the time of application as much or more than the depth of accumulated water rolling down the façade. Net wind pressure is extremely important, but keep in mind Poiseuille's Law—pushing water into small pores takes time even if the pressure differential is large.

During the period of inundation, and under the net pressure, water will be imbibed by the porous, permeable material to some depth. As the period of inundation increases, the depth of saturation increases; and as the net pressure differential increases during inundation, the depth of penetration increases. And let us not forget that energy is simultaneously being surrendered by the fabric. As the cold water is being absorbed into the fabric, residual heat in the fabric will warm the cold water. Concurrently, heat from deeper in the interior of the material will begin migrating to the now cooler portion of the material, meaning, for the most part, the saturated material. To the extent that heat is available from the interior on the building, and depending on the transfer rate of the material, more heat may begin migrating from even deeper in the material toward the exterior.

At the moment of maximum saturation, the thickness of the material at issue is thermally two distinct materials: essentially dry fabric and saturated fabric. Despite the structural similarity of the material, thermally, the properties of unsaturated and saturated fabric are completely different. The saturated fabric will have a lower thermal resistance because air, which is highly resistive, has been replaced with water that is relatively conductive. The specific heat capacity will increase because the specific heat of water is significantly higher than the displaced air. The more porous the material, the greater the change in thermal properties, because more air is displaced by the dramatically different water.

We can plot the temperature gradient for the partially saturated material thickness in the same way we used to plot the temperature gradient of two disparate materials. The temperature can be identified at every point from the interior surface to the exterior surface. In executing this analysis it is important to take into account the contribution of the overall thermal properties of the wall and the air film in immediate contact with the exterior surface. In still air, this layer is affected to a significant degree by the heat transfer from the wall to the air, and it can act in a highly insulative manner as long as it is undisturbed. This is why wind velocity is an important component in the thermodynamic equation of the assembly. The higher the wind velocity, the thinner the insulating air cushion against the surface, and, therefore, the lower the overall thermal resistance of the wall and of the material.

Returning for the moment to the material itself, let us take as an example a monolithic wall of homogeneous material, 16 inches thick. The temperature inside is 72° F; the outside surface temperature is 32° F ($\Delta T = 40°$ F). Discounting the still air effect at the outside surface, the temperature gradient is a straight line from the inside, decreasing linearly to the outside surface. The temperature at any intermediate point is the appropriate dimension into the wall, divided by the total wall thickness, times the temperature differential. If we look at the location 4 inches in from the extreme outside surface, the temperature is 32° + (4 inches/16 inches × 40°) = 32° + 10° = 42° F. Let us also assume that the unit resistance to heat transfer is 2; therefore, the total R value is (2 × 16 inches) = 32.

We now inundate the exterior surface with water so that the same four inches is now saturated with water, thereby substantially reducing its resistance; in fact, the unit resistance is reduced from 2 to 0.5 for that portion of the wall. This means that the total thermal resistance of the wall is now R = (4 inches × 0.5) + (12 inches × 2) = 26. The temperature at the boundary between the wet material and the dry material is computed as 32° + (2/26 × 40° F) = 32° F + 3° = 35° F. Every point in the wall is operating at a lower temperature than before, and the total thermal resistance of the wall has decreased by approximately 20 percent. More heat is passing through the wall per hour, and the material is colder than when the same wall was dry; but no portion of the wall is frozen.

The wind direction begins to shift, and the rain slackens, while the temperature plunges to 20° F. The water at the surface begins to freeze. At this precise moment, the wall is sealed from further liquid migration from the outside. (Liquid migration may be impossible anyway because the air is well below freezing, and any further precipitation will be solid.) Remember, it is possible for the initial precipitation to start as rain and to freeze as it passes through very cold lower air, but it is not required. If the rain passes through even 20° F air very rapidly, it may reach the surface in liquid form and immediately freeze on the surface of objects that are themselves well below freezing.

In our example, freezing rain is not an issue, because once the water at the outside surface of the fabric freezes, the frozen water provides an effective moisture and vapor barrier to the outside. As surely as more water is precluded from being absorbed, it is also true that any water trapped inside the fabric cannot migrate back through the fabric to the outside surface and drain away without fracturing the crust formed on and just inside the exterior surface. To fracture the crust, the water inside the fabric would have to exert a lateral force which is not present. The water inside the fabric at the time that the exterior surface freezes is essentially the water that will engage the freeze-thaw cycle.

The line at the leading edge of the ice formation inside the specimen we will call the *frost line*; it will progress into and through the material as long as the outside temperature remains below freezing. The farther the outside temperature drops below freezing, the faster the frost line will progress, because the temperature gradient and the rate of thermal loss through the fabric will both increase. As the frost line progresses through the fabric, water converts to ice, but there is no abrupt shift in volume at that point.

The best information on this suggests that at the precise point of conversion, the density of 32° F water and 32° F ice are indistinguishable. The reason the overall density of ice is less than that of water is because the captured ice in the core of a frozen volume is less dense than the surrounding warmer ice. It is the less dense core that enables ice cubes, icebergs, and other formations to float. If this information is correct, there is no immediate and abrupt volumetric shift at solidification; in which case, at the precise moment of solidification, there is no tendency to expand laterally against the pore wall, hence no confining pressure at the instant of solidification. The expansion occurs as the ice continues to cool below freezing. If there is an abrupt volumetric shift at the instant of solidification, there is a volumetric increase. In addition to immediately precipitating internal confining pressure, the expanding ice will push the still liquid water ahead of it, generating pressure within the mass of the water. Added pressure to the fluid will depress the freezing point of the water slightly; but the pressure is likely to be slight, and the depressed freezing point will be minor, as little as a fraction of a degree.

At the point where the thickness of the ice layer stops advancing, three are present—as well as a three-layer thermal problem. We want to know how deep the ice will form. Determining the various intermediate temperatures in this problem is simplified to some extent because we know what the temperature is at the boundary between the ice and the saturated fabric; it is 32° F by definition. Using our preceding example, we also know the thickness of the wall (16 inches) and the inside temperature 72° F. We can set the outside temperature at 22° F, which means that $\Delta T_{TOTAL} = 50°$ F. If we know the depth of saturation and the respective unit thermal transmission values of the three layers, we can solve for the depth of frost penetration. If the U value of the dry material is 2, of wet material is 0.5, and of frozen wet material is 1.0, and if the depth of saturation is 8 inches, then the total R value of the dry material is (8 inches × 2) = 16; of the wet layer, (8 inches – X”) ×) 0.5 = 4 – 0.5X; and of the frozen layer, X inches × 1.0 = X.

The sum of the R values = 16 + (4 – 0.5X) + X = 20 + 0.5X. We know that, by definition, the change in temperature from the outside surface to the boundary between the frozen and the wet material is 10° F, because the outside temperature is 22° F and the temperature at the boundary is defined as 32° F. We also know from Chapter 3 that the change in temperature across a given layer is proportional to total change in temperature, as is the thermal resistance of that layer to the total resistance of the wall. When the values are plugged in and the arithmetic crank is turned, the depth of the frost line from the outside surface is 4.44 inches, and the temperature at the boundary of the saturated material and the dry material is 35.2°. All of the temperatures across the section are below the dry temperature gradient, which is what we expect when the overall resistance of the section is reduced from the thermal response of the dry material alone.

From this we learn an important lesson: When the temperature is freezing outside, and the wall is saturated, it does not follow that the frost line progresses until all the water is frozen. There is a position at which the frost will arrest in a state of thermodynamic equilibrium, and it is not necessarily the full depth of saturation. But one surprise awaits. We must back up and start with water penetration of one inch in depth and work from there.

If the 16 inches of wall with a U rating of 2.0 (R = 16 inches × 2 = 32) is absolute, the outside ambient temperature is 22° F and the inside ambient temperature is 72° F, then the temperature of 32° F lies on a line 3.2 inches inward from the outside face of the wall. Of course, there is no freeze-thaw damage because there is no water in the wall; so the fact that over three inches of the material is below freezing is of no particular significance, except that some of the wall is very cold.

Now let's begin adding water. If we add enough water to saturate the first inch inboard from the outside surface, the thermal conductivity of the wall decreases because the resistance of the water is less than that of the dry material. The temperature at the boundary between the saturated and the dry material is 23.6° F, and all the water is frozen. If we saturate 2 inches, the temperature at the boundary will be 25.3° F, and all the water will be frozen. If we saturate 3 inches, the boundary temperature will be 27.2° F, and all the water will be frozen. If we saturate four inches, the boundary temperature will be 29.1° F, and all the water will be frozen. If we saturate 5 inches, the temperature at the boundary will be 31.3° F, and all the water will still be frozen. At a saturation depth of 5.33 inches, the temperature at the boundary will be exactly 32° F, and all the water will be frozen, but this is the limit of potential frost penetration.

Now comes the peculiar part. If we saturate 6 inches, not all the water freezes. The depth of frost penetration is 5.1 inches which means that 0.9 inches of saturated material will not freeze and 10 inches of material will remain dry. The temperature at the boundary between the frost and the saturated material is 32° F, by definition, and the temperature at the boundary between the saturated material and the dry material is 32.9° F. None of this by itself is surprising, but combined with the other results, the peculiarity becomes apparent. The frost line when the depth of saturation is 6 inches is actually less than when the depth of saturation was less. The frost penetrated deeper when the water did not saturate as far. When the saturation reaches 8 inches, the depth of frost penetration, 4.44 inches, is less than when the saturation was only 6 inches. The ice advances with the water, then it reverses. In fact, if the entire 16-inch wall were saturated, the depth of frost penetration would be only 1.8 inches. How is this possible?

In the answer lies another lesson, this one in multivariable problems. When any water enters the wall, the thermal resistance of the wall as a whole decreases because water is thermally more conductive than the dry material that was just wetted. When that same water freezes, the thermal resistance actually rebounds a bit because though the thermal resistance of ice is greater than water, it is still less than the dry material. The inch-by-inch progression of wetting and freezing, discussed above, has a parallel story— that is, what was happening to the thermal efficiency of the wall as a whole. The sequence of wetting and freezing was accompanied by a trail of decreasing thermal resistance, followed by some rebound to that decrease due to wetting alone. At a certain point, the residual dry material— meaning the material left after we designated how much of the wall was to be wetted—was diminished to the degree that so much heat was leaking through the wall the water was no longer freezing all the way through the wetted area. As the water penetration continued, we were effectively substituting wet material for dry material, which degraded the thermal resistance even further. Heat passed through the wet area even faster than the formation of ice, effectively retarding the advance of the frost line, but only at the expense of increasing amounts of heat passing through the wall. The reason the frost eventually retreats to only 1.8 inches is because the complete saturation of the wall has so diminished the thermal resistance of the fabric that the formation of ice is virtually defeated by the flood of heat passing through the relatively nonresistive wet material. What looks like a loss of freeze-thaw activity is gained at the expense of pumping vast quantities of heat into the wall and accepting the additional consequences of saturating added material.

Though obviously not intended to advocate saturating the wall in the interest of frost retardation, the preceding discussion does raise some interesting and provocative possibilities for intervention where freeze-thaw cycling is a severe problem. Suppose, for example, that we could design a wall with variable thermal resistance, or that we could retrofit a wall for variable thermal resistance. If we could, we could turn up the heat *inside* the wall at the time of impending freezing and deliberately heat the fabric, thereby defeating frost penetration.

CRACK INITIATION AND PROPAGATION

Now that we understand how and when ice forms in the material, we can turn our attention to the exact moment that the crack forms. As the ice forms and continues to cool, we know that the ice expands with a potential force on the order of 660 psi. We know from Chapter 3 that to develop that expansive pressure, the ice expansion must be confined. Inasmuch as the ice is forming inside pores surrounded by solid material, the matrix of the surrounding material provides the rigid confining reactive material against which the expanding ice exerts pressure. To the extent that the solid material can withstand the pressure, it may elastically deform, and the entire volume of the solid may expand in a manner not unlike other examples of thermal expansion, which we examined in detail in Chapter 3.

To the extent the ultimate strength of the surrounding solid material is exceeded, there is the potential for failure in the form of a crack. This crack is strain-limited by the expansion of the ice, which is finite, based on its volume and the further drop in temperature of the ice. As soon as the crack forms, the total volume of the pore increases by the volume of the void of the crack. When the increase in volume and the residual confining pressure of the solid reestablishes equilibrium with the expansive force and volume of the ice, the crack arrests, because there is insufficient force to continue crack advancement. As stated in Chapter 3, once initiated, cracks tend to run quickly and disproportionately far before self-arresting to the strain limits of the expanding material.

This review of material presented in Chapter 3 is purposeful, because here we add a twist from earlier in this chapter: the simultaneous contraction of the solid matrix due to conventional thermal contraction as the material cools. This is not a trivial issue, and so warrants further investigation. We speculated earlier in this chapter that the effective coefficient of thermal expansion of ice is negative, and approximately −0.0033 percent /100° F. We also speculated, less justifiably, that the effective modulus of elasticity of ice is approximately 1,000,000 psi. If, say, the confining material is brick, then the coefficient of thermal expansion over the same range is positive and approximately 0.00035 percent /100° F. The modulus of elasticity of brick is approximately 3,000,000 psi. Over the temperature range of maximum concern, namely 32° F to 12° F, the ice expands in one direction and the brick contracts in the other. If the ice is unrestrained, as computed earlier, the expansive potential is 0.00066 percent and the confining pressure required to maintain zero expansion is 660 psi. Over the same range, the brick will contract to close the pore dimension by 0.000070 percent and the confining pressure to resist expansion will be 210 psi. What is the location of the pore surface relative to its original position, and what is the stress in the brick at that point?

Instinctively, we want to add the two opposing forces such that the stress at the face will be 870 psi, but doing so would miss a very important point about simultaneously expanding bodies; that is, the only way that the stress in the ice can achieve 660 psi over the temperature range at issue ($\Delta T = 20°$ F) is for the ice to be *completely* restrained from any movement. If the ice is allowed to expand the full 0.00066 percent, there is no stress in the ice; it is strictly at differential expansion due to change in temperature. Volume changes without internal stress unless there is confinement. If there is confinement, and it is less than total, the ice expands some, if not to its full potential. That is, there is some internal stress, but less than what is required for complete restraint.

The same is true of the brick. Full, unimpeded expansion is 0.00007 percent. Complete restraint would require 210 psi, but the restraint is not compete. Consequently, both the ice and the brick expand, but not to the full dimension of unconfined specimens. Likewise, both the brick and the ice experience internal stress, but neither will experience the full potential of total confinement. If we knew the relative dimensions of the pore and the brick beyond the pore, we could complete the calculation for the actual stresses and actual dimensional change in each material. The missing information is relevant because the percent change in dimension, namely 0.00066 percent and .00007 percent for the ice and brick, respectively, are percentages of the original dimension, which is unknown. If L_{ICE} and L_{BRICK} were equal, we could set each equal to unity and proceed with the calculations.

We might take a guess as to the effective original dimensions based on porosity. If the porosity is 8 percent, then the pore volume is 8 percent of the total volume. That being the case, we could make an argument that the void-to-solid ratio, based on volume, is 92:8, from which we could conclude that the dimensional ratio is the cube root of the volume ratio, namely 2.3:1. This assumes a great deal, however: that pores are evenly distributed and evenly sized, and that they are cubic bubbles, none of which is either verifiable or even remotely logical. A bit more rational is the notion that pores are regularly spaced tubes passing through the volume. This simplifies the problem to a two-dimensional analysis, resulting in a spacing ratio of the square of the volume ratio, namely 3.5:1. It assumes that the pores are packed like a bundle of straws and, therefore, that there is an axis of directionality associated with the pores. This jibes with our notions of timber, but far less so for brick. If, only for the sake of the example, we arbitrarily select a dimensional ratio somewhere between these two, namely 3.0, we can at least begin to attach some numbers to the problem.

The ice and the brick are competing for the same volume, and are pushing against one another in some relationship to their respective expansive and elastic properties. Regardless of how this conflict is resolved in terms of actual movement, we know that the resulting stress at the contact surface between the two materials must be the same. If one were pushing against the other harder, and the other was not reciprocating, they would still be in motion. They are not in motion, so the unit forces must be in equilibrium. When we crank the numbers, we find that, instead of expanding 0.00066 percent, the ice expands 0.000225 percent; and instead of expanding 0.00007 percent, the brick is itself actually compressed 0.000075 percent. The resulting stress along the boundary is 435 psi, which is experienced in both materials simultaneously. Despite the apparent reduction in the stress at the contact plane from the maximum confining pressure of ice, 435 psi is not trivial in terms of crack extension pressure; in some materials, 435 psi may be enough to independently initiate a crack, particularly in cold materials.

Still, the cycle is not complete. The material must go through the thawing portion of the process. In terms of damage, the primary deterioration mechanism is the freezing portion of the cycle, because that is the period during which the material is actually stressed, and as a consequence, cracks are initiated or advanced. Nevertheless, we must also consider the thawing stage because the conditions are set for the next cycle. As previously described, once the specimen has achieved thermodynamic equilibrium during the freezing stage, there may be only ice and dry material or saturated material and dry material. As long as the exterior and interior temperatures remain unchanged, little can occur in the interior of the wall, but that is not to say that *nothing* can change.

If the thermal equilibrium is established such that there is an unfrozen saturated layer in the wall, there is a distinct possibility that even as the outside surface remains frozen, the moisture trapped by the ice seal will migrate inboard due to the continuing absorptive properties of the dry material. If, as in our earlier example, the 16-inch wall was saturated to a depth of 8 inches, and the frost line stabilized at 4.44 inches inboard of the outside surface, it is not inaccurate to analyze the situation as a three-layer thermal transfer problem. It is, however, inaccurate, as well as an oversimplification to think that over a period of time saturated material will remain completely saturated and dry material will remain completely dry when they are in contact. How long does it take a dry sponge to absorb water from a wet sponge?

Similar to the perfusion of vapor in a volume of gas, and based on the notion of differential partial pressures associated with the concentration of vapor molecules, liquid will perfuse absorptive material based on the concentration of the fluid in the source materials and the dryness of the target material. Because the volume of water is fixed as soon as the ice seal forms at the outside surface, the moisture content of the total unfrozen material will be the volume of water caught in the saturated zone, divided by the total residual unfrozen volume. In our continuing problem, the amount of saturated depth was computed to be 3.56 inches, and the dry portion was set at 8 inches. If the wall remains undisturbed for a sufficient period of time, the concentration of water in the saturated area will gradually perfuse the dry zone, and after a period of time, the concentration of moisture in the wall will follow a declining concentration from the face of the frost line to the interior surface. At some intermediate point, moisture will perfuse the dry material until measurable moisture reaches the interior surface. At that precise moment, we can draw a moisture gradient from the interior surface where the moisture content is barely zero back toward the frost line such that the water below the gradient has the same volume of water as the 3.56 inches of saturated material. Let us say that the porosity of the material is 8 percent, and that over the depth of 3.56 inches all of the pores are filled. As the moisture in the brick levels out, at the point when moisture first reaches the interior face, the absolute moisture content at the frost line will have declined to 4.93 percent per unit volume, as opposed to the saturated condition of 8 percent. In relative terms, the moisture content at the frost line has declined 38 percent.

The leveling process might continue until all of the available fluid is evenly distributed through the unfrozen 11.56 inches of wall; and if that were to occur, then the relative moisture content of that portion of the wall would average approximately 2.5 percent of the available volume or 31 percent of the saturated condition. If that were true, the moisture content at the interior surface would be just as wet as any other part of the unfrozen wall, namely 31 percent of maximum moisture content. But that is unlikely because once water reaches the interior face of the wall, it will probably begin to evaporate. We will discuss the mechanics of evaporation shortly, but for now it is sufficient to say that water is removed at the surface of the interior wall into the ambient air. This is not an absolute physical requirement, but we have substantial evidence that evaporation will occur.

As moisture evaporates from the interior surface, the moisture gradient will slowly subside until the entire unfrozen portion of the wall is dry. In the interim, there is a continuously changing moisture content at every position within the wall, meaning that the thermal resistance of the wall is also constantly changing, and it is no longer feasible to analyze the heat transfer based on layers. It is possible to solve for the temperature gradient through a changing resistance, but the formerly straight lines connecting the points of temperature at the various interfaces will now be curved. These curves are fairly simple—nothing higher-than-second-order parabolas —but the mathematics begin to have exponents and we have tried diligently to avoid those sorts of things. The significance of moisture reaching the interior surface during periods of freezing weather should be only somewhat surprising. Consider how much moisture migrates through massive masonry long after a storm has passed, and it's easy to understand why.

Another altogether different scenario has the front passing—the cold spell is broken. The sun comes out and warms the frozen wall. The furnace inside can ease up, though it still must maintain the inside temperature at 72° F. The outside ambient temperature rises to 52° F. Thawing begins. This is the rising temperature condition we analyzed in some detail at the beginning of this chapter. The exterior surface temperature climbs; heat transferred from the interior and the exterior begins melting the ice from both directions. There is a lag in the thawing of the ice inboard of the surface, and for a period, an ice dam will remain in place between a thawing and now draining zone to the outside as well as the inside. As long as some frozen water barrier exists between these two zones, they will thaw and drain independently. Once the ice barrier is dissipated, the two zones will merge, and moisture levels will gradually merge. All this takes time. Depending on the depth of frost penetration and the extreme exterior temperature low point, the complete thaw may take a day, a week, or more.

In the interim, a second cold front may sweep in from Canada and start the process all over again. The end of a complete freeze-thaw cycle is marked by one of two events: either the wall completely thaws, leaving no ice in the interior, or a second freezing period begins, which was preceded by at least partial thawing. The former is a complete cycle; the latter is as complete as it can be. If one cold front is succeeded by a second, precluding thawing between them, that comprises only one long, dreary cycle. Thawing is essential to the progressive deterioration of the fabric because the newly formed or extended cracks from the initial cycle must have an opportunity to reload with water. If the wall freezes and stays frozen, cracks may propagate, but once they have arrested, they will not extend further until more water is added to the void, and it freezes and expands accordingly.

The progressive nature of the freeze-thaw deterioration mechanism is a function of the severity of the individual cycles and, more important, the number of complete cycles. As the microcracks and fissures extend, and new ones are formed, the cracks merge into a network of interconnected fractures. When even a small grain of material is disengaged from the main body of the solid, it can become dislodged and disattached from the surface altogether. Whether and when that actually occurs is dependent on a number of variables; but once the fracture plane is complete, even if the fragment remains intact with the main body, stresses cannot be transferred across the fracture boundary. This means that the fragment is structurally no longer part of the solid even if it remains ostensibly connected to the main body.

PRESSURE RELEASE AND AIR ENTRAINMENT

So far, we have assumed that the wetted portion of the material is saturated at the onset of freezing. This is a not unwarranted assumption. It is, however, altogether possible that the material at issue is only partially saturated, which changes the situation dramatically—so dramatically in fact that damage from freeze-thaw cycling may be precluded altogether.

What does partial saturation mean? One of three things: all the pores are partially filled, a portion of the pores is completely filled, or a combination of these two options. If every pore that has any water in it is only partially filled, and the partially saturated pores freeze, ice forms in the pore; the ice expands into the remaining void without confinement, hence there is no expansive pressure. The solid simply contains a certain amount of unstressed ice inside the pores. We said earlier that the actual percentage of expansion is minimal, so very little residual void is required to relieve the expanding ice from confinement and subsequent internal stress. In short, residual unfilled volume acts as a frost break; and though the water in a partially filled pore may freeze solid, the resulting expansion, because it is unconfined, will not result in damage to the fabric.

The second possibility is that some of the pores are completely filled while others are partially filled or empty. This may seem at first thought to be impossible, but recall that very small pores take much longer to fill than larger pores. If, in the course of the inundation, the wall imbibes water to the large pores, but freezing begins before the smaller pores are filled, the result is both filled and unfilled pores close to one another in the same specimen. During the initial stages of cooling, between the temperatures of 39° F and 32° F, while everything is still fluid, the small pores provide escape chambers for the water, which is expanding in the larger pores—assuming there is at least minimal connection. Once solidification begins, pressure relief is limited strictly by the amount of time between the onset and completion of solidification. If during these crucial seconds, additional water can be pushed into the empty pores, then that much more pressure is relieved. Once the water in a given pore has turned to ice, no further pressure relief is possible, and continued expansion occurs within that filled pore, as described above, even though an empty pore may lie a micron away. From this we can deduce that the majority of damage occurs from frozen water inside large pores, if the initial inundation is too rapid for the small pores to fill. Another consequence occurs during thawing. If the majority of water is frozen in large pores, and they thaw, there will be rapid drainage. This is important because that same night the temperatures may reach freezing again, and if any water remains in the pores, there is the possibility of refreezing in small undrained pores. Of the two basic types of partial saturation, it is more likely that some of the pores will be completely filled rather than all of the pores partially filled.

In concrete, the concept of partial saturation and pressure relief is particularly relevant. Recall from Chapter 3 that concrete is an agglomerate of sand, cement, aggregate, and water. Freeze-thaw cycling can attack concrete on two levels, the aggregate and the cement. Both are vulnerable, but for different reasons. The aggregate is vulnerable similar to the manner discussed generally in this section: water freezes in confining pores, expands, exerts rupturing pressure on pore walls, and cracks are initiated or extended. Cement paste has a different structure, hence a different response to freezing temperatures. Technically, cement is a colloid, which is not so much a crystalline solid as a very stiff suspension. Colloids come in two flavors: *lyophobic* and *lyophilic*. The former are solvent-repelling; the later, of which cement is an example, are solvent-attracting. Lyophilic colloids are also called gels, which is how cement is commonly classified in the literature. Water is the solvent in cement. (In the sense that it is used here, solvent refers to water in which suspension occurs, not as the mechanism to dissolve the colloid.) The structure of the cement gel comprises a series of alternating wafers of complex aluminates and water. Though held in a stable relationship within the solid layers, the water remains mobile. If the cement is subject to freezing temperatures, the water in the gel pore space can freeze. When that occurs, the water will go through the same expansion process as described in the preceding discussion.

While the gel pore water is still mobile, and is expanding, it can migrate. If a void is nearby and of the approximate scale of the gel pore space, then the expanding water can escape into the void. This is precisely the purpose of entrained air. The entrained air in concrete is composed of

The Vagaries of Cement

There are excellent examples of enduring early twentieth-century concrete that was not air-entrained, but other factors compensated so that the concrete has held up remarkably well. These are exceptions; very little pre-1930s concrete survives a hundred years.

Other superbly formed and placed examples of concrete have been condemned because the designers used salt water in the mix, reputedly to depress the freezing point to enable them to place the concrete during cold weather. For reasons we will explore in more detail in Chapter 5, the addition of salt immediately initiates a corrosion mechanism, which essentially dooms the concrete before it is even in service. When the column bases of the famous Marlborough and Blenheim Hotels in Atlantic City were excavated to determine how much bar steel remained, workers found *no* steel left. The compression bars had totally deteriorated; the compressive loads were being born completely by the concrete.

In still other examples, the concrete is doomed by poor hydration due to inadequate water; still others, because of too much water, by inadequate vibration, or by over-vibration, hence segregation. There is a long list of ways to produce inferior concrete. Unfortunately, many are still used, even recognizing their premature deteriorating effects.

extremely small bubbles, well below the optical range. These bubbles must be chemically produced. Simply by trial and error we have determined that between 4 percent and 6 percent, based on volume, is the optimum air entrainment for concrete that will be exposed to freezing weather. It is important to remember that air entrainment will not preclude freeze-thaw damage, but that it will extend the performance of concrete exposed to freeze-thaw cycling. The fact that air entrainment was not developed until the late 1920s and was not routinely used in concrete until the 1930s explains the failure of much of the preair-entrained concrete.

Pocking and honeycomb in concrete are sometimes perceived as voids that provide a benefit akin to air entrainment. In fact, there is no such benefit. All that optically sized voids do for concrete is to reduce the effective cross-section and, therefore, reduce the structural capacity of the concrete. There are two basic reasons that large voids do not provide effective pressure relief in concrete. First, large voids are irregular in their dispersion and uneven in their relative exposure to the cement gel. The distances to a large pore are, in most cases, thousands of times greater than the effective distance that gel pore water can migrate within the time periods available to avoid expansion. The gel pores that are immediately proximate to the interface of a large void may discharge gel pore water into the void; but despite the size of the void, there is not enough locally and immediately available contact surface for large pores, so dispersion and size are detriments to large voids. Second, if water were to enter into a large void from the gel pore space, those few molecules might as well have rolled off Niagara Falls. When the frost melts and the concrete warms again, if the former gel pore water is proximate to the gel pore space, the water—which, in the case of a lyophilic colloid, is attracted to the matrix of the gel—will be reabsorbed into the gel. If, however, the water is lying at the bottom of "Niagara Falls," it is a long trip back into the atmosphere, out over the Atlantic, raining onto the Canadian forest, and draining back into Lake Ontario. Otherwise, it is has to climb back up through solid rock to the top of the Falls. Either way it is a long trip. In short, if water does escape into a large void, which it theoretically may, it is not readily available for reabsorption into the cement paste. The long-term consequence of the net loss of gel pore water is that the colloid is weakened; hence, the cement fails.

INTERVENTION IN FREEZE-THAW CYCLING

We have now established and explored the six necessary and sufficient conditions for freeze-thaw cycling, which are displayed in Table 4.4. As in Chapter 3, the necessary and sufficient conditions are arrayed vertically along the left side, and the basic approaches to intervention are listed across the top. The highlighted boxes represent the more common

Table 4.4 INTERVENTION MATRIX: FREEZE-THAW CYCLING

Necessary and Sufficient Conditions	Intervention Approaches						
	Abstention	Mitigation	Reconstitution	Substitution	Circumvention	Acceleration
General Mechanism	Accept existing damage and rate of continuing damage without intervention.					Demolish all or portion of damaged material, and do not execute repair.
1 Porous, permeable material			Replace the material with the same but undamaged material.	Replace the material with a less permeable substance.	Alter the existing material, such as through consolidation, to change the porosity properties.	Remove any loose or deteriorated fabric from the substrate.
2 Liquid water at the surface of the material		Shed water before it comes in contact with the surface.			Apply surface sealer to substrate.	
3 Lateral motive force		Positively pressurize the inside of the fabric, or pressure-neutralize the surface.				

Table 4.4 (continued)

Necessary and Sufficient Conditions	Intervention Approaches					
	Abstention	Mitigation	Reconstitution	Substitution	Circumvention	Acceleration
4 Conductive capillary size distribution					Fill or consolidate material so that capillary distribution shifts to smaller diameters.	
5 Modulus of rupture under 700 psi		Increase tensile capacity of fabric with consolidation.		Replace material with another material with higher modulus of rupture.		
6 Temperature falling through freezing		Remove obstructions to solar radiation; install obstruction to wind blast.			Install artificial heat sources in or on materials, to reduce chill and to induce thawing and evaporation.	

371

approaches to intervention of freeze-thaw cycling, primarily because they cost less. Again, the stipulated provisions of general abstention and acceleration are indicated as candidates, and should never be summarily dismissed. That said, no nuances of intervention in freeze-thaw cycling can advance the discussion of these two options beyond that of Chapter 3, so we will not reopen those two options here. For the freeze-thaw deterioration mechanism, we designate concrete as our initial example; we will deal later with more general masonry issues.

Concrete comes in two general classes: poured-in-place and precast. The peculiarities of precast concrete make dealing with it similar to large-cut stone panels. Achieving a close match of texture and color can be problematic even if the same manufacturer is making the same product. Precast concrete is difficult to mix in identical lots in much the same way that is difficult to match dye lots in the textile business.

Poured-in-place concrete is, if not unique, certainly unusual in that it is the only large-scale building material molded from plastic media, fabricated in the field, and exposed to weather. Plaster may claim the first two characteristics, but it is rarely exposed to freeze-thaw activity, and certainly not intentionally or for very long. The monolithic character of poured-in-place concrete means that it has few joints, which preclude clean removal and replacement. Removing concrete almost invariably requires chipping or cutting the material, resulting in a boundary plane between the existing material and whatever is put into the void created from the excision. Just the existence of a plane of discontinuity is, in and of itself, a source of long term vulnerability. Regardless of what is put into the void, the two materials will not be a precise match of all relevant properties. The disparity may not be great, but as it impacts on long-term repairs, any disparity is enough to generate differential behavior, thermal- and moisture-related differential movements being the leading candidates. There are also stress and strain disparities in the materials, which mean that, structurally, the patch will always be a point of relative hardness or softness in the matrix of the larger field.

As discovered from prior examples, intervention necessarily addresses the fabric, the environment, or some combination of the two. When the fabric is concrete, all currently available techniques of fabric-related intervention are tedious, expensive, and not particularly reliable. One fundamental problem with concrete fabric intervention is that of scale. When the specimen is a one-inch cube, many options are available in a laboratory setting; but when the specimen is Boulder Dam, the options are obviously more limited. Part of the problem is logistics, which include the difficulty of controlling the environment of the site during the period of treatment; erecting scaffolding around the specimen, which could entail lifting and protecting the agents, equipment, and chemicals; maintaining application control and uniformity; and, finally, costs. Another part of the concrete intervention problem is that the material is monolithic in its construction, making it difficult to deal with in discrete, identifiable pieces. Unless we treat the specimen uniformly with an applied substance, working on a select section of concrete usually draws attention to the fact that an intervention has occurred. If we need to excise a portion of rotten concrete, the scar of the excision is usually quite evident to even the most casual observer.

Several deterioration mechanisms afflict reinforced concrete, and the intervention in all of them suffer this dilemma. In the case of freeze-thaw damage, one of the specific difficulties of intervening in concrete is that the magnitude of the damage will not be uniform over the surface of the sample. This difficulty is compounded by the presence of other deterioration mechanisms, such as carbonation or corrosion. If we are fortunate to have identifiable joints that circumscribe the area of damage, we may be able to conform the excision area to coincide with the joints, but typically the area of damage is small relative to the panel size distinguished by joints, if they exist at all.

The reason freeze-thaw damage tends to occur in patches or blotches is a combination of slight variations in the exposure to weather and the local idiosyncrasies of the concrete. Even within a single pour, there are subtle variations in the water-to-cement ratio, compaction, hydration temperature, bleed-water loss, and a dozen or more other variables. Thus, what is ostensibly identical concrete will, when exposed to enough freeze-thaw cycles, experience variations in its resistance, and so the damage will vary a good deal even within a limited area.

This and situations similar raise a general intervention issue having to do with the practicalities of all of the intervention approaches that directly affect the fabric. If the material is monolithic, such as concrete, or relatively monolithic, such as large sections of cut stone or precast concrete panels, often the damage is significantly less than the area of the panels or section. That said, there is an added premium to any approach that results in the excision or alteration of the damaged material. The broader issue is often addressed in articles and texts having to do with construction or fabrication, but the peculiarities of the materials are extensions of their inherent properties, as opposed to separate unrelated aspects of production and construction. While it is altogether feasible to fabricate a concrete block wall in much the same manner as a brick wall, we cannot pour a brick wall as we can a poured-in-place concrete wall. We can cut stone, or even scavenge stones of dimensions comparable to brick, and, subsequently, build a stone wall as we can a brick wall; but we cannot make a poured-in-place stone wall, nor do we ordinarily fabricate a brick panel of 36 square feet by 4 inches thick to erect a wall.

One characteristic of concrete is that it is manufactured from a plastic medium. We can mold into smaller pieces or pour into larger molds and cure it, normally without having to add heat or water. There are not many moldable materials, particularly cold-molded materials. We mold cast iron, but the method of producing the plastic medium renders it inappropriate to molding it on-site. We can cut stone into huge blocks, but we cannot fuse the joints between the blocks to create a truly monolithic product. Stone is only in a plastic state when it is the magma of igneous rocks, which makes it not only infeasible to mold in the field, but impractical to mold in a factory as well.

Brick does go through a plastic phase during its manufacture, and within highly proscriptive limits we can mold brick, but we cannot feasibly mold it, then fire it in the field. It is the cooling requirement of brick that renders it an impractical candidate for production as large slabs. The post-molding heat requirements and the potential of cooling cracks forces us to manufacture bricks into modestly sized blocks. Both of these manufacturing limitations are directly related to the properties and peculiarities of fired clay; they are not the arbitrary decisions of either brick manufacturers or designers of brick structures.

During the intervention analysis of the most general case of deterioration damage, we must determine the proportion of the damaged area to the gross area of the element or member. When dealing with monolithic concrete, this is of particular importance, as it is a major determinant in the selection of the intervention approach. If the element is a block of stone, even a large block of stone, the option to replace the entire piece is, at least, technically feasible. If the element is brick, because the pieces are small and the damaged areas coincide more with the physical size of the pieces, approaches are much more practical, inviting replacement or substitution.

Suppose three walls each have 10 square inches of freeze-thaw damaged material, and that the degree of damage warrants intervention. Further suppose that the only factor that distinguishes the three is that the first is poured-in-place concrete, the second is constructed of cut slabs of dimensioned stone, and the third is made of modular brick and mortar. All other things being equal, the brick invites simple replacement of the affected pieces because the costs of reconstitution are low, there are few if any incompatibility issues, the technology involved is well known and reliable, and the resources to accomplish a satisfactory reconstitution, namely bricks and masons, are readily available. Reconstitution of the large stone panel is feasible, but the cost alone suggests at least investigating alternatives. Replacement pieces may be less readily available, depending on quarry operations. Even pieces from the same quarry may be unacceptably variable because of fading (frequently the case with slate). And the logistics of installing a large stone panel into an existing wall make the process itself more risky.

Many of the options in Table 4.4 carry the risks of anticipated or secondary deterioration mechanisms, due to incompatibilities or improper application. One, the application of surface sealants, is so well known to cause complications that it hardly represents an option, at all, though the situation may arise when the application of surface sealants is warranted. The specific sealers may be derivatives of silicone, acrylic, or other synthetic vehicles. Paint can be used as a surface sealer on porous substrates. The logic is compelling, and the costs are rarely prohibitive, so what is the problem? The same problem, that all materials deteriorate. Surface sealers are no exception, and when they do deteriorate, small fissures and cracks appear in the surface sealer itself. These small fissures are capillaries that can and will transport water to the thirsty, absorptive substrate that the sealers are designed to protect. As long as the surface sealer is perfect, and the only source of water is from the opposite side of the membrane, the sealer remains an effective barrier to water intrusion into porous substrates. These two conditions are, however, extremely difficult to establish and maintain. Membranes are rarely perfect; and even if they are initially, they too deteriorate.

Exterior precipitation is certainly the predominant avenue and source of water penetration of wall materials, but it is not the only source. Water can get into wall materials through roof leaks, rising damp, condensation from the interior of the building, through joints and even through materials associated with windows and doors. As soon as the membrane develops a flaw, an avenue of entry opens for fluid to pass through and saturate the porous substrate in the immediate vicinity of the flaw. Admittedly, the actual quantity of fluid may be small because most of the membrane is still intact. One might argue that there is a still a net benefit because a great deal of water is still being repelled. The problem is that the primary mechanism for eliminating the water that does penetrate the porous material is through evaporation of the fluid from the surface of the material. The rate of evaporation is directly proportionate to the available evaporative surface. Water that enters a membrane-covered surface, however small an amount, will remain in the fabric for an extended period of time because the membrane, while flawed in terms of fluid penetration, is still virtually intact as an evaporation inhibitor. Even that small amount of water cannot escape back through the pinhole through which it entered, so it perfuses the substrate. Evaporation through the interior surface may occur, but typically water accumulates in the fabric.

If the wall is perpetually saturated either globally or locally, whenever the temperature drops through the freezing point, the result will be freeze-thaw damage. Ordinarily, for freeze-thaw activity to develop, a necessary and sufficient condition is that water is present at the outside surface immediately preceding the decline in temperature. By virtue of installing an effective water entrapment scheme, the water is already present which means that there are more opportunities for freeze-thaw damage. Instead of needing both water and declining temperature, now all that is required is declining temperature, because the water is already in place. As a rule, membranes, even the best surface-applied substances, begin to develop flaws after three years. The net positive effect may continue for one or two more years, but by the fifth year, the pendulum swings. The rate of freeze-thaw damage, as well as other mechanisms of water-related deterioration begin to accelerate, and by the sixth or seventh year, the cumulative damage is greater than if the membrane had not been installed at all.

Another intervention option, removal of the friable material, is highlighted in Table 4.4 as in common use and inexpensive, but that assessment requires amplification, for it involves the removal of deteriorated material from the surface of the substrate. In many cases, this can be accomplished with the application of a stiff bristle brush, and without the assistance of paid consultants, so it can be inexpensive. The problem is, there is often considerable resistance to the notion that the way to preserve fabric is to remove it.

Materials that develop crusts of deteriorated fabric such as sandstones, granites, limestones, and bricks, and that have already lost their fire skins are candidates for this minor form of accelerative intervention. The crusts are completely deteriorated fabric, which is often hanging onto the wall by little more than habit and a trace of salt acting as a binder to the substrate. It is often so friable that the least touch will dislodge it. Sometimes, particularly with granites and limestone, the crust, though rather hard to the touch, is hollow because the crust has separated from the substrate and is no longer structurally integrated with the substrate. The common denominator between deteriorated, but not yet disengaged, crusts is that they are excellent sponges, and so hold water against the substrate that would otherwise fall harmlessly to grade. Hence, the crust provides added water, inhibits evaporation from the sound material, and, thereby, accelerates the deterioration of the substrate.

To those who resist this option, it should be explained that the removal of the crust is not removing sound material; it is removing unsalvageable, wholly disintegrated material. It is true that the removal of the crust exposes the sound material directly to the elements, but the long term advantage is that the sound material will deteriorate more slowly if the harmful crust is removed.

4.3.2 Rising Damp and Salt Decay

We have investigated the general condition of water being absorbed laterally primarily as a consequence of rain running down the outside face of the wall. As stated repeatedly, this is the most common source of laterally imbibed water but it may be introduced via other means and from other directions. For one, the water may come from below, from the base of the wall or even through the footing. If the base of the wall is saturated, probably the water will be absorbed vertically into portions of the wall that are above the level of the water source. The general conditions for transportation of fluids in porous media are the same as for laterally transported water: there must be a water source; the material must be porous, permeable, and absorptive; and the pore size distribution must be conducive to capillary action and fluid distribution.

According to Seymour Lewin, under conditions normally associated with common sandstones, the resulting vertical rise may be as high as 30 inches, or less than one meter. Absent any other external motive forces, the exact height of vertical absorption will depend primarily upon capillary size and distribution. Even allowing for variation in those factors, the vertical rise of water within common masonry materials is on the order of a meter or less. The fact that vertical transport of water higher than a meter has been known to occur suggests that there are other contributing factors, which we will also discuss in this section.

Lewin conducted his experiments so as to eliminate any other explanation for the vertical rise of fluid in the stone other than the capillarity of the stone and the fluid properties of the water. From them, he concluded, persuasively, that the initial elevation of the height of fluid rise within the stone is a function of the combination of material properties, primarily to the stone and, secondarily, to the water. If the water that rises in the wall also happens to carry with it various dissolved salts, the conditions are in place for damage to the fabric through the action called *salt decay*. The exact mechanism of the damage to the fabric by the salts is not completely understood, but the evidence is compelling that such damage is attributable to the accumulation of salt in the pores of the material, where they are transported to the locus of damage. This section, generally, examines the flow of saline solutions in porous materials and the potential effects on the materials as a result of that movement.

Previous discussions addressed the general concept of moisture content; here we expand on that, exploring the distinctions among various moisture levels within the specimen. Moisture content can be divided into three separate categories: the hygroscopic point, the critical water content, and the saturation point. The hygroscopic point, or natural moisture content, is the point at which the only moisture present is that which is retained by the surface tension located in the small interstices of the pores. This is the moisture that adheres to the walls of the pores through wetting action. (Recall that the wetting properties are the result of the specific combination of properties of the solid material and the fluid.) Wetting is affected by temperature and impurities, if the fluid is, in fact, water. Moisture movement or transfer is not possible because the film of water adheres to the surfaces of the pores and is not mobile. The levels of hygroscopic moisture are dependent upon the levels of relative humidity within the environment: if it is low, a condition we discuss below, the moisture evaporates; if it is high, moisture is likely to condense on the surfaces.

The critical water content is the point at which the capillary movement is just possible and moisture can transfer. Critical water content can be reached through a variety of methods: Water can be introduced laterally, as discussed previously; it can enter the fabric through leaks from any water source, including a leaking gutter, drain, or supply line; it can enter vertically upward from ground sources; and it can be reached through prolonged condensation. The point is, a threshold quantity is required. If the water source cannot or does not satisfy the threshold, critical water content will not be reached, and capillary transportation cannot occur. Brief exposures to water will not normally be enough to initiate a rising damp situation, simply because the critical water content is not satisfied.

The saturation point is the level of highest moisture content—every pore is filled with water. As explained, prolonged exposure to water will result in the absorption of water into the solid portions of the material as well, but from the perspective of salt damage and, generally, for freeze-thaw damage as well, the point of maximum risk is achieved at saturation. The source of water must be both prolonged and of sufficient volume to fill the voids and sustain the saturation, at the same time that fluid is evacuating the solid through gravitational forces and evaporation.

Building materials almost always contain soluble and unsoluble salts; and water is almost always present at some point, if not repeatedly, during the life of the fabric. Again, the potential sources of water are not limited to the occasional precipitation. As just stated, the bases of buildings can be exposed to ground water and other sources of water from below, which is the immediate subject of this section; water may also enter the fabric through copings and leaks in roof membranes; faulty drainage systems, and ruptures in domestic water systems. These latter sources can inundate the wall of a building and, if uncorrected, transport significant quantities of salt to the locus of damage. Parapets are particularly vulnerable to precipitation-induced saturation and concomitant salt migration.

Any significant source of water in a new structure is a result of the building process. Many building materials contain substantial quantities of water. For example, the mortar in masonry can introduce water to the wall as a whole during the curing process. In the case of brick masonry, it is not unusual for the individual bricks to be soaked in water prior to installation to preclude excessive amounts of water being absorbed out of the mortar during the curing process. And finished masonry is almost always washed down prior to delivery, and so may imbibe a significant amount of water during that process. These construction-related sources are, however, fairly limited in scope and duration. Once the construction water has migrated out of the fabric, it ceases to be an ongoing source of salt damage. That said, construction-related water is often the source of salts on the surface of buildings, and so is an object of client dissatisfaction. But the damage from construction-related water is nominal and normally self-correcting.

SOLUTIONS AND SALTS

Salts become a deterioration problem only when a water source is constantly or cyclically available, thereby allowing mobility and crystallization of the salts throughout the building fabric. Salt is a chemical compound formed by replacing all or part of the hydrogen ions of an acid with one or more positive ions of a base. Typically, this means that salts are ionically bonded ratios of metal with nonmetals. While the chemical formulae are expressed as though salts constitute discrete molecular bundles, it is more accurate to view salt formulae as ratios rather than discrete molecules. $NaCl$, for example, is the chemical expression of common table salt, sodium chloride. The sodium chloride crystal is orthogonal in structure when nested in layers in such a way that there is an equal number of sodium ions and chloride ions. The pattern of a single layer has Na and Cl ions alternating in a rectilinear pattern. The next layer overlays the first such that the chlorides of the second bear on the sodium ions of the layer below. In this manner, each sodium ion is in contact, not with a single discrete chloride, but with four chloride ions in the primary layer, plus one each in the layer above and the layer below the primary layer, for a total of six points of contact. In hexagonal packing patterns, a single ion may be in contact with six ions in the same layer, plus three each in the superior and inferior layers, for a total of 12 points of contact. The determination of the packing arrangement is a product not only of the ratios that will establish electrical balance, but the feasibility of a physical packing arrangement that is consistent with that ratio.

Salts are hygroscopic in nature; they have the property to absorb and to hold water vapor from the atmosphere. Materials containing salts exhibit a higher level of hygroscopicity, and will, by their nature, retain more moisture, thus preventing the fabric from drying out. The aspect of the saline solution that inhibits the evaporation of the solvent is one of the colligative effects of saline solutions. When a nonvolatile solute such as an ionizable salt is dissolved in a volatile solvent such as water, a specific array of properties accrue to the solution. The salt solute will depress the freezing point of the water, increase the boiling point of the water, and inhibit the evaporation of the water. The inhibition of evaporation and the hygroscopicity of salts are aspects of the same basic phenomenon; one is relative to the solvent and the other is relative to the solute.

Not all salts are equally soluble; some, in fact, are virtually insoluble in pure aqueous solvents. Others are soluble, but tend to reprecipitate almost as rapidly as they dissolve without much net effect one way or another. Soluble salts are, generally, more damaging than insoluble salts because, once in solution, they become highly mobile, which enables them to penetrate deep into the material. On the other hand, soluble salts tend to stay in solution longer, as opposed to insoluble salts which are inclined to crystallize out quicker.

Though most people are familiar with simple salts such as sodium chloride, table salt, and even perhaps slightly more complex salts, such as calcium carbonate, which is the basis of limestone and marble, or calcium sulfate, which is otherwise known as gypsum, in general, salts can be quite complicated. Not only can the numbers of elements in the crystal increase, but the packing arrangements can vary considerably, as can the various levels of possible moisture within the crystalline structure. Moreover, several types of salts can exist in either form, meaning with or without water worked into the crystalline matrix. The most common salt to exist in both states, hydrous or anhydrous, is sodium sulfate ($NaSO_4$) or thernardite. An anhydrous salt compound is one that does not contain a molecular water bond within its structure. Conversely, a hydrous salt is a compound that does contain a molecular water bond. Once inside a porous material, an anhydrous salt has the potential to recrystallize as one or more hydrated salts. Again using the anhydrous sodium sulfate as an example, the two primary hydrous forms of the compound are the decahydrate $NaSO_4 \cdot 10H_2O$ (mirabalite) and the septahydrate $NaSO_4 \cdot 7H_2O$. The deca and septa preceding "hydrate" refers respectively to the 10 and 7 water molecules within the compound. The significance of our inquiry into salt damage is that the volume, density, and other material properties of the variations of the same essential saline chemistry vary considerably simply because of the consequences of hydration on the crystalline structure.

The crystallization of salts from solution and the hydration of salts are generally accepted as the primary mechanisms of salt decay. Virtually any salt, in principle, can crystallize, yet fewer salts are capable of causing hydration damage, meaning that the salt must be able to exist in more than one hydration state. The growth of salt crystals leading to fabric deterioration depends upon three primary vehicles: the salt solution within the walls; the properties of the substrate, such as its porosity, wettability, and applied coating; and environmental conditions, especially relative humidity and temperature. Though there are numerous sources from which salts may come in contact with a building material, they fall into only two categories: extrinsic and intrinsic or autochthonous. An autochthonous, or intrinsic, salt is one that is inherent within the building material. In contrast, as the term implies, an extrinsic salt source is introduced into the material via the environment or other external vehicles. A building can be contaminated by one or the other or a combination of these two sources.

The most obvious extrinsic source of soluble salts is from precipitation, in the form of common rain. Water-soluble, or hygroscopic, salts from the atmosphere may combine with the calcium, sodium, or magnesium already present within the building materials. The sources of these elements within the building material depends on the provenance of the original materials, but many of these positive ions are inherent in the chemistry of the fabric. Atmospheric pollution may also contribute to the supply of metallic and nonmetallic ions, and can come in the form of acid rain or dry deposit. The sources of pollution associated with deterioration mechanisms are the sulphur contained in fossil fuels, and nitrogen, which is already a free gas in the atmosphere.

Fossil fuels, oil and coal, contain varying levels of sulphur as part of their original deposition and chemistry. In fact, the quality of both materials is highly dependent on the concentration of sulphur, not so much because of the polluting potential as the heat production value of the fuel. The greater the sulphur content, normally, the lower the BTU potential per unit weight. When the fuel is burned either in internal or external combustion processes, the sulphur combines with atmospheric oxygen to produce either SO or SO_2. Together, they are simply referred to as oxides of sulphur (SO_x). When SO_x enters the open atmosphere, the potential exists for these compounds to combine with water to produce transient concentrations of H_2SO_4, better known as sulphuric acid. It is more accurate to describe the production of the acid as one of statistical probability than of absolute formation. As with many solutions, the presence of the compound identifiable as sulphuric acid is more a matter of circumstance and momentary proximity. Nevertheless, what is identifiable as sulphuric acid in one instance is only a solution of water and oxides of sulphur in another, and vice versa.

Oxides of nitrogen experience a similar phenomenon. In this case, the source of the nitrogen is the general atmosphere. The mechanism of conversion to NO and NO_2, the oxides of nitrogen (NO_x) is the explosion inside an internal combustion engine. Under the pressure and heat of the explosion inside such engines, the free nitrogen and oxygen available in the open atmosphere combine to form oxides of nitrogen, independent of the fuel or other ingredients in the fuel. The process of internal combustion is inherently polluting, regardless of the specific fuel. The oxides of nitrogen do not form in external combustion processes because the pressure cannot be developed, which is necessary for the formation of the oxides. Once formed, the oxides of nitrogen dissolve in water in the atmosphere and form nitric acid.

These two common pollution-derived acids attack materials directly through their corrosive action and, secondarily, through the supply of free radicals for the formation of salts. When these substances introduce sulfate or nitrate ions into a building, the result is salt decay. Sulfate ions in concert with sodium ion depositions create sodium sulfate. Nitrates, on the other hand, may produce significantly soluble salts when combined with sodium, potassium, or calcium ions.

Salts originate from various sources regionally. In the coastal regions, sea water is the most common source of salt. Sea salt contains large amounts of sodium chloride, as well as sulfates, which can infiltrate a building through surface sea spray or ground water. Salt comes to be suspended in the atmosphere, and is subsequently transported to the surface of the building, in the form of an aerosol. The salt in solution does not evaporate; therefore, the water suspended in vapor form over a body of salt water does not contain salt in solution. The salt water suspended in the atmosphere is in the form of tiny droplets, which are mechanically produced by pounding surf and/or wind literally blowing the caps off of waves and tossing them aloft. The aerosol may attract true vapor molecules, which dilute the solution, but it is important to remember that salts do not evaporate at normal temperatures. In northern climates, deicing salts contain chlorides and sulfates, which serve as a catalyst in salt penetration. When the salt-contaminated ice and snow melts, it becomes a primary ingredient in ground water. Again, rising damp assists in its perforation into the building. (Deicing agents without a salt content are available on the market, and their use is recommended, especially on the grounds of an historic structure, because there is less chance that salts will be transported into the building materials.)

Incompatible materials placed adjacent to one another may initiate a soluble salt-related deterioration problem. For example, limestone situated above terra cotta or other siliceous stones, such as sandstone or granite, may deposit calcium bicarbonate and calcium sulfate, which may leach into the ceramic, resulting in blistering and exfoliation. Former uses of the building also may be the antagonist of salt decay. For example, in the smokehouse in Colonial Williamsburg the salt deterioration of the brick was directly related to its original use.

Autochthonous salts inherent to the building material is another cause of salt decay. Concrete, for example, contains sulfates, as does under-fired brick. Salt may have been added to the material purposefully or inadvertently. Evidence has shown that, at times, salt was added to the color glazes used on terra cotta. And a good example of an intentional addition can be found in coastal regions, where sea water may have been used in the composition of mortar and of concrete, theoretically to act as an antifreeze during cold weather operations. Beach sands used in mortar and concrete must be thoroughly washed in order to remove unwanted salt.

Modern interventions to historic materials may also facilitate salt deterioration. The alkaline salts of some modern materials will react with the ambient atmosphere or with the autochthonous salts within the historic fabric. For example, the repointing of a structure using a modern Portland cement mixture may lead to deterioration because, similar to concrete, alkaline salts are found in the material.

The most common types of salts found within building materials are chlorides, nitrates, sulfates and, to a lesser extent, carbonates. The following subsections offer a brief description of each basic salt type. Salt systems are far too complex to fully address in this book, so this discussion gives the general characteristics of the four most likely salts one would encounter in a building.

Chlorides

Chlorides are found in deicing salts, and are primarily a problem in the northern regions. In coastal regions, chlorides are distributed through maritime aerosols. Their presence is also found naturally in masonry, in particular brick masonry. Common chloride forms found in brick are halite ($NaCl$), sylvite (KCl) and antarecticite ($CaCl_2 \cdot 6H_2O$), which is a hydrated form of calcium chloride.

Chlorides are extremely soluble and hygroscopic, making the salt powerfully destructive when mobilized. Upon crystallization, a chloride's porous deposit absorbs water by way of capillary action. Because chloride crystals are able to absorb moisture from humidity, small changes in ambient moisture levels of the air cause renewed crystal growth which, theoretically, changes the pressure against pore walls. If the water source continues, the crystals dissolve once again, and because of their rapid mobility rate, move to other zones that foster crystallization more readily. In this way, their mobility acts as a catalyst to facilitate the disruptive action, and at different areas within the building material.

Calcium (Ca) and magnesium (Mg) chlorides, cannot crystallize out under normal conditions on walls because of their high hygroscopicity, even if they are present in high amounts. Rising damp and sea spray deposition are among the most common vehicles of transporting extrinsic chlorides into building materials.

Nitrates

Nitrates are found primarily in organic sources such as soils, soil fertilizers, and areas of decomposing organic materials. They can also be transported through atmospheric migration. Traces of nitrate in efflorescence (explained later in the chapter) have typically been found on church walls and tombstones due to the proximity of burial grounds.

The most common nitrate found in salt crystallization is niter (KNO_3). A constant water supply to the surface in concert with evaporation over a long period of time facilitates the outward growth of the needlelike crystals of nitrates. The disruptive effect on the pore walls is, however, generally negligible if the water source is low, even when efflorescence is present. Calcium and magnesium nitrates need specific conditions and will crystallize only when the relative humidity reaches around 50 percent.

Sulfates

The most common salt found in efflorescence are compounds of sulfates. In comparison to other salts, sulfates tend to be less soluble and mobile; still, they can be extremely damaging due to the size and the aggressive growth of their crystals. The salts most likely to be found in a wall are gypsum ($CaSO_4 \cdot 2H_2O$) and thernardite (Na_2SO_4) in differing hydration states. Upon water evaporation, sulfates precipitate on the pore walls in the form of hydrate salts, which can then become anhydrous. If the humidity is low the anhydrous salts do not dissolve, but hydrate, thus increasing in volume, which potentially can exert pressure on the pore walls. To enable the hydration of sulfates, there must be an increase in temperature. Sulfates require a slow hydration process to allow for the growth of their characteristic large crystals; therefore, a warm surface, generally a thick wall, will permit the sulfate to hydrate and allow for the execution and completion of crystal development. It is interesting to note that, in the same stone, magnesium sulfates may predominate in a rural environment while sodium sulfates generally prevail in urban environments.

Carbonates

Characteristically, carbonates have a relatively low water solubility, therefore they are not always included as soluble salts. But, in the presence of high carbon dioxide concentrations, they can be dissolved. The carbonates of calcium and magnesium are major components of certain building stones, such as limestones or sandstones, which makes them a salt source via the building material. Sodium carbonate is a predominant component of efflorescence found on interior walls, mostly resulting from ground moisture. It can form two different hydrates, $Na_2CO_3 \cdot 10H_2O$ (natrite) and $Na_2CO_3 \cdot N_2O$ (thermonatrite), plus the stable heptahydrate of the same basic salt.

EVAPORATION AND THE EVAPORATIVE ENGINE

A critical component of the rising damp and salt damage stories is the evaporation of water. Evaporation is a familiar phenomenon, whether we have analyzed the details of its physics or not. We all know that on warm days we sweat, that the sweat evaporates, hence we are cooled. We also know that boiling salt water eventually disappears, leaving a salt deposit on the inside of the pot. We know that rain puddles disappear when the sun comes out. We know that all of these events are the result of evaporation; what many people do not know is how the mechanism works.

Reiterating the fact that molecules are constantly oscillating, in the liquid state, the individual molecules achieve a level of excitation such that the crystalline bonds cannot maintain them in a long-range order; but there is sufficient molecular attraction to hold the molecules in close proximity. This is, remember, indicative of the cohesive property of the fluid. The molecules continue to oscillate in the liquid state, but they do not necessarily oscillate at precisely the same velocities; the temperature of the fluid is a measure of the average energy level of all the molecules in the liquid. But that implies that at any given moment some of the individual molecules are moving less rapidly than the average and some are moving more rapidly than the average.

There is a statistical probability that at least one molecule of the fluid will achieve a velocity sufficient to rupture the surface tension of the liquid surface, and break the adhesive link with the fluid. Whereas the temperature of the fluid is constant at the macroscale, the distribution of energy at the molecular level is considerably more varied. The oscillation of the molecules in the liquid varies, and when a molecule does achieve the energy level, hence the oscillating velocity to escape the surface of the liquid, it carries with it the heat or energy that made the escape possible. The remaining molecules redistribute the available energy, but the net loss of energy due to the escaping molecule is disproportionate to the loss of mass and volume of the escaped molecule.

One of the important consequences of a molecule escaping through evaporation and, therefore, of the reduction in average redistributed energy, is that the temperature of the fluid decreases. Evaporation has a cooling effect on the residual fluid. If the residual fluid is sweat, heat is rejected from the body through the evaporation of the fluid on the skin. This is the basis of the differentiation between "dry" versus "wet" heat. It may, in fact, be easier for the body to reject excess heat in 100° F air of 40 percent relative humidy (Rh) than in 90° F air of 90 percent Rh. Thus, the comfort level of the thermally hotter condition may be higher than the "cooler" condition. There are distinct limits to this, however: though the body may be more comfortable in the "dry" heat, depending on the physics, the body may or may not actually be rejecting more heat in the "dry" heat condition. The cliché that 120° F is tolerable as long as it's dry, is false if the net rejection rate of heat is not better than at 90° F and humid. The ensuing heat prostration in both cases will have the same symptoms.

Evaporation is, therefore, the process of molecules escaping from the surface of a fluid into a gaseous atmosphere. The necessary and sufficient condition for the escape is that the energy level of the individual molecule be enough to overcome the surface tension and, therefore, the cohesive strength of the fluid. It is also true that when evaporation occurs, the volume of the fluid is diminished by the loss of the molecule, and that the resulting average temperature of the fluid is reduced. If the process continues unabated, the fluid may evaporate completely. When this occurs at the surface of a saturated porous solid, the term *dessication* is used to describe the process of drying through evaporation of the free and the bound fluid in the solid.

The primary factors affecting the rate of evaporation are the temperature of the fluid and the contact plane, or evaporative surface, between the fluid and the gaseous atmosphere. As the temperature increases, two things occur to stimulate evaporation: more individual molecules are oscillating at or near escape velocity and the surface tension of the fluid is declining.

At some fairly predictable temperature, even the average velocity is such that almost any molecule proximate to the surface can escape. This is the boiling point of the liquid. The variables that may affect the boiling point are, notably, the pressure of the gaseous atmosphere and the constituent solutes in the fluid. An increase in atmospheric pressure will elevate the boiling point; a decrease in pressure will depress the boiling point. All other things being equal, water will boil at a lower temperature in Denver than in New York or Los Angeles. One way to harness this phenomenon is through pressure cookers. At one atmosphere of pressure, water boils at 212° F. If we put the boiling water into a container that is sealed against escaping water vapor, and continue to add heat, the pressure of the vapor increases, and boiling is suppressed, but the temperature of the residual fluid can increase to a new boiling point. If the object of the exercise is to raise the temperature of the fluid, we can do that by enclosing the whole boiling process in a sealed container capable of withstanding the resulting internal pressure.

When the fluid at issue can and will evaporate, it is considered a *volatile liquid*. To some extent volatility is relative. At 70° F, mercury is liquid, but it is far from volatile, whereas, at 3000° F, even mercury may be volatile. At normal temperatures and pressures, water is considered a volatile liquid. We also know that water will dissolve salts. The dissolved salts are no longer in a crystalline solid state, and they are, therefore, in a liquid state, albeit only because of the presence of the solvent, namely water. At normal temperatures, salts in solution are nonvolatile. While the water within the aqueous solution may evaporate, because it is relatively volatile, the salt will not evaporate, and remains bound in the solution even as water molecules are escaping through evaporation.

In fact, a solution of nonvolatile salts in volatile solvents alters the properties of the liquid in a set of phenomena called the *colligative effects*, which were mentioned above. The colligative effect relevant here is that the solution of a nonvolatile solute in a volatile solvent will elevate the boiling point of the solvent. In plain terms, this means that adding salt to water increases the boiling point, which is one reason we add salt to water when we cook. Without the salt, the water will boil at sea level at 212° F, and the average temperature of the water in the pot will remain at 212° F, until the water boils away or the atmospheric pressure changes. Remember that evaporating water carries heat with it, so we must continue to add heat to the water to sustain the average temperature at the boiling point during the process. The water cannot, on average, go above the boiling point during the process—although it may boil more rapidly if we turn up the heat.

If we add salt, however, we change the boiling point of the water; we increase the boiling point. Recall that one of the requisite conditions for evaporation, hence boiling, is proximity to the interface between the fluid and the gaseous atmosphere. In solutions, the solute—in this case the salt—is oscillating in much the same fashion as the water. At any given instant, a certain percentage of the surface film area is occupied by saline ions rather than water. The net available evaporative surface is reduced by the blocking salt ions, and escape is suppressed. The evaporative process is inhibited, and the boiling point is increased. If the object of cooking is to achieve the wetting of surfaces, such as when cooking pasta, then wetting and absorption are enhanced when the temperature is higher. The addition of salt to the pasta water will accomplish this result. If the objective of cooking is to effect a chemical reaction, such reactions are typically facilitated at higher temperatures. If the objective of cooking is to rupture cell walls and release cellular contents, higher temperature may facilitate the cell wall destruction. Salt can increase the operating temperature of the water because the boiling point is increased; therefore, when boiling occurs, the temperature is stabilized, at a higher degree than would have been the case in unsalted water.

In the end, in the case of saline solutions, that which depresses the rate of escape of the water from the liquid state is the effective reduction of the available evaporative surface. The same or, at least, similar results can be accomplished mechanically. If the liquid is in a glass container, the evaporative surface is the exposed area of the liquid. If the fluid is put into a deep but narrow tube, the evaporative surface is less than if the same quantity of fluid is put in, say, a broad-base pan. If all other factors are equal, the fluid in the broad-base pan will evaporate faster than in the tube because the available surface area from which molecules can escape is greater. If the liquid is absorbed into a porous solid, the surface of the solid is the effective evaporative surface. As liquid evaporates from the surface of the solid, it presents a locally drier point than the adjacent material. The locally drier point offers a point of relative attraction to the wetter adjacent area, and fluid is drawn to the point previously occupied by the evaporated liquid. If the evaporation rate is directly related to the available evaporative surface, and the surface of porous solids constitutes the effective evaporative surface when the solid is wet, the greater the irregularity of the porous surface, the greater the evaporative surface, and the greater the drying rate of the material. Very smooth, polished surfaces offer less evaporative surface than rough or partially deteriorated surfaces.

Throughout this discussion of evaporation, we have been specifically and exclusively referring to the absolute loss of molecules from the liquid state to the vapor state. Precise though that is, it is not the common definition of evaporation. The popular definition of evaporation is that there is an absolute reduction in the quantity of the fluid due to the evaporative process. To be precise, this is the *net evaporative effect*.

In a manner analogous to the escape of excited molecules from the liquid to the gaseous state, a simultaneously occurring possibility is that certain molecules in the vapor state will collide with the liquid surface and be captured by the fluid. As do molecules in the liquid state, molecules in the gaseous state also move at various velocities, hence carry varying energy levels. When the colliding molecule hits the liquid surface, it is highly possible that it will effectively stick to the liquid and remain part of the fluid body. The simple act of conversion from the gaseous to the liquid state, even if the gas and the fluid are the same average temperature, requires a loss of energy called the *heat of liquefaction*. The heat of liquefaction is distributed through the fluid, leading to a net increase in the average temperature as a result of captured vapor molecule by the liquid. If the vapor molecule is operating at an energy level higher than what otherwise is required to escape from the same liquid, that energy is also dissipated and distributed in the liquid, with a concomitant increase in average temperature level and, therefore, in temperature.

The conversion of a gas or vapor to a liquid is often referred to as condensation. Certainly, any form of condensation means that vapor is converted to liquid; but not every case of conversion of gas to liquid is necessarily condensation. The broadest category for such a change of state is liquefaction; the term condensation is typically reserved for conditions whereby the liquefaction occurs at or, more precisely, on a surface, and where the net exchange of liquid to gas and gas to liquid weighs in favor of net liquid formation. If the liquefaction occurs aloft, the preferred term for the condensing of liquid from vapor is precipitation; again, the term presumes a net liquid formation.

The important concept to remember is that at any contact surface between a liquid and a gas, there is a constant exchange of molecules from liquid to gas, and vice versa. If, in the net exchange, more vapor is formed than liquid, there is a net evaporative effect. If, in the net exchange, more liquid is formed than vapor, there is net precipitation or condensation. The third possibility is that the rates of exchange are equal, resulting in no net change in either the liquid or the vapor quantities. The condition that determines which of these three possibilities will occur is the degree to which the gas is saturated with vapor from the liquid. The major determinant in whether the gas is, in fact, saturated is the temperature of the air, relative to the moisture held in the gas at a given time.

The reason temperature is a major factor in the balance between net evaporative or net precipitative activity is that the amount of vapor of a particular liquid that can be supported as a block of air is highly dependent upon the temperature of that block. The variation in the capacity to hold vapor is huge. The absolute amount of water suspended in a block of air at any given time is the humidity in that block, and that quantity is measured in terms of weight, such as grains or milligrams. The maximum amount of water potentially suspended in the same block of air is the saturation humidity of that volume. If the actual humidity of the block is lower than the saturation level, then the air can potentially absorb more vapor. The ratio of the actual humidity to the saturation point is the relative humidity (Rh) of that block of air, and it is expressed in terms of percentage. An Rh of 60 percent means that the actual amount of humidity suspended in a particular block of air is 60 percent of what can potentially be suspended in that same block. As the Rh increases toward 100 percent, the block is filling up with vapor. At precisely 100 percent Rh, the block is full of vapor; the equilibrium between net evaporation and net condensation has been achieved, and no further change will occur in the quantity of either the liquid or the vapor. If additional vapor is introduced into the air volume once 100 percent Rh has been achieved, an equal quantity of water must precipitate or condense out from the air.

All of this assumes that the temperature of the block of air is constant. As soon as the temperature of the air changes, entirely new conditions exist determining what the block can potentially hold. The absolute quantity of vapor suspended in the block is unchanged, as is the absolute humidity, but the potential saturation point has shifted. If the block of air is warmer, the saturation capacity increases. If the maximum potential humidity increases while the actual humidity remains constant, the relative humidity decreases. If the air cools, the maximum potential humidity capacity decreases. If the absolute humidity remains constant, which it will, the relative humidity will increase as the air cools. Consequently, as air cools, at some point, even though the suspended quantity of water is unchanged, the capacity will decrease to a point at which the relative humidity reaches 100 percent; we refer to this as the *dew point*, meaning quite simply that precipitation will occur.

If the relative humidity is low, the rate of net evaporation will increase; if the relative humidity is high, the rate of net evaporation will be low. A minor, but important, factor affecting the rate of net evaporation is the temperature of the liquid. If two blocks of air are of equal temperature in relative humidity, the block over the warmer liquid will absorb vapor faster and, therefore, potentially reach saturation faster.

Now suppose that the material at hand is a porous masonry wall, and that it is saturated with water. The air proximate to the exterior surface is relatively dry—that is, the relative humidity is below saturation level—so the potential for additional net evaporation exists. As water evaporates from the wetted wall surface, the surface begins to dry and the air becomes increasingly saturated. At this point, two processes set into motion. The relatively dry masonry surface means that there is a local moisture gradient within the material. Moisture will migrate from the wetter interior of the material toward the relatively dry surface. At the same time, the increasingly saturated air at the surface will effectively establish a moisture gradient in the air block, and the concentrated water vapor at the surface will disperse within the larger block of air, thereby reducing the relative humidity at the surface. The combination of continued reabsorption of water by the solid at the surface, and the dispersion of the evaporated vapor away from the surface, means that as long as the relative humidity of the air block as a whole remains below the dew point, water will migrate through the porous solid toward the surface, where it will evaporate. If, at the same time, the surface of the wall is also being warmed by the sun, the rate of net evaporation will increase, because the liquid molecules will become more excited and, therefore, edge closer to escape velocity.

If the supply of water to the interior of the material is through a maze of capillaries, the net evaporative loss of moisture at the surface will act as a siphon, and will generate a vacuum effect to the upper end of the capillary; that is, the continued net evaporation at the surface of a wall can provide the motive force to move fluid through small capillaries from an inferior position to a higher elevation in the wall. This potential for water movement within a porous, permeable material is a form of engine driven by the siphoning effect of net evaporation at the wall surface.

EFFLORESCENCE AND SUBFLORESCENCE

Salt decay is exhibited through two primary visual effects: *efflorescence* and *subflorescence*.

Efflorescence

Efflorescence occurs when the masonry material is saturated with water, in which there is a substantial amount of soluble salts. As evaporation proceeds at the surface of the material, the concentration of salt increases to the point that precipitation of the crystalline salt occurs at the site of evaporation; and the formation and growth of crystals will often appear on the surface. The basic process is as follows:

1. When a salt in solution comes in contact with a gradually decreasing ambient relative humidity, water will evaporate from the solution, making it more concentrated.
2. When the ambient Rh reaches the equilibrium Rh, the salt solution will have become saturated.
3. If the ambient Rh is reduced further, it will cause the remaining water to evaporate, and allow the salt to crystallize out, producing the familiar efflorescence.

In short, the rate of evaporation must be slower than the rate of migration of the solution from the inside of the masonry material to its surface.

Efflorescence may be found in almost every part of a building where water penetrates the porous material. Cellars as well as areas near the roof are vulnerable to the growth of crystals, whose formation is most common during periods of low relative humidity. As such, it is more likely that efflorescence will appear in the wintertime and generally disappear in the summertime, meaning that crystals will appear when the air is cold and dry rather than warm and humid. The less soluble salts, such as sulfates, are more likely to form the characteristic efflorescence, as opposed to the highly soluble salts such as nitrates and chlorides, that tend to remain in solution longer. Efflorescence by itself does not physically damage the material because it is merely a deposition on the surface of the material. Its presence does, however, compromise the aesthetic integrity of the fabric and signal the potential for damaging salt formation below the surface. Once formed, efflorescence can undergo various transitions depending on environmental conditions; it can age, hydrate and dehydrate, and dissolve and recrystallize. In the hydration and dehydration of salt minerals, the hydrated phase crystallizes first, which facilitates the same damage as produced by salts unable to hydrate.

The transformation procedures usually cannot occur if salts or salt aggregates do not dissolve and recrystallize, at least partially. This happens when an undersaturated solution from any source is supplied to the salt crystals, or when a salt is mobilized as the result of a temperature change. This can be damaging, in that once the dissolution process is in place, in concert with the aging process, loose salts will diminish in size, thereby producing single crystals that will compact into larger aggregates and crusts.

Subflorescence

Subflorescence is clearly damaging to the physical fabric of the structure even though its exact mechanism is not totally understood. We do know that subflorescence is the result of salts that crystallize or precipitate under the surface of a building material, potentially resulting in flaking, spalling, sugaring, and/or pitting. Its genesis is the converse of efflorescence. To occur, the rate of net evaporation of the solution must be faster than the capillary network can supply moisture to the surface. The water, in effect, escapes before the solution actually reaches the surface. Whereas efflorescence is relatively easy to remove, subflorescence is difficult because it is formed within the fabric.

Other Salt-Related Decay Mechanisms

The presence of salts may also initiate or exacerbate other decay mechanisms. Most salts, especially sodium chloride, are more susceptible to the freeze-thaw mechanism, even though the freezing point of the solution may be depressed by virtue of the salt in solution. At the point near freezing, the maximum sodium chloride concentration achievable is approximately 25 percent. At that concentration, the freezing point of the solution is depressed to approximately 20° F. If the salt concentration in the wall is at or near saline saturation when freezing occurs, the water does freeze. Even though the temperature is depressed, the combined expansive effect of the freezing water/salt combination is greater than that of water alone at 32° F.

When the saline solution freezes, depressed though that point may be, the ice crystals exclude the salt. As the ice forms, the salt concentration in the residual solution increases. Depending upon the salt concentration prior to the initiation of ice formation, the salt concentration will sooner or later reach saturation and begin to precipitate salt crystals embedded in the body of the ice. The volumetric sum of crystalline water plus crystalline salt is greater than the same combination in solution. The difference is that the two sums are not great; and they are often assumed to be additive, but in most cases the volume of the solution is actually less than the sum of the separated fractions.

The condition of embedded crystals can occur only where the salt is confined along with the water, such as in the pore of a porous solid. In the preceding section we calculated that the expansive potential of confined ice inside a pore might be over 3,700 psi, although a more likely figure is an internal pressure of less than 700 psi. In any case, if the solution is saline, and it freezes, these figures will increase due to the trapped salt crystals as they form.

If the excluded salt can remain in solution, it will. In polar ice packs, the ice forms at a temperature less than 32° F. If sea water is a 5 percent saline solution, the freezing point is depressed approximately 2° F. The calculation is complicated by the fact that the salt in sea water is not a single salt, although it is predominantly sodium chloride. When the saline solutes consist of more than one salt, the properties of the solution can be derived only by laboratory analysis because the properties of the individual salts are not additive, multiplicative, or any other known theoretically deducible algorithm of the individual properties. As the ice pack forms, the salts are excluded; but instead of being trapped inside a confined pore, they are pushed back into the surrounding ocean.

SALT DECAY

There is no doubt that these crystal formations resulting from crystallization and hydration cause damage within porous materials, but the mechanisms that exert the internal pressure requisite for crack formation are not fully understood (see Figs. 4.7A and 4.7B). In short, little is known regarding what is going on in situ. Many theories have been advanced to explain the damage process, but no one has definitively identified the operating mechanism.

The majority of the current research on the damage process suggests that easily hydrated soluble salts enter the pores of a material and crystallize within the pore, filling its void. This, in turn, creates stress on the pore walls, thereby damaging the structure of the pore. This explanation does not, however, take into account that salts that do not or cannot transform in hydrated phases, such as sodium chloride and gypsum, can also cause damage. To further complicate the matter, rarely is there homogenous salt content throughout a wall. Usually, there are a variety of salt minerals, including sodium, potassium, magnesium, sulfates, nitrates, and chlorides within one building. Furthermore, the soluble characteristic of salts strengthens the probability of interactions and reactions among the different salts, thereby creating a complex salt system.

Relative humidity, and more specifically, the equilibrium Rh of a salt (which is the relative humidity of air in equilibrium with a saturated solution of a particular salt) are important factors in the initiation of the hydration and crystallization cycle. Another complication is that different salts have different equilibrium Rh values. Therefore, interactions between two salts can radically affect the crystallization behavior of either salt. This means that several salts contained within a building will not crystallize independently based on their individual equilibrium relative humidities; instead, they will crystallize and hydrate based on a range of equilibrium relative humidities, making it difficult to determine a relatively safe Rh. The presence of multiple salts could drastically alter their individual solubility, thus usually increase their potential deterioration effects. It has also been demonstrated that salt concentration and hydration levels may change from area to area, even on a small scale, so that certain closely adjacent zones may contain different salts and different salt properties.

To date, the search for the precise mechanism behind salt damage tends to focus on a source of expansive behavior in the salt itself, which follows this logical sequence: salt crystals form in the pore; the crystal grows in an anhydrous form; it is subsequently hydrated and reforms into a hydrous structure; the hydrous form expands, thereby exerting pressure on the pore wall. The logic is, unfortunately, incomplete, hence misleading. There is no reason the initial crystallization will be anhydrous. It is more likely or at least as likely that it will be a hydrous form. Further desiccation may result in reconfiguration to an anhydrous form, resulting in lower volume, not the opposite. If simple rehydration occurs, the increased volume can only reassume the original volume; it cannot exceed it.

A more persuasive, albeit complicated, explanation for the expansive effect of salt crystallization is that complex salt solutions precipitate at varying rates and relative humidities. While some crystals are forming, others are still in solution. The earlier-forming crystals are more likely to be hydrated crystals, but as the later-forming salts crystallize, the earlier-forming salts are dehydrated by the subsequent generations of saline solutions. The result is that the pore can be completely packed with crystalline solids, and some of them are anhydrous crystals. Upon a subsequent application of water, instead of reversing the process, the anhydrous salts recrystallize before the later-forming salts reenter the solution. Consequently, the entire volume increases, and, as it is confined, exerts lateral pressure on the pore walls.

FIGURE 4.7A The damage here is largely from salt decay. Idiosyncracies among stones, bedding orientation, and the prevailing elements result in seemingly random degrees of severity.

FIGURE 4.7B The primary mechanism here is also salt decay, but the mechanism is exacerbated by the corner location.

Salt crystallization either is a sufficient lone variable resulting in crack formation at the pore perimeter or it is not. If it is, then we do not completely understand the process. If it is not, then the examination of salt formation as a lone variable cannot and will not surrender the complete explanation of the mechanism.

Perhaps part of the process attributed to salt expansion is actually differential movement of the porous solid relative to the crystalline formation of the salt. If, as temperature decreases, the saline solution tends to precipitate at the same time the solid is contracting, the mechanism may be a function of the one solid squeezing the crystalline plug until it cracks, as opposed to the salt plug expanding until the solid cracks. There may also be a contribution from the hygroscopic contraction of the porous solid at the same time, with similar constraining consequences.

To minimize or eliminate salt decay, removing the salts—or, in other words, desalinating the building—may be desirable. Before desalination, the infiltrating water source must be identified and eliminated. Following elimination, if the salts are not removed, the wall will gradually dry out, further triggering salt precipitation and damage. It is difficult to assess the effectiveness of desalinating a building, but there are some techniques that seem to adequately remove the salts. As previously stated, a small amount of salts within a building material may not cause harm; therefore, its desalination is unnecessary. For example, a chloride content of less than 1.5 percent by weight in a brick may not initiate damage. On the other hand, a chloride content in reinforced concrete that is less than .04 percent may cause the corrosion of the iron. In short, different materials contain different amounts of *safe* salt content.

Recall that efflorescence is easier to remove than subflorescence. If loosely attached, as noted, the easiest and least-invasive technique to remove efflorescence is to simply brush the crystals off the surface. Although the crystals are not particularly harmful, their removal ensures they will not reenter the fabric; and this type of intervention improves the appearance of the material. However, the salts will remain within the porous material, and may continue the crystallization and hydration processes. Another option is to remove the contaminated material and replace it with a new material. Of course, if dealing with historic fabric, this is neither desirable nor encouraged. It is important in most cases to maintain the original fabric of a historic building; unfortunately, replacement may be the only solution for the reduction and/or elimination of salts.

When removal of the original fabric is to be avoided, but eliminating the salts is essential, several techniques are available to attempt the task. One way to desalinate the building is to repeatedly flush the walls with water. In this treatment, a fine mist is sprayed onto the surface then allowed to dry, thoroughly. (The use of deionized water could increase the solvent action of water.) To reduce the cost of this treatment, water is collected, filtered, and recirculated through a mixed bed of ion-exchanged resins to deionize the water again. This procedure is repeated until the presence of new efflorescence is minimal. The risk of this process is that the water pressure may increase salt mobility, thereby forcing the salts further into the wall. Moreover, the evaporation rate may be too slow, trapping the salts into the wall, which may cause further damage to the fabric in the form of subflorescence.

Another method of desalination is to apply a sacrificial coating on a masonry surface. This technique is called *poulticing*. In this treatment, a slow-drying material, such as an expansive clay or a paper poultice, is applied to a dampened wall. (Clays typically used are attapulgite, bentonite, celite, fullers earth, and kaolinite, which are mixed with water to form a stiff paste. The fibrous materials used are generally cotton, wool, and wood paste.) Because the poultice dries at a slower rate than the salt solution, the salts move toward the surface and crystallize in or on the poultice. After the poultice dries, it is removed, or it crumbles away from the building, along with leached salts. This process may need to be repeated several times to achieve the desired results. This procedure, too, carries a risk. Similar to the first method here, using water to dampen the wall may force the salts deeper into the material, which may stimulate further salt decay.

To minimize the potential of mobilization of salts caused by the poulticing method, an alternative, called *injection poulticing*, can be used. In this treatment, pressure transport is used to desalinate porous materials. Holes are drilled into the mortar, and water is injected into its center, to provide a constant source of water. By forcing the water out through the mortar's face, a pressure gradient is created. As in the poultice method, a clay or paper poultice is applied to the surface. The rate of transportation by capillary movement to the poultice is greater than the rate of evaporation from the poultice. Subsequently, the salts crystallize within the poultice and are leached with its removal. This process has been claimed to desalinate a building in one to four weeks. But note, although this process does not force the salts further into the material, the dampness it produces in the wall may engender other types of decay.

A more sophisticated process, often used by architectural conservators, is *electromigration*, also called *electroosmosis*. In this process, electrodes with an electric potential are inserted, or are otherwise connected, to the masonry. Depending on the electric potential applied and the specific characteristics of the materials, the rate of desalination will vary. Studies indicate it may take a long time to completely desalinate a building, in some instances up to 25 years. Therefore, at this time, the treatment is questionable as a practical means of desalination of large structures, and will probably find greater use as a laboratory technique applicable to smaller objects.

RISING DAMP

By now it should be apparent that salt damage and salt migration are not inextricably connected with rising damp. Salt migration is a likely coincidence of water migration through almost any porous, permeable material. Salt sources are pervasive enough to generalize water migration as essentially simultaneous salt migration. We discuss them in tandem precisely because they are so often coincidental, but it is important to emphasize that they are two independent phenomena.

Rising damp refers to the general class of vertical water migration through masonry. The necessary and sufficient conditions for the rising damp condition to occur are:

- Porous, permeable masonry
- Continuous supply of bulk water at or near the base of the wall
- A vertically continuous network of water-transporting capillaries.

The basic mechanism is that water is absorbed at the base or at least a lower portion of a masonry wall and transported via capillary action to a higher elevation in the wall. Along the way through the wall, salts may, and usually do, collect and deposit in upper portions of the wall through evaporation.

The water may also attack building materials directly, as opposed to damage resulting from efflorescence or subflorescence. If the water reaches interior plaster surfaces, the water will dissolve the gypsum, transport the dissolved plaster to the interior evaporative surface, and produce efflorescence on the interior of the wall. This sort of damage is characterized by bubbling and blistering of the interior plaster surface, and normally occurs below the painted surface. Unlike exterior efflorescence, which may well be the relatively harmless precipitation of transient salts, interior deposition of dissolved plaster indicates that the plaster is severely damaged and that the mechanism is still active. The mechanism may be active only during periods of rain, tide, or other water inundation, and is, therefore, sporadic, but the damage is cumulative, the mechanism is continuous, and the damage is progressive. While the interior plaster may be in jeopardy only during periods of prolonged saturation, the damage that occurs is irreversible.

Any timber pieces in the line of travel are similarly at risk, especially embedded items, such as joist and beam ends, which are lodged in pockets. The ends of the timber members are cut across the fibers of the timber, which exposes the end grain, which exacerbates the problem. The pockets themselves also contribute to the problem in that the masonry and mortar are typically packed tight to the timber surfaces. This has the effect of transporting the water to the immediate surface of the timber, then holding the moisture against the wood for prolonged periods of time.

In fact, many materials are adversely affected by prolonged exposure to water. Gypsum drywall, mortar, calcareous stones of all varieties, cement, metals generally and iron especially, along with timber and plaster, are all inherently vulnerable to water, independent of salt migration in solution in the water. All of which is to say that water movement in porous, permeable solids is associated with a very large class of deterioration mechanisms. Rising damp per se, however, is not a deterioration mechanism. It is a *method of delivery* so commonly associated with the general class of consequential mechanisms that it is often treated as a deterioration mechanism itself. Making the distinction may seem like quibbling, but it is important to keep in mind that water movement in the material is not by itself a deterioration mechanism, because no damage is directly and exclusively attributable to the water alone. No identifiable source of deterioration has the single necessary and sufficient condition of the presence of water.

It is true, nonetheless, that water is one of a set of necessary and sufficient conditions for a deterioration mechanism. Because of the prevalence of water, and because of its prevalence as a necessary condition for deterioration, water is often maligned as the fiercest enemy of the built environment. Water is not the enemy; it is only a natural element with which we must cope as building pathologists. Viewing the presence of water as the source of all building problems, instead of completing the list of necessary and sufficient conditions, precludes an accurate analysis of the problem. Rising damp is one such quasi-mechanism that tends to divert attention away from the actual deterioration mechanism and, therefore, away from cost-effective intervention.

Let's revisit at the requisite conditions for rising damp, and invoke the results of Lewin's experiments. In a monolithic sandstone column, under laboratory conditions (which means the stone pillar was placed in a tub of salt water to a height of saturation approximately 22 centimeters). The height did not (and does not) change with fluctuations in relative humidity, although there is some rise in height with the increase in salt concentration. The degree and rate of formation of efflorescence versus subflorescence may change depending on relative humidity, but not the height of the evaporative surface. After purging the salt at least from the surface of the pillar, Lewin was able to induce a second evaporative surface as much as eight centimeters higher than the first, but still only 30 centimeters above the saline source, and with a significantly more concentrated solution. It is tempting to conclude that further applications may have climbed further until, upon an infinite number of saturations, the evaporative surface would climb an infinite distance. Only the hyperbolic extension of such a conclusion exemplifies the fallacy. It is, nonetheless, widely accepted that rising damp (as defined earlier, as the capillary rise of moisture in masonry construction) can easily achieve elevations of three meters, and, in Venetian structures, occasionally reaching 10 meters or more. That said, there are other and simpler explanations for the observed phenomenon, as follow:

- No closely controlled experiment, including those by Lewin, has ever induced water to rise more than one meter even upon multiple applications of inundation at the base.
- No single masonry capillary or capillary network can come close to the monolithic pillar Lewin used in his experiments. A disruption in the capillary will disrupt capillary action and, therefore, the flow of fluid in the capillary.
- The time required for capillary flow as quantified by Poiseuille's Law means that the water supplied to achieve even modest rises through capillary action must be prolonged, and the volume of available water must be virtually unlimited.

The design of the critical experiment to determine maximum potential moisture rise is not difficult to imagine:

1. Either build a test wall or take an existing wall and saturate its base in a solution of water and radioactive iodine.
2. Plot the fluctuations in radiographic responses over time.

We may even be able to achieve measurable results using simple dyes, but to date no reported experiments have been conducted on large scale constructions. The interpolations of Lewin's results to apparent salt depositions several meters above a known bulk water source are not conclusive or even logical.

Capillary action is not an engine. The idea that severing a capillary in which water had risen a meter at half a meter up its length would result in an endless stream of fluid out of the end is simply incorrect. Perhaps capillary action is sometimes confused with the phenomenon we observe when we cut a plant stem above-grade and fluid continues to ooze from the severed plant. Despite the apparent analogue of small inorganic capillaries and organic cells in plants, the phenomena are only vaguely connected. The flow of water in the stems of plants may in part be attributable to capillarity, but to no greater a height than a meter or so. As suggested earlier, there is the added boost of the evaporative engine through the leaves of the plant, but that can take effect only after the fluid reaches the leaf. The potential of an evaporative surface cannot draw the fluid to the evaporative site. In other words, the evaporative engine is a compelling argument for sustained fluid flow, but it is inadequate to explain flow initiation.

Initial fluid transportation in plants is primarily explicable through the fourth of the colligative factors: osmotic potential. In solutions of volatile solvents and nonvolatile solutes, there is the potential of osmotic action across selective membranes. This means that fluids will permeate osmotic membranes when the solute concentrations on either side of the membrane are different. The movement will occur in the direction of the lower concentration. The fact that the solute may be a simple organic compound, such as xylose or fructose, as opposed to inorganic salts, does not alter the phenomenon. Water moves, at least initially, by being passed from cell head to cell tail through osmotic transfer. Capillarity alone does not get the water to the leaves of a tall tree, and capillary action alone does not get water three meters above the water source.

The capillaries in a masonry wall are highly fractured and disjointed. Arguably, the closest approximation of the continuous capillary necessary to account for any rise of fluid in the material will occur in the mortar, not the stone or brick. Even the continuity of the mortar is dependent on the tightness of the jointing and the compaction of the voids in the wall with mortar. What is far more likely is that the face of every brick or stone is a break in the capillarity of that material, and every void in the mortar is a disjunction in the capillary network of that material. Taken together, the prospect of a continuous, unbroken capillary net from the water source to a superior elevation in the wall is remote.

The movement of fluid through small capillaries is slow. Lewin's experimental runs were six to eight hours of continuous emersion of the stone pillar in a constant supply of solution. Any intermittent water supply may be enough to initiate capillary action, but it is limited to the period of inundation. Sporadic sources may, however, be able to saturate earth that can act as a reservoir for the subsequent application of fluid to the wall, but this does not diminish the fact that sporadic sources cannot directly account for a phenomenon dependent upon a prolonged, continuous supply of fluid to accomplish even a one-meter rise. Remember, Lewin was unable to induce water above 30 centimeters even after leaving the stone in the water for two weeks.

Where does this leave us? If rising damp is defined as a transportation device dependent upon capillary action, there is little if any evidence that the phenomenon can account for the vertical transportation of water higher than approximately one meter above the water source. That water source must be continuously available and rechargeable regularly. What difference does it make how we define the phenomenon if we know that water is in the wall and damage results from its presence? It makes a difference in the approach we take to intervention, which we will discuss in depth shortly. In short form, however, this means that the source of the offending water may be misdiagnosed and, therefore, inappropriately addressed.

All this is not to imply that rising damp does not exist. It does; but the magnitude of the vertical rise is significantly lower than is often identified as attributable to the phenomenon. Even if limited to a meter or less, there is still a great deal of vulnerable material within that dimension. If capillary action is fairly limited to one meter or less above the water source, the water that wets the first-floor joist ends when there is a full basement is probably not coming up through the footings, which are two to three meters below the joist pockets. Probably the source of water that is rotting the joist ends is a meter or less below the joist pockets, which means in most cases that the water is moving laterally into the wall at or immediately below-grade. The water is probably moving laterally up and down the entire surface of the wall, wetting the entire region, plus approximately one meter above-grade.

The entire basement wall may be saturated, as we discussed in Chapter 3, but the portion of the structure subject to damage because of rising damp, namely the joist ends, is limited to less than one meter above the highest source of the water (see Fig. 4.8). If the object of intervention is, in the end, to protect the joist ends, then intervention can be limited to that zone alone. Specifically, if the joist ends are only slightly above-grade or even at-grade, the intervention at the perimeter of the building can be limited to one meter below-grade. This may be accomplished by trenching along the wall to a depth of approximately one meter, and installing a drainage board against the wall. When water comes in contact with the drainage board, the water will immediately be transported to the base of the drainage board. The soil at the base of the drainage board may be saturated, and that water may very well penetrate the masonry basement wall, but the migration back up the wall to the joists is beyond the reach of capillary action. Therefore, the joist ends are protected; and if that was the object of the intervention, the intervention was effective.

Obviously, this intervention does not address other occasions of dampness or humidity in the basement, but that was not defined as an objective of the intervention. Some observers may argue, not totally without justification, that stopping short of a more comprehensive intervention is short-sighted and wasteful, but it is extremely important that the pathologist not be presumptive in his or her intervention. Treating only the upper meter of a wet wall is a sound approach to protect joist ends, but it probably is not an effective intervention to cure a humid basement. This does not mean, however, that a limited treatment is wasteful or inappropriate. If humidity in the basement is not demonstrably damaging to the building, and the client does not want to incur the added cost, anything more aggressive is what is wasteful.

If the objective of the intervention is to protect joist ends and the pathologist assumes that rising damp initiates as far away as three meters,

FIGURE 4.8 Rising damp is evident here by the discernibly wet (darker) portion of the wall.

then the intervention may result in the complete excavation of the basement wall, as was described in Chapter 3, with all the attendant costs. The point is, appropriate intervention begins with the rigorous application of diagnostics, combined with a precise explication of the intervention objectives. Rising damp is not the same as lateral saturation; thus, the intervention approaches used to deal with these two conditions are not identical. This is not intended to advocate avoidance of complete intervention of lateral water migration through basement walls. It is to argue that such treatments must be precisely aligned with the appropriate attendant pathologies, and not be chosen out of misunderstandings about water or the physics of water migration.

Let us take another example. Damp basements are at times combated with dehumidification devices. As explained in Chapter 3, this means that, in some instances, more moisture is effectively sucked through the wall than would otherwise be able to migrate through the material on it own. Now suppose that rising damp starts with water at the base of a three-meter-deep basement wall. Theoretically, in this case, dehumidification could pull the water out of the wall before it reached the ends of the joists at the first floor, and so save the building.

Now, putting aside that additional water would be pulled from some source through the wall, suppose that the primary source of water affecting the joist ends originated at- or near-grade. In this case, the dehumidification of the lower portion of the wall would pull water from all available sources into the wall at an accelerated rate. The water at-grade, if of a relatively unlimited volume, would continue to wick up through the wall even as additional water was diverted downward toward the dehumidifiers.

There would be no net reduction in water available to the joists at the same time that water was being suctioned through the wall. This misapplication of intervention effort is due to misidentifying the source of the water because of a presumptive notion of how and how far rising damp alone can wet materials.

If rising damp acting alone is limited to relatively short vertical distances, why then does water *appear* to rise much higher in walls? There are two ways to answer the question. The first is that water is, indeed, rising higher in the wall than can be readily explained by capillarity alone. The second is that water is not, in fact, rising from a remote lower elevation that the question presumes.

Much of the literature regarding salts (and the reason that rising damp and salt damage are linked together in this chapter) focuses on the hygroscopic capacity of salts to pull water up into materials and, thereby, establish an evaporative engine that will perpetuate the migration of water through the fabric. The idea can be analogized to a sort of absorptive bucket brigade. The salts in contact with the water source absorb the water and pass it onto adjacent salt concentrations, until a fluid channel is open to the evaporative surface and the engine is operational. If the conditions are such that evaporation does not occur until the bucket brigade reaches 10 meters, that is where a salt line appears; hence, we have rising damp to that elevation.

The factually supportable part of this theory is that many dehydrated salts are hygroscopic; solid salts will indeed absorb water to the point at or near hydration. Salts can absorb moisture out of the atmosphere to the point of hydration, provided the relative humidity is high enough and remains high enough for a protracted period of time, such as may occur on the interior of a saturated masonry wall. Lewin's experiments support the notion that at least as far as subsequent wettings are concerned, the salt train may advance the evaporation line eight centimeters. The evidence is inconclusive, to say the least, that continued reapplication of the solution would or could advance the salt line even to a meter, much less three or 10 meters above the water source.

More credible is that water is introduced into the open vertical channel not from capillary action alone, but externally from some other source. Suppose that rainwater enters the wall laterally, as described in Chapter 3, and percolates downward through the capillary and pores of the materials until many of the channels are filled all the way to the bulk water source. The rain stops and evaporation begins at the wall surface. The evaporative engine does not have to overcome resistance of pulling water up through the labyrinth of empty capillaries, because they are already primed. Despite the appeal and logic of such an explanation, there is no confirming experiment or study to support the hypothesis. Nothing in this theory even suggests that such an engine can reach a particular altitude.

The second answer is that the water does not rise more than a few centimeters. The lines of salt are real, but they are the products of perched water sources. If water enters a wall laterally and collects in the pores of the materials above shelf angles, flashings, or simply a succession of impermeable window heads, that water may subsequently rise in the wall, evaporate, and produce the characteristic salt line at the perched elevation. If water migration occurs at an already elevated position, and the source of the water is, in fact, lateral and not vertical, the fact that there is a visible salt line on the exterior surface does not prove the existence of rising damp much less rising damp to that elevation.

In conclusion, the examination of two closely affiliated phenomena, salt damage and rising damp, leaves us with far less in the way of definitive intervention strategies than we might wish, primarily because our understanding of the phenomena is limited. To date, in regard to the problems of rising damp and salt damage, there is far more opinion than sound science. Depending on the client and the situation, we can engage in prophylactic measures such as inserting or injecting capillary breaks in walls or installing complete perimeter drainage systems. These interventions have the definite advantage of isolating the water source, but at considerable expense and disruption to the fabric or, potentially, to archeologically significant terrain. Simply put, because we do not totally understand the mechanism of the problem, we can not be sure we are solving it either efficiently or economically.

ACID RAIN AND OTHER POLLUTANTS

Generally, air pollution which contains acid rain and other pollutants in the form of particulates and aerosols, is transported by numerous vehicles, all of which fall into two broad categories: man-made and natural. Common to both are primary particles that are emitted directly, and include, for example, fly ash, soot, and dust. Secondary particles are those created by gas-to-particle transformations, such as oxides of sulfur, nitrates formed of oxides of nitrogen, and hydrocarbon vapors that generate secondary organic compounds. The extent of air pollution is influenced by meteorological factors, as well as the amounts and levels of pollutants actually produced. The determinants of pollution concentrations include, but are not limited to: temperature, precipitation rates, fog and condensation rates, wind speed, population density, industrial productivity, and urban and rural settings. Because there are so many variables, pollution cannot be understood comprehensively, regardless of all the attention paid to it. We do know, however, that pollutants will degrade *all* building materials.

Temperature is a principle determinant of particle formation and deposition. Urban climates are regarded as milder and moister than rural environs; temperatures in these areas tend to be approximately 10 percent higher than those of surrounding, more open, terrains because of the numerous heat sources, such as industrial activity, and the concentration of transportation heat producers, such as trucks and automobiles. This is despite the presence of smoke and haze, which can absorb up to 15 percent of the reflected light and infrared radiation. The mean temperature is also a function of the size of the urban population; and precipitation is about 10 percent greater in urban area, due to the abundance of condensation nuclei from dust and smoke. Urban humidity is, however, lower, by about 5 percent, than in rural areas. The average temperatures are lower outside urban heat domes, and greater transpiration over agricultural lands and open water provide added humidity. Rural-area transpiration generally provides more moisture than in urban environments.

Pollution concentrates in small areas when the wind speed is low; in contrast, strong winds dilute pollution, making it less concentrated, but spread it over a larger area. The particle matter transported by winds at normal speeds, range from about 100 to 10 micrometers or microns (μm), which is one-thousandth of a millimeter or one-millionth of a meter. Extremely small particles, which form a thin smoke, or haze, do not fall at all, because they are inherently susceptible to collisions with molecules of air and water and to the weak pull of gravity. Small particles may be carried considerable distances before they fall to the ground; otherwise, their rate of fall increases with the square of the logarithm of their size.

Particulates and aerosols are solid or liquid particles small enough to remain suspended in air, as opposed to particles such as dust, sand, or grit, that settle in still air. The diameter of airborne particulates is also measured in micrometers or microns. Although wind can transport particles as large 100 μm to even 10,000 μm in size, even smaller particulates ranging from 15 to 20 μm typically settle near their point of origin in still air.

The evaporation and condensation phenomena result in the accumulation of atmospheric particles of varying sizes, which can coagulate; and, in foggy weather, their concentration usually increases. Fog droplets can absorb pollutants in higher concentrations because of their large specific surfaces and very slow fall velocities. In the summer, fog is approximately 30 percent greater in cities than in nonurban areas. In the winter, however, the percentage of fog will nearly double. This increase in pollutants lowers the pH of fog, making it more acidic than rain. By contrast, rain cleans out the atmosphere by precipitating the pollutants to the ground. A steady rain falling at the rate of 1 millimeters per hour (mm/h) for 15 minutes will eliminate 28 percent of the particles measuring about 10 μm. This effectiveness decreases with the size of the particles, meaning that smaller particles are more likely to remain on the surface material. Rain will typically not eliminate particles measuring less than approximately 2 μm.

Aerosols can be taken from the air by falling raindrops at rates determined by the height of fall from the cloud base to the land surface, in conjunction with the diameter and speed of the raindrop. The absorption equilibrium of rain is approached much faster by slowly falling small droplets with a larger specific surface area of absorption than by large drops falling at greater speed. Large raindrops can absorb a maximum of 10 parts per million (ppm), whereas smaller fog droplets can absorb a maximum of 60 ppm or more. Urban winter fog can therefore be very corrosive and toxic.

Extensive studies of the composition and optical properties of atmospheric aerosols, have shown that soot causes the black coloring of the submicron fraction (particles smaller than 1 micro) and is responsible for the blackening on painted surfaces. The black pigmentation is usually due to a number of factors, such as "smoke" particles in the size range 1 to 1,000 μm diameter. Smoke includes all products of combustion ejected into the atmosphere, including domestic, commercial, and vehicular sources. Coal, oil, and diesel internal combustion sources predominate in modern settings. Mixed in with these particles are organic and inorganic particles from a variety of sources. In short, the soiling of façades by exposure to pollutants is the result of a sequence of occurrences starting with emission of the pollutant, followed by atmospheric transference, and, finally, deposition onto the façade surface.

Soot in the atmosphere is the result of incomplete combustion. It is present in the submicron size range (0.1 to 1 μm) and, consequently, remains in the atmosphere for a substantial time. The considerable amounts of soot present in the atmosphere are more responsible for the soiling of material surfaces than any other coarse particles. Soot particles are aggregates of primary particles of graphitic material. Automobile aerosol, which is principally soot, consists of aggregates of several hundred primaries of .05 μm, a geometric mean diameter.

Submicron soot particles are deposited onto surfaces mainly via precipitation. Soot concentrations in wet deposition are about 200 ug per liter, but this type of deposition onto façades, does not lead to appreciable soiling. However, an important and degenerative interaction takes place between soot and gypsum on building surfaces: it is the reaction of the building material with SO_2 and acidic sulfates. Materials with low porosity and good tensile strength are the most resistant to this type of decay; for example, well-fired bricks or tiles are appreciably resistant, whereas limestone, sandstone, and lime mortars will be deeply attacked.

Atmospheric constituents are characterized as airborne particulate matter. As mentioned, pollutants may be either naturally occurring particles, such as dust, or man-made pollutants. The majority of pollutants are anthropogenic, such as soot, industrial chemical emissions, and vehicle exhaust emissions. Modern urban atmospheres also contain gaseous pollutants, which are significant as agents of stone decay. The intensity of air pollution is reflected in the quantity of total suspended matter. The number and size of particles decrease with height, but increase with rising Rh.

These toxins are removed from the atmosphere on a daily basis. Aerosols are removed through dry fallout, in the form of dust or soot, or are washed out through precipitation. Gases are eliminated by washout through rain adsorption; the decomposition in the atmosphere by a reaction with other gases, which results in the form of aerosols or smog; adsorption by vegetation and stone decay; or the escape of some light compound into the outer atmosphere.

Concomitant to the decline of coal burning as a cause of pollution is the increase of diesel fumes as a major source of pollution, which are now nearly double those of coal emissions. Diesel exhaust is a rich source of a very fine carbon particle known as the particulate elemental carbon (PEC). While they are chemically inert, the particles are very small, approximately .02 μm in size, and are very sticky because of their hydrocarbon content. PEC is less water-soluble than a suspended soil particle, which is generally washed off the surface. Acidic gases such as sulfur and nitrogen oxides can, in turn, become easily absorbed onto the surface; once deposited on the soiled stone surface, they may act in conjunction with the diesel particles. Also, PEC, due to its electronic structure and electron system, may aid in the catalytic conversion of such acidic gases to the sulfate and nitrate forms, respectively.

The general observation that surfaces facing downward are blackish corroborates the aerodynamic properties of soot-containing ambient particles, which are part of their deposition characteristics. Consequently, the soiling patterns of façades reflect the eroding action of rainfall runoff that reacts locally with this uniform soot deposit.

The dense coating of soot upon a surface is the result of an agglomeration of carbon atoms reacting with certain chemicals within the material, and enables the surface to adsorb large quantities of gaseous hydrocarbons and other pollutants from the air. A coat of soot can form a tight film, which can protect the stone surface even as the interior of the stone decays.

Atmospheric sulfuric and nitric acids also affect the building fabric. Once sulfuric and nitric acids are present on the building fabric surface, they may react, for example, with limestone, which may foster accelerated erosion rates. The calcium carbonate that comprises limestone and marble is technically a soluble salt, but in pure water the rate of dissolution of calcium carbonate is quite slow. If the solution is acidic, however, the rate of dissolution of calcium carbonate increases dramatically. If the acid dissolved in the rainwater is sulfuric acid, the calcium carbonate is effectively replaced with calcium sulphate, which is relatively easily ionized and dissolved. If the acid is nitric acid, the carbonate is effectively replaced with nitrate. The calcium nitrate is even more leachable than calcium sulphate; and because the emissions of nitrogen oxides are increasing within the ambient atmosphere (over 40 percent of the national emissions of nitrogen oxides are vehicularly derived), the rates of building erosion are similar to when ambient levels of pollution from sulfur dioxide were much greater.

There are two primary reactions between the external influence of the atmosphere and the deposition of the pollutant on a building. First, dry depositions are made of reactive gases and of water-soluble solid materials such as sea salt; second, wet deposition of rainwater can bring the dry-deposited products into solution. The wet deposition can attack the façade material itself, and supply the wet condition mandatory to facilitate corrosion. The wet deposition can also transport deposited products on the façade or even into the surface itself.

By its nature, rain is acidic, with a theoretical mean pH of 5.6. This is based on the present equilibrium of the CO_2 content in the atmosphere with rainwater, forming the weak carbonic acid. The reaction is $CO_2 + H_2O - H_2CO_3$. Doubling or tripling of the atmospheric CO_2 from near .034 to .07 percent lowers the pH to 5.4. At .1 percent CO_2, the pH is near 5.3. In addition to carbonic acid, polluted atmospheres inside or near urban centers contain variable amounts of sulfur dioxide produced by the burning of sulfur-containing fuel. Sulfuric acid is strong enough to cause the deterioration of several minerals, such as carbonates and silicates, at a higher rate than water containing only carbonic acid.

Sulfuric acid and carbonic acid are not the only deteriorating mechanisms formed in the atmosphere. As stated, the action of pollution is not fully understood because of its numerous variables, all of which affect the atmosphere and building materials. First, other pollutants besides sulfur dioxide and carbon dioxide are consistently present, some of which can cause the formation of other acids, which also foster a corrosive action. Second, the sulfur dioxide attack on materials can follow different paths, reacting with a variety of minerals or chemicals. Under these same atmospheric conditions, the carbonates of calcium and magnesium can be transformed into bicarbonates, and slowly dissolve.

Relevant, too, are the mass and size of distributions, because the composition of the various size fractions differs greatly due to the individual contributions of aerosol sources to particle size fractions. The mass/size distribution of the atmospheric aerosol is usually dimodal, from about 0.1 μm to several μm and beyond. The mass in the coarse particle type can be dominant in the atmospheric aerosol mass, typically under conditions of dry, windy weather. The smaller particulates are primarily composed of sulfates, nitrates, ammonia, soot, and organic material. Coarse particles

resemble soil composition to a large extent. In coastal areas sea salt particles can contribute to the coarse particle mode, and to a lesser extent to the
submicron regime.

There are four primary, and rather broad, constituents of coarse particulates: inorganic airborne dust particles, such as soil, coal, ore, and "fly
ash" particles; organic airborne particles, including plant remains and
pollen; inorganic precipitants such as gypsum; and organic growth in or on
the crust, such as bacteria, fungi, and bird fecal matter.

Dust particles from both organic and inorganic sources can range in
size from 1 to 100 μm and severely impact the appearance of a building.
Generally speaking, anthropogenic aerosols can be more destructive than
natural sources; of natural sources, sea salt in the atmosphere is the greatest contributor to building deterioration. Salts in the atmosphere can infiltrate masonry and material up to a distance of 300 miles from coastal
waters, bursting bubbles of saltwater to form aerosol-size water droplets.
The evaporation of the water droplets further reduces their size.

Sulfate is the strongest of the corrosive pollutants in rain because it can
hydrolyze either to sulfurous acid from unoxidized SO_2 or to the more
corrosive sulfuric acid from SO_3. The supply of chloride and sulfate is
abundant near seashores and deserts, as well as around major industrial
areas. The flux of chlorides near the coast is more than 100 times greater
than elsewhere.

Typically, bricks and stones come under attack by rainwater or by
condensation and the corrosive action of polluted atmospheres. This can
be considered as a type of stress-corrosion process, in which chemical attacks are combined with mechanical stresses. Silicates of calcium and aluminum are the primary components of ceramic products such as brick
and tiles and they are mostly water-insoluble. As such, they should be fairly
resistant to rainwater, but their slow reaction with slightly acidic water is
not fully understood; we do know that it is destructive to clay products.

Limestones and marbles are considered delicate stones composed of
calcium carbonate, which is highly sensitive to acids. As limestone weathers, the outside layer becomes weaker and more porous. The weakness renders the rock more susceptible to decay. Sulfur dioxide, which is one of the
primary constituents in the acid attack of calcareous surfaces, dissolves
readily in water to form sulfurous acid (H_2SO_3). Sulfuric-based acid combined with calcium carbonate forms calcium sulfate, which upon crystallization becomes gypsum ($CaSO_4 \cdot 2H_2O$).

Because gypsum is quite soluble in water, heavy rain will remove the
gypsum particulates and, thereby, the residue of the deterioration and the
base material. By contrast, in regions of high pollution, not exposed to rain
expulsion, the particles will become trapped on the surface. In the process,

the "skin" formed on the stone becomes increasingly less permeable, and the buildup of particles becomes apparent as a dark crust. In urban environments where pollution is high, the encrustations can build to a thickness of 0.25 to 0.5 inches (6 to 12 millimeters). Less polluted areas will generally reveal a brownish surface crust. Different limestones will react in varying ways depending on their coarseness. For example, more durable limestones may be able to retain their skin under chemical attack, whereas more susceptible, finer-pored stones may form blisters, which will eventually split to reveal a deteriorated, flaking, powdery stone.

Sandstone tends to be much more uniform in its deposition than limestone. The surface accumulates the grime, but a variegated discoloration remains because of inhomogeneities of the surface; this gives a patchwork appearance to some buildings. If surfaces are left rusticated, some parts of a block may be clean while others are dirty. For dry deposition on a monument detail, sandstone surfaces can accumulate crusts as thick as limestone. Generally, however, on sandstone buildings, continuous rain-washing leaves crust-free areas, as there is little opportunity for deposition to accumulate. A more complex deposition can be seen where there are horizontal or subhorizontal streaking of black crusts. These are the result of localized airflow around buildings (they may appear on limestone buildings as well).

Architectural details tend to crust where they interrupt surface water flow. On details, particularly on sculptures and balustrades, crusts can be very thick and sharply defined. In these places, the buildup is produced by airflow (dry deposition) and may be several millimeters thick. Historic building materials differ in the way they hold grime and how susceptible they are to damage during cleaning because of the wide variation in their physical properties (porosity, permeability, hardness, strength, water absorption, texture, and structure), in chemical composition, in surface finish, architectural form, and orientation. Local environmental pollution is significant because many historic buildings suffered dirty conditions even before industrialization. Differential crust formation can often be seen even where there appears to be a uniformity of features.

4.3.4 Hazards of Cleaning

Exposure to the environment will weather and soil all traditional and contemporary building materials. Initial light soiling usually causes fairly subtle changes to the appearance of a building, and can be complimentary to the architecture. As the soiling advances, details and coloring begin to lose their sharpness, and become muted. Once the soiling becomes heavy, much of the building's surface becomes uniformly dark. At this stage, many owners, observers, and practitioners believe that cleaning is necessary to improve the durability and/or enhance the aesthetics of the structure. They believe that removing the particles will renew the building's appearance, as well as halt the damage caused by destructive pollutants. There is some truth to both of those assumptions. That said, debates continue over the viability and appropriateness of cleaning, primarily focusing on the importance and value of patina and age. Patina may be a legitimate issue for older buildings, but not for modern buildings, which are viewed only as dirty. Practical debates regarding cleaning rage as well, raising questions such as: Does cleaning really create a more durable surface? Will cleaning accelerate deterioration through other mechanisms? Does cleaning really remove the particles? The answers to these questions are not fully known; much more research is still necessary to fully understand the effects of the cleaning on masonry.

The cleaning industry itself is a complicated topic. Industry professionals must have a comprehensive understanding of building materials in addition to a high level of skill. They must master cleaning techniques to know and appreciate their outcome. These requirements are not easy to meet, because at present there is no single known method of cleaning that safely, and totally, removes established patterns of soiling without altering or damaging the stone, its tooling, or detailing. Furthermore, there seems to be no established, reliable system under which the long-term consequences of cleaning can be predicted with confidence. By contrast, numerous short-term consequences of the three principle cleaning methods —water, chemical, or mechanical—are known. These will be discussed in greater detail later in this section.

In most cases, cleaning does involve the modification of the surface structure. Whether by applying chemicals, blasting the surface structure, or even applying water to the fabric, the surface material will be changed. If the change results in the removal of the original skin, cleaning is a negative. The ideal cleaning method would affect only the soiling, but it remains an ideal, not a reality. Consequently, the following discussion focuses primarily on the application of water as a cleaning method, because it is thought to be the least invasive of the three methods. But "least invasive" does not mean water cleaning cannot cause damage. It can.

To reiterate, the decision to clean a surface is based primarily upon two reasons: actual damage associated with soiling, and aesthetic viewpoint. Generally, the primary purpose of cleaning is to improve the durability of a surface. Unfortunately; it is difficult, if not impossible to remove soiling without affecting a building's masonry substrate. Soiling is frequently divided into two types: nonbiological and biological. Soiling not only exists as a superficial layer; it can be closely associated with the mineral elements of the original surface, and extend inward from the surface layers. Usually, soiling will have markedly different physical properties from the base masonry. This is why it is considered a source of deterioration that needs to be ameliorated. Deposited soiling will contain soluble salts (pollutants) that can be transported into the pores and fissures of the stone during wetting and drying cycles, where they may cause ongoing deterioration.

Despite the deterioration caused by soiling, cleaning can be the most severe experience to which a building fabric will ever be subjected. In the strict sense of the word, cleaning is intended to remove the outer layer of incrustations, soot particles, and microorganisms that are presumably damaging the building material. Cleaning, then, is the separation of a particle from the substrate. Thus, cleaning can be divided into two specific operations: the detachment of the particles from the substrate, and the transportation of the detached particles from the surface. The process of detaching the dirt particles depends on the type of bonding between the particles and the substrate; therefore, there are numerous processes for particle removal. Choosing a method must be done based on the bonding relationship between the particle and the substrate.

PARTICLE ADHESION AND REMOVAL

Before describing the specific cleaning methods, it is necessary first to explain particle adhesion and removal. Particles are deposited onto building's surface through a combined action of diffusion, sedimentation, inertia, electrical charges, and local electrostatic fields. These factors can be characterized by three phenomena that lead to particle bonding: molecular interaction, electrostatic interaction, and capillary condensation. Remember, particles are not uniform in size, but exhibit a wide range of shapes and sizes.

Molecular interactions are based on van der Waals's theory of dispersive interactions, which suggests that atoms are instantaneous dipoles, and the induced dipoles in neighboring atoms are summed over all atoms. More simply, this means that the molecular attraction depends on the first power of the particle diameter. By contrast, the mechanical removal forces are represented by "mass acceleration," and therefore depend on the third power of particle diameter. Thus, very high accelerations are required to remove small particles. A good example of this is the adhesion of small particles on the windshield of an automobile. Though these particles are in no way mechanically attached to the glass, some withstand rain hitting the windshield in excess of 60 miles per hour (others do become dislodged as is apparent in the difference in particle concentration above and below the line swept by the wiper blades).

Particle adhesion is also dependent upon the surface to which it clings; that is, the surface or substrate can either strengthen or weaken the bond.

For example, microsurface roughness—meaning that the cavitations are smaller than the particle—reduces the mass of the contact plane which results in a less adhesive force. The opposite is the case if the surface roughness is larger than the particle.

For particles measuring larger than 50 μm in diameter, the attraction is electrostatic forces. Electrostatic interactions increase particle adhesion in two ways. The first is due to the relative attraction of the two different materials. The two materials form a contact that has a maximum electrical potential value of 0.5 V. This is produced from the outer layers of the particle and the surface. The second way is the electric charge on the particle or the surface. This effect creates an electrical image of the particle in the substrate, which facilitates an adhesive force. And, this force will change over time as a result of leakage.

Capillary condensation will also increase adhesion. It is the process by which water vapor forms between the particles and the surface. The meniscus between the particles and the surface attracts the vapor, due to surface tension, and alleviates the pressure of the liquid, which thereby generates an attractive force. This force, however, can be negated. It is thought that when a particle and substrate are fully immersed in a liquid, the electrostatic forces become negligible, and the van der Waals forces are drastically reduced because of the shielding effects of the liquid. This is the theory behind using low-pressure water spray as a masonry cleaning technique. Relative adhesion forces up to several million times the force of gravity have been speculated for particles in the submicron range.

There are two concepts, in addition to immersion, that may determine the feasibility of particle removal from a surface: *drag force* and *"law of wall"* combined with *particle "hideout."* A drag force is imparted upon a stationary particle when in a moving fluid stream. It is dragged along because of the pressure exerted by the moving fluid, and the friction between fluid molecules and the particle skin as they flow around the particle. The concept behind particle removal is to have the drag force exceed the force of adhesion. This can be accomplished through the increase of fluid density and local velocity.

Local velocity is dictated by the "law of the wall." This law states that as a fluid passes over a surface, the velocity at the surface immediately approaches zero. But, at some distance above the surface, it will increase to a maximum, called the *free stream velocity*. Consequently, at some distance above the surface, the velocity acting on the particle will be quite different. The effect of particle "hideout" is established in the attempt to remove micron-sized particles that adhere to the substrate where the velocity approaches zero. Small particles exhibit a greater adhesive strength, which exemplifies the challenges of using a water-pressure cleaning approach. That said, if the particle size is large enough, the relationship of pressure required for removal could be determined as a selection process for this type of method.

Some particles cannot be removed because the forces needed to do so are just too great, and would most likely destroy the physical fabric. The intermolecular forces on a 1,000 μm (equals 1.0 mm) particle are comparable to one, times the force of gravity; the force on a 5 μm particle is more than two million times the force of gravity, which illustrates the difficulty of removing micrometer-size particles. The bonding strength and subsequent shear stress required to sever the bond between particle and substrate will determine removal feasibility. (This makes something of a mockery of the notion that asbestos particles clinging to an air duct will be picked up by the moving air stream and distributed into the ambient air. The size of asbestos fibers that represent the alleged health hazard would require millions of times the force of gravity to remove them from the duct surface, meaning that the air velocity would have to reach ultrasonic velocity.)

CLEANING METHODS

Before taking on a cleaning project, consideration must be given to the specific masonry substrate and its condition; the nature and degree of soiling; the effect of cleaning on the building's appearance; the effect of the cleaning at a microscopic level and in relation to the future weathering of the building; the safety of the operatives, members of the public, and occupants; and the effect of the processes on the environment.

Before beginning a cleaning project it is vitally important to conduct a small test program on locations throughout the building. Sandstone is a good example to illustrate the need for careful preparations. Sandstone is often used as a general description for a group of rocks that are detrital, quartz-rich sediments with grain sizes varying from 2.00 to .0625 mm. That description does not, however, begin to reflect the wide variation of mineralogical, physical, and chemical characteristics found in these stones. This inadequate characterization has in the past led to the implementation of many inappropriate cleaning methods and remedial treatments resulting in discoloration and accelerated decay rates. To avoid making a similar mistake, an analytical investigation should include a microscopic view of the materials to be cleaned, comprising their constituents, details of their condition, and the reasons for their deterioration and breakdown. A simple visual inspection is rarely sufficient to make certain cleaning decisions.

Three methods of cleaning are typically used:

- Water in the form of a spray, steam, high-pressure vacuum, and soft brushing

- Abrasives, either by blasting or mechanical means, with or without water
- Chemical process, involving acids, alkalis, and/or solvents.

Dividing the methods into three separate types may seem pointless, considering that two, and sometimes all three, are used in conjunction with each other, and that water is almost always an ingredient—except in dry-grit blasting.

Water Cleaning

As mentioned, we will focus on water cleaning, the least expensive and simplest method. It can be broken down into three submethods:

- Low-pressure water over an extended period of time
- Moderate- to high-pressure wash
- Steam.

Water plays a key role in cleaning because of its capability to soften, dissolve, and dislodge the particles. Simple, direct water application is used for removing light soiling, though brushes are often used in conjunction. Two applications are used for surface cleaning with water. The first phase, gently sprinkling or spraying the water, softens the particles. This presoftening action facilitates the washing away of the particles. The water layer surrounding the particles diminishes the mutual attraction between the particles and surface of the façade. This is the immersion process discussed earlier. The action can be accelerated by adding a surface active agent, a surfactant or detergent. The dirt may actually absorb water, increasing the diameter of the particle, which means that less force will be required to dislodge it from the substrate. In the second phase, the bond between the particles and the substrate is broken. This can be accomplished either through the brushing off of the particles or by applying a low-pressure stream of water to the surface. Water mists are highly efficient in dissolving soot crusts on stone because the tiny droplets suspended in air have a large specific surface, and create a large interface when they are deposited on the stone.

Water can also sometimes dissolve the outer skin of the surface of the façade, thereby destroying the bonding agent of the particle to the substrate. For example, in pure rainwater, hardened cement paste may dissolve slightly. (Chemicals are applied based on the same dissolution principles.) Typically, water is not expected to be able to dissolve a portion of the surface. This points out that a water application is not truly an innocuous cleaning method. Water can also dissolve the cementing products of dirt.

For example, soiling particles often adhere to the surface by plaster (gypsum, $CaSO_4 \cdot 2H_2O$), which has been formed through the reaction of SO_2 from the air and the lime located within the substrate. Because plaster is partially soluble in water, it is possible to dissolve the plaster. The dirt is loosened along with the surface.

Water to be applied to the building must be collected at the base of the wall and be directed to rainwater drains so as not to saturate the foundation and lower portions of the wall. Normally, washing is done from the top to the bottom, as to get maximum use of water. (In contrast, chemical treatments should be applied from bottom to top, to avoid double treatment, resulting in pattern stains.) Water is delivered through *lances*, using either high- or low-pressure. The shape of the lance nozzle is an important consideration. "Pencil" nozzle jets should not be used; fan-spray tips of a minimum 30 degrees are generally recommended for architectural cleaning. The cutting action of the water lance is useful in removing stubborn patches of dirt; however, they must be used with care, as it is easy to cause loss of the masonry surface or joint material. Light levels of loosely adhered atmospheric soiling and moderate to high levels of organic soiling on limestone can, at times, be removed using just low-pressure water lancing. High pressure is reserved for severely or deeply soiled sandstone. In most cases, a basic softening must be done before any soiling can be removed.

The erosive cutting power of the jet of a pressure washer depends on the pressure produced by the pump, the flow rate, the diameter and shape of the nozzle, and the distance between the nozzle and the surface. The equation is PQ/a: P is the pressure at the pump, Q is the flow rate of water, and a is the area of the masonry in contact with the jet of water. Used at up to 20 to 1,500 psi, the pressure should be only as high as needed to soften the particles. Acceptable maximum pressures will depend on the type of stone involved, its condition, and the condition of associated pointing and masonry. But note that pressure that may be acceptable to masonry units may be unsuitable to the surrounding mortar joint, which although weathered, may be otherwise in acceptable condition.

To complicate matters, high and low pressure are defined differently by various industry professionals. For example, a representative of the building cleaning industry might consider high-pressure water cleaning to be anything over 5,000 psi, reaching as high as 10,000 to 15,000 psi. Water under this much pressure might be necessary to clean industrial structures or machinery, but it would destroy most historic building materials. Industrial chemical cleaning commonly utilizes pressures between 1,000 and 2,500 psi. A conscientious cleaning of a historic structure could be conducted within the range of 20 to 100 psi at between 3 to 12 inches from the substrate. Cleaning at this low pressure requires the use of a very fine (00 or 0) mesh grit forced through a nozzle with a one-fourth-inch opening.

That said, some professional cleaning companies that specialize in historic masonry buildings use chemicals and water at a pressure of approximately 1,500 psi; while others recommend lower pressures, ranging from 200 to 800 psi, for a similar project. Some historic structures can be cleaned properly using a moderate pressure (200 to 600 psi), or even a high pressure (600 to 1,800 psi) water rinse, but they are generally exceptions to the rule.

To reiterate, then, the main problem in using pressure-related cleaning methods stem from the differences in objectives of the various commercial, industrial, and architectural building cleaners, as well as the nature of building materials at issue. No one cleaning formula or pressure is suitable for all structures and so decisions regarding the proper cleaning process must be made only after conducting a careful analysis of the building fabric, and sample testing. It has even been suggested by some that classifying by numbers the proper water pressure for masonry cleaning is irrelevant.

When steam is used as part of the water-cleaning process it is generated in on-site equipment then directed against the masonry surface through a low-pressure nozzle. The steam and condensed water soften and swell the dirt deposits, which can then be flushed from the surface by the pressure of the jet stream. Low volumes are used to prevent or at least minimize staining.

Steam has proven to be a viable choice for cleaning highly carved surfaces without mechanical damage. It is also a useful method for removing greasy or tarry deposits, chewing gum, wax crayon, and for killing mold and algae on damp surfaces. Steam is not useful for simply removing dirt.

Steam cleaning is generally not harmful to stone, unless the stone is particularly soft, but there is an ongoing controversy regarding how the temperature of steam may affect the stone surface. Often this is not a problem, but before steam is used on valuable or sensitive surfaces, its effect should be investigated.

Overall, water cleaning is the simplest and least harmful to the building fabric, but large quantities of water could damage or instigate further deterioration to the whole building. With that in mind, before instituting a water-cleaning program, all joints, including mortars and sealants, must be determined to be sound to minimize water ingress to the interior. Faulty joints and fissures could allow water into the building, thereby potentially facilitating rusting of hidden iron cramps and fostering the propagation of dry rot in masonry or where the water has come in contact with timber end beams, bond timbers, and timber paneling. In addition, porous masonry may absorb excess amounts of water during cleaning, causing damage within the wall or on the interior surfaces. An increase in water volume or pressure could induce interior flooding, damage to the interior lining of the building, or the electric wiring. In dry climates, the water may evaporate inside the masonry, leaving the salt just below the surface to activate the subflorescence mechanism.

Decayed stone, especially small-scale details, are very vulnerable to washing with water sprays, as the process can dissolve sulfated surface areas. Where the sulfated skin is thick, this can mean unwanted losses. The spray water technique should not be utilized in winter, because it could activate freeze-thaw cycling, and damage the fabric through spalling and cracking. Frost may also occur and impart up to a 10 percent expansion in pore volume. Moreover, because a wall may take several days to a week to dry, no water cleaning should be permitted for several days prior to the first predicted frost date—even earlier if local forecasters predict cold weather. The use of hot water raises the risk of inducing thermal shock to some masonry, especially at extremely high temperatures, which could affect glazing putty and plastic fittings.

One of the most damaging consequences of water washing is the activation of the salts within the building fabric, which usually is the result of excessive water saturation. The force of the water could also push the existing salts further into the building material, which would cause severe damage once the salts dehydrate. Following hydration, salts beneath the surface will generate the loss of the building fabric through spalling and cracking, that is, subflorescence. The result is the deposition of both efflorescence and subflorescence; and unless the salts are removed, the process will be on-going, exacerbated by the wetting/drying cycle.

Overuse of water to clean can produce brown and orange stains that appear as the surface dries out. It is believed that the primary cause of this type of staining is the high iron content within the water. Stones typically affected by this type of staining include limestone and white marble. Subsequent rainwater washings can, but do not always, reduce staining. The brown staining can appear in irregular patterns, dictated by the previous soiling pattern or natural variations in the stones. The staining of limestone surfaces may relate to the take-up by individual stones of soiled water that is allowed to run down from the surface being wetted above. This is particularly problematic at the early stages of a water-cleaning operation. Thus, it is not always easy to achieve a uniform appearance over a light-colored limestone building that has been cleaned by water washing.

Extended periods of water saturation on sandstone surfaces can produce staining to a considerable depth, a result of mineralogical color changes within the stone. Usually it is not possible to remove such soiling without also removing excessive amounts of the masonry substrate. Frequently, the depth of stone affected by the residual staining will mean that even surface removal by abrasive cleaning will not eradicate enough of the stain, and the marks are considered permanent.

Postcleaning staining of white limestone and marble stone surfaces is frequently attributed to the dissolution of the surface or subsurface soiling, its migration into the stone in liquid form, and its redeposition on the stone surface during the drying-out period. The staining is not always superficial or water-soluble, and can be difficult to eliminate.

Copper stains are usually dilute solutions of copper sulfate or copper chloride that have run off from a cuprous surface (copper, brass, bronze) above. When such a solution comes in contact with limestone or mortar, the reaction produces a blue-green copper salt, which permanently stains the porous building material. Rust from reinforced concrete or runoff from iron or steel typically produce dark, intense stains. Both types of stains are difficult or even impossible to remove without damaging the substrate.

Abrasive Cleaning

When the water-cleaning method is not sufficient to remove hard, encrusted dirt, mechanical, or abrasive, techniques are employed. Mechanical cleaning directs particle abrasives onto the surface through a directed force of air and sometimes water. Typically, it is called *grit blasting;* it operates by abrading the dirt off the surface of the masonry, rather than reacting with the dirt and masonry as in water and chemical methods. The abrasives do not differentiate between the dirt and the masonry, so some erosion of the masonry surface is inevitable, hence this cleaning method has a very restricted application on both limestones and sandstones, and should be used on brick only in exceptional cases.

Erosion and pitting of the building material by abrasive cleaning creates a greater surface area on which dirt and pollutants collect. Needless to say, the building fabric thereafter attracts more dirt, and requires more frequent cleaning in the future. Also, the roughened surface will henceforth have a slower water runoff rate, meaning that water that would previous to cleaning have left the surface will now most likely be absorbed into the masonry, to activate numerous mechanical failures.

Despite these potential hazards, abrasive cleaning can be a viable choice in certain circumstances, as long as the risks are understood and mitigated. Metals, for example, particularly steel, can be quite satisfactorily cleaned with blasting techniques. In the past 20 years, many new types of grits have been developed that are more selective in the range of particles they can lift from a surface. One technique that uses glass beads projected at such a high a velocity that they shatter on impact and, thereby, protect the substrate from overabrading.

But whenever abrasion is used, a system of tight controls and careful supervision must be instituted. The technique is inherently abrading, and the level of erosion of the substrate is directly related to the time spent on a given location. Therefore, duration of the cleaning cannot be based on results; the exposure to the abrasive must be time-limited. The results are whatever can be accomplished in that period. If the abrasive is directed too long on a tough spot, it will erode a hole or pit in the substrate. For these reasons, many in the industry seek to eliminate abrasive cleaning from the inventory. In fact, the Secretary of Interior's *Guidelines for Rehabilitation* flat-out discount the use of abrasive cleaning, with no exceptions. (Remarkably, however, those same guidelines allow the use of chemical cleaners with concentrations as high as 40 percent phosphoric acid.)

The cleaning action of abrasives is accomplished by the impingement of the particles, by which a part of the surface layer also may be pulverized and carried off together with the dirt. The dislodging of the dirt deposits thus takes place by the breaking-up of the surface layer beneath the deposits to a potential depth of several millimeters. (The removal of a superficial dirt layer may take off a fraction of a millimeter of the stone, but the removal of grime-impregnated stones could require the removal of several millimeters.) Such a treatment could lead to massive destruction of the substrate if the crust is thick. The protective skin of the fabric will be removed, and reveal the softer interior, which will make the cleaned surface more vulnerable, as noted earlier. When the surface hardness is not equal over the whole area, the softer parts will erode, leaving an uneven-textured appearance. In short, the stone is damaged and unsightly.

Dry blasting may cause pitting of the surface if the stone is too soft or if the wrong abrasive is used. Moreover, erratic movements of the "gun" leave a mottled effect, especially with wet blasting. Disc cleaning may leave scour marks or wavy arises, and so requires a very high level of skill to be done successfully. The impact of the grit particles tends to erode the bond between the mortar and the brick, causing cracks or enlarging existing cracks, where water can enter. This then can cause joint shadows or loss of a joint detail.

By using a wet-head gun, dry abrasive cleaning may be adapted to a wet process, which introduces water into the air and abrasive stream. Water may also be introduced inline. This type of cleaning generally operates under the same parameters as dry blasting. Wet blasting, like dry blasting, potentially will remove a portion of the surface. The addition of water also raises similar risks as those discussed above, including oversaturation, the ingress of salts, and production of efflorescence and subflorescence, freeze/thaw cycling, and so on. The primary additional consequence is that the slurry generated from wet-blasted masonry surfaces makes this type of cleaning very messy and fairly undesirable. To eliminate the slurry, copious amounts of water may be needed to continuously rinse down the surface.

Historically, wet blasting had been considered the more sympathetic means of abrasive cleaning. The addition of water into the air stream was thought to reduce the impact of the abrasive force, minimizing surface loss and pitting effects. In fact, the inverse is true. Evidence mounts that proves the addition of water into the air stream increases the cavitation action of the jet cleaning rather than reducing it.

Another mechanical, abrasive technique, called dry microblasting or the pencil abrasive technique, is based on the same principle as grit blasting, but the abrasives used are considerably smaller than the typical grits. Fine powders of materials such as aluminum oxide and glass microbeads are used. Because these particles have a lower specific gravity, and are softer, they can achieve a milder mechanical cleaning action. The nozzle used is very narrow, similar to a writing instrument (hence the term pencil abrasive). But this method is time-consuming, and requires expert skill, so it has, to date, largely been applied in the museum conservation setting, specifically for sculpture cleaning.

All of these methods require an operator to control the nozzle; therefore human error and inconsistency are often the principle causes of damage. The operator determines distance, air pressure, water pressure, and/or abrasive level, so for example, if the operator steps too close to the surface, the blast will give an uneven appearance to the fabric.

Chemical Cleaning

The third alternative to a simple water application is the use of chemical solutions. Chemical agents dissolve or destroy the cohesion of the dirt to the surface. Acids transform the dirt from a fixed state to a movable state, which facilitates washing away with water. In any case, the façade must be saturated with water beforehand, to minimize the ingress of the chemicals too far into the surface. Following application of the chemicals, large amounts of water are necessary to thoroughly rinse the surface of any remaining chemicals. Any left behind may cause the appearance of efflorescence, which will impart salt decay.

Conservators seem to prefer chemical cleaning processes, even though there are marked risks associated with their use, not only to the substrates at issue but to other parts of the building, notably exposed metals and, more importantly, embedded metal anchors and ties. Acidic cleaners or alkaline cleaners are the chemicals of choice. The most common acids used for cleaning are hydrofluoric acid, ammonium bifluoride, hydrochloric acid, phosphoric acid, sulphuric acid, and acetic acid. But not all acids are suitable for all substrates, and no acid is suitable for contact with metals. Professionals must investigate carefully to find the least dangerous acid.

Some acids contain salts, and others may react with salts already present within the substrate. For example, hydrochloric acids may produce chlorides that are particularly hygroscopic, hence will increase the moisture level of the fabric. The most common alkalis used for cleaning are sodium hydroxide, potassium hydroxide, ammonium hydroxide, sodium carbonate, and sodium bicarbonate. All of the alkalis have the potential to deposit salts into the masonry.

As stated, before applying a sufficiently diluted chemical cleaning solution, the surface must be wetted, to provide a protection for the masonry. (Note we say "a protection," but it is far from absolute.) The chemical solution should be applied immediately after soaking. Once applied, the chemical solution sits on the surface and acts upon the soiling, theoretically without reacting with the substrate. "Dwell time" will vary based on the various chemical solutions, in conjunction with the substrate at hand. Following the dwell time, a thorough rinse of the solution is mandatory. The recommended rinse time is, generally, much longer than most contractors want to give, so supervision and quality control are imperative at this stage. As stated, if the chemicals are not thoroughly rinsed away, their residue may activate varying chemical and mechanical deterioration mechanisms.

LONG-TERM EFFECTS OF CLEANING

A study conducted in the early 1990s in London revealed that the resoiling of previously cleaned buildings was rapid. The light reflectance of a surface at an urban site can fall to 60 percent of its initial value only five years after cleaning. Subsequent studies designed to measure soiling rates within the urban environment have generally established that soiling appears to be very rapid during the first period of exposure after cleaning (a decrease of 32 percent of the original reflectance, within seven days). Suspended road dust and soil-derived particles may be deposited on materials, but are more likely to be removed by rainfall than deposited diesel particles. Calcareous stone buildings in urban environments, for example, may need to be cleaned every 5 to 10 years. In rural areas, the period may extend to 10 to 15 years or more.

These findings imply that the benefit of cleaning may have a concomitant disadvantage of simultaneously exposing the substrate to accelerated redeposition of particulates especially. We can speculate as to some of the reasons for this. The original surface of the substrate, whether it be stone, brick, or concrete, was the least defective surface of the material. In some cases, the soiling process may develop a self-limiting crust; in other cases, the irregularities of the surface and the exposure of unprotected substrate may become filled with particles. When the crust and particles are attached to the building, chemically malignant though they may have been, they are relatively inert; the chemical activity has been partially or completely neutralized by interaction with the substrate. The crust, while not totally benign, is not on the whole necessarily erosive or damaging. In some cases, such as with the freeze-thaw action, the crust may act as a wet sponge against the substrate, which can exacerbate frost penetration and salt migration. We know that some crusts include concentrations of dehydrated sulfuric and sulfurous acid, which can activate when they are hydrated.

Simply put, it is not conclusive that removing the accretions and depositions on buildings brings an unmitigated benefit. The net long-term advantages of cleaning are far from proven. The substrate when reexposed to the elements, is not, obviously, the same pristine material it was originally. And the surface is degraded, regardless of the technique employed in the cleaning process, thereby increasing the rate of redeposition, and making the substrate more vulnerable to continued attack. So why do it?

As pathologists, our job is to offer *all* the options to our clients (not only those that are politically correct or professionally advantageous), to explain the consequences and costs, to make recommendations, and then to execute *their* decisions to the best of our ability, even when those decisions are not the ones we would make.

4.4 Systemic Pathology

We've defined the vertical closure system as a selective barrier. As a system, the vertical closure must exclude weather, but allow for light to enter the occupied zone when desired. It must safeguard against unwanted intruders, but be welcoming to guests and inhabitants. It must repel heat and humidity, but not trap those elements. It must exclude pollutants, but admit fresh air. In short, a vertical closure system must meet a rather large set of seemingly conflicting objectives. However, the conflict is more apparent than real, because all those objectives rarely must be met at the same time. The vertical closure system, more than any other functional system, must be able to adjust, to accommodate a constantly changing list of criteria. This means that certain of the criteria must temporarily be compromised in any given situation.

This section examines those aspects of this set of objectives that can or most likely will result in the progression of a deterioration mechanism of the system and, therefore, of the building as a whole. We begin by discussing three broad and very important aspects of vertical closure: water exclusion, heat modulation, and vapor transmission. We proceed to develop a typology of walls, which will allow for the analysis of any wall or closure system. We will find that the issues associated with openings in the system are no different from those associated with the nonoperating portions of the system. We close the section by addressing a couple of surface-related issues unique to the vertical closure, namely vegetative growth and surface coatings.

4.4.1 How Walls Leak Water

Understanding how walls leak, water in particular, requires that we examine walls typologically rather than encyclopedically, because no two walls are exactly alike, precluding a catalogue approach. To start, it is important to emphasize that walls do not leak because of the way they were designed, detailed, or constructed. They do not leak because they are made of this material instead of that material. They leak because water passes through them from the outside to the inside in a manner determined to be potentially damaging to the fabric or the contents, or that impacts the comfort of the occupants. For a leak to occur three conditions must be in place: water at the exterior surface, a motive force to move it through the wall, and sufficient time to transport the fluid. The necessary and sufficient conditions for a leak to occur include those three conditions, plus a path from the wet side to the dry side.

We will concentrate initially on the transportation of water through the solid portions of the wall because that is the path that leads to the greatest potential damage to fabric. Most water enters the building through joints between materials or at junctures of change in geometry or of the pieces comprising the construction. But whether the water enters through the fabric of the wall or through a construction joint, the necessary and sufficient conditions are the same: water plus path plus motive force plus time equals leak. All wall systems attempt to mitigate or eliminate one of those four factors. The success or failure of the system to achieve the desired effect is, therefore, not necessarily inherent in the design or the construction. A leak may even be an acceptable compromise in the face of the cost to preclude the leak.

The preceding is not to say that we, as an industry, categorically do not know how to prevent leaks in buildings. Our problem as building pathologists is more one of figuring out how to unravel the cause of leaks of earlier, in some cases even archaic, designs. Obviously, we cannot, economically, disassemble every wall that has a leak in order to determine the construction detailing, much less to track the path of the leaks. Indeed, many of the leaks we face come through materials that have become historically significant; thus, destructive investigation adds cost in terms of lost fabric.

We need a theoretically based model to assist designers in the future to prevent leaks and to aid pathologists to detect and deal with active leaks. The model must start with the water and motive force. Earlier in this chapter we inventoried the motive forces that move water through a porous solid. These forces are the same without qualification or exception. Water cannot traverse a plane, even a porous plane, without a motive force, with a component vector perpendicular to that plane.

Of all the potential motive forces that can theoretically move water through or into porous solids, the dominant one is differential pressure. In fact, at the macroscale of the wall as a system, there is no other significant force. Consequently, every approach to water exclusion that we will examine is, essentially, an effort to defeat differential pressure and, therefore, water transportation through the barrier. Indeed, the types of walls we discuss are sorted based on the approach taken to defeating differential pressure. In the end, that approach will have more to do with the success of defeating differential pressure than the selection of specific materials or construction technologies.

4.4.2 Wall Types

There are two general classes of walls relative to defeating differential pressure: single-layer and multiple-layer systems. Each comes in two varieties: for single-layer systems, they are thin and thick; for multiple-layer systems, they are drained cavities and pressure-equalized cavities. All of the varieties of wall construction share the functional objective of water exclusion. And the typology presented here emphasizes as a means to achieving that objective establishing an intermediate objective: to defeat the motive force that moves water through the barrier. We will address the water exclusion aspect first. Each variety is characterized by the approach taken to exclude water from the interior volume, not the specifics of the construction. Often, from the standpoint of materials employed and construction details used to bond the materials, very little distinguishes one wall from another. The very small details, however, can and do radically distinguish how to defeat differential pressure and, therefore, how to exclude or not, in some cases, to exclude water.

> **Note:** Although we focus on vertical closure primarily from the perspective of the wall, it is important to recognize the other major components of the vertical closure system. At a minimum, those include doors and windows. Following our exploration of wall systems, we will return to these subelements, at which point we will discover that they are not so different from walls except in scale. The same principles of water exclusion apply to large or small walls, which just happen to be called windows or doors.

SINGLE-LAYER SYSTEMS

All single-layer wall systems use the skin of the wall itself to defeat differential pressure and, therefore, the intrusion of water. If water penetrates the outer layer of the wall, the water enters the "occupied zone," or, at least, the interior finishes of the occupied zone, because there is only one layer to the wall, in which case the system has failed to provide a barrier against water (see Figs. 4.9A and 4.9B). Though an interior collection device may be in place, generally when water breaches a single-layer system, the system is considered to have failed.

Again, single-layer systems come in two varieties, thin and thick. Thin systems, in turn, also come in two varieties: permeable, which are more suitable for temporary structures, and impermeable, which are more popular for permanent structures. The best example of a thin, permeable, single-layer system is the tent. We will not dwell on this system, because tents are seldom the site of construction pathologies. In contrast, the thin, impermeable, single-layer system acts as a "perfect" membrane, and as such need not be any thicker than is required for fabrication, installation, or to satisfy some other selective barrier function, such as physical security.

FIGURE 4.9A Water rolling over the shallow gutter (at arrow) has saturated the substrate, removing the clay lime mortar, and initiating a wholesale collapse.

FIGURE 4.9B   The same wall from Figure 4.9A displays a complete disengagement and lateral displacement of more than an inch. Catastrophic collapse is imminent.

Examples of common perfect membrane walls include glass curtains, aluminum or porcelain enamel panel systems, or other impermeable panelized systems. Any impermeable material satisfies the category requirement; for example, most window systems, as we typically understand them, are thin, impermeable, single-layer systems.

Thick Single-Layer Systems

Thick single-layer systems may be deceptive in that they may be constructed of several layers of material. For example, solid masonry walls are considered thick single-layer walls even though there may be more than one wythe of masonry that comprise the construction of the wall. Regardless of how many layers of brick or stone make up the wall, it is the solid mass of the masonry that stands between the water source outside and what we hope is a dry interior. The point is, if the solid masonry, however thick it is and however many wythes are employed in its construction, is all there is between the interior and the weather, then it is a thick single-layer system.

Thin Single-Layer Systems

Thin single-layer approaches to water and weather exclusion comprise some of both the oldest and the newest walls in the inventory. We have been putting animal skins and woven fabric between us and the weather for a long time in the form of tents and lean-to's. More recently we have been manufacturing large panels of thin impermeable materials and joining those panels to make continuous sheets of durable membranes. The pathology of such impermeable thin systems is relatively simple, but economically significant because they are pervasive, and they tend to be found in large constructions. And, they also all leak eventually.

Permeable Single Layers

Recall that permeable means that the material can and will allow the passage of both gases and liquids only if there is a motive force to cause the movement. In the case of gases, the motive force can be as slight as a differential in the concentration of a given gas across the barrier. The differential partial pressure of the gas from one side of the permeable barrier to another is enough to cause molecules of the gas to pass from the higher concentration to the lower concentration. Even though the apparent force to cause such a movement is slight, it is, nonetheless, required for movement to occur.

The same is true of fluids. The fact that one side of the fabric may be dripping with water does not mean that the system has failed as a barrier to water. Consider shower curtains: Although they are not so popular as they were before the development of inexpensive impermeable sheet goods, there was a time when shower curtains were made of woven cotton or linen cloth. Today, we tend to use these as decorative outer curtains, but not so long ago they were *the* shower curtains. A cloth shower curtain works because the force of the shower spray is not enough to traverse the porous properties of the fabric. The water may wet the curtain, even saturate it, but the water stays inside the tub or stall, which is the objective. Ironically, a cloth shower curtain is even more effective as a barrier when it is wet than when it is dry to because the porous space is actually smaller when the threads are saturated and, therefore, hygroscopically expanded. The adhesion of the water to the surfaces of the thread also acts as a film on the surface of the fabric as a whole. And at each of the individual pores, the water acts as a partial block to the pore and a passage through the pore. If the curtain is actually dripping down the inside face, that sheet of fluid can theoretically act as a barrier to vapor passage, as well as inhibiting fluid passage through the curtain.

We can learn a lot about leak prevention from the cloth shower curtain. First, the fact that the curtain gets saturated does not mean it has failed. Second, the barrier need not be a perfect membrane to exclude water. These principles can be extended to tents. Long before nylon and Goretex, tents were made of canvass and other woven fabrics. In the nineteenth century, we began coating the canvass to produce oil cloth, which moved us closer to true membranes. The point is—and it is an important one—porous materials, even in the presence of water for protracted periods of time, do not necessarily spring leaks. Something more is required: the motive force to traverse the plane of the fabric. With shower curtains and tents, there is insufficient differential pressure from one side of the curtain to the other (thin though the fabric may be) to suck the water through the cloth. Shower curtains operate in environments in which the air pressure on one side is the same as on the other, because there is, generally, a large connecting path between the inside of the tub or stall to the outside. Thus, the air pressure is equal on both sides. Tents, on the other hand, are quite drafty. Although there may be external reasons for the potential development of unequal pressure from the outside to the inside, that potential is quickly dissipated by the capability of the standard tent to immediately admit any passing air current and, thereby, equalize the pressure inside with the pressure outside. If there is no pressure differential, there is no motive force, thus there is no leak.

The shower curtain principle does, however, have its limits. If we use screening material as a shower curtain, we will get wet, because the pore space and diameter are large enough that, upon impact by a drop of water, the momentum of the drop striking the screen will be enough to propel some of the water through the screen and onto the window sill or on us. Screening material is effective as an insect barrier, not as a rain barrier, but only because of the spacing of the pores. If the space of the inherently impermeable wires were drawn down to a very small pore space, to become fine openings through which vapor, but not liquid water could pass, then we would have an effective water screen.

If we can weave a fabric with a sufficiently small space between threads, and those threads are themselves hydrophobic, then a fabric that is permeable to vapor can repel liquids. Goretex is such a fabric. It is constructed by wrapping a Teflon filament around a thread of another fiber, such as nylon or rayon and then weaving the wrapped threads into a tightly woven fabric. The resulting product is highly flexible, impermeable to water, permeable to vapor, and economically feasible. The long-range problem of the material will be recovery and recycling, not disintegration.

Though we have not as yet erected a building made of Goretex, we are wrapping buildings in a material called Tyvek, which performs in a similar manner. It is a pressed synthetic material that is porous. Tyvek is permeable to vapor and only semipermeable or impermeable to liquid—or at least that is the claim. Assuming Tyvek can perform as advertised, why would we want to wrap a building with it, particularly given that it is not free? Theoretically, a selective membrane such as Tyvek allows the building to "breathe." The idea is that the selective barrier will allow vapor to pass through the membrane and retard liquid passage. If water does get inside the barrier for whatever reason (presumably other than through the fabric), it can evaporate back out through the fabric, which *is* permeable to vapor. Any liquid that reaches the fabric from the outside is repelled, so less water requires evaporation. Unfortunately, the best thing that can be said for the sale and application of the selective barrier wraps is that the promotion departments of the manufacturers of such products are more effective than their products. Either the water is getting into the wall or it is not. If it is, then buying more building wrap will not save the situation; if it is not, then buying more expensive building wrap is a waste of money.

Impermeable Single Layers

Thin, impermeable, single-layer systems comprise a huge class. We do not build whole structures using this approach; rather we use it for parts of buildings. The theory is simple. Outside is rain, wind, sleet, and the gloom of night; between the outside elements and the inside is a *perfect*, impermeable barrier to both liquid and vapor. As long as the barrier is perfect, it matters not an iota that there is a pressure differential between the outside and the inside. The pressure on the inside can drop as low and as far as will support life, and as long as the membrane is perfect, not one single molecule of water will enter our domain. Not only is the theory simple, it is as flawless as the membrane.

The theory acknowledges that pressure differentials exist; it acknowledges that water may be running down the face of the membrane for eternity. But it does not matter, because when the membrane is perfect, there is no connecting path by which the water can traverse the plane. Impermeable membranes defeat differential pressure by eliminating the pathway.

Sadly, there are two essential flaws with the perfect membrane approach. One is a short-term, even immediate, problem; the second is a longer-term problem, in some cases measured in only months. The immediate flaw is that any system predicated on perfection requires more than perfection of materials. Perfect, impermeable systems also require perfect design, detailing, fabrication, delivery, and installation. Inasmuch as these perfect systems are also thin, there is limited dimension in which to work to achieve all this perfection.

Anyone who has watched a trade draw a sealant bead knows how vulnerable perfection can be. If the sealant does not bond properly, or is too thick or too thin, or too cold or too hot, the perfection is compromised. If the panel is warped or the edge is crimped or gets hammered into position, perfection is at risk. If the spacing is greater than the gasket can bridge or the gasket has not been cleaned with the proper solvent or the trade's glove had some grease on it, perfection takes a break. The only good news is that these sources of imperfection are often detected before the building is delivered, but not always. The flaw exists from the day the system is put into service. It may not manifest immediately, although it will, generally, manifest quickly.

Once flaws are detected, they can be repaired. This triggers two issues, one unique to perfect membranes, the other, to detection, in general. The general issue of detection we will defer to a subsequent discussion on the difficulties of tracking leaks once they are manifest. The more immediate and direct issue relates to the repair of perfect membrane systems. To repair a flawed perfect membrane is to introduce a point of discontinuity in the system that was neither intended nor designed. The only way to repair a flawed system predicated on perfection is to disassemble the wall and start over. If the "wall" is a single eight inch by eight inch pane of glass, we may very well do just that. If, on the other hand, the flaw is the unbonded corner of a gasket in a 20-foot-square display window, we will squirt something with the word "Magic" in the brand name into the gap and stare away. Repaired flaws are weaknesses in any system if for no other reason than the introduction of discontinuities into the fabric.

The fundamental flaw of all systems predicated on perfection is the presumption that the system as installed, perfect though it may be initially, will never change. To repeat the mantra of this book: All materials deteriorate; all joints move; all systems change. Sooner or later, perfection weathers. Well-designed membrane systems take this failure into account, and allow for replacement of deteriorated components and reconstitution of the membrane.

A common example of such a component is sealant. There is not a sealant on the market that does not deteriorate over time. There is not a sealant manufacturer who claims otherwise. Designers tell clients that sealants must be maintained on a schedule not unlike that for painted surfaces, meaning that at some point they must be replaced. But there is not a building owner on record as maintaining and replacing sealants on that schedule. As sealants weather, they typically become harder and less elastic; they crack or debond from the substrate. The resulting fine-line crack is a perfect channel for water to enter if a pressure differential exists across the crack. And because the system was designed without regard to pressure differential, there is a distinct prospect that such a condition will develop, at which point all the necessary and sufficient conditions are present except the time it takes the water to traverse the crack, which for optically sized cracks can be met in a few seconds after the other conditions are in place. A cracked sealant in a thin membrane wall is a leak waiting for the rain.

As the pieces within the assembly move, due primarily to thermal expansion, they work joints, which means stressing gaskets, adhesives, and seals. Given long enough, the basic materials will deteriorate. Metal panels will pit and corrode. Glass will crack. There is simply no way to put a perfect membrane into an open and hostile environment and expect anything other than deterioration to occur. As soon as the perfection is compromised, the system is vulnerable to infiltration and water intrusion.

Thick Single-Layer Systems

While the development of thin, impermeable systems is relatively new, we have been building single-layer walls, more commonly referred to as solid walls, for a long time. Solid walls are typically constructed of masonry, either brick or stone, but solid wall principles also apply to chinked log, rammed earth, or adobe constructions. We have been piling up blocks and rocks for centuries, in the name of providing shelter, and many of those systems are variations of thick, single-layer construction. The materials and the joinery are, however, far from perfect membranes, so how do they work to defeat differential pressure? They do it through the manipulation of the time element. Inherently imperfect materials connected with imperfect joints cannot begin to act as true impermeable membranes, but they can be built to leak slowly enough to spare the inside of the building from cascading water.

Understanding how thick, solid walls work is not difficult, as long as we do not equivocate over semantics. Solid walls will leak, just not always through to the inside. Because they are generally successful in preventing water from entering the occupied zone, it is tempting to say that solid walls are not inherently inclined to leak. In fact, solid walls can be designed only to minimize the occurrences of bulk water intrusion, but it is really just a matter of time until the conditions are right for a leak to activate.

Currently all solid wall constructions are imperfect barriers. They all have pore paths that connect the outside to the inside. These paths vary considerably in size, and if they are small, it can take a considerable amount of time before any fluid is pulled through. If a rainstorm lasts less time than required to pull the water through the pore path, the wall begins to dry via evaporation at the very surface from which the water was supplied. Inasmuch as the interior surface may also provide an evaporative plane, it is not unusual for solid walls to wick lingering moisture to the inside surface, and for that water to evaporate at that surface long after the rain has dissipated. Thus, one of the aggravating aspects of solid walls is that the damage is not manifest until after the motivating origins have been removed. Occupants may complain about plaster damage on solid walls at the same time they deny vehemently that the wall leaks.

When evaluating solid wall construction, therefore, it is important to remember that while it rains, the wall will soak up water, and though it will slowly dissipate and evaporate, it will be wet inside the wall for some period of time. As a protection barrier for the interior, the solid wall performed its function, nevertheless, it is wet. For this reason, historically (especially in damp climates) solid walls have been constructed of materials that are relatively resistant to decay when they get wet. Brick and stone constructed with lime- and cement-based mortars satisfy that criterion quite well. Chinked log construction does not work as well, but given the economy of using logs at the time such construction was popular, the cost of the occasional rotted log was little more than walking to the woods and cutting down a replacement. All compacted earth and unfired mud construction, such as adobe, is less than ideal, too, but it has the redeeming qualities of low cost and high fire resistance. And as with any thick, single-layer system, there is some delay between the onset and intrusion of water. In earthen constructions, the race is actually more against intrusion into and disintegration of the fabric itself. Following seasonal rains in areas where adobe construction is prevalent is a period of wall repair. Still, the damage to the solid wall is an acceptable price relative to importing brick from abroad. In short, any solid wall can be made to work if the limitations of the materials and the environment are taken into account relative to the available resources to entertain other options.

During the period in this country when common construction consisted of assembling stones in random-rubble walls using a mortar that was a mixture of clay and coarsely ground lime, the masonry was generally not coursed, because the surface was not intended to be displayed or to resist weather directly. This class of solid wall construction was always intended to have a weather-resistant surface, generally a cement-based stucco or rendering. The presence of clay-lime mortar and random rubble means almost certainly that the building was originally coated with an exterior plaster, render, or stucco. Any building of this construction still standing in a damp climate after 200 years has not survived on the strength of the clay-lime mortar when exposed to water.

For similar reasons, many chinked log buildings were clad or stuccoed with a second, less vulnerable, weather surface. In some states, such as New

Jersey and Pennsylvania, some houses or portions of houses are still constructed of chinked logs, but are covered with clapboard. Unless the clapboard is removed to expose the log construction, there is little evidence that the solid wall lies below the surface.

The practice of applying a weathering surface to an otherwise relatively vulnerable solid wall continued unabated until late 1945. Shortly after World War II, and as a direct consequence of the widespread expansion in the use of Portland cement and aluminum production, particularly the latter, the cost of applying a second surface to standard rowhouse construction dropped so low that Americans began to clad our cities with cement stucco, disguised to look like stone, and aluminum siding, disguised to look like clapboard. The trend continues to this day, with the addition of pressed vinyl and synthetic stucco to the inventory of surface covers.

Though many in the industry denounce these claddings, they do work. Of course, any cladding, clapboard and cement-based stucco included, can be installed incorrectly, resulting in damage to the substrate, but the failure lies in the design and application, not the material.

Of cement-based applications intended to resemble stone masonry, for generations, the most widely used product was Permastone, in use so long and pervasively that the term has entered the vocabulary as a nonproprietary descriptor of the generic product, permastone. More recently, synthetic stucco applications called "brick face" have replaced permastone as the render of choice. Like cement-based stucco applications, the addition of permastone does not change the classification of the thick, single-layer wall.

Synthetic, "brick face," stucco is not cement-based, and is neither porous nor permeable. The vehicle of such products initially were epoxy resins, but today may be any number of styrenes or other thermoplastic derivatives. The application of synthetic stucco *does* change the classification of the wall from a thick, single-layer system to a thin, single-layer system. Whether intentional or not, as soon as an impermeable material is added to the exterior surface of a porous, permeable substrate, the approach to defeating differential pressure shifts from a delay tactic, characteristic of thick systems, to a path destruction tactic, characteristic of thin, perfect membranes. To understand how walls work, it is essential to remember that what distinguishes one system from another is not the name or the description, but how the system defeats or damages property. (We will discuss the details of such systems in a subsequent section dedicated to coatings and paints.)

Aluminum and vinyl siding are an altogether different story. They are not restricted to use on brick and stone; they are applied as second surfaces to wood-clad buildings as well. Furthermore, they are not restricted to use as add-on applications; thousands of new houses are being constructed with aluminum or, more likely, vinyl siding as original materials.

Wood clapboard, aluminum siding, and vinyl siding can all be installed in one of two ways, as thin, single-layer systems or as part of multiple-layer systems. (The mechanics of the multiple layer system follow in the next section.) In the single-layer application, the siding is sealed at all its edges and laps, to form a thin, impermeable, perfect membrane. This is rarely the intention of the installer, but it happens, nonetheless. Sidings generally perform better as part of a multiple-layer system for reasons we will explain in the next section. The point here is the same point we have made numerous times: Nothing perfect remains perfect forever. Eventually, a flaw forms in the system, and through that flaw water enters with virtually no evaporative surface behind it through which to exit the systems after the rain stops.

Numerous solid masonry walls have been constructed, some with finishes applied directly to the inside face of the masonry, and they occasionally leak, precisely because they were not good membranes, and they were not thick enough, given their imperfection, to delay the leak long enough to get past the rainstorm. But leaks occur in otherwise competently built solid walls for environmental reasons. The wind may blow particularly hard, causing an unusually high-pressure gradient across the wall that tends to draw the water through at a faster rate. This factor alone can account for leaks through a wall that has never leaked before and probably will not do so again until another hurricane passes through town.

Or the wind may be less than hurricane force, but lasts for an usually long period of time without changing direction. An advantage of rainstorms is that they do not ordinarily last for prolonged periods of time, and as they pass by, the direction of the wind tends to change. Though the wind velocity and, therefore, the absolute pressure differential may be constant, the surface subjected to positive pressure changes. So even if a storm lasts three hours, no single surface will be subjected to net positive pressure for much more than a third of that time. When, however, a storm continues or, for some other reason, the wind blows from a single direction, a leak may be activated simply because the wall ran out of time.

A third environmental reason competent solid walls may leak is that the volume of water sheeting down the exterior surface of the wall is unusually heavy. The higher the volume of water on the surface, the wider the flaw it can bridge. For water to be sucked through the wall, the wet end of the "straw" must be immersed in fluid. Similarly, we cannot suck milk through a straw unless the bottom end of the straw is in the milk. Neither can water transport across a crack or pore unless the water running down the wall covers the outside end of the channel at the same time that the positive pressure occurs at that point. If the volume of water is slight, only very small flaws or pores will be immersed. Recall that the rate of passage of fluids through small-diameter pores is a function of the fourth power of the diameter. The larger the pore, the more rapidly it can transport fluid—but the outside end must first be immersed in the liquid. If the volume of water is greater, the diameter of the immersed end can increase, which, in turn, decreases the time it takes for water to get into the building.

For all of these reasons and some infinite number of various combinations of these reasons, a leak may be activated in one storm, then not again for 15 or 20 years. And let us not forget that the wall itself will change over that time, rarely to become *more resistant* to water. New cracks form; old cracks increase in width; the thickness of the wall decreases. Everything that can contribute to reduced water resistance is exacerbated. Moreover, gross changes may occur that are human-induced, such as the installation of a window or a door. Although such changes do not necessarily result in leaks, it is much easier to construct a water-tight window installation in a new construction than in a retrofit installation.

In summary, then, thick, single-layer approaches to water exclusion are inherently vulnerable to leaks due to differential pressure when the differential is large or prolonged, or the volume of water is particularly high; and they seldom, if ever, improve in this regard with age or modification. As we will see shortly, solid walls are also prone to air and vapor infiltration. But they do have one somewhat redeeming feature: given the same quantity of material, solid walls are stiffer than multiple-layer walls. A 12-inch solid wall has a moment of inertia of 1,728 inches^4 per linear foot of wall. If the same amount of material is used to construct a two-layer wall of four inches and eight inches, respectively, the combined moment of inertia of the two layers is 576 inches^4 per linear foot. The solid wall is three times as stiff as the two layer wall; therefore, a given height has three times the critical load capacity of the two layer wall. If in addition to excluding water, the wall must also provide substantial load-bearing capacity, the designer may strike a compromise in favor of solid construction knowing full well that part of the price may be a decreased water resistance.

This balancing act has been part of the building process for some time. Fortunately, in the last hundred years or so, the relative costs of materials declined to a point at which we could afford to construct multiple-layer systems *and* add enough material to compensate for the reduction in stiffness. Today, if we need the strength provided by a 12-inch solid wall, we simply build it and add the other 4 inches for the benefit, as we will soon see, of a multiple-layer system.

In new constructions, there are ways to detail solid walls to significantly improve their water resistance, but at an added cost. If the solid wall is constructed of two discrete, but bonded, layers so that they act as a single-layer system, a water-proofing material can be installed between the layers to act as a capillary break inside the wall. Thus, the differential pressure is defeated by eliminating the connecting path through the solid wall. The price, obviously, is the cost of installing the water-proofing in addition to the cost of having to install masonry reinforcement in a wall that otherwise could be bonded with the masonry alone. Needless to say, the search for better and less expensive wall systems is ongoing.

MULTIPLE-LAYER SYSTEMS

If thin, perfect, membranes are not possible or, at best, are unreliable, and if solid walls are prone to leak or, at best, to perpetual dampness, how can we reliably, permanently, and economically keep water out of buildings? We must begin by accepting that we are dealing with inherently porous materials and/or construction joinery—it does not matter which. Water will be present sooner or later in sufficient quantity and differential pressure to pass through any conceivable single-layer system. From that acceptance, we can begin to invent an alternative approach.

That alternative approach operates in two stages. The first stage intercepts the water, and the second stage protects the pathway, open though it may be, from the intercepted water. This approach requires at least two distinct layers, physically separated by a space or void sufficient in its width and detail to preclude water from bridging the distance, to prevent any contact between the outer layer, which stops the water, and the second and dry layer, which otherwise divides the inside from the outside elements.

There are two fundamental ways of interpreting this design. One is to accept that water will traverse the outer layer, but recognize that because it is unable to bridge the cavity between it and the second layer, it will drain to the bottom of the cavity where it will collect and, typically, return to the outside. The other fundamental approach does not accept admission of water into the cavity as a given. Using the cavity as an extension of the outside air pressure, the pressure differential across the outer layer is defeated by equalizing the pressure on both sides of the outer layer. Thus, a pressure differential exists between the cavity and the interior, but there is no water on the surface of the second layer; therefore, there is nothing to be pulled through the second layer.

Both approaches are valid, and both will work. But both systems can be corrupted, at which point both will fail. The following subsections examine these two approaches in detail, to see how they may be applied to large- and small-scale construction.

Drained Cavity Systems

The first multiple-layer system, the drained cavity system, is in the United States, the more widely implemented of the two multiple-layer approaches.

The drained cavity system has five essential components: the outer layer, the inner layer, the cavity itself, the gutter that collects the water, and the drain. We will describe each of these components and its function. Then we will assemble the pieces and critique that process. Finally, we will look at the pathology of a constructed drained cavity system.

An advantage of the drained cavity wall is that the outer layer can be constructed from almost anything (almost the only functional requirement of the outer layer is that it not fall off the building). Because an operating assumption of the drained cavity wall is that the outer layer will leak, there is no reason to construct it as though it will not. A dry cavity in a drained cavity system is more likely to be the result of pressure equalization than because the membrane of the outer single layer is perfect. In fact, sound outer layers with few connecting pathways will enhance *any* multiple layer wall, whether it is a drained cavity wall or a pressure-equalized wall. The less water entering the cavity, the less water will flow into the system.

Achieving a sound outer layer is, however, no mean task and some premium added to its construction may be justifiable. As stated repeatedly, in solid walls, it is accepted that some water will penetrate the material, and that when water does penetrate materials, particularly porous, permeable materials, it heightens the risk of damage from a variety of deterioration mechanisms. We tend to mitigate those mechanisms by providing large evaporative surfaces on the exterior and interior of the wall. Historically, we have also leaned toward accelerating the evaporation, at least in cold weather, by providing heat to the wall. In addition to other potential advantages, this increases the partial pressure differential of the water vapor on the inside of the building thereby accelerating the evaporation process. The heat retards the formation of the ice, and thaws a frozen wall more quickly.

To the outer layer of a drained cavity wall, the benefit of internal heat is diminished; simply, the outer layer is more vulnerable. In the absence of contact with the inner layer to effectively insulate it, the outer layer is figuratively and literally divorced from the rest of the wall. If the only difference in two walls is the presence of a cavity in one of them, the outer surface and the outer layer, generally, the one with the cavity will weather more rapidly. The solid wall may leak, and it may pull water deeper into the fabric, but the outer layer of the cavity wall will weather more rapidly. For this reason, some designers view the outer layer as sacrificial, and deliberately design it to deteriorate and to be replaced, while others do the opposite, and invest as much durability into the outer layer as the client will buy.

For reasons related more to habit than science, the materials usually used for the outer layer of drained cavity walls are porous, permeable solids such as brick and stone. Sometimes they are laid up in rather conventional methods associated with masonry construction. This means, however, that as a matter of construction facility, outer layers of brick must be nominally four inches thick; if they are constructed of stone, six inches is preferred.

These thicknesses are not required to execute the function of the outer layer, but because it is exceedingly difficult to build a wall out of relatively small pieces of brick and block.

Such outer layers have a couple of other disadvantages. Despite the depths of four or six inches, these constructions are still considered thin walls, and so must be braced regularly to raise them even one floor level. (We discuss wall reinforcing in the next section.) Another disadvantage is one of real estate. Buildings are often constructed from the property line inward. Every square inch that goes into the wall is a square inch that cannot be counted as rentable space. In a commercial office building 200 feet on a side, and four stories high, the difference in a four-inch outer layer and a two-inch outer layer is the lost rent on 45 square feet every year.

All things considered, however, building an outer layer out of masonry is a sound, durable solution if one takes the sacrificial approach to drained cavity construction. The materials are relatively inert chemically; they can withstand impact; they are fire-resistant; they are known technology; and they are not particularly expensive. But the decision to construct the outer layer with masonry comes with some caveats. To illustrate we will recall the age of the 8 × 8 brick, which lasted approximately five years, from 1976 to 1981. The bricks came in the usual assortment of colors and finishes, and were manufactured with the hollow cores running in one direction; the outer surfaces parallel to the cores tended to be smooth and finished in the same manner as the large flat faces of the interior and exterior surfaces, which were identical. This meant that the brick could be laid with cores vertical or horizontal, with no visual difference on the exposed surface.

The difference was that if the cores were turned horizontally, there was no cleating effect of the mortar in the bed joints oozing into the cores and locking the cells, hence the bricks, together in the plane of potential buckling. This meant that more wall reinforcement would be necessary to stabilize the outer layer. If, on the other hand, the cores were turned vertically, there was an advantage at the bedding joints, but the eight-inch-long head joints had no other bonding than the finished surface of the header faces of adjoining bricks.

Most designers of the time elected the latter course as the apparent lesser of two disadvantages, giving greater priority to the relative structural advantage of vertical cores. Once in service, however, these walls leaked, some disastrously, some catastrophically, some only annoyingly; but they all leaked. With the advantage of hindsight, we can learn much from the 8 × 8 experience. The failure of the 8 × 8 brick in the late '70s had less to do with the brick than with the collective lack of understanding of how drained cavity systems work. The outer layers all passed water, as is predictable with drained cavity walls; but the designers did not expect water to pass through the outer wall, and so did not detail the remainder of the system to accommodate water accumulating in the bottom of the cavity. In every case, water entered the occupied zone primarily because the remainder of the system was not designed or not constructed to handle the water.

The amount of water that came through the face brick was shocking. It became immediately apparent that the water could not be going through the body of the brick (although the brick manufacturers certainly had to accept their share of the blame). The bricks themselves were often denser and generally of as high or higher quality than any brick we have used since then. The mortars were the same as those used in hundreds of modular brick walls that were *not* leaking. In situ testing of the walls demonstrated unequivocally that the water was coming through the wall between the mortar and the surface of the brick at the vertical joints. The mortar debonded from the vertical face, resulting in a long vertical separation on one side or the other of the vertical mortar joint.

In a modular brick wall, there are $6\frac{3}{4}$ bricks in every square foot of masonry, with 5 linear feet of horizontal mortar joint with a nominal length of 9 linear feet of contact length; the adjacent brick affects some or all of the horizontal edges of the $6\frac{3}{4}$ bricks, and the edges or joint segments of the equivalent of $13\frac{1}{2}$ bricks. There are 2 feet of vertical mortar joint, with a contact length of 3 feet, affecting vertical edges of the same $6\frac{3}{4}$ bricks in $13\frac{1}{2}$ segments. The typical length of a horizontal segment is a nominal 8 inches, whereas the typical vertical segment is approximately $2\frac{5}{8}$ inches. In contrast, in the typical 8×8 wall, there are only $2\frac{1}{4}$ bricks with only 2 linear feet of horizontal mortar joint, with a nominal contact length of 3 feet, with the same amount of vertical joint as horizontal joint length and contact length. There are equivalent $4\frac{1}{2}$ segments in each direction. Based on these factoids, the larger unit has significantly fewer inches of horizontal joint and fewer inches of horizontal contact length in fewer segments. This means that the opportunities for failure of the horizontal joints are significantly reduced. The total length of the vertical joints and the vertical contact lengths are the same in both cases, but the number of segments is decreased so that instead of the average length of a vertical segment being approximately $2\frac{5}{8}$ inches, the average vertical segment is 8 inches long.

The reduction of horizontal vulnerability was a reduction of a negligible risk. Masonry joints seldom crack, and even more rarely debond along bedding joints. The weight of the wall itself is sufficient to compress the joints and to maintain sufficient compression during mortar curing to reasonably preclude cracking and debonding. Reduction in a negligible risk cannot result in a substantial risk reduction because there is not much to reduce. Ostensibly, maintaining equivalency of the vertical joint would have neither increased nor decreased vulnerability to joint deterioration, but that proved not to be the case.

Vertical mortar joints, unlike their horizontal cousins, are under very little, if any, external compressive stress. If, during the curing period, the mortar contracts, tensile stress develops within the body of the mortar joint and at the bonding plane between the mortar and side face of the adjacent brick. If the mortar is sufficiently elastic, with a sufficiently high modulus of rupture, then the mortar itself will not crack. If it does not, a crack will form along the vertical center line of the joint. If the adhesive strength of the bond between the mortar and the brick is greater than the tensile stress, debonding will not occur. If it does not, a debonding crack will occur along one, but not both, of the contact faces of the joint. Along which of the two possible faces the debonding will occur is strictly a function of which of the two debonding planes is the weaker. This is true whether the joint is $2\frac{5}{8}$ or 8 inches long.

The intersection of the vertical joint with the horizontal joint beneficially affects the prospect of a fracture debonding. The primary reason cracks or debonding occur in mortar joints is lateral shrinkage of the mortar during curing, and the primary reason for shrinkage is premature dehydration of water from the mortar. The mortar dries out from evaporation before the curing of the constituent materials in complete, hence, before full strength and elasticity can develop. The loss of water results in the absolute shrinkage at the same time that the adhesive bond and strength of the mortar are impaired. The presence of a horizontal intersection provides a source of potentially beneficial water precisely when such water could mean the difference between a failed and a competent vertical joint. Although the crack may initiate at mid-height of an otherwise under-hydrated mortar, the crack will at worst propagate to the end of the joint segment. It is also highly possible that the crack may not occur at all due to the infusion of water at the critical moment in the curing process; but if it does, it may well terminate upon encountering competent mortar near the horizontal intersection. The greater the number of such crack inhibitors, the greater the prospect of total numeric crack reduction and total crack length reduction.

The essential weakness of the 8 × 8 brick is the unbroken distance of the vertical joints. Evidence showed that the percentage of failed vertical joints was greater than its modular counterpart; even more notable was that the total length of the failed vertical joints was six times greater than a comparable modular construction; and the maximum crack widths were much greater in the 8 × 8 than in the modular equivalent. Again, recall that the capacity for fluid transportation increases to the fourth power of the pore width. In short, much larger quantities of water were coming through the 8 × 8 bricks at the vertical joints than were reasonably expected.

That lengthy explanation is not an excuse for the failure of a system supposedly designed to accept water coming through the outer layer. The failures of the 8 × 8 brick, drained cavity walls were not simply due to the unanticipated increase in water infiltration caused by the vulnerability of the vertical joints. We repeat: the primary reason for the failures of the 8 × 8 brick walls was the collective misunderstanding on the part of the designers as to how the walls were intended to work versus how they needed to be detailed and constructed to work as drained cavity systems.

We will revisit the saga of the 8 × 8 brick as we explore the balance of the components of the drained cavity system, because there was at least one notable failure of each component once the water penetrated the outer layer.

Let's return to the theme of the dimension of the outer layer. As one response to the minimum thickness of masonry outer layers, the cut-stone industry developed attachment details and systems to allow for the installation of panels of stone that were two inches thick. The size of an individual panel is limited to its structural capacity to carry the weight of the stones stacked above it without bracing, other than at the attachment points, which are normally at the edges of the panel. Such installations offer the aesthetic quality of stone, combined with the practicality of a drained cavity wall, without the weight of stones twice as thick. (Note: The weight is not really a structural issue, but it can make construction a more difficult undertaking.)

Remarkably, even using these panels, many designers still inject vast quantities of sealant to pack the joints, evidence that they still do not trust the system to collect and drain the water. Certainly, the reduction of a gross water stream has some advantage, but any attempt to convert cut-stone panels into a perfect membrane are futile, unnecessary, and risky to the stone. And the risk to the stone is the sealant itself. More than one fine stone specimen has been ruined by goo in the joints.

More recent systems include the use of panelized sheets of stone to decrease the stone thickness to three-fourths of an inch. The brick industry has developed similar techniques for mounting relatively thin sections of brick onto significantly sized panels: the panels are raised as units into place.

One problem with most panelized systems is that the backing frame that holds the thin sections of stone or brick in place itself has some dimension, which intrudes into or through the cavity, often compromising the effectiveness of the cavity and, therefore, the entire system. Once the cavity is compromised, the exterior panel systems become little more than single-layer membranes made of porous, permeable materials highly fragmented with joints.

When the panels are large slabs of glass or metal, the same problem exists, except that the materials are more appropriate for membrane construction, if that is the effect of the assembly. It is possible to detail a glass curtain wall or an aluminum panel wall as a drained cavity system. The point is, it is not the material of the exterior layer that determines the wall type; it is the full composition of the pieces, and how they work together, that defines the system. The craft and, ultimately, the art of exterior wall design requires both the selection and consistent development of the approach to defeat differential pressure.

The inner layer of the drained cavity wall, often called the backup wall or just the backup, has a number of functions, beginning with dividing the occupied zone from the cavity—that is, from the outside elements. Typically, the backup is the structural component of multiple-layer systems; it is the last layer of defense when other parts of the drained cavity wall fail.

The structure of the backup layer has three identifiable components: wind resistance, seismic resistance, and load-bearing resistance. Primary among these is the lateral resistance component. A typical drained cavity wall has an outer layer, which may be as thin as a fraction of an inch up to four inches. The inner layer backup is often eight inches, although many are six inches thick. If the composition of the two layers comprises four inches of facing material and eight inches of backup, the combined moment of inertia of the wall is 576 inches^4. That value is the sum of the moments of inertia of the two layers, which are calculated independently because they operate in bending situations as connected but unbonded layers. The moment of inertia of the outer layer in this case is 64 inches^4 which is 11 percent of the total moment of inertia of the wall, despite the fact that it comprises 33 percent of the material thickness of the wall. This means that 89 percent of the stiffness of this particular drained cavity wall is provided by the backup layer based on the thicknesses of the two layers.

For all practical purposes, the capacity of any panel of a drained cavity wall to resist lateral loads perpendicular to the plane of the wall is limited by the structural stiffness and strength of the backup. Wind pressure at the outer layer cannot be assumed to equalize with the pressure in the cavity, and, therefore, is a net pressure at the outside surface. That pressure is transferred to the inner layer by devices specifically designed to act as short axial elements capable of transferring tension or compression from the outer layer to the backup.

These elements are typically part of somewhat larger assemblies that also reinforce the outer and the inner layers. In their simplest forms, they are flat straps of metal embedded in the bed joint of the outer and inner layers. This means that the coursing of the face material and the backup must be reasonably aligned. The *ties*, as they are called can absorb only limited misalignment, if the item is designed to absorb compression. If the tie is bent to adjust for a stagger in the joint alignment, the tie can still be placed in tension, and operate effectively; but if it is placed in compression, then the line of action is such that the strap, instead of developing in compression, will operate primarily in bending, for which it is ineffective.

Typically, these simple flat straps have rippled, or crimped, ends, which is why they are often called crimped ties. They are the tie of choice in many situations, including many solid wall applications. These ties can be modified on one end or the other to integrate with other components and/or other methods of attachment to the layers. Their disadvantages are that they are, indeed, flat and, therefore, relatively weak in compression even when they are placed horizontally. The compressive weakness is often reinforced by adding a dollop of mortar directly on top of the tie, between the two layers. When the mortar hardens, it acts as a plug between the layers, which is supported from falling into the cavity by the tie. (We will discuss this in more detail later.) The ties also provide bridges for water to traverse the cavity (which we will also discuss more in the following section).

The outboard ends of these small posts are often connected to wires embedded in the bed joints, and run along the length of the outer layer. The purpose of the joint reinforcement is to stiffen the outer layer and, essentially, to collect the forces exerted on the outer layer and redistribute them to the posts. Once collected, the forces in the posts are directed to the backup wall. In the bed joint of the backup, the post or strap is connected to additional wires; it then redistributes the axial forces in the connecting posts or straps to the backup as a whole. The wires in the backup are often configured into small trusses, or box patterns called *ladders*. These trusses and ladders can bridge the cavity and provide the connections from the backup to the outer layer directly; they can be coupled to the outer layer with the straps, as described above; or they can be connected to the outer layer and its embedded wire with hooks.

The hooks come in many flavors, but essentially they are welded to the horizontal wire embedded in the bed joint of the outer layer. The eye or loop through which the hook fits is part of the ladder or truss in the backup. There are advantages and a disadvantage to this family of hook ties or, collectively called *two-piece ties*.

One advantage to two-piece ties is that the backup layer can be constructed independently of the outer layer. The backup is laid with the loops projecting from the outboard face of the backup layer. The outer layer can then be raised at a later time, rather than at the same time, which had been the practice until the invention of the two-piece ties. (Prior to the advent of two-piece systems, the backup could also be raised prior to the outer layer, but that meant the projecting ladders or trusses had to be bent out of the way of the facing layer when it was raised. In the interim, the ladders and trusses, generically referred to as *joint reinforcement*, were subject to damage.)

Another advantage of two-piece ties is that they can accommodate more joint misalignment than solid ties or straps without compromising the compression capacity of the assembly.

The disadvantage of these ties is that they have been the object of much controversy. In the late '70s and early '80s, two-piece systems were criticized as dangerous to passers-by walking below such installations. Since then, the ties have been, generally, vindicated. The substance of the criticism was that the hook, a bent wire, could not provide the same tensile or compressive capacity as a solid axial wire of the same diameter. The bent wire would necessarily deform more than the same wire axially loaded. Though factually correct, the argument was absent quantification.

Hook ties with a single hook and eye are specified less often than they might be otherwise because many professionals simply do not want to re-open the controversy. In any case, in the interim, new concerns have emerged, which may mute the entire question of single hook ties. One is the issue of seismic design and protection. In most communities, the structural requirements for drained cavity wall design dictate the capacity of the wall reinforcement to meet seismic criteria. For all intents and purposes, this means that all hooked-wire ties are obsolete. There are two-piece ties that effectively replace the bent wires with bent plates of metal, which then slip through loops in the ladders or trusses in the backup. There is a cost differential between the wire hooks and the so-called seismic-rated ties, but it is a small price to pay to gain the advantages of two-piece ties.

The significance of the load-bearing capacity of the backup layer depends upon whether the backup, in addition to bracing the exterior layer, is a major component of the structural system. As is often the case with solid walls, heavy backup layers may have considerable bearing capacity and may be used as the perimeter load-bearing element. There is nothing inconsistent or incompatible about a solid or a backup layer being part of the structural system at the same time that it is part of the vertical closure system. However, there is a practical implication to the pathology of such elements. If an element, such as a load-bearing backup layer, operates in two systems, then that element is subject to the potential deterioration environments of the mechanisms associated with *both* systems. Furthermore, because that deterioration does not distinguish among functions, once incurred due to a failure in the closure system, it immediately and directly affects the structural system. Clearly, this increases the vulnerability of the buildings as a whole.

The backup is the defining element of drained cavity systems, separating the outside from the inside. Whether it is actually part of the inside or of the outside of the building is an issue we will resolve shortly. We can say definitely that the inboard side of the backup is inside the building, and that probably insulation and finish materials remain between the inboard face and the interior of the building. A simple way to test inside versus outside is to query whether water penetration is significant or insignificant. If water penetration is insignificant, then that location is outside; if water penetration is significant, then that location is inside.

The location of the boundary is important because that is the preferred location for the air barrier. *Every* wall requires an air barrier. Somewhere in the cross-section of the wall there necessarily is a point at which the block of air on one side is defined as outside, and, as such, is subject both generally and specifically to, and is determined by, the weather. To the inboard side of that line, the block of air is to a greater or lesser degree influenced by the outside conditions, but it is also immediately and specifically influenced by the occupant through various environmental controls.

In a drained cavity system, the distinction from the solid wall takes place the backup layer, so the outermost location of the air barrier is the backup, in contrast to a single-layer system where the air barrier is at or within the single and, therefore, the outer, layer. The function of the air barrier in a drained cavity wall is one of thermal efficiency; it does not affect the performance of the wall itself. In pressure-equalized systems, the air barrier is critical to the performance of the system as a whole, as we will explain shortly.

The drained cavity wall is designed generally to preclude rainwater from reaching the surface of the backup, but because the backup is normally outside the protective envelope of the insulation and vapor barrier, condensation is another source of invasive moisture. Some designers argue that water vapor will not permeate the cavity because the cavity is not specifically connected to the source of moisture; but that precludes that water will penetrate the outer layer. If water is drawn through the outer layer because of a pressure differential between the outside and the cavity, the conditions for vapor penetration are present and operating. Condensation forms in the cavity of the drained cavity wall; it forms on all surfaces at or below the dew point. This means that condensation can and will form on the exterior surface of the backup. It is recommended, therefore, that this surface be treated with damp-proofing material, to minimize damage from condensation, and to improve the air barrier properties at the same time.

In general, the separation of the inner layer and the outer layer can be breached by a variety of elements that bridge the cavity and provide a water path from the intentionally porous outer layer to the inner layer, which is intended to be dry. Despite the assumption that the primary pressure drop occurs between the outside and the cavity, across the section of the outer layer, it is unreasonable to conclude that no pressure differential exists across the backup. The pressure drop through a drained cavity wall is more accurately described as follows: The primary line where the pressure differential is defeated is the backup, not the outer layer; the pressure differential that occurs across the outer layer and that is, therefore, responsible for moisture being sucked into the cavity, exists less because the cavity is part of the interior nearly than that it is inefficiently connected to the exterior. The pressure differential between the exterior and the cavity exists because of the time delay that occurs when the exterior pressure changes abruptly. Time delay in pressure-equalized walls is inherent to that type of wall system, so we will defer further discussion of the phenomenon until that section. As the time delay relates to drained cavity walls it is important to remember only that there is a significant pressure differential between the cavity and the interior of the building; therefore, any moisture that reaches the surface of the backup is subject to being pulled into or through the backup material into the interior volume.

Other bridges across the cavity include virtually any tie that connects the outer and inner layers. Hook ties tend to defeat the water bridge feature by providing the built-in drip of the hook itself. Certain rigid ladders and trusses have small crimps in the wires that cross the cavity, to act as drips for any water that clings to and traverses the wire. As minor a detail as this may seem, it is an annoyingly difficult problem to remedy if it turns out to be the source of a leak later on.

Solid plugs of mortar in the cavity also act as bridges for water transmission. Recall that such plugs are used in conjunction with ties, which perform well in tension but not in compression. The plugs act as spacers, and maintain the dimensional tolerance of the cavity in positive pressure situations. If the water passes through the outer layer and wets a mortar plug, however, the porous, permeable mortar acts as a wick to the backup. If the backup layer is faced with damp-proofing, such items are seldom an issue; but if the backup surface is an exposed porous surface, this can be the source of a remarkably significant amount of water.

Far more common than plugs in the upper reaches of the cavity are mortar droppings (unfortunately referred to as snots) in the bottom of the cavity. They can accumulate in the bottom of the cavity and foul up the cavity. If they pile up high enough, they will wedge between the outer layer and the backup, and wick water in the same manner as the plugs just mentioned. At the very bottom of the cavity, snots are more likely to be wedged between the outer layer and the flashing, and therefore are not likely to pass water into the backup. As we will see, however, they may have a devastating effect on the operation of the cavity itself. How they get into the bottom of the cavity is a function of the construction of the cavity, which we will explain later.

The materials used to construct backup layers are typically less expensive and technically inferior in many respects to what are used in the exterior layer. Backup materials, particularly when faced with damp-proofing, are unlikely to become saturated. Even when subjected to freezing temperatures, freeze-thaw action is not likely to occur, due to the absence of water. Salt damage is a possibility, but again, the relative absence of water in the backup minimizes that prospect. The backup material of choice today is the concrete masonry units (CMU), which is molded concrete. The concrete mix is steam-cured in a variety of shapes and sizes. The CMU is extremely porous and permeable, though its compressive strength is not particularly impressive compared to brick or stone. A major advantage is that it's very inexpensive. On the other hand, it's quite heavy, but still manageable by an accomplished mason.

Laying a standard size CMU is equivalent to laying a dozen bricks, but a great deal faster and cheaper. The crushing strength of a CMU is not nearly as valuable an attribute as its capability to achieve a high moment of inertia at a low price. The stiffness of the backup is more readily achieved with depth than with inherent material properties. Accepting that there will be compromises associated with a reduction in the enclosed area and volume, CMU backup can be increased in increments of four inches to produce rather deep backup layers.

If area and volume are a consideration, and additional stiffness is required, CMUs lend themselves, as do all blocks with large cores, to being grouted solid and to the incorporation of reinforcing rods into and through the units. One company has developed a CMU that has slots cast in the webs to facilitate reinforcing the block both vertically and horizontally. Walls constructed with these units have been referred to as the "poor person's reinforced concrete wall" because the block performs the function of form work without the need to actually construct or strip the forms.

As the CMU is filled with grout and reinforcing, the pathological consequences increase. The hollow CMU is subject to water and vapor passage, but it is relatively resistant to weathering. When the cores are filled with concrete grout, the CMU essentially becomes a shell wrapped around a core that is significantly stiffer and stronger than the shell. This places the CMU material at risk. If, in addition to the concrete in the core, there is reinforcing steel as well, the situation is compounded because of the corrosion potential of the rods.

Unfortunately, the CMU is occasionally used as an exterior material. It was not designed for exterior use, but its low cost and the relative ease of construction is too tempting for some to resist. When the CMU is used on an exterior, it is normally painted, which does tend to minimize the open pores of the material. Nevertheless, it is still an imprudent choice of exterior material.

Other backup materials include common brick and structural clay tile, though common brick is used infrequently because the drained cavity system became popular after structural clay tile was made readily available, and at a significantly lower price than brick per square foot of wall. It is more common to find brick backup in solid walls than in drained cavity walls.

Structural clay tile was the backup material of choice until World War II, when the CMU captured the market. Structural clay tile is a superb material; it is lightweight thanks to the high void or core percentage, and as dense and strong as brick. In its heyday, a wide variety of shapes and sizes were available.

The major disadvantage of clay tile is that it is brittle, and its cores are so large relative to the solid portions of the tile that an addition to or modification of structural clay tile usually winds up destroying the integrity of the unit. For example, it is difficult to anchor into simple structural clay tile because probably the anchor will penetrate a core.

Conceptually, the drained cavity wall is designed without much regard to where the positive pressure drop between the inside and the outside will occur. As opposed to thin, single-layer systems and pressure-equalized systems, pressure equalization is not a great concern, and the detailing of many drained cavity walls reflect that attitude. The cavity itself typically is viewed as though it is connected back to the outside, but only to the extent that water that collects at the bottom of the cavity drains back to the outside. From the standpoint of pressure, implicit in the notion of the drained cavity wall is that water will penetrate the outer layer, and that there is, therefore, a pressure differential across the outer layer. Thus, by logical deduction, the cavity is part of the inside of the building, at least in terms of pressure equalization.

As stated, the cavity in a drained cavity wall is typically two inches wide, for only one reason, which has to do with construction feasibility. Whether the outer layer is put up with the backup or after the backup, the mason usually works from a scaffold mounted outside the building. (In some situations, the mason works from the floor of the interior of a building, but normally walls are laid from the outside.) If the outer layer is brick, after the mason has pressed the face brick into the mortar bed, he or she must tip the trowel down into the space between the brick and backup and scrape the snot off the back side of the outer layer without knocking it into the cavity. This is the source of droppings in the bottom of the cavity. This task can be made easier with the use of a narrow strip of wood that is pulled up through the cavity as the work advances. The so-called drip board is cleaned off every two feet when the wall reinforcement is laid. The board is placed on top of the protruding reinforcement and the process is repeated. The drip board is a good device, but it takes time to use and vigilance to ensure effectiveness.

Regarding the two-inch dimension of the cavity: It has no other function than construction feasibility, despite some occasional assertions that it is somehow related to water penetration. Functionally, the cavity of a drained cavity wall performs exactly the same whether it is a half inch or four inches, which is one of the features that distinguishes drained cavity walls from pressure-equalized walls. The only difference is that shallow cavities are difficult to build and wide ones are somewhat wasteful of building volume. (A development in cavity design that did affect the constructed cavity dimension was the introduction of rigid insulation into the cavity, which began as a response to the energy crisis of the '70s.)

The bottom of the cavity must have a collection device capable of gathering and channeling the water to the appropriate exit port. Of all the parts of the drained cavity wall that fail, this probably heads the list. The common name for this device is *flashing*, which unfortunately belies its function and importance. Flashing is one of the least precise terms in the construction lexicon, as it is applied to almost any process that involves the closure of a gap or turn that might otherwise admit water and that is not closed with a sealant. Flashings line valleys and ridges; they can be found around vents, stacks, and chimneys, and along parapets in horizontal closure systems; flashings cap parapets, tops of window and doors; they exist below sills, and at the bottoms of cavities at every horizontal break in the cavity.

In the drained cavity wall system, flashing is an impermeable membrane of metal or synthetic fabric that lines the bottom of the cavity. Typically, it is carried up the face of the backup and runs into a convenient bed joint eight or more inches above the bottom of the cavity. The outboard edges run below the bottom edge of the outer layer and end at the extreme outside face of the outer layer. The theory is that water will sheet down the cavity side of the outer layer and collect at the bottom of the cavity without penetrating the flashing. The flashing will direct small quantities of moisture along its surface horizontally back through the bond break between the flashing and the outer layer that is bearing on it. Larger quantities of water will be carried laterally across the bottom of the cavity to a local exit port called a *weep*, which we will discuss very shortly.

One reason flashings are such common sources of failure in drained cavity wall design is that there is a flaw in the theory of their operation. Water does *not* work back out of a drained cavity by following the boundary plane between the flashing and the bedding of the outer layer or the bottom edge of the outer layer itself. (Water has been known to wick back through the bedding mortar at the base of the cavity, but not often.) To water working through fine cracks or capillaries there is no difference going one direction or the other. If pressure is required to push or pull water through fine fissures and pores, it takes time. The drained cavity system is designed so that when a pressure differential exists at the same time that water is present on the exterior surface, water that may be pulled through the outer layer is collected in the cavity.

For water to be moved back through a more limited area of capillary action, the pressure must be reversed; therefore, the time required to achieve the result is much greater. An implicit presumption of most drained cavity wall details is that the water will build up inside the cavity, and it is the hydrostatic pressure that supplies the motive force to push the water through the joint. This means that water must collect in the cavity to a depth sufficient to provide such pressure, at which point it will flow laterally to an exit port. If there are no such ports, or those provided are ineffective, very little water is required to fill the cavity to reach above the maximum height of the flashing at the backup. At that point, the water crests the flashing and is just as likely to move through the backup into the occupied zone as out through the outboard material at the base of the cavity.

In the one case where we saw clear indications that water was moving through the bedding joint between the flashing and the bottom course of brick (the aforementioned 8 × 8), the depth of the water in several measured locations was over 10 inches deep inside the cavity. The claim was that the brick was faulty because a great deal more water was coming through the outer layer than could reasonably have been anticipated. Supposedly, ordinary quantities of water would have and could have leeched through the bottom bedding joint. This was and is incorrect. The only reason water was migrating through that joint was because there was a 10-inch head of water pushing it through. If the quantity of water entering the cavity had been significantly less there would have been no comparable hydrostatic pressure and no water would have come out through the joint. To follow the logic of the claim of cause, the only way the drained cavity wall could work was if no water would ever enter the cavity, which as we know is completely contrary to the purpose of the cavity.

In another case also involving the now infamous 8 × 8 brick, the architect did not like the appearance of the flap of stainless steel flashing protruding at the shelf angle, so he had the flashing trimmed back and sealant injected at the toe of the shelf angle. The effect was to bury the flashing lip behind an impervious and impermeable line. When the water accumulated in the cavity, it flowed laterally along the flashing until it passed through an open lap in the flashing. At that point, the water might have worked back to the exterior between the horizontal leg of the shelf angle and the flashing, but the toe of the shelf and the flashing were sealed at the horizontal joint. Instead, the water built up behind the flashing until it crested the shelf angle, then poured onto the window head and into the interior of the building.

Here, too, the claim was that the 8 × 8 brick was flawed because so much water had gotten into the cavity. True the brick admitted more water than was customary for reasons previously discussed, but the ultimate failure of the wall was not because the water got into the cavity. Technically, the failure occurred because the primary path back to the outside was sealed, the lap joints in the flashing were not sealed, and the occasional weeps were clogged with mortar droppings. The more fundamental reason the wall failed was because the architect did not understand that drained cavity walls are supposed to collect water.

This example illustrates another aspect of cavity flashings worth discussing. The unsealed lap joint between segments of flashing is a common oversight brought on, we suspect, because such details are drawn in section through the bottom of the cavity. This suggests to the designer that the primary path of water movement will be over the surface of the flashing and out to the exterior surface along the entire length of the detail. That is not, however, how the water moves in the cavity. If water accumulates in the cavity to any measurable depth, there is a very good chance that it will not flow anywhere. It will accumulate on the flashing in a long slender puddle because the designer viewed the detail only in section. If the designer called the detail a gutter, and not a flashing, he or she would know immediately what to do with the water.

This brings us at last to the drains (the weeps) themselves. Simply put, many if not most, do not work. Cavity drains are placed at the bottom of the cavity for good reason: that is where the water will collect. (We state the obvious here because we have seen designs with the cavity drains *above* the bottom of the cavity.) The cavity drain of choice is a vacant head joint in the exterior layer, two feet on center. Such an opening connects the cavity, which is presumed to be charged with water with the exterior of the wall. The water flows from the cavity through the open head joint and out of the wall altogether. This, obviously, is applicable only when there are head joints two feet on center that can be voided, as in a brick outer layer. In stone and other materials, the closest comparable spacing is recommended.

The drains can be farther apart than two feet; they can be spaced any distance apart, in fact, but there are consequences. If the drains are 10 feet apart, the water that may have accumulated in the cavity to a depth of half an inch when the joints were two feet apart may be as high as two and half inches. If there is a pinhole leak in the cavity gutter, the hydrostatic pressure across the flaw is five times as great. Even though the water is moving toward the drain, in the latter case, the entire bottom course of brick is below water, and the depth of water at the drain is the full height of the drain itself. Typically not that much water will be in a cavity, but the path of that logic leads to trivialization of the problem and to a discounting of the premise upon which the drained cavity wall is predicated: water enters the cavity.

One substitute drain configuration for open head joints is the pencil hole, or its high-tech cousin, the $\frac{3}{8}$ inch diameter, transparent plastic tube.

The pencil hole derives its name from the diameter, which is a derivative of the manner in which it is fabricated, namely by ramming a pencil through the green mortar at the time of construction. Even if the diameter were sufficient to perform as a drain, there are problems with the pencil hole. The mortar is still plastic enough to admit a pencil or, more likely, a 9-gauge wire through the mortar. When the wire is retracted, the soft mortar contracts around the hole. In some cases, the cavity end of the drain closes entirely, leaving a cosmetic imprint of a drain hole on the outside and a sealed end on the cavity side. If a snot falls behind a pencil hole, it is certain to clog. The same dropping behind a vacant head joint is far less likely to close the drain. A pencil hole is also a favorite nest for a number of insects, particularly those with a first name of Carpenter. Carpenter wasps scour the planet searching for drain holes of this size. When they locate one, they build a dwelling inside the hole, defeating the function of the drain.

Holes that have tubing inserted into an already below-standard diameter hole are only less likely to work than their slightly larger surrounding perimeter. They may project away from the face of the wall, which may be an attempt to throw the water away from the face of the building—at which point any idea of disguising their presence is fairly well defeated—or they may be cut flush with the wall surface. In either case, they are ineffective and a general waste of time and money.

One of the more alarming cavity drain details in the inventory is the sash cord detail, which comes in two flavors. The first is a variation on the pencil hole theme. A length of rope, traditionally sash cord, is built into the wall at the future locations of the drain. It is built right into the wall as the mortar is laid. When the mortar begins to set, the mason walks along the wall and jerks the sash cords out of the wall, leaving a well-formed pencil hole. The advantage is that, unlike the pencil hole, there is no displaced mortar to collapse into the hole, so if small holes must serve as drains, questionable though that decision may be, this is the preferred method of fabrication. In the second variation, instead of pulling the cord out of the wall, the mason cuts the cord flush with the face of the building, leaving exposed the end of a piece of sash cord set in mortar. Enough said.

Pressure-Equalized Systems

Pressure-equalized systems are not new; they are, in fact, quite ancient, as we will demonstrate in Chapter 5 when we apply the principles to the analysis of roof systems. What is relatively new is how pressure equalization is applied to walls. The pressure-equalized wall is an accepted part of the European and Canadian construction repertoire, but it is less widely used in the United States. There is no major functional difference between a drained cavity wall and a pressure-equalized; both systems will, when properly executed, defeat the pressure differential and, therefore, preclude water from entering the occupied zone. The pressure-equalized system has a slight cost premium for additional vents and for baffling, approximately $1.00 per square foot of wall, excluding windows and doors. For common masonry walls, that comes to approximately a 2 percent premium. The benefit of that premium is enhanced durability of the exterior surface material and added assurance against leaks.

The operational difference between the drained cavity wall and the pressure-equalized wall is that whereas water is expected to enter the cavity of the drained cavity wall, it is not expected to enter a pressure-equalized system as far as the cavity, though the exterior surface of the outer layer will be wetted. A major premise of the pressure-equalized system is that there is no motive force to pull moisture deeper into the fabric than will occur by passive absorption by the wetted exterior surface. As explained without external motive force, the penetration by absorption alone is slight and slow.

Pressure-equalized walls defeat pressure differential by physically separating the plane where the pressure differential occurs and where the water is present. In contrast to the drained cavity system, the pressure-equalized system operates on the premise, that the pressure drop occurs across the inner layer, not across the outer layer. The outer layer serves as a bulk water interceptor and is wetted at its outer surface. The pressure in the cavity is always the same as the outside air; therefore, there is no motive force across the outer layer, and no water is drawn into the cavity. Because the cavity remains dry, there is no risk of water at the surface of the inner layer. There is a pressure differential across the inner layer, but no water, so even if the air barrier of the inner layer is less than perfect, no leak will occur because there is no water to leak.

The elegance of the pressure-equalized system is that none of the components must be perfect and the materials will actually remain drier than other systems even when performing optimally. In contrast, the thin, single-layer system requires a perfect outer surface as a membrane; and the thick, single-layer system requires a perfect environment (meaning that the system will work as long as the weather is never bad enough to reach the assumed penetration depth). The drained cavity system is much more forgiving than either of the single-layer approaches, but it still requires a perfect cavity and drains.

The essential components of a pressure-equalized system are:

- An outer layer, also called a screen or rainscreen
- An inner layer, with an incorporated air barrier
- A pressure chamber
- Baffles
- Vents

Although not essential to the concept of a pressure-equalized system, a gutter is also necessary to the system, for reasons we will explain.

The outer layer of a pressure-equalized system is not distinguishable from the outer layer of a drained cavity system; neither is required to be more than a barrier to rain, and neither must be impermeable. Ironically, more attention is paid to the detailing of the outer layer of a typical drained cavity wall to defeat water at the outside surface than to pressure-equalized walls, even though it is assumed in the former that water will penetrate the layer but will not in the latter. Why the discrepancy? Remember, the drained cavity system relies upon the perfection of the gutter and drains, so even in a well-conceived, -designed, and -detailed drained cavity wall, it is prudent to minimize the load on the parts that must perform perfectly. The pressure-equalized system, by contrast, works precisely because the outer layer *is* flawed; literally there are holes punched through the outer layer. A few more flaws one way or another will not affect the performance of the system.

As just noted, the outer layer of pressure-equalized systems have been dubbed screen or rainscreen, a term that is sometimes used to refer to the entire pressure-equalized system. In fact, the outer layer of a pressure-equalized wall is no more a screen against rain than is the outer layer of a drained cavity system, but the term has become exclusively identified with pressure-equalized systems. The screen may be of any material, thin or thick, light or heavy. The only functional requirement is that it intercept rain, or at least the vast majority of it. (It is premature at this point to discuss the significance of partial rain interception until we have explained the entire system, so we will leave this subject for now.) As a matter of practicality, for reasons such as durability and economy, rainscreens tend to be constructed of familiar wall surface materials. They can be made of stone or brick, and curtain wall materials will serve as well. Often, because the system is tolerant of flawed materials and fabrication, designers are tempted to allow lower-quality construction of the rainscreen. However, the rainscreen has its own technical demands, which suggest this is an unwise tactic to take.

As with drained cavity walls, the outer layer must be braced against buckling under self-weight, though lateral load from wind on the outer layer is significantly less in a pressure-equalized system than in a drained cavity system. This point is often misunderstood and sometimes openly challenged, so we will digress here to explain the physics behind pressure equalization.

The abstract concept of pressure differential is not difficult to understand. And we know that fluids can be and will be drawn through porous, permeable materials under the motive force of differential pressure. The difficulty arises over those two points in conjunction with the absence of lateral load on the exterior of the outer layer in pressure-equalized systems.

To begin to understand, consider how it feels to walk in a strong wind. We feel pressure on our bodies; our senses tell us there is external pressure. Then why is the outer layer of a rainscreen not affected in the same way? Because at precisely the windward *surface* of a wall, the air is not moving; it is relatively static. Let's extend the analogy. A person standing in a windstorm is analogous to the building as a whole. The person is wearing glasses. If the force of the glasses against the person's face was proportionately the same as that of the wind on the structure as a whole, the glasses would be jammed into the face so hard as to imprint on the person's face. The glasses are a rainscreen for the eyes. The external air pressure is absorbed by the eyes, not the glasses; but if rain is mixed in with the wind, the glasses will intercept the rain.

More formal proof can be found in a common bit of logic: When A occurs, then B occurs; therefore, if B does not exist, then A cannot exist. Here, the analogue of A is a differential pressure across a solid plane, and B is a net external force. The initial proposition then becomes: When there is a differential pressure across a solid plane, there must be a net external force on the solid plane. This is precisely the motive force that pulls fluid through the outer layer of a drained cavity wall. This is the force that pulls water through a fine-line crack in a single-layer system, generally. In those situations, there is a pressure differential across solid planes; therefore, there is a net external pressure on the exterior surfaces of those solid planes.

The second part of the proposition uses the inverse of the first part to disprove the existence of the first condition. If there is no pressure differential across the solid plane, there is no net external force on that plane. A classic use of this same logic can be found in mechanics: A body at rest remains at rest unless acted upon by a force. If the body is, in fact, at rest, there is no force acting on the body. The absence of force is proven by the absence of motion. If we can prove that there is no pressure differential across the outer layer of a pressure-equalized wall, we will have proven that there is no net external force on the surface.

At the scale of the building as a whole, we can diagram this phenomenon using lines of equal pressure, called *isobars*, as they would appear if we take a section through the ground plane in the path of an air stream. If air is moving over a smooth surface, there is some minor friction with the surface; but the air is essentially moving with little turbulence parallel to the surface. The velocity of the air, though not pertinent, we'll say is moving at 40 miles per hour. The air *pressure* is one atmosphere at sea level; the velocity of the air does not increase its pressure.

If an obstruction is placed perpendicular to the path of the moving air, the molecules of the air pile up at the face of the obstruction, and exert a force on the obstruction. The air stream glides up and over the obstruction, compressing the stream at the crest of the obstruction, and increasing the velocity of the stream as it rushes over the obstruction. When the stream clears the obstruction, the velocity of the stream redistributes uniformly across the section of the stream. At the windward and leeward faces of the obstruction, the velocity of the air in the direction of the overall air stream is at or near zero. What movement there is at the windward face is typically vertical upward; at the leeward face, vertical downward. In neither case is there much, if any, air movement perpendicular to the face of the obstruction. Although the air movement is minimal, the pressure at the windward surface is greater than ambient atmospheric pressure; and at the leeward surface, the pressure is lower than ambient atmospheric pressure. The vectoral directions of the positive pressure on the windward side, and the negative pressure on the leeward side are the same, and the sum of these two forces is the total force applied to the obstruction.

These observations are profound relative to the pressure and movement of air at the surface of a rainscreen. If the rainscreen is on the windward face of the building, which is perpendicular to the stream of moving air, the *movement* of the air at the surface is negligible. Any movement that does occur is parallel to the surface, not perpendicular. This is true whether the stream is perpendicular or oblique to the obstruction. The net force vector of the moving air at the surface is negligible perpendicular to the surface.

If the force vector perpendicular to the surface is zero or near zero, the dynamic force is zero or near zero, even though the pressure is greater than one atmosphere at the positive face and less than one atmosphere at the negative face.

If, say, the obstruction is an empty box, and the pressure inside the box is equal to one atmosphere, the pressure on the windward surface is greater than the pressure inside the box, and the pressure on the leeward surface is less than one atmosphere. If the sides of the box are flexible membranes, the windward surface will bulge inward, and the leeward surface will bulge outward. If we pump air into the box until it is equal to the pressure on the windward surface, the net force on that surface will be zero, but the force on the leeward surface will be the sum of the increased inside pressure and the negative air pressure. If we pull air out the box, the effects will be reversed. The point is that the force of the pressure of the moving air against an obstruction is constant, based on the velocity of the air and the character of the obstruction. The net pressure on a specific surface, however, is dependent upon the combination of the external pressure and the internal pressure if the obstruction is hollow.

Of interest to us when water is present on the exterior surface is the specific combination of external positive pressure relative to the internal, because that is the condition conducive to leaking. Water may be present at the exterior of the negative pressure surface, but that water will not penetrate the building through a fine-line fissure or crack. In fact, it is possible to observe buildings "blowing bubbles" through water that is running down a negatively pressurized surface. The exact inverse is occurring on the other side of the building.

We should also point out that it is also possible to develop relatively positive pressure on an exterior surface by means other than moving air. Some buildings are relatively negatively pressurized through mechanical means. Either deliberately or inadvertently, air handlers in a building can pull air out of a volume faster than it can be made up through infiltration or open venting. This is done deliberately when something is inside the building that we prefer to keep inside such as microbes or fumes. We can filter the air as we pump it out; and unless we do, it is possible for the noxious air or microbes to seep out through the cracks. We can reasonably preclude this happening by reversing the air flow through those cracks and fissures so that the air is always moving from the outside to the inside. Needless to say, negatively pressurized buildings are particularly challenging to keep from leaking. Buildings deliberately designed to be positively pressurized include medical facilities, where the objective is to keep the noxious air and microbes outside. The same explanation applies to positive pressurization, but in reverse (however, positively pressurized buildings rarely leak).

Despite the best intentions, under the wrong conditions, it is possible to overwhelm the pressurization system, causing a localized or temporary external positive pressure. This happens in taller, more tubular-shaped buildings. As the air inside the building warms, for whatever reason, it rises within the enclosure of the building as a whole. The air pressure in the upper part of a high-rise building is typically greater than the air pressure at the lower level of the same building. This is called the *stack effect*, derived from smokestacks, which operate on the principle that hot air rises. (In fact, hot does not really rise; it is pushed up by denser, cooler air. It displaces it by the effect of gravity on the denser air mass, so in the end, warmer air winds up at a higher elevation.) In tall buildings, the warmer air at the top tends to positively pressurize the upper part of the building. Consequently, tall, tubular buildings rarely leak at the upper levels. The surcharge of air that accumulates at the top of the stack must, however, be replaced from somewhere, and that somewhere is typically by infiltration through the walls of the lower levels, which are negatively pressurized because of the effective evacuation created by the rising warm air. The result of this is that the lower levels of tall, tubular buildings tend to leak from all sides when it rains.

The overarching point here is that pressure differential between the inside and the outside of a building may develop for several reasons at the same time. In the first place, wind is not the only source of pressure differential. And where the source is not wind, the pressure differential will be continuous and uninterrupted, which means that the building will be more inclined to leak when it rains because the wind direction is not as much a determinant. This can be an advantage as we can deliberately design structures that are positively pressurized mechanically. We simply use the outside air and a big fan to inflate the building; and if we are very clever, we can selectively inflate the building only when it is raining.

More commonly, however, pressure differentials will develop as a natural consequence of wind and other factors, even if the differential is slight, momentary, or fluctuating. For that differential to exist, there must be segregation of the air masses, which we assign as outside and inside. The separation of those two air masses must be maintained by a barrier that is impermeable or relatively impermeable to air. This is the air barrier; and in pressure-equalized systems, this function is assigned to the inner layer of the system.

The impermeability of the inner layer of a pressure-equalized system is more important than in the drained cavity system; the impermeability of the outer layer is less important. It is always assumed that water is in the cavity of the drained cavity system, and as long as the water drops to the bottom of the cavity without wetting the cavity side of the inner layer, it is not important that the inner layer be an effective air barrier. In the pressure-equalized wall, however, for the air pressure inside the cavity to be reliably equal to the outside air pressure, it is necessary not only to deliberately connect the cavity with the outside air, but to ensure that the air not leak through the inner layer and lose the pressure in the cavity.

The impermeability of the inner layer is important and functionally necessary, but it need not be perfect. In fact, the pressure-equalized system is effective as long as the permeability of the inner layer is 10 percent of the permeability of the outer layer. The rationale behind this experimentally verified percentage is that a major determinant of the performance of the pressure-equalized system is the time it takes to achieve inflation of the cavity relative to the air loss through the inner layer. In practice, it is fairly easy to accommodate the 10:1 permeability ratio of the two layers. That said, we point out that the more permeable the inner layer, the more likely it is the system will perform as a drained cavity system, whether that was the intention of the designer or not. A good air barrier is necessary to achieve the pressure differential between the inside and the cavity. It is also of value relative to air infiltration whether it is raining or not. (We will discuss infiltration in more detail when we discuss how walls leak air.) Good air barriers also have the belt added to the suspenders of the pressure-equalized wall. As in drained cavities, condensation can form anywhere in the pressure-equalized system. Further, the presence of a reasonably moisture-proof inner layer can preclude the migration of liquid and vapor into the more vulnerable materials of the inner layer and of the inside generally.

Effecting the inner layer air barrier can be as easy as applying damp-proofing to the backup block, adhering membrane roofing vertically, or using any other approach that results in a fairly air tight inner wall. It is not necessary that the air barrier be as resistant to exterior conditions as the outer layer, because it will not be as subjected to hostile weathering processes as the outer surface of the rainscreen.

The pressure chamber of the pressure-equalized system is not significantly different from the cavity of the drained cavity system. Its essential characteristic is that it be a void, which is not that difficult to achieve. Again, in the pressure-equalized wall, we assume that water is not present in the chamber; therefore, there is less concern about bridging the chamber. That said, as a practical matter, it is prudent to apply similar, if not equal, rigor to chamber integrity and clarity as drained cavity systems, for construction purposes. For masonry outer layers, that means the minimum size be two inches from the back of the outer layer to back of the chamber. The surface at the back of the chamber may, in fact, be the surface of the air barrier; or, as we will find shortly, it may be the face of rigid insulation. The width of the pressure chamber is not a function of the nature of the back surface; it is driven by the fabrication requirements of the mason or builder.

Aside from fabrication requirements, the smaller the pressure chamber, the better. The performance of the pressure-equalized system is driven by the speed at which pressure equalization can be reached. Remember, time delay is a function of the magnitude of the pressure differential, the orifice area between the chamber and the outside air, the rate of leakage across the air barrier, and the volume of the chamber itself. It is the last variable that is the subject here. All other things being equal, a small chamber will equalize pressure faster than a large chamber simply because fewer cubic inches of air are moving. Where there is no construction requirement for a two-inch chamber width, as when the outer layer is a panel system, there is considerable advantage to reducing the pressure chamber width to less than two inches. The absolute width is not relevant.

Volume-to-orifice ratio is important to consider, but it has not been experimentally determined. If the chamber is very small then, within limits, the orifice connecting the chamber to the outside air can be small. The question is, how small is small. Obviously, if the chamber volume is large, the total orifice must be large. When the vertical or horizontal dimensions of the chamber become great, there is greater concern over the compressibility of air. Suppose the volume of one chamber is 1,000 cubic inches, the dimensions of which are 10 inches on a side, and the orifice is one square inch in a cross-sectional area located at the geometric center of one of the faces. Next suppose that the volume of a second chamber is also 1,000 cubic inches, and that the orifice is also one square inch in area. The second chamber, however, measures 2 inches by 2 inches by 250 inches long, and the orifice is located on one of the small 2 by 2 faces. Under the same pressure differential, the second chamber will take slightly longer to equalize than the first, because air is a compressible material. This means that, in both cases, as air is pushed into or pulled out of the chamber through the orifice, there will be a time lag before all the air in the chamber is pressure-equalized. The farther from the orifice, the longer the delay, all other things being equal. It follows, then, that not only do we want the orifice to be of sufficient size in terms of area, we also want the orifice to serve a limited area or to be divided into multiple orifices so that the chamber is rapidly and evenly equalized.

Partly for this reason, chambers are divided into compartments, which limits the volume and dimensions of any one chamber. This is accomplished using baffles, which act as partitions to subdivide the larger chamber into smaller, more rapidly responsive chambers. The baffles can be simple strips of rigid insulation, which "choke" the chamber into smaller compartments. But, note, each compartment must be vented, ideally with a single vent. A single vent is because of a second and far more important reason for partitioning the chamber into compartments: If a compartment is vented with two or more orifices, it is possible that a stream of air moving laterally across the face of the wall will generate a cross-current, or draft, within the chamber. That draft acts, essentially as a leak within the chamber, and can cause a pressure differential across the face material, especially the area immediately adjacent to the intake orifice.

The air seal between compartments does not have to be perfect any more than the air barrier at the rear of the chamber does. The baffles must, however, be resistive enough to leakage to defeat cross-currents. Thus, they too must display less than 10 percent of the permeability of the exterior face.

The vents or orifices may be of any number of forms. In a brick wall, they may be empty head joints distributed over the surface of the wall at approximately 24 inches on center, vertically and horizontally. For a typical two-inch-deep chamber, this is an orifice-to-volume ratio of 1:1,200. Based on experience, ratios as high as 1:2,500 have been successful; but as noted, more controlled experimentation is required to verify this parameter.

And because there are sources of moisture other than rain, namely condensation, it is wise to provide a gutter at the bottom of the chamber, and to connect that gutter to the outside in the same manner as in a drained cavity system.

OPENINGS AND DETAILS

There is a general tendency in the study of building and construction technology to treat each component of a building as conceptually different. In fact, the principles that determine wall design are the same for all components. Moisture is drawn through fine-line cracks regardless of whether the crack is a through-body fracture of a solid material or the line between a gasket and a pane of glass. From the standpoint of water coming through suboptical fissures and pores, all cracks are created equal.

We will examine two specific examples of what are typically thought of simply as details: the double-glazed window, and the joint at the jamb between a window or door and the rough opening. The physics underlying large segments of exterior wall are just as applicable to doors, windows, or the joint between a window frame and the rough opening.

The typical single-layer window is, by definition, a membrane. The glass itself is a near-perfect membrane material, but that may or may not suggest that the system into which it is set is an equally viable thin, single-layer system. In a common divided-light, double-sash situation, the panes are set in a rabbet and sealed there with a glazing compound, also known as putty. The seal between the glass and the glazing compound forms a remarkably tight bond, which often endures for 20 or more years if properly installed.

The single most common problem with single-pane glazing involves preparation of the adhesion surfaces. If the glass is dirty, the adhesive bond between the glass and the compound may not develop or may only partially develop. This is more likely to occur during reglazing a broken pane. If, or when, the compound disengages from the glass the resulting crack may not be more than a thousandth of an inch, but water can enter through the fine-line crack and freeze; the subsequent expansion of the ice will increase the width of the crack, possibly even breaking the glass. If a pressure differential exists across the plane of the window, and if the crack completes a path from the outside to the inside, and water is present at the outside end of the crack, the window will leak, as surely as any thin, single-layer membrane with a flaw will leak. The fact that the element with the flaw is a window and not the wall is irrelevant. The principles described relative to single-layer and multiple-layer systems are applicable at any scale.

Windows are good examples of pressure-equalization, because their function is not unlike that of the field of the wall, in that it must provide a barrier to the elements. In most cases, however, the window is expected to function as a barrier, yet be transparent, to let light in and for visibility. These two functions, though coincidental, are not inherently related. (There are, of course, translucent or even opaque windows, though arguably they are not.)

A window's two functions complicate the design process, but they do not alter the fundamental goal of maintaining the physical separation of the inside air mass from the outside air mass; nor do they alter the fundamental approaches to defeating the pressure differential, which is the underlying motive force responsible for water and air infiltration.

This discussion has been based on the simple, single pane, set in glazing compound in a wood muntin and frame. That combination has many variations, from glazing compound and wood frames to neoprene gaskets and aluminum frames to silicone beads and vinyl frames. For all, the physics remain the same; they are all examples of thin, single-layer systems. In their basic configurations, these variations have at least four components: transparent glass, the frame, the seal between the light and the frame, and the seal between the frame and the rough opening. The presence of multiple panes only adds to the number of repetitions of the same set of basic components.

Likewise, with most thin, single-layer walls, the window materials are essentially impermeable. And many of the materials comprising the components will deteriorate to the point of failure, but the system is vulnerable primarily to the deterioration of the adhesive bonds between the components. In the simplest four-component, fixed, thin, single-layer systems, leaks around the glazing seal are fairly easy to detect and repair. Leaks around the frame/rough opening seal are less obvious, but if the seal is a single layer, the leak will not be difficult to detect.

The problem is, even simple installations are not all that simple. The glazing gasket is likely to be between the glass and a fixed glazing stop on the outside and a wet seal between the glass and the applied glazing stop on the inside. The glass itself will be set up from the lower frame on shims, producing a small void around almost the entire perimeter of the glass. This continuous void may be small, but if it is not vented back to the outside, a pressure differential probably will develop between it and the outside. If there is a flaw in the outer gasket, and a pressure differential between the outside and the small void, water will be drawn into the small void, to accumulate until the water covers a flaw connecting the small void to another zone of lower pressure, such as the interior of the building. The water will, commonly, work its way past the frame via the attachment holes, into the rough opening surround or spandrel below the window. In curtain wall systems, the frame itself is essentially a tube, and water can collect in the void of the hollow frame, through the attachment holes or, more commonly, through the joints at the corners where the material is cut.

Mitered or box cuts at the corners of frames may also provide avenues of entry, not only to the hollow void in the frame, but directly to the inside of the building. If the glass is exterior glazed, the mitered glazing stops can provide a path into the small void around the glass, which then connects to the frame void or to the interior. The corner joints at the bottom of a window are particularly vulnerable, but cuts at the top of the frame can draw water as well. Seals at these locations are extremely important; but they are impossible to inspect once the frame is installed. In higher-priced aluminum and steel frames, the designer can specify fully welded and ground corner joints, which not only provide excellent sealing of the joints but enable easy identification of substandard manufacture. In contrast, when the corner joints of any assembly are butted and sealed, quality assurance is difficult.

In aluminum windows it is common for the frame to consist of a primary frame, for receiving the glazing, and a separate subframe, which is attached to the rough opening. This approach allows the primary frame to expand into slots provided by the subframe, to accommodate the thermal expansion characteristic of aluminum assemblies. This means, however, that the subframe is fixed to the rough opening, and the corners of the subframe are likely to be sealed in the field. There is yet another perimeter seal between the subframe and the primary frame, which is a potential point of seal failure. Given the construction and fabrication complexities of modern window assemblies, even allowing for advances in gasket technology, it is extremely difficult to fabricate a durable, pure single-layer window assembly.

Most curtain wall systems combine the thin, single-layer approach with a pressure-equalized approach, in a form of a hybrid system. The glazing, gaskets, and frames are used as the impermeable materials that they are, and the voids formed through their joinery are treated as pressure-equalized chambers, including the gap between the frame or subframe and the rough opening. If these voids are connected back to the exterior with sufficient vent area, they will operate at the same pressure as the outside air. The requirements for a hybrid system are identical as those for pressure-equalized walls, except the scales and particularly the chambers are typically smaller. If the small void around the glass, for example, is connected to the void in the hollow frame, and if that void, in turn, is connected back to the outside, then both voids act as pressure chambers. This is true even for the gap between the frame or subframe and the rough opening. If the joint is detailed such that there is an outer barrier near the outside surface, and a separate line of filler toward the rear of the joint that serves as an air barrier, the void between the two lines of filler can perform as a pressure chamber, if the void is positively connected back to the exterior. Despite the common apprehension that such connectors will open paths to water infiltration, the reality is that, as long as pressure equalization is achieved, water does not pass across the outer barrier.

Much of the credit for the development of the hybrid curtain wall goes to Swiss engineers working in the early 1950s for the aluminum consortium Alusuisse. The influence of those early Alusuisse designs can be seen more than 50 years later.

A comparatively recent development was the introduction of double-glazed systems, the most common of which is the *insulated glass unit*, or IGU. Designed primarily to reduce consumption of energy in our buildings, double glazing, generally, and IGU, specifically, are thermally more efficient than single glazing, but not enough to justify the cost, based on energy costs alone. But double glazing also offers greater creature comfort by reducing drafts at or close to the plane of the glass. This is the result of a warmer inside pane of glass, which means there is little if any chilled air descending the inside surface of the window. Double-glazing also offers acoustical advantages to single glazing, which we have come to expect and enjoy.

Double-glazing comes in three forms: the aforementioned IGU, interior vented systems, and pressure-equalized systems. Each has its advantages and limitations. The IGU consists of two panes of glass bonded together with a spacer strip and a perimeter gasket or seal. The air between the two layers of glass is sealed into the unit at whatever ambient pressure existed at the time of manufacture. If the pressure on either face is greater than the pressure of the trapped air, there is an implosive force on the unit. If the pressure on either face is less than the pressure of the trapped air, there is a net explosive force on the unit. Though there are no reports of units that have actually imploded or exploded, the units do undergo constant changes in pressure on both faces and the seals. These pressure fluctuations, combined with other weathering effects on the seals, typically result in microcracking or debonding to the seal and one of the layers of glass. Water vapor is pumped into the unit; when it cools, it forms condensation inside the unit. The condensate remains inside the unit, which gradually clouds over; and eventually the unit must be replaced. The typical IGU will fail in 10 to 15 years; specifically, the seal will fail, and the unit will become fogged with condensation. If the replacement costs of IGU are factored into the total cost, rarely can their use be justified. Still, they are convenient: they arrive at the assembly point as a unit, ready to be inserted into the frame; consequently, the glazing process is familiar and relatively simple. In the event of impact damage or fogging, the reglazing process is not substantially different from reglazing a single light. As a single unit, IGUs perform as single-layer systems relative to pressure differential; therefore, they leak similar to single-layer systems, at and around the perimeter glazing seal.

The interior vented system, is the apparent cousin of the drained cavity system, although with a major distinguishing difference, which negates the similarity. In the interior vented system, not only is the cavity at the same pressure, hence of the same air mass, as the interior volume, the cavity is drained to the interior as well. The outer layer of a drained cavity system is a permeable water barrier, slowing the water penetration enough so that it collects in the cavity and is drained back to the outside. In the interior vented system, the outer barrier is a thin, single-layer system, designed to defeat any pressure differential relative to both air and water. The gap between the layers is serving other purposes altogether, because the outer layer does all the work of a barrier. As with IGUs there is some antidraft benefit to the interior vented system, as well as some nominal operating efficiency. The disadvantage of the interior vented system is that the weather integrity is no better than with any other thin, single-layer system, and it costs more.

The interior vented approach is used more in retrofit situations than in initial installations, most commonly exterior mounted storm windows. The most frequently occurring pathological error associated with exterior storm windows is caused by failure to clearly designate the primary pressure barrier. Exterior storm windows are often loosely fitted to the frames of the existing windows without gaskets or seals. This suggests that the chamber between the two layers is intended to be part of the outside air mass, hence a pressure-equalized system; but there is no specific connection other than incidental infiltration, and there is no specific provision for drainage of incidental moisture or condensation back to the exterior. If the seal between the original window and the interior of the building is less efficient than the seal between the storm window and the frame, the chamber will operate as part of the interior air mass, at or near the air pressure of the interior. This means that the pressure differential exists across the outer layer, which is, therefore, a thin, single-layer system, with all of the frailties of that approach. Exacerbating the situation is that most exterior storm windows are neither manufactured nor are they detailed as membrane systems.

If the window is detailed from the outset as a double-glazed system, with the outer panel as the pressure panel, the chamber is connected to the interior, and the inner panel has little practical value except as draft reduction. If the interior panel is, however, inherently vulnerable to vandalism or other damage, then the exterior pressure panel may simultaneously provide physical protection to the interior light, as well as single-layer protection to the interior. Examples of such a condition are stained glass, prism glass, leaded glass, generally, or other glazing of a precious nature such as historically significant or antique glass.

In many such cases, the protection is a later addition, which can compound the difficulty of achieving the desired level of air barrier at the exterior surface. There is an alternative approach to exterior protective layers, which we will discuss shortly, but this discussion focuses on the exterior pressure panel approach, which assumes that the pressure differential is defeated at the exterior layer. If the protective panel is part of the original design, the designer will detail that panel as a thin, single-layer system. The protected glass or layer rides inside, absent pressure or weather. The remaining design issue, and the point of some pathological significance, is whether to vent the air space between the layers, and if so, how to vent the air space.

If the exterior panel is sealed, and an air barrier and the air space is not vented at all, then the glazing is effectively an insulated glass unit using a piece of precious glass on the inner light. The air between the two layers is trapped at whatever static pressure was present at the time of glazing. As the exterior air pressure varies, the exterior panel presumably is capable of absorbing and distributing the relative positive or negative pressure. As the interior pressure varies, however, the relative positive or negative force of that fluctuation will act on the inner panel, which is precisely the panel that is supposed to be protected. Even if the inner layer is not precious or vulnerable to lateral forces due to pressure differential, there is another problem.

Recall that one of the disadvantages of IGUs is that the perimeter seals fail and the unit fogs up with trapped condensation. If the panels are not detailed to accommodate that eventuality, the risk of failure of the perimeter seal increases. If the inner layer is constructed of divided lights, there is the distinct possibility that the seals around the individual lights will fail, especially if the unit is subject to the lateral force of a pressure differential. The trapped condensation may be of greater concern than the optical fogging of the glass. The accumulation of moisture in the air space could become a source of deterioration of the materials themselves.

For this reason, the air space is typically vented. If the exterior layer is the pressure panel, obviously, the only way to vent the space is to the interior. If the air space is vented to the interior, only a few basic vent configurations are possible: top, bottom, side, and combinations of these options. For chambers vented to the interior, the selection of vent locations is not particularly important, except to account for vents placed on more than one edge. Top and side vents will probably collect more dust in the chamber than bottom vents, but any one of these is preferable to vents on both top and bottom. Despite assertions as to the efficacy of opposed vents relative to evaporative advantages, opposed vents will allow cross-currents to circulate in the chamber. While this may, indeed, accelerate evaporation of water in the chamber, this arrangement allows the formation of drafts and currents across the inside surface of the exterior pressure panel; and these drafts, of relatively warm, moist interior air, will produce the very condensation that must subsequently be evaporated. It is elimination of this very condition that we set as an objective when we raised this issue. If the application of the exterior protection is not part of the original installation, or is a demountable application, if the window is a double sash or casement, it is often relatively simple to open the window slightly, which will unequivocally connect the air space to the interior volume. If the window is a double sash, then only the lower sash should be opened.

The mounting of a panel, whether the primary purpose of the panel is to defeat pressure differential or to provide physical protection of the inner layer, is a somewhat controversial issue. Some practitioners advocate venting the air space to the interior, as described above, while others advocate venting the air space to the exterior. If the air space formed by the application of an outer protection layer or storm window is vented to the exterior, the pressure in the air space will track exterior air pressure. This means that the space is, in fact, a pressure-equalized chamber. There is no inherent problem with the creation of such a chamber on an existing window, but some unanticipated problems may occur. Proponents of exterior venting have good reasons for their position, namely a concern for condensation, and the relative difficulty of developing a thin, single-layer system on a retrofitted panel to prevent leaks.

Venting the chamber to the exterior, thereby creating a pressure-equalized chamber out of the air space, can and probably will reduce the risk of water infiltrating to the occupied zone; but it also means that the interior layer is the pressure panel and will, therefore, absorb the lateral force of the pressure differential between the outside and the inside. If the inner panel is, in fact, designed to absorb such a load, then exterior venting is, generally, the preferred option. If, however, the original purpose of applying the exterior panel was to relieve the inner panel from lateral force, which may be the case with a divided light or stained-glass panel on the interior, then venting the chamber to the exterior may not be advisable.

The options discussed so far have the added panel or storm window on the outside with the option of venting to either the inside or the outside. Another option is to place the added panel on the inside. In the case of stained glass or divided lights this has the obvious consequence of placing the vision priority on the outside. If that is the preference, then the storm panel can be mounted to the inside relative to the original, or priority, panel. If the air space is vented to the inside, the outer panel is the weather panel and the pressure panel, which is to say, simply, that the outer layer is a thin, single-layer panel. If the outer layer was the intended beneficiary of the additional panel, venting the air space to the inside is not consistent with the objectives.

If the air space is vented to the outside, then the interior light is the pressure panel and air barrier. Even though the outer layer is the panel exposed to the weather, it will operate without the lateral force of the pressure differential. When the relatively vulnerable light must be mounted on the outside, because the priority of vision is from the outside back toward the façade, then placing the pressure panel on the interior and venting back to the outside is the optimal solution.

In summary, in general, adding a second light to an existing window presents a number of options as to panel location and venting direction. Each has appropriate applications, but also potential disadvantages. Determining which to use will be directed by the objectives of adding a second layer. The most common cause of untoward consequences of adding storm sashes and protective glazing is failure to articulate the objectives. When the objectives do not exist, it is impossible to adopt an appropriate technique. It is not unusual to wind up with systems of mixed, and at times ineffective, details, which are ultimately traceable to the absence of clear objectives.

Another location of interest is the joint between the jamb of the window frame and the rough opening, but this is more problematic than either the head or the sill, primarily because designers are more cautious with these elements. Heads of windows tend also to be the bottom of a cavity above, and are, therefore, flashed accordingly. Often, sills are flashed for the wrong reasons, but the flashing below the sill typically benefits the window whether the detailer's motives were well considered or not. Undersill flashing is generally assumed to be a back-up to the sill, which is typically horizontal, jointed, and/or porous. The horizontal component is by itself sufficient to satisfy the requirement for motive force across a barrier, because gravity alone will pull water through a vulnerable sill. The orientation will also tend to collect water to a higher average depth than a comparable sloping member. If the material is inherently porous, and water is present on a horizontal sill, then all of the necessary and sufficient conditions for water penetration are present, and the sill will leak. Through-body joints will accomplish the same result although the development of the fissure may take some time. Multiple-piece stone sills are notable for leaking through the joints. Sills composed of multiple pieces of metal with the joints filled with sealant will eventually fail, either through the body of the sealant or at the boundary of the sealant and the metal. If the sill is made of wood, even painted or coated wood, there is the chance that the wood will check or rot, thereby opening an avenue through the material. The paint or coating will crack, providing a point of entry for water to accelerate the deterioration process relative to the wood.

Recognizing that, for any number of reasons, the sill may fail to exclude water is a sound reason for flashing below the sill. The section through a sill is best viewed as a horizontal, drained cavity, and should be detailed accordingly. If the sill is a single, solid, albeit porous, member, then a pressure-equalized approach is viable, but the chamber must still be wept back to the outside. If the sill is a continuous, impermeable member, such as a one-piece vinyl or metal sill, then flashing is theoretically not required based on this analysis; however, water penetration through the sill is only one of two major reasons for undersill flashing.

The second reason for installing undersill flashing, to defeat jamb joint failure, is present regardless of the composition of the sill. Arguably, this is the more important of the two reasons. When water penetrates the joint between the window frame and the rough opening, it does so in one of three locations: through the body of the filler, or across the boundary plane between the filler material and either the rough opening or the frame. By now, the reasons for water penetration across fine-line cracks and fissures should be committed to memory: pathway, water, and motive force, normally in the form of pressure differential. The distinguishing feature of jamb leaks is that the trajectory of the leak is not strictly horizontal. If the joint is, in fact, a small, vertical, drained cavity, then water will traverse the outer layer, collect in the narrow, vertical cavity, and drain toward the sill. If the water is not intercepted by the undersill flashing, it will continue down the wall until it pools somewhere.

One such somewhere is the floor line below the window. It is not unusual in steel frame construction for the floor slab to pierce the line of the backup block, and for the edge of the slab to be supported by a spandrel beam. If the cavity is choked at each floor and flashed to the outside, then the jamb water is managed the same way as any other water collecting in the cavity. (This assumes that the narrow cavity behind the outer layer of the jamb is clearly connected to the larger wall cavity.) A similar possibility exists with pressure-equalized systems, namely that the small chamber behind the outer layer of the jamb joint is positively connected to a part of the larger wall chamber. Pressure-equalized jamb joints are highly effective, and can be constructed whether the field of the wall is a pressure-equalized system or not.

If the wall is a single-layer system, any water collecting in the jamb joint will either be pulled through the wall altogether or collect at the sill of the rough opening below the window. It is, therefore, particularly important in single-layer systems, whether they are thin or thick, that the undersill flashing carry across the bottom of the jamb joint, turn up and into the wall on the outboard side of the joint, and drain positively to the exterior. A rather old, but sound, sill detail that took all these features into account was the one-piece marble sill. The material was relatively impermeable, although not perfectly so. The sill was shaped such that the portion below and outboard of the window was slightly scooped, to form a raised rim on the ends and inboard edges. The piece fit into the rough opening such that the ends extended beyond the rough opening; the rim was higher than the exposed part of the sill, and the window sat on the back rim with a slight drip over the edge of the rim into the shaped basin of the sill. The one-piece sculpted marble sill was, and still is, one of the most technically advanced pieces of building design in history.

WHY LEAKS ARE DETRIMENTAL

This section addresses those aspects of leak formation that are responsible for so much of the difficulty of detection and intervention—at least insofar as they occur in walls and components of walls. The dominant reasons for the difficulty of detecting leaks start with people: we misunderstand the physics of leaks, specifically, certain aspects of their physics that make leaks more difficult to detect. But by becoming knowledgeable about those few points, we can improve our chances of finding the source of leaks and of designing effective intervention approaches.

We begin by reiterating the basic leak mechanism, which means accepting and understanding the necessary and sufficient conditions: leaks occur only when all necessary and sufficient conditions are present; when they are, a leak will always occur. If any one of the necessary and sufficient conditions is removed, even momentarily, the leak will stop. As soon as they all are present again, the leak will reactivate.

Moreover, once a leak is manifest—even if for the first time in the history of the building—one of the necessary and sufficient conditions is established as a fact, namely a viable pathway, and that pathway will usually remain present. Typically, the fluid capacity of the pathway will increase because such pathways that develop in aging buildings are the result of advancing and progressive deterioration.

Previous sections have examined the fundamentals of capillarity, fluid properties, differential and hydrostatic pressure, porosity and permeability, absorption and adsorption, adhesion and cohesion, Poiseuille's and Bernoulli's Laws, evaporation and condensation, solution and precipitation, and the properties of water generally. These principles, in place and in force at all times, if not immutable, represent our best understanding of the reality we call the world. To the extent that they are valid, they are valid for all buildings and all building designers. They are valid for tenements and monuments. And they are valid all the time.

Once a leak has manifested, it probably will recur. A single event is enough to establish proclivity. Of course, there are leaks that occur only under extraordinary conditions, but we cannot begin from that position. The leak occurring during a hurricane is strong evidence of an extraordinary circumstance, but by itself the hurricane is not enough to establish the rarity of event. We need also to know the history of the velocity and direction of the wind, the rate of precipitation, and the duration of the storm. From these facts we can begin to construct *event scenarios*, which could lead to the same leak. We may find that the ordinary noreaster has the velocity and rainfall, but not the duration, to match the circumstances, or that a passing warm front has the duration but not the velocity and rainfall. In other words, the circumstances of the initiating event may have been unusual, but also quite similar to other events carrying different labels. If the properties of the pathway degrade only slightly, we may find a number of events that can reactivate the leak.

Poiseuille's Law, that filling a capillary to transport water through a small channel takes time, forces us to take into account the duration of any set of external pressures and precipitation rates. It is possible that a capillary route was, in fact, filled for the first time as a result of the rain delivered for a protracted period under a high level of differential pressure. It may be that such circumstances were necessary to initiate the leak. But now the capillary is filled, and unless the fluid in the capillary dissipates, the reactivation of the leak will not require the same set of circumstances. A saturated pathway can reactivate under much less severe circumstances. We can think of such leaks as the aftershocks following a major earthquake.

In all probability, the leak pathway existed prior to the initiating event; the point is, that pathway has now had water pass through it. The passage of water through a capillary normally has an erosive effect on the capillary itself, meaning that the size of the capillary has increased. We know that the fluid transportation capacity of a small capillary is a function of its diameter to the fourth power. We also know that any salts dissolved from the capillary surfaces near the orifice were transported to the interior evaporative surface. This increases the size of the capillary at the orifice ends, and moves the potentially absorptive salts to the evaporative end. Both of these actions increase the capacity of the capillary in the future.

Clearly, then, reassembling the sequence of events preceding the activation of the leak must take into account, not just the circumstances present at the time of notice of the leak, but of those prior. Given the time delay from initiation to recognition, it is possible that the capillary began filling much earlier. It is not illogical to believe that a leak is connected with the most proximate event, but it is just as logical to view leaks as a cumulative phenomenon and that the precipitating event is no more than the last step in the sequence that results in awareness.

With that said, we point out a condition associated with the establishment of the pathway that may occur only when the differential pressure reaches a threshold, and when the threshold remains at an effectively high value. If the pathway forces the trail of water vertically upward as it traverses the wall, there may well be a durable minimum velocity of wind or other source of differential pressure which can activate the leak. The pressure resulting from a stream of moving air against a surface perpendicular to the direction of the stream is a function of the square of the velocity.

That value, however, can be deceptive, in that it represents the force resulting from the pressure differential at any given point. If the obstruction is a slab of monolithic concrete, for example, the pressure differential is measured across the thickness of the obstructing slab from the windward surface to the leeward surface. These values are represented in Table 4.5 as the total pressure differential associated with a particular wind velocity. The pressure differential can also be represented in terms of inches of a vertical column of water that can be sustained by such a pressure (these values appear in the third column). At rather elevated velocities, therefore, the potential vertical rise may reach several inches.

In buildings, however, these maximum potential values of vertical rise are not likely because buildings are not monolithic slabs; they are hollow boxes. If the internal pressure of the hollow box is neutral, then the pressure differential on the windward side of the building is positive, and half the value of the total potential pressure differential. The pressure at the leeward side is negative, and half the value of the total pressure differential. Positive pressure tends to pull water and air into the volume, negative pressure, out of the volume. The façade that is vulnerable to water infiltration is the positive side. Air infiltration will also, obviously, occur at the positively pressurized surface; in contrast, exfiltration, which is the loss of air, can simultaneously occur at the negatively pressurized surface. In terms of thermal loss, exfiltration is as much a concern as infiltration.

Table 4.5	WIND VELOCITY AND FORCE		
Velocity (mph)	Total Pressure Differential (psf)	Total Pressure (Inches of Water)	Half Pressure (Inches of Water)
5	0.123	0.024	0.012
10	0.492	0.095	0.048
20	1.969	0.38	0.19
40	7.873	1.51	0.75
60	17.71	3.41	1.70
80	31.49	6.06	3.03
100	49.2	9.46	4.73
120	70.8	13.6	6.80
150	110	21.2	10.6
180	159	30.6	15.3

To maintain the pressure-neutral status of the internal volume, infiltration must equal exfiltration; otherwise, the box becomes either positively or negatively pressurized. If the building is not pressure-neutral, then one surface will experience more than the half pressure of the pressure-neutral volume. For example, suppose that, during a storm, we open all the windows on the leeward side of the house and close all the windows on the other three sides. As the storm passes, we track the leeward side and open and close windows respectively to keep the leeward side open. The pressure inside the building will equalize with the negatively pressurized side, the leeward side, and the pressure differential at the positive surface will move close to the value of total pressure differential shown in the table. That means that the potential vertical climb of water on the windward surface will increase, which will significantly increase the risk of leaks; but it will reduce the risk of a far greater catastrophe.

If, in such a circumstance, the building remains sealed shut, the internal pressure relative to the wind is more likely to remain neutral relative to the windward and the leeward surface pressures. The internal pressure is not, however, zero; it is some absolute value. As the hurricane passes directly overhead, the atmospheric pressure at its eye drops precipitously with the risk that the internal pressure in the building is suddenly much higher than the outside atmospheric pressure. That differential can be enough to explode the building, a fate considerably more grievous than water infiltration.

To appreciate the power of differential pressure relative to potential lifting of water, imagine that the water in a small fissure or pathway is analogous to a straw with the lower end immersed in water and the upper end connected to the interior of the building. The pressure differential in a straw is typically supplied by creating a partial vacuum at the upper end. To get a column of water moving through and out of an eight-inch-long straw requires a differential pressure of approximately 4.6 ounces per square inch, or the equivalent of approximately 41.6 pounds per square foot. If the same pressure differential is supplied by an increase in outside air pressure, as opposed to the creation of a vacuum, then the overpressure, which will achieve the same eight-inch rise in the straw, is approximately 0.018 atmospheres of pressure. If the source of the external pressure is wind, the velocity required to lift fluid up an eight-inch straw is approximately 92 miles per hour.

If water is sheeting down the surface of a wall, and occludes or covers the extreme end of a potential pathway to the interior, just the wind blowing at 92 miles per hour does not assure that the water will be pumped vertically as much as eight inches. The pressure differential of 0.018 atmospheres must exist between the outside and the inside for a sustained period of time, to allow the fluid to flow through a constricted capillary at a rate consistent with Poiseuille's Law. During that period of time, the actual pressure differential between the two volumes may be less than the theoretic pressure achievable by the moving air stream, due to general pressure equalization between the inside and the outside. The old rule in hurricane country was to run around the house opening the windows on the leeward side of the storm, to pressure-equalize the chamber, which in that case, was the house as a whole. In more customary situations, simple infiltration will reduce the net pressure differential between the inside and the outside such that the wind velocity required to effect the eight-inch rise is actually greater than the theoretic value.

The time delay between sucking on a straw and having fluid come out of the straw, is in effect with a wall leak, although the delay is probably longer. If during this delay, the pressure drops, the height of the fluid in the straw will recede. Even though the peak velocity of the wind may reach the threshold required to theoretically cause a leak, if that pressure is not sustained for the required initial period of time, the fluid will recede and the leak will never be realized even though the pressure differential reached the threshold. Fluctuating velocities can reach peaks well in excess of the theoretic threshold, but if they are not sustained for the requisite initiation period, the leak will not activate.

Once the fluid in the capillary reaches the maximum altitude, or climbs and breaches the point of maximum requisite pressure differential, the fluid in the pore begins operating as a siphon. The pressure required to sustain the flow may actually drop. Whether that is the case depends on the elevation of the exit location relative to the exterior orifice. If the outlet is at the same elevation as the point of maximum climb, eight inches in our example, then the differential pressure is the same as that required to pull the fluid to that elevation. If the interior outlet is higher than the inlet, but lower than the elevation of the maximum climb, a net positive differential pressure must be sustained to perpetuate the flow of fluid through the pathway; but that pressure is less than what was required to breach the high point. If the outlet is actually lower than the inlet orifice, the fluid flow will be self-sustaining, as long as the supply of water at the inlet is continuous; this is a true siphoning situation.

Siphoning is an important element in understanding leaks. This principle says that a significant sustained pressure differential may be required to initiate a leak; but once the leak is initiated, it may essentially be self-perpetuating—that is, not stop even after the pressure differential drops back below the threshold value. Once initiated, siphoning will continue as long as the water supply, (read: flooded gutter or roof pond) continues to supply water. This means that leaks may continue even after the rain stops and the wind has long since subsided.

Another major reason leaks may go undetected, hence unaddressed, is that we don't usually see leaks as and when they occur, primarily because this requires standing in the rain. Simply put, if we really want to track leaks, we must get wet. We must gather data from the site. If we, the experts, cannot afford the time—or, more likely, our clients are disinclined to pay us to sit and to wait for the leak to reappear—we must find ways of gaining the relevant information within the financial means of our clients. This takes time.

To gain the time to monitor a building for leaks requires the forbearance and patience of the client, who may be expecting a definitive and quick solution, in which case he or she must be educated to the nature of leak detection. Having gained the compliance of the owner to monitor the site of a leak, we must design the monitoring program less to gain information than to *eliminate* options. The essence of diagnostics is, ideally, to reduce the possible explanations to a single mechanism, and to then effectively and definitively intervene in that process. It is tempting, during monitoring to attempt to replicate the leak scenario through full-scale testing. Properly executed, full-scale field tests can be immensely informative and definitive. But improperly executed, field tests can cause more harm than good. That said, because testing costs money, it is not always feasible.

Full-scale tests executed according to ASTM standards are fairly representative of the typical conditions during a rainstorm. The wall in question is sprayed fairly evenly for a prolonged period, and differential pressure is developed by constructing a negatively pressurized chamber inside the building opposite the area being saturated. The volume of water, the duration of the test, and the level of negative pressure can all be varied and controlled. There are, however, some limitations to this type of test. It can be executed only over a relatively small area at any one time. If the investigator wants several areas tested, the testing rig has to be dismantled and reassembled at each location. The tests, moreover, are necessarily limited to rooms and walls where the negative pressure chamber can be assembled on the inside. This means that leaks that originate at, exit from, or pass through the intersection of a floor and the wall cannot be effectively tested with this technique because the pressure chamber cannot be erected through the floor.

As just mentioned, full-scale wall tests are not cheap. A testing crew for one day of testing costs between $2,000 and $4,000, including a report on the findings. This can be enough to dissuade the client from conducting such testing even when he or she is convinced of the efficacy. Thus, full-scale testing must offer enough potential savings to offset the costs. That is usually possible only with systemic problems in newer construction, not when the leak is isolated, the building is older, and/or the value of the damage is low.

4.4.3 Pathology of Vertical Closure Thermodynamics

Heat migrates in both directions through walls in the way and for the reasons discussed at the beginning of this chapter. The fact that heat, or the absence of heat, fluctuates within walls and the materials comprising those walls is well known; and we have seen some of the thermal consequences of that fluctuation on materials. This section details the pathological concerns and consequences of those same basic principles as they apply to the larger context of vertical closure systems.

THERMAL INSULATION

Thermal insulation is the term used for a large class of materials whose function is to resist the flow of heat through them. Generally, as noted previously the substance of which the insulation is made is less important than the amount of air the material can trap and hold, as it is the still air that provides the insulative advantage. Air trapping works in two ways: the air is encapsulated in small bubbles within the matrix of the substance, which we call *closed-cell insulation*, or the air can flow through the fabric of the substance, which we refer to as *blanket material* or, simply, *blankets*. This second type of insulation works not because it literally traps the air, but because the flow of air is so impeded and delayed by the complexity of the passage through the material that the air is essentially trapped. Closed-cell insulation is typified by expanded or extruded polystyrene; absorbed blanket insulation is typified by fiberglass batts. Both broad classes of insulation are effective at resisting heat transmission, and the use of both can result in the pathological deterioration of buildings.

The building-threatening conditions associated with insulation are of two general types: the first is the result of an untoward shift in the thermal response of exterior materials; the second is the result of condensation occurring in untoward locations and quantities within the wall. The former is the result of the insulation working in the manner intended; the latter is the result of insulation *not* working in the manner intended. Both are decidedly more likely to occur in retrofits to existing buildings than in original designs.

It is only fairly recently that the physics of thermodynamics have been applied to building design. Earlier, the technical design of buildings was largely a matter of repeating what had worked in the past and avoiding what did not. A major difference in current building technology and the technology of 200 years ago is our increased reliance on mechanical devices, a subject we will address in Chapter 6. When buildings were heated by burning timber in open fireplaces, and cooled by ventilation alone, the wall sections tended to be more adapted to operating passively in the ambient environment. In the absence of insulation in the wall, what heat was generated in the occupied zone usually dissipated through the walls more rapidly than we consider acceptable today. On the other hand, the loss of heat through those walls also tended to sustain the porous, permeable materials. The relative initial cost of a brick was higher 200 years ago, so the use of interior heat as a form of material protection and maintenance was good resource management. Though it is tempting to discount such combinations of assemblies and operations as the result of limited options, it is incorrect. There *were* other known options and construction techniques. Which combination was selected was based on what was holistically optimal given all the circumstances of construction of the day. Over time, naturally, circumstances change, and we have lost the operating knowledge of the fabric of those older buildings. But they endure, and we tend to expect they will adapt to those changes. When, for example, we insert insulation into an existing wall, we are altering the composition of the entire assembly. When we do so, two general results accrue. Generally, there is an untoward shift in the thermal response of exterior materials, and there is a significantly higher risk of condensation occurring in untoward locations and quantities within the wall.

When the thermal resistance of the wall increases, the rate at which heat will migrate through the wall in either direction decreases. The reason to increase thermal resistance is to decrease the cost of maintaining constant or nearly constant climate conditions to the interior of building. The issue here is that one of the contributing factors to achieving interior environment control is to increase the thermal resistance of the envelope. When we do this, we must accept, as a consequence, that the extreme exterior materials will be less subject to modification by the interior conditions and more subject to the exterior conditions. The temperature and moisture content of the exterior materials will "float" with the ambient exterior conditions. If this is assumed during the original design phase, we can select the materials with those temperature and moisture fluctuations in mind. We may even decide on another approach to water exclusion altogether. When we alter a wall that is already in service, we cannot change the brick to granite or add a cavity where one does not exist. The range of adjustment to existing conditions is severely limited by the mass and tangibility of the existing conditions. And when we alter one variable, we affect all the variables. In our effort to increase thermal efficiency, we necessarily subject the envelope to harsher conditions. If those conditions result in the accelerated deterioration of the fabric, then part of the cost is the energy debit to the fabric. That debit is ultimately payable in the form of increased intervention and/or higher intervention costs, which are, in turn, translatable into energy, an operating expense.

The message is this: Alterations of integrated construction systems, such as walls, roofs, or support systems, can have untoward effects, even if only a single variable such as thermal resistance dominates the alteration scheme. The structural and the contributing systems were not developed or designed unidimensionally, so there is no reason to expect that they will respond systemically and holistically to single-variable modifications. Quite the contrary. Regardless of the virtue of the objectives, any alteration in a system has systemic ramifications. Failure to acknowledge the intrinsic integration of buildings, and their limitations, can cause detriment to the building. Buildings cannot learn. They can, however, absorb modification and change with varying degrees of collateral impact. To what degree is to some extent dependent upon the original character of the design. But we, the agents of change and modification, can also unwittingly damage buildings with simplistic, incompatible, and unintegrated alterations.

INFILTRATION AND AIR SEALS

Generically, the term seal refers to an impermeable junction of two materials. In many cases, we cannot achieve an impermeable junction directly, so we invoke an intermediate material to bond first to one of the pieces and then to the other to produce two seals. An example of the two-stage seal is the head gasket of an internal combustion engine, where the objective is to join the head to the block without the loss of compression in the engine cylinder. We can torque the head bolts until they shear off, and we will not be able to clamp the head directly to the block tightly enough to keep the gases in and the coolant out of the cylinders.

We have a similar situation with window frames in rough openings. We probably could find ways to bolt a frame directly to the rough opening, but probably not in a way that would preclude water or vapor from passing through the joint. So here, too, we invoke an intermediate material, to effect a seal between itself and the frame, and alternately, between itself and the rough opening. These intermediate materials and devices are often the weakest link in the sequence of materials that we rely upon to protect the interior from the exterior; and they are often overlooked as materials in their own right. Not only are they vulnerable to direct deterioration, but their mere presence doubles the risk of seal failure.

For purposes of this discussion we will refer to the class of intermediate materials used as double-sided seals as either *gaskets* or *sealants*. The term gasket is normally reserved for rigid or flexible materials that have a prescribed shape prior to installation, as opposed to sealants, which are extruded from a tube into the joints at the time of installation. A close cousin to gaskets and sealants are *adhesives*, which have a specific structural task of physically joining the base pieces. Adhesives may or may not be required to exclude water and vapor. When they are, then they too, join the great class of double-faced seals.

Fundamentally, gaskets and sealants are low-modulus plastic materials or low-modulus elastic materials with plastic or even viscous surfaces. In plain language, this means they will conform to the irregularities of the rough surface, filling voids and conforming to protrusion at the suboptical level. Once set and cured, they are all inherently impermeable and non-porous, and the bond formed with the substrate is so closely adhered that even vapor under pressure will not pass between the gasket or sealant and the substrate.

Gaskets and sealants fail in two fundamental ways. The material itself can fail in ways familiar to us by now: differential expansion, temperature embrittlement, and so on. As a class, gaskets and sealants are vulnerable to degradation by ultraviolet (UV) radiation. As a mechanism, UV degradation is generally associated with the fracture of long polymeric chains (we devote attention to UV degradation in Chapter 5). Because of recent knowledge of UV vulnerability, most gasket and sealant formulations on the market today are much more resistant to UV, but improved resistance and invulnerability are not the same thing. Most gaskets and sealants are still vulnerable to UV to some degree. In addition, many are vulnerable to ozone; some are vulnerable to acids; all are vulnerable to thermal expansion and contraction.

The last vulnerability, thermal expansion and contraction is complicated by the fact that the seal is wedged between two other materials of rather disparate properties. Gaskets and sealants must contend not only with the prospect of being sheared on one face, but may be at risk of double-shear, depending upon the respective properties of all three materials. As a general rule, when gaskets and sealants are being sheared due to declining temperature, they are also probably being put into tension. When the seal itself is in tension and/or shear across its width, the boundary conditions are as well. The bond between gaskets, sealants, and substrates are less vulnerable to atmospheric pollution and UV degradation simply because of its minimal exposure to the environment; in contrast, the contact layer is by definition in immediate and intimate touch with the substrate. The substrate may react chemically with either the sealing material or the adhesive layer, where there is yet another material used as a bonding agent between the substrate and the material of the seal. This commonly occurs with gaskets, less so with wet sealants.

When they fail, gaskets and sealants do so either through the body of the seal or along the bonding plane with the substrate. Through-body failures are the result of the reasons cited above and in other sections on fundamental material-based deterioration. Bond failures are more likely due to some condition that existed at the time of initial installation, of which there are two types: design errors and preparation errors. A common design error is a miscalculation of the optimal contact area at the bonding plane. If the area is too small relative to the elasticity of the seal material, as the joint opens while the seal material is contracting, under falling temperature conditions, the bonding surface may not be sufficient to maintain the adhesive force required to overcome the total elongation of the seal. This same error can also result in a fracture through the body of the seal. The idea is to balance the total bonding capacity with the strength of the seal material.

The second general class of design errors is incorrect gasket or sealant selection, made more prevalent by the market-flood of new gaskets, sealants, and adhesives, and their attendant promotional hype. Exacerbating the potential for error is that some of the products have basic chemical incompatibilities which, by and large, are not disclosed by the manufacturers.

Preparation errors include improper storage of materials prior to installation, improper cleaning or surface preparation of the substrate, improper bead size, and improper curing conditions. Many glazing seals fail simply because the glass was not cleaned first. Likewise, many masonry seals fail because dirt or dust was on the contact area.

All of this discussion assumes that gaskets and sealants are necessary. In fact, whenever a seal has deteriorated, the pathologist may legitimately query what the effect would be to eliminate the seal. Contrary to intuition, the building may or may not develop a significant leak. In fact, in many details, notably a window surround, the joint is filled from the back *and* the front with different compounds. The back of the joint may be covered with a trim piece, which by itself, is an imperfect but not insignificant seal; the face is filled, let us say, with sealant. Immediately behind the sealant is a backer rod, which is pushed into the gap to prevent the sealant from filling the entire void. Behind the backer rod is a void. This void may be open, or, as is increasingly commonplace, it will be filled with a foam extruded from a canister into the gap, where it solidifies into a rigid, air-entrained filler. The joint is packed solidly from front to back, from outside to inside, with one substance or another. The materials themselves are relatively impermeable, but the adhesive bond with the substrate is weak, and often debonds. This means that there is a fine-line pathway through the wall without the benefit of a cavity or chamber. The conditions are ripe for water intrusion.

Even when the sealant is removed in such a circumstance, it will make little difference in terms of excluding of water from the interior. If, however, the same joint is covered with the trim at the interior face, behind which is a compression tape in the joint itself, and there is a sealant and backer rod at the exterior face, then there is an open void between the backer rod and the compression tape. This void is a potential pressure equalization chamber. If the void is connected back to the exterior to achieve pressure equalization, there is a greater chance of excluding water from the joint than if the same detail is packed solidly with expanding foam or other filler material.

If, on the other hand, the primary purpose of the seal is to exclude air rather than water, the quality of the air barrier is significant. If the detail is intended primarily to defeat water penetration, a pressure-equalized detail will perform as long as the permeability of the air barrier is less than 10 percent of that of the outer layer. If the primary purpose of the detail is to defeat air infiltration, the relative permeability of the two layers is more or less irrelevant. The effectiveness of a barrier against air infiltration is a measure of the minimum total permeability in any single layer, not the arrangement of, number of, or pressure equalization between, layers.

This means that a pressure-equalized system may be effective at defeating water intrusion, even when the air barrier is permeable as long as the relative permeability of that air barrier is less than 10 percent of the permeability of the outer layer. Depending upon the absolute air permeability of the air barrier, this same system may be totally unsatisfactory as an anti-air infiltration device. By the same token, a thin, single-layer system may leak water at an unacceptable rate, and still provide acceptable air integrity.

This simple difference is often discounted or overlooked when assessing and evaluating barrier integrity. Walls that let air pass are often assumed to also pass liquid in unacceptable quantities; conversely, those that let liquid are assumed to let air pass unacceptably. In fact, neither assumption is correct. Effective air barriers and water barriers are essentially and physically different, and the capable design and detail of one does not ensure the capability of the other.

VAPOR BARRIERS AND CONDENSATION

Related to the notion of air barriers, at least in concept, are vapor barriers. They are typically membranes placed below the surface of either the interior or, in some cases, the exterior finish surfaces. Vapor barriers may act as air barriers—and certainly will if they are effective vapor barriers—their purpose has less to do with infiltration than with condensation control. As discussed earlier in the book, the purpose of a vapor barrier is to maintain the segregation of warm, moist air from a plane of condensation. Ordinarily, this means that the vapor barrier is placed at or near the interior surface of an exterior wall, the presumption being that the air inside the building is warmer than the exterior ambient temperature, and that it is relatively humid as a result of interior moisture conditions. For buildings in temperate zones and in cooler climates, this presumption is reasonably well founded. Certainly, while relatively warm, moister air may be outside the structure, even in cooler climates, on balance, placing the vapor barrier at the interior surface is valid. In more consistently warm climates, however, the vapor barrier may be more effectively placed at the exterior surface, particularly if the building is air conditioned.

This subject is raised in this context because the functions of air barriers, insulation, and vapor barriers often overlap, hence become confusing. The risk is that the wall will have two vapor barriers, increasing the risk that water will get trapped between the two barriers without an accessible evaporative surface. The accumulated water will subsequently find ways into the fabric of the structure, resulting in the now familiar array of negative consequences.

In buildings with insulation in the drained cavity or in the pressure equalization chamber, that insulation is typically a form of expanded or extruded foam product. Those materials are impermeable, which means that when they are applied to the backup block, they form a vapor barrier at that plane. The rigid insulation may not be a perfect vapor barrier, because the seams are rarely taped or sealed, but the effect is that the impermeable insulation acts as an exterior vapor barrier, at least relative to the backup. In some installations, a damp-proofing or a waterproofing membrane is applied directly to the backup, so even if the insulation is absent or ineffective as a vapor barrier, the water-repellent material will certainly perform the function. In such cases, it could be highly detrimental to then finish the interior with a nominal vapor barrier, such as polyethylene plastic or foil-faced batt insulation.

The point is, anything impermeable can act as a vapor barrier whether it is called that or not. Simply, if the barrier is impervious to vapor, then it is a vapor barrier. If the location of that barrier is such that it duplicates another vapor barrier somewhere in the wall, then a double vapor barrier exists, regardless of what the various layers are labeled.

As a general rule, when a wall contains rigid, impermeable insulation, no other vapor barrier is required and, in fact, should be avoided regardless of where the insulation is installed within the cross-section. Rigid, impermeable insulations should also have the seams taped and sealed. A vapor barrier, deliberate or not, should perform as a complete vapor barrier, not as a partial or imperfect barrier.

A common flaw of all vapor barrier installations is that the seams and joints are left open, through which vapor can penetrate and flow around such gaps and openings. This compromises the effectiveness of the vapor barrier, which can lead to condensation, infiltration, heat loss, drafts, and leaks. Unfortunately, the proper placement and integrity of the vapor barrier are often sacrificed. Joints are not properly sealed, and sometimes holes are punched through the vapor barrier at receptacles, by perimeter piping, and at fan coil makeup ducts, to name but a few. If the only function of the vapor barrier were to act as an air barrier, modest compromises might reduce effectiveness but only of the compromised area in proportion to the overall area. Recall that air barriers are effective even when compromised; and their reduced effectiveness is a linear function of the magnitude of the compromise.

In contrast, the compromise of a vapor barrier is not a linear reduction of effectiveness. The reduction of effectiveness of a punctured vapor barrier may be negligible in a climate where warm, moist interior air is unlikely to condense in the wall thickness. Even in cooler climates, where condensation may form as a result of the migration of relatively warm moist air through the vapor barrier, the consequences are not necessarily disastrous; the construction of the building may be such that nominal amounts of water in the wall are inconsequential. It is also possible that the water, while undesirable in an absolute sense, will evaporate quickly enough as to have inconsequential effects. But in structures where accumulated water is inherently detrimental, faulty vapor barriers can have disproportionately negative effects. Timber frame structures with blanket-type insulation tend to collect condensate in the depth of the insulation, where it percolates to the bottom of furring space, to rot the timber. In metal structures, the consistent supply of even small quantities of water in the wall space can be extremely detrimental to steel members and connectors. And because the damage is slow, incremental, and not readily visible, the mechanism is detected when it is at the catastrophic stage.

A relatively recent development in the general category of vapor barriers is the use of building wraps, impermeable membranes, typically polyethylene films or asphalt-impregnated felts, applied to the outside surface of the exterior sheathing. The wrap, if perfectly executed, can theoretically provide a thin, single-layer protection to defeat the pressure differential. As described in the previous discussion, a less-than-perfect installation will result in water collecting in the sheathing and in the furring space, causing damage to the structure and the interior surfaces. In construction where wraps are employed, it is common to install foil-faced batt insulation in the furring space, following the axiom of construction practice to place the vapor barrier "on the warm side." In fact, this means that the wall has a double vapor barrier—the traditional foil on the inside surface of the batt insulation and the impermeable building wrap on the outside of the structure. Any water that does get between those two layers will not evaporate, and will remain in the materials and the space for protracted periods of time.

Wraps also raise the risk of condensation forming in the wall due to violations of the inside vapor barrier; this compounds the already detrimental conditions in the wall. The origin of this highly flawed construction combination is the independent development of insulation, the traditional vapor barrier location, the development of the building wrap associated with recent construction practices, and the failure of anyone involved in the process to identify the wrap either as a flawed single-layer system or as a second vapor barrier.

Some builders who recognized the problem of two vapor barriers began using a wrap that is permeable to vapor but impermeable to water. One popular version of this product is Tyvek, which is designed to repel water that might migrate through the outer layers of the construction, but allow any errant water that does penetrate the wall to evaporate back out and through those same outer layers. The theory behind the product is fundamentally sound, but the technical issue is one of rates. Tyvek is not inexpensive, although that may change with the expiration of the patent rights, and many builders are more penurious than they are wise. Ironically in buildings on which Tyvek is used, the general quality of the construction is high enough that the marginal benefit of the material is probably not as significant as its marginal cost; and in construction where the marginal value might be high, the builders are less likely to use the material.

THERMAL RETROFITS OF WALLS

Chapter 3 delved into the issues of thermal expansion and contraction of walls. The principles explained there do not change when insulation is included in the design and construction of an exterior wall. Keep in mind that, as a result of the wall being insulated, the exterior skin of the building will undergo more thermal fluctuations of greater amplitudes than a comparable uninsulated wall. If that level of fluctuation is incorporated into the original design, the consequences of the increased levels of thermal movement can be and, generally are, mitigated to inconsequential proportions. This is accomplished in two basic ways. Either reinforcement, typically in the form of steel wire or rods, is added such that the tensile capability of the wall is increased to a level high enough to endure the internal stress of thermal movement; or joints are installed in the wall, to relieve the stress as it develops when the wall moves. The former effectively absorbs the forces generated by thermal expansion; the latter relieves the stress before it even develops. Both methods are acceptable, and both are routinely and successfully employed in wall design.

When insulation is installed in existing walls, however, the situation is altogether different. The exterior skin, once constructed, is one of the least adjustable of any of the systems of a building. The capacity of the wall to endure thermal movement, either restrained or unrestrained, is built into the fabric of the wall itself. If the amplitude of the thermal shift or swing increases significantly over the life of the skin, in particular, expansion cracks may appear where none existed before.

Such sudden changes can occur in the external microclimate of the wall. Let's say a nearby evergreen tree, which provided shielding from cold wind, or a deciduous tree, which provided shade from the sun is cut down. The wall is now exposed to greater external thermal stress. On the other hand, if other portions of the same wall are performing satisfactorily, without ever having had the benefit of a buffer, the exposure of a relatively protected segment of the same wall will not ordinarily result in thermal stress damage.

A far more likely source of significant thermal stress increase is the introduction of insulation in a previously uninsulated wall, and one never designed to be insulated. When insulation is added to such a wall, rarely is the fabric of the skin either reinforced or has joints added to relieve the stress. In other words, too often, the change is one-dimensional, and does not take into account other aspects of comprehensive wall design and integrity.

4.4.4 Vertical Closure Pathology

Vertical closure pathology is very much the pathology of leaks; and the general patterns of vertical closure pathology are implicit in the discussions of the various types of walls, their components, and materials. The classes of pathology can be grouped along those very lines: material issues, assembly issues, and design issues.

Material pathology issues may manifest as common leaks, but the origin of the failure lies in the deterioration mechanism of the fabric of a component of the field of the wall. If a wall leaks for the first time after several years of satisfactory service, and after exposure to severe weather, the odds are that the immediate problem is the gradual deterioration of a fabric component.

A common example of this type or class of leak occurs thick, single-layer walls that have performed well for several years without any particular problem. Then, rather suddenly, a leak appears, often right through the solid field of the wall. This type of leak is usually traceable to mortar bond failure or some other material deterioration. Often the leak in a wall, particularly a thick solid wall, originates at a higher elevation than the site of appearance, such as a parapet or, more specifically, a failed coping joint. The point is, the delay in manifestation is relevant information, particularly if the delay is a matter of several years.

If the leak appears relatively soon after construction of the wall, it is more likely that the leak belongs to a class associated with an assembly or a design problem. Design-related failures are often traceable to the design documents themselves. A surprising number of architects do not know how to keep water out of a building. Tracing design flaws is not particularly difficult because they are usually clearly visible in the drawings. If, for example, the drawings call for a porous, segmented coping without through-wall flashing below the coping, the leak may even occur before construction is completed. If, however, the coping is set well, the quality of the installation may disguise the design failure until after occupancy, but probably not very long after occupancy.

Generally speaking, assembly problems are the most difficult to analyze. As with design flaws, assembly flaws become apparent relatively soon after occupancy. Inasmuch as any leak is dependent to some extent on the peculiarities of meeting the complete set of necessary and sufficient conditions, the leak event may not appear for weeks, months, or even years after occupancy. Unlike with design errors, there is no document of the construction to assist in the analysis. Assembly problems occur because the building was not built according to specifications and drawings; therefore, the design documents will offer little assistance in finding the flaw. If, for example, the leak is a result of an unsealed seam in the flashing at the bottom of a cavity, there is no way to determine the culprit from the design documents, which clearly call for all such seams to lap four inches with fully bedded adhesive for the entire width of the lap seam. If the specification had been followed, there would be no leak.

A clue to determining if a leak is the consequence of a design error or an assembly error is whether the problem is pervasive or localized. If it is a design error, a leak is more likely to occur at every location where the error occurs. Assembly errors are more likely to occur where the mechanic forgot to apply the adhesive, or some other specific oversight, but this is by no means a reliable indicator. If the design documents call for adhesive at every lap seam, but the mechanic did not follow the specifications, then every lap seam on the job may leak and still not be a design error.

5 Horizontal Closure Systems

5.1 Functional Definition

The functional definition of the horizontal closure system is to collect, channel, and divert water, from any source, from the time the water comes in contact with a given property until it is properly and appropriately discharged from that same property. The system includes the building and the site, and from that standpoint, the horizontal closure system is notably different from the structural and vertical closure systems we have examined thus far. However, it is not unique that the system extends beyond the perimeter of the foundation. As we will find in Chapter 6, the environmental modification system includes not only the immediate site, but the surrounding terrain as well.

As stated throughout the book, the most common source of water is precipitation, either solid or liquid. Other sources include surface water from an adjacent site or flooded water course, ground water, storm surge, and tides. In fact, water from any source other than domestic supplies is the functional object of the horizontal closure system.

In this book, we will concentrate on the building-related components of the system because they are the dominant sites of deterioration and failure within the system. We will also, to a lesser degree, address how the site itself and vegetation, in particular, cause system malfunction and, thereby, impact performance.

5.2 Components and Description

If we define the horizontal closure system strictly in terms of the building, the largest single building-related component is the roof. Taking the system as a whole, however, the largest component is the site drainage subsystem. In total, the basic components relative to the building include the roof, the valleys, ridges, gutters, leaders, and flashings associated with the cover of the building proper. The horizontal closure system continues to function until the water collected and diverted by these components is directed to conduits, catch basins, retention ponds, storm sewers, or drainage lines, and is disposed of off the site in a controlled and proper manner.

We will examine the horizontal closure system following the pattern set in Chapters 3 and 4: describe the materials-related issues associated with horizontal closure, followed by a discussion of system-related failures. But in addressing horizontal closure, generally, and roofs, specifically, because the materials and the gross geometry are so interrelated, we must discuss the typology of roofs and the materials associated with that typology before we delve into the materials (see Figs. 5.1A and 5.1B).

5.2.1 Roof Types

There are three broad classes of roofs: single-layer, multiple-layer, and hybrids. Their distinctions, in terms of components and materials, are driven by the initial decision as to which basic configuration will be employed to collect and channel the water. In many cases, the geometry is guided by an initial material selection; in other cases, the geometry "leads" and the material selection "follows." In no other system is the very concept of the system such an integral part of the materials selection, and vice versa.

Single-layer roof systems are, by definition, membranes, with all the requirements for perfection that we discussed relative to thin, single-layer walls. The single-layer roof is an exercise in extending an impermeable sheet over the occupied zone and relying on the absolute impermeability of the sheet to repel water.

Multiple-layer roofs are necessarily pressure-equalized systems, because there is no equivalent to the drained cavity system for roofs. Recall that the presumption of the drained cavity system is that water penetrates the outer layer. In walls, having penetrated the outer layer, the water descends vertically into the cavity, where it is collected and returned to the exterior. If such an approach were employed with roofs, the water would be descending directly onto or into the occupied zone. If, on the other hand, the presumption is that water does not enter the chamber, but the screen is inherently permeable, then the system is considered pressure-equalized whether it is labeled as such or not.

The layers of the multiple-layer pressure-equalized roof are the outer layer, or screen, and the air barrier, similar to pressure-equalized walls. But one significant difference between pressure-equalized walls and pressure-equalized roofs is the potential size of the chamber. In pressure-equalized walls, the depth of the chamber is typically a few inches, and it is fairly uniform because the screen and the air barrier are normally parallel. Though this is possible as well in pressure-equalized roof systems, it is not the only option. The screen and the air barrier may be parallel, in which case the depth of the chamber is also typically a few inches and uniformly so. Where there may be an impermeable layer below tiles or shingles, it is the size and shape of the chamber that is altered, not the basic physics. It is also possible, however, for the air barrier *not* to be parallel to the screen. The void formed by such a relationship may be, for example, an attic. When properly detailed as such, the attic becomes a large pressure equalization chamber, a fact that, historically, designers understood well. More recently, however, with the rising popularity of hybrid roofs, some designers neglect one of the basic functions of the attic, and instead treat it more as an extension of the occupied zone rather than as a necessary part of a properly functioning multiple-layer roof. This oversight has resulted not only in poorly designed new construction, but, more regrettably, the corruption of otherwise functional older attics and roofs.

The third type of roof, the hybrid, is a relatively recent entry into the roofing scheme. As its name suggests, the hybrid roof shares some

FIGURE 5.1A The ideal roof penetration from the standpoint of water integrity is top, dead center, of the ridge; however, this may interfere with the ridge pole or riband.

FIGURE 5.1B The less desirable location for a roof penetration is in the field of the sloping plane. This inherently produces an upslope crotch, which is a constant source of compensating detail and risk even when well tended.

characteristics of both the single- and multiple-layer systems. Both the single- and the multiple-layer approaches are somewhat material-specific, but some interchange of the materials is possible between them. In contrast, the materials associated with the hybrid roof are unique to the approach; they are not applicable to either the single-layer or the multiple-layer approaches. The dominant material in hybrid roofing is the asphalt shingle (although it is arguable that cap sheets on built-up roofing (BUR) serve much the same effect). The asphalt shingle roof system was popularized in the 1950s, and it is currently the dominant residential roof found on American homes. Surprisingly, then, this system is not well understood by designers, who tend to rely on roofers for details and specifications for asphalt shingle systems.

If the roofing approach is a single-layer system, the material options are limited to constructions and fabrications that can be adapted to continuous sheets or membranes. Among the earliest membrane roofs were those made of metals. That said, we must point out that, though there are materials that are perfectly impermeable, such as metals, this does not mean that every perfectly impermeable material can be used effectively to produce a single-layer membrane roof. The success of metals used in membrane roofs depends much more on the success of the joints than the properties of the metal. In fact, metals are not good candidates for single-layer roofing.

Another widely used material has been coal tar or asphalt bitumen applied directly to a substrate such as stone or wooden boards. If assiduously maintained, a roof made thusly can be rather successful. To this day, it is customary in Britain to apply liberal quantities (20 millimeters or more) of bitumen directly to concrete substrates. This general approach was improved dramatically with the implementation of roofing felt, which developed from rolls or sheets of paper impregnated with coal tar or asphalt used in the ship-building industry. The felts were laid on hot layers of compatible binders, then mopped with more of the same melted adhesive substance. The layering was repeated as many time was desired. We have not improved on this BUR system very much in recent times except to substitute the base material of the felt with glass or other more durable fibers. (We will discuss BUR in greater detail in a subsequent section.)

A more recent, and major, development in the field of roofing are roofing membranes operating under the deceptively simple name of single-ply roofing (which serves only to distinguish it from the multiply approach of BUR). One of the earliest entries into this market was Hypolon, which was developed in the 1950s by Dupont as a reservoir liner. Although an excellent roofing material, Hypolon was never a big seller due to its price. Another Dupont product was Neoprene, which was not developed as roofing but which was adapted for that purpose. It too languished on the market due to price.

But there was a tough, cheap sheet material available, namely rubber. The fundamental problem with rubber, either natural or early synthetic varieties, was its vulnerability to ultraviolet radiation and ozone. As long as the substance was encased in a tire, it performed well as an inner tube; installed on a roof, it would not have endured well. That problem was solved with the development of ethylene propylene diene monomer/terpolymer, a polymeric synthetic rubber. Better known as EPDM, this material has quietly revolutionized roofing. Though it has taken a few years, EPDM has become the single most widely used material for single-layer roofs, ever since the emergence of other synthetic membranes.

Having a reliable, durable membrane, regardless of its material composition, allows the designer to develop roof configurations that are more or less sloping. Notice we didn't use the term flat as it applies to roofing. That is because "flat" roofs are not flat; they are sloped, however slightly, to enable gravity to pull water off the roof and toward the collection and channeling devices. Typically, a flat roof slopes vertically no less than 1:48 (one-fourth-inch vertical drop per foot of horizontal run or approximately two centimeters vertically per meter horizontally). Usually, this is at a pitch sufficient to avoid pooling of water on the roof and to preclude a condition, called ponding, which we will examine in more detail in the section on systems pathology.

The material choices for multiple-layer systems are inherently greater than for single-layer systems simply because neither the materials nor the assembly need be impermeable. Certainly, impermeable materials, which is to say membrane sheet goods, can be used in multiple-layer systems, but doing so has the effect of negating the added costs of developing the geometry and configuration of the multiple-layer roof. Better is to simply use the panels or sheets of metal as though they were not impermeable, and develop them as multiple-layer, pressure-equalized roofs.

Because multiple-layer, pressure-equalized roofs are assumed to be permeable, they must employ a cascading runoff approach, which is another way of saying that all multiple-layer roofs are sloped. Whereas membrane roofs may be sloped or nearly flat, pressure-equalized roofs must be sloped. The angle of the slope is determined by a number of factors, which we will explore in detail later; here, suffice it to say that the angle is far from the 1:48 pitch of proverbially flat roofs. The lower range of acceptable pitch in a sloped roof is 16:96 (2:12) and increases from there.

Once we decide that a roof is to be sloped and pressure-equalized, we can relax the assembly of the pieces. If the materials need not be impermeable, then neither does the assembly. If the pieces are arrayed such that the water traverses the roof laterally as it descends vertically, we can use gravity as the motive force to move the water off the roof, in the same manner as flat roofs, but without the necessity of the pieces themselves being perfectly joined together. The development of the shingle roof is coincident with the need to use something other than membranes to protect the interior of the building. Shingles, sloped roofs, and pressure-equalized systems form a nexus of materials and assembly that is as viable as it is old. In contrast to flat roofs that require membranes, sloped roofs do not require shingles. It is more accurate to say that sloped roofs "allow" shingles.

5.3 Materials Pathology

The typical deterioration mechanisms associated with the horizontal closure systems are largely the consequence of the inherent hostility of the environment coincident with the system, which, as defined, is essentially a water collection and diversion device. As we have seen over and over again, water is itself an erosive agent, or it is associated with deterioration mechanisms of extreme aggressiveness. To this list of aggressive mechanisms we add corrosion and galvanic action, which we will discuss below.

In addition to the occasional inundation by water, much of the system, and specifically the roof, undergoes the most extreme thermal shift of any part of the building. In addition to water and heat, roofs are particularly vulnerable to ultraviolet degradation, another deterioration mechanism we will discuss below. When we combine the peculiar deterioration mechanisms associated with the horizontal closure system with the mechanisms discussed in Chapter 4, it is not surprising that roofs fail. It is more surprising that they work at all.

The materials used to produce the horizontal closure system, and the roof in particular, are closely related to the basic approach to repel, collect, channel, and transport the water, as described above. This section organizes the various materials used in the horizontal closure system according to the primary deterioration mechanism, rather than as an inventory of all the materials available. We take this approach for two reasons: First, it is beyond the scope of this book to inventory every material employed in the construction of horizontal closure systems. Moreover doing so would be a waste of time since the list would change before the book was published. Except for sealants and adhesives (which are major components of contemporary roofing technology), no area of passive building systems changes more rapidly than that of roofing and drainage. Second, a primary objective of this book to change the way professionals view building deterioration; that is, to move away from a focus on accumulations of material and/or specific flaws to a general approach based on deterioration mechanisms that can be applied to the subject at hand.

We will address the most common mechanisms in more or less the order of their historical development and continuing economic importance.

5.3.1 Differential Thermal Movement

The physics of differential thermal movement were covered in sections of Chapters 3 and 4. These physics are the same for roofs, but the orientation of the roof is typically more horizontal; or, more accurately, it has an added horizontal component. Combined, thermal expansion and contraction in roofing is even more of a two-dimensional issue than is the case with walls, in particular, as regards the materials comprising the wall. There is little, if any, cause for concern about thermal effects perpendicular to the plane of the roof. We will, however, discuss one notable exception, that is, the tendency of fasteners to "back out" of a roof, which is an action perpendicular to the roof plane and is the result of thermal expansion.

By its orientation, a roof's materials are generally exposed to incident solar radiation for longer periods, often at less oblique angles, and with less obscuration or screening than walls. If a roof is flat, it begins to incur solar impact within minutes after sunrise, especially intense during the summer months. Though the roofing will reradiate much of that energy back into the atmosphere, the net energy gain will cause the temperature of the roof to rise throughout the day. In the summer in temperate zones, the peak solar altitude is more than 70° above horizontal. At midday, the angle of incidence on the south wall is only 20°; therefore, the solar impact is only 12 percent of the maximum gain of a similar plane oriented perpendicular to the incident radiation. At the same time that the south wall is subject to only 12 percent of the incoming solar energy, the roof is subject to 88 percent of the radiation. Furthermore, until approximately 10:00 A.M. in a temperate zone, the south wall is in or nearly in shade. In contrast, by 10:00 A.M. on the same summer day, the angle of incidence on a roof is approximately 50°, and the impact is almost 60 percent of maximum solar gain; by that time the roof has been receiving solar load for several hours.

The situation is different on sloped roofs, because the angles of incidence are different from those of flat roofs; in addition, a sloped roof may be "folded," which results in a bifurcated panel problem. When a roof is flat, the angle of incidence is the same as the solar altitude; when the roof is sloped, two variables must be factored into the problem at the outset: the angle of the slope and the orientation of the pitch. If, for example, the slope of the roof is 20° above horizontal, that translates to approximately a 4:12 pitch, meaning that for every 12 inches of horizontal run, the vertical dimension increases four inches. If the axis of the pitch is oriented to solar south, then at noon on a hypothetical midsummer day in a temperate zone, the roof will receive 100 percent of the maximum solar load, as opposed to 88 percent for a flat roof. If the axis of orientation of this same roof is toward the north, the maximum angle of incidence is 50°, and the solar impact is 60 percent of maximum. If the orientation is due west, the solar impact is 80 percent of maximum. The point is, slope can increase solar load at noon, if the orientation of the slope is to the south; if, on the other hand, the orientation is other than due south, the gain may be reduced for that peak altitude, which can reduce solar load. The downside is that when the sun swings around to the axis of orientation, the gain may be higher at that hour than for the same roof oriented south. In short, a roof tilted toward the sun will absorb more energy than one tilted away from the sun.

Roof slope and axis of orientation are two very important variables in the diurnal thermal fluctuation of a roof and, therefore, of the deterioration of that same roof. For this reason, on, for example, folded roofs with one panel oriented to the south and the other to the north, the south slope weathers faster and usually fails first. Solar impact is the principal contributing factor to temperature change and, therefore, to thermal expansion.

In addition to orientation and pitch, roofs and roof materials tend to have relatively high emissivity values. Emissivity, a material property, is a measure of the tendency of a surface to absorb and to reradiate radiant energy. If a material is highly absorptive of incident radiation then the emissivity value approaches 1.0; if it reflects radiation and, therefore, does not absorb radiation, the emissivity value approaches 0.0. The factors that affect emissivity are color, texture, and specular aspect.

The color of a material is an indication of a range of the visible spectrum that is not being absorbed. What we perceive as the color of an object are the electromagnetic waves reflected from the surface that are detected and interpreted by the human eye as the color of the object. The wider the range of rejected light, the lighter the object; the darker the object, the higher the degree of absorbed light, hence the greater the degree of absorbed radiation. Objects that reflect the entire range of visible radiation appear to us as white; objects that absorb the entire range of visible radiation and, therefore, reflect no light appear to us as black.

In addition to the visible spectrum, the object may accept or reject radiation from the nonvisible spectrums above and/or below the visible range. The radiation immediately below the visible range is called the *infrared* (below red) radiation; below the infrareds are various bands of radio ranges. The range immediately above the visible range is *ultraviolet* (above violet) radiation; above ultraviolet are the X ray and gamma ray bands. All of these bands are collectively referred to as the *electromagnetic spectrum*, and they can all impart energy to an object if they are absorbed. The fact that an object is seen as white does not mean that all radiation is reflected;

it is altogether possible for an object to reflect the entire visible range of the spectrum and still absorb radiation in the infrared (IR) and/or ultraviolet (UV) ranges. It is equally possible at least theoretically, that an object can reflect IR and or UV rays and still absorb part of the visible range.

An object that absorbs all radiation regardless of wave length is referred to as a *black body*. As radiation of whatever wave length is absorbed, the atoms within the molecules and/or crystalline solids oscillate at increasing amplitude, and the absorption of additional energy is manifested in an increase in the temperature of the solid. The oscillating atoms retransmit radiation, which means that all matter is constantly emitting radiation because all matter consists of oscillating atoms. The rate of retransmission is always in balance with the rate of reception and absorption.

The magnitude, wavelength, and frequency of the retransmission of radiation from the solid are all functions of the temperature of the solid, which is in no small part a function of the absorbed energy of incoming radiation. As the object increases in temperature, the frequency of the retransmitted energy increases as well. This is why very hot objects begin to glow. The heated object initially retransmits energy in the suboptical range; as the object continues to gain heat from the source at hand, the frequency of the retransmission increases through the IR range until the retransmitted radiation enters the visible range, beginning with the red part of the visible spectrum. As the object continues to increase in temperature, the breadth of the retransmission range increases, as does the intensity of the retransmission. To the reds are added yellow and, eventually, green. But note, we never perceive the green retransmitted light from a very hot object because the greens, and later the blues, are additions to the yellows and reds. Instead of changing color in the upper range of the spectrum, the object simply becomes brighter, turning more white as the width of the visible retransmission band widens. When the object is hot enough to retransmit the entire visible spectrum, it appears to emit a white light. An example of such retransmission of this so-called white light is the filament of a high-intensity lightbulb.

The color of the retransmitted energy is strictly a function of the temperature of the object; it is not related to the nature or structure of the material. White-hot steel is the same temperature as white-hot salt, approximately 4100° F (2300° C). That the steel is a liquid at that temperature does not alter the phenomenon relative to temperature and the spectrum of retransmitted radiation. The spectrum of energy transmitted from a radiant body is so predictable that we can determine with great precision the temperature of an object by measuring the width of the retransmitted spectrum. What varies among substances is the amount of energy required to raise the temperature to the point at which it retransmits energy at all. This is reflected in the specific heat of a substance. So, though it may be true that white-hot steel and white-hot salt are the same temperature, if we discount the effects of heat of liquefaction it is also true that the heat required to get the steel to that temperature is significantly less than the amount of heat required to get salt white hot because the specific heat of salt is greater than that of steel.

The balance of incoming radiation, retransmitted radiation, and temperature is called the Stefan-Boltzmann Law, named for Josef Stefan, who experimentally verified the phenomenon in 1879, and Ludwig Boltzmann, who mathematically justified Stefan's findings in 1884.

Because of the exposure and orientation of roofs, the consequence of incident radiation often determines the general performance and durability of the entire system. Incoming radiation can and, generally, does raise the temperature of the roof well above ambient air temperature. On a bright day, with the roof oriented into the sun, the surface temperature of the roof may reach 140° F (60° C) or more, depending upon the cooling effected by the ambient air. At night, the still-warm roof, in the obvious absence of the sun will emit radiation at a faster rate than it will absorb radiation. Depending on the clarity of the night sky, the loss of radiant energy from the roof may be greater than can be replaced by the surrounding ambient air through convection; thus, is possible that the surface temperature may drop below ambient air temperature.

One consequence of the Stefan-Boltzmann Law is that materials subject to intense radiation and that reradiate that stored energy into an unobstructed atmosphere can cycle through temperature ranges well in excess of the high and low air temperature range in an ordinary day. During a summer day when the air temperature may reach a high of 85° F (29° C) and a low of 60° F (15.5° C), if the atmosphere is clear, the roof may cycle from 140° F (60° C) to 50° F (10° C). Even in very cold weather, the surface temperature of the roof may experience large swings in surface temperature.

The Stefan-Boltzmann Law is not restricted to roofs. In a somewhat less dramatic manner, walls and other surfaces experience the effects of accumulated and subsequently released radiant energy. In porous, permeable materials, the effects can be rather destructive. If, for example, in the evening and later in the night, the material cools to below ambient air temperature, which drops below the dew point of the surrounding ambient air, condensation will form in and on the material. Thus, materials become wet even in the absence of rain. We emphasize the consequences of dew on porous, permeable materials because this class of materials, as described in Chapter 4, is vulnerable to freeze-thaw cycling. If, subsequent to deposition, the dew freezes, there is a risk of surficial damage from freeze-thaw even though the source of the water is other than precipitation. Although the depth of frost penetration may be slight, and the expansive force very low, it is still a deterioration mechanism, and the damage is cumulative.

Another aspect of the Stefan-Boltzmann Law is called the greenhouse effect. When warm bodies radiate energy, the dominant wavelength of that radiation is generally below the visible spectrum. It is, in fact, only at relatively high temperatures that the radiation from a warm body includes enough radiation in the visible range to be detectable. The floor of a room near a south-facing window, for example, may be exposed to substantial direct solar radiation. The spectrum of that radiation includes the complete visible range, plus IR, UV, and both higher and lower wavelengths. The glass in the window, even if it is clear glass, is a partial filter of that radiation. In fact, glass is fairly opaque to IR and low wavelengths, so that portion of the solar radiation, largely reflected by the glass, never reaches the interior surfaces. The radiation that does reach the floor, raises the temperature of the floor; subsequently, the floor begins to emit radiation back into the room, because the temperature of the floor, while higher than ambient air temperature, is still well below the threshold for emitting visible radiation. The wavelengths tend to be precisely those reflected by glass. Consequently, that visible light passes through the glass and heats the surfaces of the interior, including the floor; but the reradiated energy is reflected back into the space by the very glass that allowed the energy into the room. All the surfaces in the room absorb the low-frequency radiation from the heated surfaces, thus the ambient temperature of the entire volume increases. Anyone who has parked a car in the sun has experienced how hot such surfaces and interiors can become as a result of the greenhouse effect.

The immediate concern of the greenhouse effect relative to roofs and other exterior components of buildings is the potential 100° F (38° C) or greater swing in the temperature of the materials that comprise these components. If the roof is a shingle system, the consequences of such large temperature swings will be mitigated by the fact that the individual pieces are small and relatively unconfined. As they expand or contract with temperature, they are released at their respective boundaries, resulting in little if any internal stress. If, however, the roof is a membrane—which is to say a continuous sheet—the cumulative change in dimension due to thermal movement may be considerable. If, for example, the roof is sheet copper, with a long dimension of 80 feet, that undergoes a seasonal thermal change from 140° F (60° C) in the heat of an overhead, summer sun to −10° F (−23.3° C) in the chill of the coldest night of the year, the unrestrained change in dimension is $1\frac{3}{8}$ inches (35 mm). While that change is only $\frac{14}{100}$ of 1 percent relative to the total length, the absolute dimension is huge. If the change were to be accommodated on only one side of a single sheet, the copper would pull itself out of its cleats or rip along one or more seams. The prospect of having to accommodate that amount of change in dimension is daunting, but it is not unrealistic. Attempting to restrain dimensional changes of this magnitude risks the very integrity of the material.

This leads to an important issue associated with membrane systems, thus all flat-roof approaches. All membranes must accommodate significant dimensional change due to thermal movement, and they must do so within the plane of the roof itself. If the shift is restrained, the thickness of the material normally is insufficient to absorb the force, as would be the case, for example, if we simply nailed the edges to the deck. One way to successfully restrain the roof is to adhere the membrane to the substrate. Because the thermal movement is restrained at every point, the shearing stress is distributed over the entire area of the roof. If the modulus of elasticity of the roofing is low, the restraining stress will be modest.

Suppose that, for a three-ply asphalt roof, the modulus of elasticity at depressed temperatures were on the order of 500,000 psi and the coefficient of thermal expansion were 0.005 percent per 100° F (0.009 percent per 100° C). Under the conditions of a 140° F (60° C) change, the shearing stress at the bonding plane between the roofing and the substrate would be approximately 420 psi (2,900 Paschals). Shear stresses of this magnitude are significant; the bond could fracture at the adhesive plane or in the body of the roofing material itself. A mitigating factor in this situation is the possibility that the roof rarely undergoes a net dimensional shift over a range of a 140° F (60° C)—not because such shifts do not occur, but because the roof was not originally set all the way at one end of the thermal range.

Suppose the roof were originally adhered under conditions such that when the hot asphalt set, the ambient temperature was 80° F (26.7° C). If the asphalt set up so that the baseline dimension was determined at 80° F (26.7° C), the dimensional shift should be based on a positive or negative shift from that baseline. This means that the maximum positive shift is more likely to be on the order of 60 to 70° F (15.5 to 21.1° C), and the maximum negative shift, 80 to 90° F (26.7 to 32.2° C).

A second mitigating factor, particularly in asphalt roofs, is that asphalt is an amorphous solid, so as the asphalt increases in temperature, there is a decrease in the modulus of elasticity. If the modulus of elasticity is, in fact, 500,000 psi at 0° F (−17.8° C), then at 80° F (26.7° C), the modulus of elasticity may be less than half of that value. At 150° F (66° C), the modulus of elasticity may cease to have any significance at all because the material will have softened to a point that the modulus of elasticity is zero, as the material at that point is more plastic than elastic.

A third mitigating factor, which relates to the combination of the asphalt being a very low-modulus substance and BUR being laminar in construction, is that shear deformation can occur between the layers of felt. Some of the stress is relieved because one ply can slide, however minutely, relative to the layer below. On the downside is the fact that, as asphalt cools, it becomes more brittle, and the elastic coefficient increases. As the material becomes stiffer, it also become less resistant to cracking. (In a subsequent section we will add to the list of detriments of asphalt roofing the effects of UV, which is the terminal deterioration mechanism of the materials.)

A fourth mitigating factor is that, as the membrane heats, so does the substrate. If the coefficient of thermal expansion of the substrate is remotely close to that of the membrane, the two materials will expand at more or less the same rate. The substrate may also provide a thermal sink for the membrane, thereby drawing off some of the heat from the membrane itself. This factor should be kept in mind when placing a membrane on top of insulation. The insulation does not draw heat away from the membrane, so the membrane probably will reach higher sustained temperatures than if the same membrane were laid directly on a thermally massive substrate.

The point here is that, as the membrane expands and contracts, forces are created due to thermal expansion and contraction. For large expanses of roofing, these forces may be beyond the capacity of the material to restrain the movements. If the material is bonded to the substrate, these forces are absorbed over the entire area of the roof such that the shear stress at the bonding plane is distributed over the entire plane of the roof. The shear stress per square inch at the bonding plane is necessarily lower than if the same magnitude of force were concentrated along a line or at a point. In asphalt roofing, several mitigating factors can reduce a theoretical or potential shear stress of 420 psi (2,900 Pa) to a more manageable level of a couple hundred psi or less. Even at reduced shear stresses, however, over time and many cycles of thermal change, the bond with the substrate may weaken and fail. If the bond fails, the roof begins to migrate in a less restrained manner, and the remaining points of contact will necessarily absorb more of the lateral shear forces.

If the membrane material is metal, such as copper, the percent change in dimension is as great or greater than for BUR, and the modulus of elasticity is significantly higher. That modulus is not mitigated by softening of the material as it warms, and is, therefore, relatively constant over the entire operating temperature range. The modulus of elasticity of copper is 16,000,000 psi (1.1×10^{11} Pa). If the sheet metal is copper, the theoretical shear stress over a 100° F (37.8° C) range at the bonding plane—if the metal were fully adhered—would be almost 900 psi (6.2×10^6 Pa). If the membrane were sheet steel, the theoretical shear stress at the bonding plane would be over 1,100 psi (7.6×10^6). Even though the materials, copper or steel, are less than 4 percent of the thickness of the BUR, the moduli of elasticity are so much greater than that of asphalt that the theoretical shear stresses are several times that of asphalt. There are also far fewer mitigating circumstances associated with sheet metal than with BUR. There is no accommodation due to amorphous solidification or interlaminar slippage. If a metal roof were adhered to the substrate, the actual shear stress would approximate the theoretical values, and the bond would be extremely vulnerable to fracture, particularly given the relative difficulty of achieving long-term durable bonds to materials with metals. Once the adhesive bond is fractured locally, the metal would migrate and redistribute the shear stresses through the roof plane until the roof would unzip from the substrate in a progressive failure of the adhesive.

For this very reason, metal roofs are not adhered. Instead, the general approach for mounting metal membrane roofs is to release the movement at the edges and at specifically constructed joints in the roof plane. This has the effect of allowing lateral movement without admitting water through the roofing. Such roofing control joints are not difficult to construct; and once installed, they tend to overcome most metal migration issues, at least at the gross movement level. In metal roofs, however, the failures are more likely to occur at a lower level of construction. The fabrication of sheets of metal is limited to some extent to the width that a machine can reliably roll the material. But even if the metal could be rolled to great width, no roofer could maneuver it onto a roof. The stiffness of the material restricts installation to relatively narrow widths and somewhat limited lengths. Therefore, metal roofs must be fabricated from pieces of metal that are then joined together in the plane of the material to form continuous, watertight joints.

Given the difficulty cited earlier in achieving durable bonds between metallic materials, this jointing is accomplished in two general ways: one is to seal the joint together with another metal, typically a lead alloy or solder; the other is to fabricate the seam such that the mouth of the joint is not in the plane of the roof and, therefore, not below the level of the water as it sheets across the metal roof. The former approach is the flat seam, the latter, the standing seam. Both are reasonable accommodations to the vicissitudes of metal membranes, but both are vulnerable to failure.

The flat-lock seam is formed by folding the two sheets of metal together in what amounts to a roll, then flattening the roll, which results in an interlock of the two sheets. The solder is melted and "sweated" into the seam. (The same principle applies to sweating a pipe fitting or joint.) The two pieces of metal to be joined are heated until the solder, when brought into contact with the hot sheet metal, melts and is drawn by adhesion into the fine-line crack of the joint. A properly sweated joint is filled with solder. Unfortunately, all too often, the base metal is not hot enough, and the solder melts not on contact with the base metal but in the flame of the torch or on the hot iron. The solder may cover the joint but it does not necessarily flow into the joint (see Figure A through E on page 554). A well-sweated joint does not require a large deposit of solder showing at the joint exposure because the solder doing the work is not visible at the surface. The difference is analogous to glue between two pieces of paper as opposed to tape covering the joint between two pieces of paper. Solder should be *in* the joint, not simply covering the joint.

The capillarity of the meandering crack between the two sheets pulls the solder into and through the interlock such that the "glue" attaches not only at the surface but well into the joint. This is more likely to occur if the sheet metal is hot enough to melt the solder without actually having to apply heat directly to the solder. If the solder melts onto basically cold sheet metal, it will congeal at the point of contact, not flow into the crack nor bond well with the adjacent sheet metal. This is the subsequent source of many solder joint failures.

Soldered joints fuse well with copper, lead-coated copper, and terne metal; nevertheless, they can fail, in two ways. The most common failure is a result of the improper soldering technique just described. As the sheet metal expands and contracts perpendicular to the joint, the "taped" solder joint is put into tension across the joint. A cold and, therefore, brittle joint will crack along the length of the joint and through the body of the surface-applied solder. When a joint is properly sweated, the force of expansion and contraction is absorbed not at the surface of the joint, hence, by the solder. The two plies of metal are interlocked, and the solder is in a much lower average state of stress.

The second way a solder joint can fail is along the length of the joint. This, too, is much more likely to occur if the solder is sitting *on* the joint rather than *in* the joint. The coefficient of thermal expansion of the lead-based solder is more than 50 percent greater than that of copper and more than double that of steel. As the joint expands and contracts along its length, the bond between the solder and the base metal is stressed. Under conditions of dramatic temperature change, the bond will be stressed, and may fail. (See Chapter 3 for an extensive discussion of differential expansion and the resulting crack propagation in composite materials.) Such a failure is more likely to occur not only if the solder is improperly applied, but also if the surface of the sheet metal is improperly prepared. Waxes, oils, dust, or other surface films can weaken the bond between the sheet metal and the solder. The fine-line crack resulting from the debonding of the solder from the base metal is an even finer crack than the original joint between the two sheets of metal. When there is a pressure differential across the failed joint, as the rain rolls across the roof, a leak is inevitable, because all of the necessary and sufficient conditions for fluid transportation have been satisfied.

Standing seam roofs join the pieces of metal in such a way that the joint is flattened perpendicular to the plane of the roof. The seam is made by turning the edges of the two pieces up and then folding them together. This places the mouth of the joint above the plane of the metal roof surface. It is commonly believed that a standing seam will not leak unless it is completely immersed in water. This no more valid than saying that water cannot be drawn through a straw unless the straw is completely immersed in the fluid, meaning that the fluid level is at the top end of the straw. For standing seams, the tip of the "straw" begins at the bottom of the down-standing flap—that is, where the mouth of the joint is located. Once the mouth is immersed in water, and a pressure differential exists, unless the joint is sealed with solder or other material, a leak will ensue.

If the membrane is EPDM, all the phenomena discussed above hold true. The temperature shift will cause the material to expand and contract. Because the modulus of elasticity is low, however, the restraining force at the bonding plane of an adhered system is low. Single-plies can, within limits, be laid loose and allowed to migrate. In fact, the construction of the expansion joints, to the extent they are necessary, are little more than up-standing loops of roofing working as expansion loops in a steam line.

When the material must be seamed, double-sided adhesive tapes that are part of some single-ply membrane kits are watertight and durable. Though all materials deteriorate, so little EPDM has failed so far, that the essential mechanisms are not yet known. If we were to hazard a guess where the long-term failure of the system would be, it is reasonable to suppose that the adhesive at the joints would fail before the EPDM itself. It is also reasonable to presume, that, though EPDM and other single-ply materials may be specifically designed to resist ultraviolet, acids, and ozone attack, they are not completely impervious to these sources of deterioration. And because we are still in the first generation of single-ply roofs, we also must consider how, when these systems begin to deteriorate, they will bond with adhesives. We may find that cutting a new penetration into an aging EPDM roof is not nearly as easy as doing so in fresh material. Currently, single-ply technology is very promising, and as economies of scale are achieved and more experience is acquired, we are likely to see even more entrants into the market. But, as stated throughout this book, history of materials is unrelenting in one respect: they all fail. If the cost and duration until failure are satisfactory, then the material can be considered to have performed well. To date, single-plies seem to be favorable options, but, again, we are only in the first generation of the technology.

If a substance is effective at repelling water and acting as an impermeable membrane when it is laid flat or nearly flat, that same membrane should be equally impermeable when laid on a significant slope. Impermeable is still impermeable. Sheet metal, for example, is often placed on a sloped plane, and arguably, sloped metal sheet is a more appropriate use of the material than laying it flat. The vulnerability of the seam, however, is still an issue, although mitigated somewhat by the reduced depth of water likely to be at the seam at any one time if the roof is relatively steeply sloped. Reduced depth means reduced hydrostatic pressure, which means less pressure differential across the crack. EPDM, too, can be laid on a slope. If the slope is shallow (1:12), the rolls can be rolled out across the roof perpendicular to the slope. If the slope is greater than 1:12, the rolls can be run up and down the slope, in which case the seams are parallel to the slope. Though seemingly more vulnerable, in practice, seams in the latter approach are sound and do hold. The question, of course, is for how long.

Asphaltic bitumin also can be placed on a slope, as long as the type is properly matched to the slope. The types are ranked I to IV, to differentiate them by the temperature at which each softens. Type I softens at the lowest temperature and is, therefore, most likely to "slump" or run when exposed the heat of the sun. Type IV reaches the highest temperature before softening, and so can be applied to a significant slope. The main disadvantage of all four types is directly related to the same property; that is, when the bitumin softens, it can "heal" a minor crack. The disadvantage emerges from the inverse of that property: the higher the softening temperature, the more brittle the material. Thus, though Type IV has the highest softening temperature, it is also the most likely to crack and to not be able to heal itself.

Coal tar bitumin, also classified by softening temperature, is less stable. Even the type with the higher softening temperature can only hold a fairly shallow slope without slumping.

Taken as a class, the problem associated with using these materials on slopes is more a matter of the practicality of placement than an issue of their capability to repel water. They can all repel water, but putting them and keeping them in place is another matter.

Nail Popping

Nail popping is a special thermally related materials issue. The name of the mechanism describes the situation: a driven fastener, typically a nail, seems to spontaneously work its way out of the substrate and to stand proud of the roof plane. Nail popping is not restricted to roofing. It also occurs in drywall that is attached with driven nails.

Nail popping can occur in any roof that includes in its assembly driven nails, whether they are used to fix the layer exposed to weather or to fix an underlayment. If the nails are under a single-layer membrane, obviously this behavior can jeopardize the membrane. Nail popping in shingle roofs is far less of a problem, because, when it does occur, a protruding nail under a shingle may not even reach the underside of a superior shingle to lift it; and even if it does lift the shingle slightly, usually, little if any harm is done to the shingle.

Nail Popping (*continued*)

Nail popping occurs because of a two-part thermal cycle. First, the nail elongates because of thermal expansion greater than that of the material into which it is embedded. The restraining force of the base material in the form of friction along the shank of the nail is not sufficient to restrain the elongation of the nail, so the frictional bond breaks. The nail elongates either into the sheathing, out of the sheathing, or both. Migration into the sheathing requires a force greater than the frictional force holding the nail in place to begin with, otherwise the nail would never have gotten into the sheathing at all. In its passive state, the maximum force that can develop in the nail is the frictional force along the shank. If a greater force than that is required to move the force in the nail hole, then it must come from outside the nail-nail hole set. Thermal expansion may satisfy this requirement, but it can do so in either direction, and the direction will be the one of least resistance, which in this case is *out* of the hole, not further into the sheathing.

For the nail to fracture the frictional bond with the sheathing, the force created by the thermal elongation of the nail must be greater than the restraining force of the clamping action of the nail hole. This means that nail popping is more likely to occur when these five factors are in place: the temperature differential is great, the surface of the nail is smooth, the density of the sheathing is low, the diameter of the nail is relatively small, and the depth of the original set is relatively short. Once the frictional force has been exceeded, the nail backs out of the hole ever so slightly. For a two-inch-(five cm) long roofing nail, the initial elongation over a 40° F (22° C) range is $\frac{1}{2000}$ of one inch (0.013 mm), which admittedly is a minuscule measure. If, however, this happens 100 times a year which is entirely possible, then in one year, the nail can potentially elongate a cumulative length of $\frac{1}{20}$ of an inch (1.25 mm); in two years, $\frac{1}{10}$ of an inch (2.5 mm). If the direction of cumulative movement is in one direction, and the necessary and sufficient conditions are present for overcoming the original clamping force, the nail can break the plane of the sheathing in two or three years. If, however, when the nail cools, it sinks back into the hole, nail popping is at least potentially reversible.

If the nail elongates and breaks the frictional clamping force, it will emerge from the hole, but it will not reverse. When the nail breaks the plane of the sheathing, there is less nail embedment than before the elongation, meaning that the total clamping force is less than before the elongation. When the nail cools, the tip will retract toward the head, because less force is required to effect that movement than for the nail to reseat itself to its original depth. This is the second half of the cycle. Not only must the nail surface from the hole, the tip must retract toward the head when the shank cools. This sets up another cycle of elongation and retraction such that the cumulative effect is in one direction.

Thermal expansion is far less of an issue with multiple-layer systems, simply because the pieces are small and essentially prejointed. The segments that comprise a shingle roof may be thermally expansive, but the movement is not restrained. That is not to say that shingle-based fields don't have their points of vulnerability; they do, but differential thermal movement is not one of them. Multi-layer systems reveal vulnerability to thermal movement where they are not, in fact, multi-layer systems at all, rather, but small pieces of single-layer systems, namely at flashings and other transition points that require the limited applications of single-layer technology.

5.3.2 Corrosion and Galvanic Action

The subject of corrosion could be legitimately addressed in any chapter of this book, because there are metal components in every system. As long as we continue to expose metals to open environments, we will have corrosion: metal structural members corrode; metal wall components corrode; metal pieces of mechanical systems corrode. We include it in this chapter, however, because arguably no system exposes as much metal to harsher conditions than the horizontal closure system. Indeed, there are entire horizontal closure systems built out of metal, totally exposed, yet expected to protect all the other systems. In this section, we will explore the structure and nature of metals. We will examine the advantages and disadvantages associated with their use; and, of course, we will discuss the peculiarities of using metals as all or part of the horizontal closure system.

As materials and substances, metals are defined in a distinctive and unequivocal manner:

- Metals are absolutely impermeable and nonporous.
- Metals are ductile.
- Within their respective elastic ranges, metals are linearly elastic.
- Generally, metals are malleable and will fracture ductilely.
- Metals can be drawn into wire or rolled into sheets and plates.
- Metals are fusible and, therefore, weldable.

These properties are valid for virtually the entire class of materials we refer to as metals. Of course, specific metals exhibit particular aspects of these properties, which lend them to specialized uses and application.

For example, copper has particularly good thermal and electrical conductance properties; and it is highly malleable and can readily be drawn into wire. Although all metals corrode—including, by the way, stainless steel—the properties that tend to inhibit corrosion in metals are more apparent in copper than in any other metals in its availability and cost range. Metals that corrode less rapidly than copper are prohibitively expensive. Metals more readily available than copper corrode far more rapidly. Aluminum is a popular metal because its density is low compared to its strength. Sheet lead is highly malleable; tin, when melted, adheres quite well to other metals; and zinc enjoys a special relationship to iron, which we will discuss in considerable detail.

Far and away, though, the economically most important single metal is iron. It is difficult to imagine life remotely close to anything as we know it without iron. Simply put, without iron, there would not have been a Industrial Revolution. Not only is it a truly magnificent material, it is plentiful and relatively inexpensive to recover. Iron is stiff, strong, and cheap. Fabricated iron costs about $1 per pound compared to aluminum which is $3 per pound or copper which is $5 or more per pound. If we could substitute metals on a volume-for-volume basis, we could normally break even with aluminum relative to iron, but a copper item would still run more in price than the same volume of iron. If the task required stiffness, aluminum fabrications would run at least 9 times greater, and copper 15 greater, than their iron counterparts because of their relative costs and stiffnesses.

There are metals more conductive than copper—gold for one, but it hardly need be said that gold is 350 times more expensive than copper. There are metals lighter than aluminum, such as manganese and magnesium, but they also tend to burn in an open atmosphere. There are even metals stiffer than iron, such as tungsten, which unfortunately is rare and relatively brittle, and platinum, which is 6,400 times more expensive than iron per pound. (If platinum were more plentiful, it would wipe many other metals off the market. It is as strong as iron and stiffer than iron, less corrodible than gold, more conductive than copper. It is, however, very dense and not very malleable.)

THE PROCESS OF ALLOYING AND REFINING

Another advantageous property of metals is that they can be mixed with other metals to form metal blends. The process is called *alloying*. Alloys should not be thought of as solutions, however, rather as true mixtures. To understand and to appreciate alloys and to more easily recognize the vulnerabilities of metals, it is necessary to become knowledgeable about the formation and refining of metals, generally.

A pure metallic solid is a crystalline structure held together by metallic bonds, which do not exist in other substances. The metallic bond is distinguished by the fact that electrons are not shared locally, as is the case with ionic bonds, nor are they shared specifically with other atoms, as is the case with covalent bonds. In metallic bonds, free electrons are shared collectively over and through the entire crystal, and they are concentrated more at the surfaces of the crystal. This enables the electrons to flow quite freely throughout the metal crystal. In contrast, the electrons in other forms of crystals are virtually immobile compared to even the most resistive metal. This attribute is present in all metals, although the electron flow is more fluid in some metals than others.

The mobility of electrons within and through the metallic crystal is, primarily, what gives metals their electrical conductivity. Moreover, if a metal is connected to an electrical generator, additional electrons can be "pumped" into a metallic crystal. If a metal is connected to an electrical sink or ground, the metal will discharge to the sink. If the metal loses electrons, the metal is positively charged; and if there is a surfeit of electrons in the crystal, it is negatively charged. A metal has a much stronger propensity to discharge electrons to an available ground than to spontaneously attract electrons, so the general character of a metal is to be positively charged. Thus, at any given instant a specific metal nucleus at the surface of a metal crystal may be individually ionized. Consequently, not only is the body of the crystal negatively charged as a whole, but there is a statistical probability that a specific metal atom is negatively charged. That said, we must note that the occurrence of a identifiable negatively charged metal atom at the surface of a metallic crystal is a short-lived event. The localized negative charge will quickly attract a passing electron, thereby neutralizing the atom which, of course, results in the exposure of another atom somewhere else in the crystal being negatively charged due to the collective shortage of electrons. The next atom to be shorted an electron may not, however, be at the surface, and this is an important element in the corrosion saga.

Suppose that near the occasionally positively charged metal atom (ion) at the surface of the metal crystal is a droplet of water. (Recall from prior discussions that the water molecule is a polarized molecule, which means that it is a miniature magnet relative to electrically charged surfaces and particles.) If at the contact plane of the water and the metal, one of the metal atoms becomes momentarily charged, the negatively charged poles of the water molecules will instantly rotate in the direction of the positively charged particle, and exert a magnetic pull on the charged ion. The collective pull of a few hundred-million water molecules pulling in one direction on a single charged metal atom will pull the metal atom out of the crystalline matrix, difficult though that is to do, and put the newly released metal ion into solution. The surface of the metal crystal is eroded by the loss of the atom.

The story is not over. Next suppose that in the solution is also a dissolved negative ion, say a chloride ion or a sulphate ion. It is possible for the metal ion to form an ionizable salt and to precipitate onto the surface of the metal as a metallic salt. If the positively charged metal ion encounters a free oxygen radical, the metallic ion and the oxygen radical can and probably will combine to form an oxide of the metal. (Free oxygen radicals are always present in water, inasmuch as water routinely ionizes into charged oxygen and hydrogen radicals, only to recombine into a stable molecule an instant later.) If the concentration of metallic salts is sufficiently high as to precipitate, crystals will grow on the surface of the base metal. Similarly, the oxides of most metals are quite stable and precipitate upon formation. The emergence of these crystals on the surfaces of crystalline metals is corrosion. That is all there is to corrosion. In the preceding example, the corrosion crystals formed as a result of the ionization of the metal because of gross electron loss and the immediate proximity of water and negative ions in solution in the water. But corrosion can also occur as a consequence of the water and negative ions acting without the benefit of externally, positively, charging the metal by grounding. The water can form a small electrical cell on the surface of the metal. Given the transient flow of electrons within the metal and a similar counterflow in the electrolyte-laden water, it is statistically possible for a metal atom to become positively charged without an overall loss of electrons from the metal crystal. The loss occurs locally within the cell; and at that precise moment, the ever-vigilant polarized water molecules do what polarized water molecules do: they "lift" the metal ion out of its matrix. This sort of corrosion at localized cells often results in the pocking, or pitting, of the metal surface. The formation of these small depressions has the effect of increasing the surface area of the water droplet relative to the volume of water, and accelerates the rate of corrosion at the site of the cell. Metals often develop the familiar dimpled surface appearance due to this pitting phenomenon. The effect on sheet metals is that, instead of corroding in a uniform manner, the corrosion works through the sheet in discrete points or pin holes.

Certainly, corrosion can occur in the absence of water altogether; but then the process requires very high energy levels, and so is normally associated with very hot metals. In this case, the energy level of the metal causes the metallic bond to be weakened, by the extreme oscillation of the metal nucleus independent of any external attraction. This so-called dry corrosion can be observed as steel is being rolled. The red-hot steel reacts with airborne ions to form corrosion products on the surface of the steel, which are promptly pressed and rolled into what is called mill scale. Mill scale is no less a corrosion product than other forms of corrosion, but because it is not economically significant, we discount it. And, because mill scale is brittle, most of it is fractured and dislodged by the rolling process, and remains behind on the mill house floor.

When aluminum is extruded, it is less than red hot, but because it is quite reactive, it corrodes in the open atmosphere immediately upon exposure to oxygen. The aluminum oxide, unlike mill scale on iron or steel, is a small crystal, which bonds very tightly to the surface of the raw aluminum. The result is that the aluminum oxide, which is chemically rather inert, forms an instantaneous protective wrapper on the surface of extruded aluminum. The depth of the aluminum oxide is thin and transparent, so the appearance of the newly extruded aluminum is considered the natural, bright finish of bare aluminum. In fact, bare aluminum is only achievable in an open atmosphere, after cooling, when the aluminum oxide is abraded off the surface. The formation of the aluminum oxide is a corrosion process; however, because it is inconspicuous and self-limiting, we leave it alone. Most people are not even aware it exists.

Aluminum is not the only metal to produce self-limiting corrosion protection. To some extent, many other metals, if undisturbed, will corrode in such a way that the corrosion product forms a relatively inert protective film or cover for the metal. For example, wrought iron will produce such a coating if it is buried; copper turns an attractive green if the atmosphere is properly polluted, with the requisite sulfates forming the ideal corrosion product. When we regard the corrosion product as attractive, we refer to it as patina; when we do not, we call it tarnish or rust.

Of course, not all corrosion products form tight crystalline depositions nor are they self-limiting or protective. The corrosion product of cast iron is a gross crystal form that does not bond well with the iron. It cracks easily, which provides perfect paths for water intrusion and cell formation. The corrosion product tends to slough off the face of the cast iron, thereby exposing more iron to the corrosive elements. In other cases, the crystalline form is small, but never bonds very well to the metal surface. As a consequence, the corrosion product simply washes away with scarcely a trace, such as lead oxide.

These examples of corrosion mechanisms may lead to the conclusion that the corrosion of metals is an easy process. It is not; the metal crystal is tough, and the corrosion of dry surfaces requires extreme heat. The difficulty of fracturing the metallic bonds mechanically is precisely the quality of metals as a class that we prize so highly. That metals corrode in the presence of water with such apparent facility is not a measure or indication of the weakness of the metallic bond, rather a testament of the power of the polar attraction of water. Water, too, is powerful stuff.

This introduction to corrosion should clarify why pure metals are rare in the undisturbed environment. If out of the molten magma, lenses of pure crystalline metal formed, over geologic time those formations corroded. Large geologic formations of economically recoverable corrosion products are called ore. Metal ores are prized on this planet as raw materials second only to fossil fuels. Corrosion of metals in service, the subject of this section, is no more than the return to a similar corrosion state from which the metal was extracted and refined.

RATES OF CORROSION

Not all metals are equally vulnerable to the discharge of electrons. One of the necessary and sufficient conditions for the corrosion mechanism to progress is the loss of electrons and, therefore, the creation of positively charged metal ions. If there is no loss of electrons, there is no corrosion. If the loss of electrons is slow, then the progression of corrosion is slow. The degree and rate of discharge is related to the propensity of a given metal to throw off, or shed, electrons. That propensity varies among metals and can be measured and displayed in a comparative series called the electromotive series. This scale is arrayed in either ascending or descending order, to illustrate that metals have different potentials for electron discharge (see Table 5.1).

In one of the standard tests for such *electromotive force* (emf), hydrogen is taken as a base. The hydrogen atom has one electron in its outer shell, which has the potential to carry two electrons. Whether the hydrogen gives up its one electron and becomes a positive ion or attracts an electron and becomes a negative ion is a good measure of the electromotive potential of

Table 5.1 ELECTROMOTIVE SERIES (NORMAL HYDROGEN SCALE)	
Metal	**Electrical Potential in Volts**
Magnesium	−2.37
Aluminum	−1.66
Zinc	−0.76
Chromium	−0.74
Iron	−0.44
Cadmium	−0.40
Cobalt	−0.28
Nickel	−0.25
Tin	−0.14
Lead	−0.13
Hydrogen	0.00
Copper	+0.34
Silver	+0.80
Gold	+1.50

a given metal. For this reason, hydrogen is assigned the electromotive potential of 0.00, and the electrical potential of metals is compared with the relative attraction or rejection of electrons compared to hydrogen. Elements that have negative values are those that will discharge or "sacrifice" to the hydrogen; elements that have positive values will attract electrons from hydrogen. The magnitude of the number in either case is a measure of the relative pull of that potential.

The metals high in the table will tend to discharge or sacrifice electrons to the metals low in the table. The metals receiving electrons are "pacified," which means they are not inclined to form ions and, therefore, are less inclined to corrode. Based on this table, given any combination of two metals, it is theoretically possible to predict which one will sacrifice to the other and, therefore, which one will corrode more or less rapidly. This phenomenon, called *galvanic action*, is not without a price, and the price is that the "sacrificial" metal will corrode at a more rapid rate, because it must not only cope with its own corrosion activity, but also compensate for the electron loss of the other metal, which is referred to as the more "noble" metal.

There is, unfortunately, a problem associated with the normal hydrogen series: the test is conducted in a pure aqueous environment. This means that the entire series is predicated on the premise that the electrolyte that forms the cell with the two metals is pure water. The prospect of having two metals in contact is very great—and of significant interest to us as pathologists—however, the likelihood of such an important deterioration mechanism occurring in a pure aqueous environment is strictly hypothetical. When the material properties are exposed to more realistic environments, the series begins to shift. It does not invert or anything else so dramatic, but the shifts in electromotive potential are considerable.

Table 5.2 arranges metals according to electromotive potential in a saline environment, which is a much more likely arena than a pure aqueous environment. Note that, due to the nonavailability of values of a few metals (chromium, cobalt, and gold), not all the metals in the first table appear in the second, and one other (brass) is added.

The important point to focus on in Table 5.2 is the reversal in the positions of aluminum and zinc. In the laboratory, using pure water, aluminum sacrifices to zinc; however, in the more likely situation involving salt water, the zinc will sacrifice to the aluminum. Also note the dramatic changes in the positions of nickel and lead: nickel moves down while lead moves up.

If we look at an abbreviated table of metals in other solutions, we find other interesting and potentially important shifts. Table 5.3 lists four metals in four different environments. The absolute values for the figures in the saline column in this table are slightly different from those in Table 5.2 due to variations in the sources; but, with the exception of the value for iron,

Table 5.2 ELECTROMOTIVE SERIES (5.85 PERCENT NaCL OR CALOMEL SCALE)	
Metal	**Electrical Potential in Volts**
Magnesium	−1.73
Zinc	−1.00
Aluminum	−0.85
Cadmium	−0.82
Iron	−0.63
Lead	−0.55
Tin	−0.49
Brass	−0.28
Copper	−0.20
Stainless steel	−0.10
Silver	−0.08
Nickel	−0.07

the numbers are close enough to accept without concern, and the order remains the same.

The difference in value for iron in a saline environment in the two tables in not explicable, based on the information available, but it may point out an important lesson: the absolute and relative values of electromotive potential are highly environmentally sensitive. A minor shift in concentration of the solute can alter the values; in some cases, the changes can be significant. Table 5.3 teaches other important lessons. In the saline environment, as we noted earlier, aluminum and zinc flip positions, relative to the hydrogen scale; they revert to the basic order in the chromate solution, and become virtually equal in the nitric acid solution. Because these two metals are prevalent, and economically significant, this flipping around is important. However, it pales in comparison with the aluminum and iron relationship. In all the environments shown in Table 5.3, aluminum will sacrifice to iron except in an acidic environment. In a nitric acid solution, the iron will sacrifice to the aluminum. We may surmise, albeit at no small risk of error, that other acids will have similar results. A similar prospect exists for salts other than sodium chloride, specifically that aluminum will sacrifice to iron in other saline environments. Because saline and acidic en-

Table 5.3	SOLUTION POTENTIAL OF SEVERAL METALS IN VARIOUS SOLUTIONS			
Metal	**Sodium Chloride**	**Sodium Chromate**	**Sodium Hydroxide**	**Nitric Acid**
Magnesium	−1.72	−0.96	−1.47	−1.49
Aluminum	−0.86	−0.71	−1.50	−0.49
Zinc	−1.15	−0.67	−1.51	−1.06
Iron	−0.72	−0.16	−0.22	−0.58

vironments are quite prevalent, the chance that two extremely important metals will reverse electromotive potentials is problematic, to say the least.

The general rule for bimetallic conditions is that the sacrificial metal will corrode at the same time it pacifies the more noble metal. If the relative electromotive potentials do reverse, what happens? In the ideal world, the former sacrificial metal would simply reconstitute itself; but, in reality, that would require a net increase in order and, therefore, in energy. Absent an external source and mechanism for increasing the potential energy, the rule is that energy always dissipates. Aluminum corrodes in the presence of iron in a saline environment, and that iron corrodes in the presence of aluminum in an acidic environment. They both corrode if the environment fluctuates from saline to acidic. It is not inconceivable for the solution to be both saline and acidic. If the concentration of each solute is precisely of the nature in Table 5.3, it is more likely that the aluminum will incur the greater damage. Notice that the electromotive differential between aluminum and iron in a saline environment is 0.14 volts, whereas the differential in the acidic environment is 0.07 volts. This suggests that the greater potential discharge will occur from the aluminum to the iron. In fact, a slightly more sophisticated view of chemical reactions is that both reactions will occur—that is, the sacrifice of both metals will occur—but the aluminum will corrode slightly faster.

THE CONVERSION PROCESS

Our discussion of corrosion began with the objective of understanding how we produce metals, which led naturally to the fact that metals rarely appear in nature in pure crystalline form because they corrode. The corroded state for every metal is a lower-energy state than that of the pure crystalline state; therefore, all metals will corrode. In the natural setting, we recognize large concentrations of corroded metal as ores; and if the concentration of metal content is adequate, we mine the ore and convert the corroded metal into a pure metal crystal. The process of conversion, in most cases the refining process, requires tremendous amounts of energy, so the story of metals manufacturing is often a discourse on the transportation logistics it takes to bring the vast tonnages of ore into proximity with the vast tonnages of fuels necessary to execute the refinement process. One major reason Pittsburgh became the steel capital of the world was because the great iron ranges of Michigan and Minnesota were reachable by lake and river transport, and the great Pennsylvania and West Virginia coal fields were reachable by canal and rail. The point of optimal convergence was at the confluence of the Monongahela and Allegheny Rivers.

When the metal is valuable enough, we take the ore to wherever is required to extract the metal. Take, for example, Jamaican bauxite, the primary ore from which we derive aluminum. Over the last hundred years, a great deal of aluminum ore has been transported to Europe and North America for refining. For zinc, lead, and antimony, the energy demands are sufficiently lower than for iron, copper, and aluminum, therefore, those metals can be refined closer to the mine head, thus reducing total transportation costs.

Once the ore reaches the point of refinement, it is crushed and sorted. By any number of means the ore is heated to a point at which the ore crystal fractures and the nonmetallic "slag" is removed, leaving behind elemental metal in economically viable quantities. In some cases, this is a two-stage sequence. Iron, for example, typically is reduced for iron ore in a carbon-rich environment. Then the carbon is removed from the carbon-rich iron in a second heating, which oxidizes the carbon in the form of carbon dioxide. The carbon-rich stage of iron production is called pig iron, named for the shape of the mold on the floor of the blast furnace into which the iron was poured, which resembled a row of suckling piglets lined up on a sow.

Dissolved within pig iron is a high concentration of carbon, which, if left in without much alteration, can become *cast iron* (the name is indicative of the relative ease of pouring the carbon-rich cast iron into molds). The carbon concentration of cast iron is approximately 5 percent by weight. The presence of this much carbon mixed into the matrix of the iron crystal means that the melting point is substantially reduced, which is why casting is more economical than pouring a metal with a higher melting point. Simply, it takes less energy to liquefy the metal. The unfortunate consequence of carbon concentrations as high as 5 percent is that the carbon does not necessarily disperse in the matrix in an absolutely uniform way; it may vary within the body of the iron crystal. Iron-carbon mixtures of different concentrations display different properties, and so are given distinguishing names, such as austinite and martensite, depending on the level of carbon content. Where the carbon concentration is high, at or near the 5 percent level, carbon may crystallize into zones, or nodes, of nearly pure carbon, that is, graphite. Graphitic iron is grainy and quite brittle; when chilled, it may fracture under impact, and when cyclically loaded it is far more likely to crack than iron mixtures with lower carbon content.

The commercially available metal with the lowest concentration of elemental iron is wrought iron. To form wrought iron, the ore is reduced in a carbon-rich environment not unlike that for cast iron. Instead of molding it into "pigs," however, wrought iron is heated until the slag is removed and the carbon is burned off in what amounts to a continuous operation. The iron content of wrought iron is 99.98 percent. Iron purer than wrought iron is achievable only in small quantities under laboratory conditions.

The term steel is reserved for a wide range of iron-based alloys. The most common commercial alloy is called *mild steel*. The more formal designation is defined by ASTM standard A36, "High Carbon Steel," which is somewhat amusing in that high carbon steel means approximately 1 percent carbon. High carbon steel is still 98 percent iron. The remaining 1 percent is partly composed of the residual impurities left from the original ore and from the "flux," a generic term for compounds added during the refining process that bind the nonmetallic inclusions in the ore into a relatively lightweight complex. This complex of bound silicates, carbonates, and other irreducible crystals float on the top of the molten iron or steel; from there they are drawn off, and once cooled, are collectively referred to as slag.

Slag accumulating around iron and steel mills for the past century or more has probably polluted more streams and rivers than any other environmental hazard. The reason slag has been and continues to be so dangerous is that one of the major impurities carried off by the flux is sulphur. The ore itself does not contain large quantities of sulphur, but the fuel, namely coke from bituminous coal, used to produce the pig iron, contains a great deal of it, which is dissolved in pig iron along with the excess carbon. Whereas the carbon and some of the sulphur can be reduced simply by reheating the iron, most of the sulphur is carried off with the slag, which, historically, was often dumped into the local stream or along a river bank. (At the time of the great flood at Johnstown, Ohio, in 1889, the width of the Conemaugh River was a fraction of its preindustrial width as a consequence of slag dumping.) Water percolating through sulphur-enriched slag combines with the sulphur to form sulfuric acid, which has had a highly debilitating effect on any aquatic life downstream of the great slag piles. Arguably, the regeneration of the great rivers of the eastern United States is far more the result of the emergence of foreign steel and the concurrent decline of the eastern steel manufacturing centers as any environmental legislation.

Of greater concern to the topic at hand, however, is the use of slag in construction applications. Even if we discount the sulphur content as an issue, slag is virtually worthless. It is of very low density, and has very low compressive capacity. Traditionally, it was used extensively as fill, in lieu of gravel, below slabs on-grade as a drainage layer. It was used on flat roofs between the roof slab and the BUR to provide slope and pitch. It was occasionally used as aggregate to produce reduced-weight concrete. It was also occasionally used as ballast on flat roofs. What all of these applications did was put sulphur and, therefore, eventually sulfuric acid, near metals, notably reinforcing steel and metal roofing items such as anchors and cleats.

When mild steel corrodes, the corrosion product results, along with the predominance of ferrous and ferric oxides, in inclusions of carbon. The resulting crystalline structures occupy approximately four times the volume of the displaced iron, and they are relatively friable when they are unconfined. Confined crystalline growth can exert expansive pressures of several thousands pounds per square inch. In certain circumstances, the crystalline growth occurs in tight crevices, such as between adjacent plates of the steel, which means that the crystals may actually deform the plates resulting from the expansive lateral force. Crystalline growth below base plates can lift columns; corroding reinforcement can explode the concrete encasement.

The crystalline product of mild steel is also brittle and friable. As it grows and expands in an unconfined environment, the crystal will spontaneously crack and slough off the from the base metal, opening avenues of water intrusion and exposing undecomposed metal to continuous and progressive deterioration. Other metal corrosion products may behave in comparable ways, or they may mimic the fine-grained, tightly bonded nature of aluminum oxide.

One specific alloy of steel, best known by the brand name Corten, but generically called *weathering steel*, was developed to corrode such that the resulting corrosion product would accumulate on the surface in much the same way as other self-limiting corrosion crystals. In weathering steel, the texture of the weather surface is grainy and dark brown once the "patina" is complete. Before reaching to a stable crystalline structure, however, the weathering steel sheds a substantial quantity of ferric oxide, which washes off the metal. The resultant "staining" precedes the stabilized stage. The patina can be retarded by any oil, including that from the palm of a human hand, which can be visible for several years after the material is put into service. Furthermore, weathering is not inviolate; it can be reinitiated chemically by acids and by cracking caused by flexure of the steel member.

The corrosion patterns and the mechanism properties of metals can be engineered by the process of alloying, which, conveniently, brings us back to how we began this section. Restating, alloys are mixtures of a base metal with other metals or nonmetals. The mixture is then cooled and solidified. If the alloy is cooled too slowly, depending on the respective heats of fusion of the alloy material and the base material, the alloy material may crystallize independently of the base material. If the alloy material crystallizes first, it will form nodules or grains within the matrix of the still-solidifying base metal. If, on the other hand, the base metal solidifies first, and slowly, the alloy material may be excluded from the base metal matrix and, essentially be pushed into pockets of pure alloy material. One way alloy mixtures are "frozen" to produce more homogeneous distributions is through rapid cooling, and one of the concerns in tempering materials is the risk of recrystallization of the alloy materials.

Theoretically, alloys can be formed by combining any two or more metals; certainly, most combinations have been tried at one time or another. Most alloys retain the same label as the base metal, so that there are dozens of alloys of iron all of which are called steel, followed by a specific ASTM designation. A high-carbon alloy is A529; two high-strength and low-alloy steels are A441 and A572. There are even more alloys of aluminum, which are designated by tempering and alloy numbers such a TU-254. These names are not particularly descriptive, nor are they familiar to anyone other than metallurgists, fabricators, and structural engineers.

There are, however, some alloys that, for a variety of reasons, have been given monikers that make them seem unique, such as brass and bronze. But as metallurgy has become far more sophisticated, even these labels have come to refer more to classes of alloy than to a specific blend of metals. Brass, for example, is copper-based; but, historically, the alloying metal was tin and/or zinc and occasionally other metals. Today, the term is reserved for copper-zinc combinations, but the range of blends is wide. Similarly, the combination of copper and tin is now called bronze, again, accepting a range of concentrations. Other commercially significant alloy classes are terne, monel, and pewter.

Terne refers to alloys with a lead base to which is added approximately 20 percent tin and some antimony. As such, terne is closely related to pewter, which originally was an alloy of tin and lead, but in its modern incarnation is an alloy of tin, antimony, and copper. Monel, named for an American business man, Ambrose Monel, comprises a group of alloys consisting of approximately 68 percent nickel and 30 percent copper.

The copper in a penny is actually an alloy predominantly copper; pure copper would be too soft to work well in a coin. Gold is rarely used in its pure form, the relative purity of which is designated as some "number of carat" gold. A carat is a designation of percentage: 100 percent is 24-carat; therefore, 12-carat gold is 50 percent gold and 50 percent other metals.

One of the most important alloys or group of alloys in the building business is the aforementioned stainless steel. This metal consists primarily of iron with approximately 20 percent chromium and other metals added to balance properties. It is expensive due to the chromium content; otherwise, we would probably use it for almost every application calling for a metal. Stainless steel, despite the inclusion of substantial quantities of alloying metal, retains many of the structural properties as mild steel, namely a modulus of elasticity of approximately 29×10^6 psi (2×10^{11} Pa) and yield strengths well above 36,000 psi (2.5×10^8 Pa). It is harder than mild steel, but somewhat more brittle and less malleable. The property for which stainless steel is valued is, of course, its relative resistance to corrosion.

Recall from Table 5.2, which listed the electromotive potentials of various metal in a saline solution, that the electromotive potential of stainless steel was identified as –0.10 volts, which put the value of stainless steel in the same range as that of silver, but significantly below that of copper for the same environment. In short, stainless steel exhibits very low electromotive potential in many corrosion environments. This does not mean that stainless steel will not corrode—it will—but it does exhibit a resistance to corrosion comparable to the most noble of the metals. It achieves this resistance through the clever use of galvanic action to pacify the base metal.

The Sacrificial Process

To reiterate from earlier in the chapter, regarding sacrificial metals and galvanic action, we can deliberately blend two metals so that the so-called sacrificial metal will corrode, but in the process will inhibit the corrosion of the base metal through pacification. It does not matter where the sacrificial metal is physically located relative to the more noble metal, as long as there is a viable electrical conduit connecting the two metals. The sacrificial metal may be in any one of three fundamental locations relative to the base metal: adjacent, proximate, or alloyed.

An example of an adjacent relationship that has been used from time to time is to connect a block of magnesium by an electrically conductive wire to an underground cast-iron *sewer*, which in this context refers to a particularly vulnerable piece of metal. Though it is exposed to hostile environments both from the interior and the exterior, it is hidden from view; access for maintenance or even for inspections is very difficult. The industry has developed techniques for inspecting such items from the interior, using small television cameras attached to cables, which send back images of the interior, but the efficacy of such inspections is relatively low, particularly when the outside is not reasonably accessible for comparable inspection, much less meaningful intervention. Certainly, sewers are well coated before placement, but corrosion will occur eventually.

If we attach a block of magnesium with a fine copper wire to the sewer, the magnesium will sacrifice electrons through the wire to the sewer every time that the iron discharges electrons. Over time, the magnesium block may corrode, but the iron sewer remains protected. If the magnesium block corrodes to the point of diminished effectiveness, we can simply disconnect the old block and install a new one. The point is that, as long as the sewer discharges to ground, the magnesium will discharge to the iron, and the iron is pacified. This process is, obviously, only a small step from connecting a block of sacrificial metal to a more noble metal to connecting a direct current generator or a battery to the base metal. This variation is employed with underground piping, including sewers. The main difference between using the magnesium block and the direct current generator or battery is that the magnesium block will discharge when a voltage drop occurs, but only to the magnitude of the voltage differential between the block and the sewer, because the current is predicated of the demand, which is established by the discharge of the sewer. The generator or battery supplies current whether the sewer is discharging or not. Excess current will discharge to ground, which means that the current and, therefore, the cost is constant and is supply-based.

If the sacrificial metal block can be attached by a conductive wire, there is no reason the sacrificial metal cannot be applied directly to the surface of the base metal. Returning to the example of the magnesium block connected to the iron sewer via an electrically conductive wire, that same magnesium could, theoretically, be attached directly to the iron, thereby eliminating the need for a connector wire. If we expect the applied sacrificial metal to supply electrons to the more noble metal, the electrical contact must be positive, and the contact surface itself cannot be subject to corrosion. Because the voltages involved are very low, a very small amount of corrosion deposited between the sacrificial metal and the more noble metal will break the contact and, therefore, the flow of current. Clearly, the contact plane between the two metals must be flawless. Thus, the connection cannot be accomplished with any kind of adhesive or binder because they will break the contact. If the sacrificial metal is applied using mechanical fasteners, the contact plane between to two metals is inherently less than full contact, and corrosion at the plane is almost certain.

The most common way to meet these objections is to coat the more noble metal with the sacrificial metal. This is most expeditiously done when the sacrificial metal is in liquid form and the more noble metal is a solid, which introduces another level of detail to the issue. The base metal must be extremely clean. This is usually accomplished with an acid wash called *pickling*. Once the base metal is clean, it is often dipped into a vat of molten sacrificial metal. The melting point of the base metal must be significantly above that of the sacrificial metal. Even when that is reached, however, there is a risk that the base metal may warp or twist due to differential expansion brought on by the asymmetry of the piece or fabrication. When all the conditions are met, the *hot dipping* of metal pieces in molten sacrificial metal is fast and relatively inexpensive.

Hot dipping, unfortunately, is not always feasible. The melting point of the sacrificial metal may be too close to that of the base metal to be done safely. Or the absolute value of the melting point of the sacrificial metal may simply be so high that the cost becomes prohibitive. The same is obviously true if the sacrificial metal is particularly rare or valued for other reasons.

When hot dipping is not technically feasible, there is another way that sacrificial metal can be applied to a base metal. The process is called *electrolysis*, better known as *plating*. The way plating works is that two electrodes are immersed in an electrically conductive fluid or electrolyte. When a direct current is passed from one pole to the other through the electrolyte, one of the poles is necessarily discharging electrons into the solution, and the other pole is necessarily collecting electrons from the solution. The pole that discharges electrons is called the *cathode*, and the collecting pole is the *anode*. (The names of the poles are understandably often confused, so it may be helpful to remember that cathode and catapult derive from the same Greek prefix, *kata-*, which means "away from"—the electrons are moving away from the cathode. *Ana-* is the Greek prefix for "back"—the electrons go back to the anode.)

If the electrolyte solution contains metallic ions, those ions will be pacified at the negative pole, or cathode, because the electrons will discharge from the electrode to the positively charged ion. If the temperature of the solution is above the melting point of the ionized metal, it will crystallize from the solution and deposit itself on the surface of the cathode. The cathode, therefore, will be coated with a metal. If the electrode is itself a metal, such that the electrode is the base metal and the plating metal is coating it, there is the possibility that one metal will have a higher electromotive potential than the other. If there is a net electromotive potential between the plating and the base metal, the plate metal may sacrifice to the base metal. Another term for this approach to pacification, that is, the application of a layer of sacrificial metal to the surface of a base metal, is called *anodic protection*, because the base metal is the metal attracting back electrons, and the sacrificial metal is catapulting electrons to the anodic metal.

Meanwhile, back at the electrolytic bath. Specifically, at the other pole, the anode, in the electrolytic bath, the negative ions in the solution are stripped of excess electrons and neutralized. If the negative ion is a chloride, chlorine gas will be produced at the anode. If the negative ion is a sulfide, sulphur will accumulate on the cathode. Note that several of the preferred electroplating salts are cyanides, so there is the risk of polluting the immediate environment with cyanide gas.

Commercially, plating is primarily used in finishes, for purposes of appearance more than anodic corrosion protection, which is typically reserved for items that are fairly small and relatively precious, such as hardware. The process of plating is somewhat costly, and it is time-consuming. To reduce costs and the required time, plating is normally deposited at the minimum depth, which will accomplish the objective, especially if the objective is cosmetic. If, on the other hand, the objective of plating is corrosion protection, the thickness must sacrifice material and still be viable, so the plating depth must increase.

Whether the applied metal is hot-dipped or plated, anodic protection is the most common relationship to the base metal. To review, this means that in most, if not all, environments, the applied metal, the metal we can actually see on the exterior of the piece, will sacrifice to the more noble metal below. Even if the metal on the surface is damaged or cracked, it will continue to provide effective anodic protection as long as there is approximately 10 percent area coverage of the base metal (see Fig. 5.2).

By far, the most commercially important applied anodic protection is the hot-dipping of steel or iron into molten zinc. This process is called galvanization, and the term means precisely that: molten zinc on steel or iron.

Another common anodic combination is lead-coated copper, usually in sheet form. This particular combination offers the advantage that, in addition to the anodic protection, the presence of the lead provides excellent fusion with solder applications. Both processes are feasible, because the sacrificial metals liquefy at temperatures significantly below those of the respective base metals; both sacrificial metals remain anodic to the respective base metals in most environments; and both are inexpensive compared to the corrosion reduction.

Another approach to bimetallic corrosion reduction that is *cathodic protection*, wherein the sacrificial metal is the same item as the object to be protected, and the more noble metal is between it and the weathering

Note: The terms galvanization and galvanic action derive from the name of an Italian physiologist, Luigi Galvani (1737–1798) who discovered, among other things, a connection between certain chemical reactions and electrical discharge, and vice versa. It is not difficult to understand how his name came to be associated with galvanic action, but it is curious how it came to be applied to a process which, specifically, does not involve an electrical/chemical interaction.

FIGURE 5.2   This is all that is left of an outboard slung metal gutter, in this case made of galvanized steel.

medium. At the outset, it may seem inexplicable, having made such a case for anodic protection, to suggest that we put the sacrificial metal below the more noble metal. To understand, imagine that the item we want to protect is a steel casement window frame. We could dip the steel frame in zinc (galvanize it) to protect it anodically, and that would certainly work. But let's also say we do not like the look of the exposed zinc, and prefer a metallic finish without paint; in fact, we like the look of weathered lead. Lead, however, is higher on the electromotive series than iron (steel), and in most environments the iron in the steel will sacrifice to the lead. While coating the steel with lead is technically feasible with regard to fusion temperatures, the combination does not provide good corrosion protection. Recall, however, that for the corrosion process to progress, the corrosive environment must be in direct contact with the corrosion surface. If we anodically protect the steel, the sacrificial metal will always discharge to the steel, which will pacify the steel and inhibit corrosion whether there is an exposed surface on the steel or it is entirely encapsulated by the sacrificial metal. Of course, this is at the expense of the sacrificial metal, which is not only exposed to the corrosive environment itself, but discharging to the steel at the same time, thereby, accelerating the corrosive effect on the sacrificial metal.

With all that in mind, we now point out that it is possible to reverse the relationship of the two metals and put the sacrificial metal on the inside of an envelope of more noble metal, to which it, the encapsulated metal, will discharge. If the inside material is still steel and is still, let us say, the structural member and, therefore, the item we want to protect, this might seem to be a counterproductive approach to protection. Though convoluted, it is not necessarily counterproductive. Why not? If the more noble metal is exposed to the corrosive environment, it will discharge electrons to that environment and, therefore, potentially ionize and corrode, except that the encapsulated steel will immediately discharge to the coating and pacify it against corrosion. The question is, obviously, why doesn't the steel itself ionize and corrode? The trick is to keep the steel away from the corrosive environment, and that can be accomplished only as long as the coating is *perfect* in its encapsulation, or to ensure that wherever the steel is exposed there is no corrosive medium. If those two conditions are not met, the steel will, indeed, corrode; and it will corrode rapidly, as it is the sacrificial metal.

Although cathodic protection may sound somewhat risky, it is feasible. This example was based on the assumption that zinc was an unacceptable surface, and that lead would be far more desirable. The question was, and is, whether we can employ a metal coating on steel and expect to achieve effective corrosion protection. The answer is that we can, provided that the steel is completely coated and protected from the corrosive environment. If we can satisfy those conditions, the benefit of cathodic protection is that the surface exposed to weather is pacified and, therefore, protected from corrosion. If we had used zinc as anodic protection, we would protect the steel, which was the primary objective, but at the expense of the diminishment of the exposed zinc surface. By using cathodic protection, the exposed surface is protected and remains in good condition as well.

There are limits, however, to the capability of cathodically protected metal to supply electrons to pacify the exposed surface metal. If the sacrificial metal is, indeed, completely encapsulated, a finite supply of electrons is available to accomplish the pacification. When that supply is exhausted, the pacification process stops and the exposed surface metal proceeds to corrode as it would whether a sacrificial metal is present or not. Silver plate is a good example of how this process works. As opposed to sterling silver, which is as pure a grade of silver as is commercially feasible, silver plate is deposited on a base metal by way of electrolysis. Including a core of far less expensive metal than silver, to cathodically protect that core and to retain the brightness of the plated finish against corrosion allows us to reduce the cost of the item. In most environments and situations, the actual silver on silver plate will corrode less than the silver surface of sterling silver in comparable exposures.

The core of the plate too has limits on its capacity to pacify the high amount of exposed silver. When the electrical capacity of the core is exhausted, the silver tarnishes, which is what we call corrosion of silver and other precious metals. This raises the question with regard to silver plate, specifically, and cathodic protection, generally, as to whether cathodic protection is not simply a one-time event, and after the pacification potential is exhausted, corrosion commences. Can the core, which is also the sacrificial metal, ever recharge? Fortunately, it can and does recharge. In the case of lead-coated steel casement window frames, the source of recharge is fairly direct and continuous. In the case of silver plate, recharging is somewhat more sporadic. The ultimate source of the replacement charge is ground. Most of the time we think of ground as the ultimate receiver of electrons, which relative to our perception of the globe, it is; but it is also the ultimate electron "donor." If the combined electrical charge of the plated object or the casement window is positive as a result of total discharge—that is, the sum of the electrical status of the core and the plate or of the frame and the coating—then there is a net positive electrical attraction. If the object is, in fact, grounded, the object will effectively recharge from ground, as surely as negatively charged objects will discharge to ground. In the case of lead-coated steel frames, the building itself acts much as ground. The silver plate is different, in that it may rest on an electrically isolated shelf for protracted periods, during which it discharges to capacity, and tarnish occurs. Theoretically, silver plate will tarnish less if it is stored on a grounded shelf, whereas sterling silver will tarnish more if it is stored on a grounded shelf.

The third basic way galvanic action can be constructively harnessed to inhibit corrosion is to bury the sacrificial metal inside the body of the base metal. If zinc on the surface of the steel can anodically protect the steel, what happens if we embed the sacrificial metal inside the steel alloy? Now, the zinc will sacrifice to the steel (iron) cathodically because the sacrificial metal is encapsulated by the more noble metal; moreover, because the zinc is perfectly encapsulated by the surrounding steel, it cannot come in direct contact with corrosive environment and will, therefore, not corrode. The fundamental metallurgical problem is, however, that the melting point of the zinc is so far below that of the steel and iron that, as the steel crystals form, the still-liquid zinc will be excluded from the crystalline formation. Simply put, it is very difficult to accomplish an alloy of metals with very disparate temperatures of fusion. The metal with the lower melting point essentially floats out of the alloy as the metal with the higher melting point congeals. What we need is a metal that has a similar melting point and is less noble than iron to inhibit corrosion through alloying. The metal that most closely fits this description is chromium, and the alloy is called stainless steel. Because the basic electrical mechanism is cathodic, stainless steel will perform better when it is grounded.

To repeat yet again, to dispel the common perception that stainless steel will not corrode, *all* metals will corrode. Even the currently popular "magic" metal, titanium will corrode.

5.3.3 UV and Ozone Degradation

We address ultraviolet (UV) degradation and ozone degradation in the same section for two reasons, neither of which has much to do with the respective deterioration mechanisms. First, ozone and UV are related to each other in that the production of ozone is much the product of light, generally, and UV, specifically, passing into and though the Earth's atmosphere. Second, practitioners often, incorrectly, connect the two mechanisms in their minds as related or reinforcing of one another. Though addressing the two subjects in this section may tend to support the second notion, we do so to *emphasize the differences* between these two processes.

In a very loose and general sense, ozone degradation, or ozonation, is analogous to corrosion, whereas UV degradation is analogous to heat damage. Ozone is particularly threatening to the class of materials called rubber, both the natural and synthetic varieties. Depending on the chemistry of the substance, ozone may damage gaskets, sealants, tires, surgical tubing, rubber bands, elastic waistbands, adhesives, and other polymers consisting of or related to natural or synthetic rubber. As damaging as ozone is to this large class of items and components, it is *not* damaging to the majority of materials, including polymeric materials. The chemistry that renders rubbers vulnerable to ozone "corrosion" is very specific to the chemistry of the class, and should not be generalized beyond that class.

UV degradation works on a principle having to do with the fracturing of bonds at the molecular level. In hydrocarbon polymers, C:C and C:H bonds are normally among the toughest in the polymeric chain, and the energy required to fracture these bonds is quite high. When energy is imparted through irradiation to a polymer or any other substance, the atoms within the molecule begin to oscillate at increasing amplitudes and velocities, as discussed earlier in this book. At the same time they increase their respective rates of oscillation because they are absorbing radiant energy, the atoms also generate radiation, which reduces the net rate of energy absorption. If energy is absorbed faster than it can be rejected, the temperature of the solid rises. If this process is carried to an extreme degree, it is conceivable that the temperature of the macrostructure of the polymer will rise to a point at which the various bonds will fail at both the short-range as well as the long-range order of the molecule and the crystal, respectively. The solid will melt if the long-range order is disrupted. If the source of radiant energy is limited to exposure to the sun, the prospect of fracturing the C:C and/or C:H bonds is remote, because the rate of energy transfer within the solid makes it difficult to concentrate the energy required to fracture the bond without it dissipating to adjacent areas of the molecule.

One way to turn the trick is to hit the region of the bond accurately, with enough energy, to fracture the bond before the energy can transfer and dissipate. One of the few sources of energy with enough wallop to do the job is UV light with wavelengths between 400 and 300 nanometers (nm). Wavelengths much below this level do not pack the energy necessary to fracture the bond, and wavelengths shorter than 300 nm, which might otherwise be able to fracture the bond, are filtered out of the Earth's atmosphere before they can do any damage.

It is at this point that the ozone and UV stories come together. Ozone is produced primarily by the introduction of light, specifically the UV and deep UV wavelengths, into and through the atmosphere. The process, called *photolysis*, depends in part on the initial concentration of nitrogen dioxide in the path of the incoming UV light. In very simplified terms, the UV and deep UV wavelengths fracture the nitrogen dioxide molecule, producing nitric oxide and a free oxygen radical. The free oxygen radical moves into and out of equilibrium states with O_2 molecules, resulting in transient and unstable molecules of ozone (O_3).

The atmospheric concentration of ozone at any location and time is dependent upon the concentration of the raw material, nitrogen dioxide, the concentration and duration of UV, and the dispersal or neutralization rate of the ozone. Since the formation of the planet, ozone has been produced in the upper atmosphere. More recently, humans have introduced to the surface new sources of highly concentrated oxides of nitrogen, in the form of exhaust gases from internal combustion engines. When fuel is mixed with air in the combustion chamber of a gasoline or diesel engine, it is inevitable that a certain amount of nitrogen will be introduced into the cylinder at the same time. Upon ignition, the atmospheric nitrogen in the fuel-air mixture inside the combustion chamber is partially or completely oxidized. Previously, we discussed this phenomenon as a source of atmospherically generated nitric acid; it is also responsible for ozone concentrations 10 to 25 time higher in urban areas than ambient concentrations in rural areas and the upper atmosphere. Ozone concentrations are also higher in the summer than at other times of the year, not because of heat, but because the angle of the sun relative to the Earth's atmosphere causes more UV to penetrate deeper into the atmosphere; more UV light is, therefore, available to react with nitrogen dioxide.

Ozone, when it comes in contact with rubber, attaches to specific sites on the rubber polymer, and essentially oxidizes or, more accurately, ozonates the site. The rubber polymer may fracture, or cross-link, which describes what happens when the fractured polymer fuses to another site. The net effect is that the polymer is reduced in length and in elastic capacity, and is increasingly brittle. That this occurs at the surface of a rubber item is relatively incidental and self-limiting, and is analogous to a corrosion product inhibiting further corrosion. In the case of rubber, however, the progression of the mechanism is directly related to the elongation of the material. If the element is not stressed and, therefore, not elongated, then the ozonation is surficial and self-inhibiting. If the material is elongated, fissures develop as the ozonation occurs. Being brittle, the ozonated crust cracks, thereby exposing fresh material to the environment. The mechanism is, therefore, progressive, as long as the affected material is elongated. The degree of elongation required to reach a threshold of progressive degradation varies among specific formulations. For natural rubber, the threshold elongation for self-perpetuating ozone degradation is 3 percent. For certain synthetic rubbers, the minimum threshold elongation is greater: acrylonitrile- butadiene rubber, 8 percent; for chloropreme rubber, 18 percent; isobutylene-isoprene rubber, 26 percent.

Although the mechanism is different, the effects of UV embrittlement are similar in some respects to ozonation. Not every hydrocarbon polymer is vulnerable to UV degradation, but for those that are, the result of the fracture of C:C and C:H bonds is that the polymer chains are abbreviated. When they are mechanically elongated, instead of displaying the elastic properties typically associated with relatively long, intact chains, the molecules break into fragments. What begins as a molecular fracture coalesces into identifiable fractures at the optical level.

One of the most dramatic examples of UV degradation is in hot asphalt roofing. Asphalt is an amorphous solid, so there is no long-range, ordered crystalline form. As asphalt cools, its viscosity increases until, at or around 90° F (32.2° C) it becomes "solid." In effect, the fluid at or around 90° F is thick enough to "stand," not flow away. As the temperature continues to decline, the asphalt becomes increasingly hard and brittle. When asphalt fractures after solidification, the fracture plane is characterized by swirls and sweeps in the fracture pattern. This is because there is no crystalline cleavage plane along which to fracture cleanly. We are familiar with this fracture pattern in glass, which is also an amorphous solid.

The polymers that we collectively call asphalt are not uniform by any means. Asphalt is more or less the residue from petroleum distillation after the more valuable fractions have been removed. It is a "brew" of residual petrochemicals collected from the bottom of catalytic crackers for marketing, not as the indeterminate sludge it is, but as a paving binder, a roofing adhesive, and sealing product. These are economical ways of disposing of an otherwise obnoxious by-product. Coal tar has a similar distinction in the coal distillation process.

Over normal temperatures—which is to say down as low as 0° F (−17.8° C)—fresh asphalt retains the elasticity requisite to avoid spontaneous fracture. But the longer the material is exposed to sunlight, the more fragile it becomes because of the cumulative molecular fractures. As asphalt cycles through warming- chilling cycles, the chains are stressed as described in Chapter 3. Eventually, optically identifiable cracks form in the surface of the asphalt body (see Fig. 5.3). The cracks themselves are, as we have discussed, geometric stress raisers; and, as we have also discussed, the propagation of cracks is far easier than their initiation. In asphalt, in particular, the pattern that develops from the merging of the optically identifiable cracks is called "alligatoring," to describe the seeming similarity to the hide of the animal.

FIGURE 5.3 This is an unusual instance of structural steel plate being used as the substrate of a roof. The roofing is hot-applied bitumen, which has delaminated and failed.

On an extremely general level, the approaches to inhibiting ozonation and UV degradation share some similarities. The vulnerable sites can be physically protected: wrapping items vulnerable to ozone, shielding items vulnerable to UV light. The more common and practical approach in both cases is to alter the chemistry of the compounds to mitigate or eliminate the consequences.

There are three basic approaches to mitigation of ozonation. First, something can be added to the rubber to inherently alter the vulnerability of the compound to ozone. These additives act effectively as molecular re-inforcement. The links and cross-links that form render the chain less vulnerable to ozone. Second, chemicals can be added to sacrifice to the ozone. These *antiozonants* react even more quickly with ozone than the rubber, and they, therefore, scavenge the ozone before it can attack the base material. Third, chemicals can be added to the material to react with the ozone resulting in the production of a durable protective layer of ozonated material that is flexible enough to protect the surface from further contact with atmospheric ozone.

Shielding and masking surfaces from UV damage is not as infeasible as wrapping ozone-vulnerable items, and should always at least be listed on any intervention matrix associated with the reduction or elimination of UV damage. The general approaches to UV protection beyond shielding include additions to the chemistry of the UV-sensitive material. These additives or alterations either directly absorb the incoming energy before the vulnerable site is struck or quickly absorb the energy resulting from impact at the site, but before the threshold energy requisite for fracture occurs.

5.4 Systemic Pathology

Keeping in mind that the function of the horizontal closure system is to collect and divert water off of the premises, we must now elaborate to say, there may be some sharing of functions among the components of the system. For example, the roof is a major component of the horizontal closure system, but it also often provides structure to the building, particularly the stabilization of the otherwise free edges of the tops of the walls. Similarly, the function of the vertical closure system is to provide a selective barrier, but the walls, too, often provide structural support. The organization of a site into systems implies that any single component may simultaneously provide functional contributions to more than one of these systems.

Thus, there is a redundancy in any accounting of components. We must assess the roof, for example, as it performs as part of both the horizontal closure system and as part of the structural system. The obvious question is why not simply address the site on a component-by-component basis? Because of the way buildings deteriorate. And how they are designed relates directly to this issue. The component approach to building accountability, whether as part of the design or diagnostics process, is the conventional way of viewing the building and the site. The standard specifications for buildings are organized into a 16-division format, the titles of which are an index of the components and the materials comprising those components. The presumption is that once all the pieces identified in the drawings and specifications are, in fact, delivered and installed, the building will be complete, and will perform as expected. Even allowing for the customary tuning that goes on with a new building, this is a reasonable way of constructing buildings; we are not suggesting otherwise.

Buildings actually get built through a contracting and subcontracting process, which is roughly parallel to the 16-division specifications format, which was developed and institutionalized by the Construction Specifiers Institute (CSI) within the past 40 years, long enough to shape our thinking about buildings based on the format.

But relative to building pathology, there are two problems with the component approach to building inventories and assessments: the first is conceptual, the second pragmatic. The conceptual problem with assessing buildings following the 16-division CSI format is that buildings may be contracted and built according to the format, but they do not perform according to the format. Buildings are not the compilation of pieces. Buildings are integrated entities, and once the pieces are in place, how a building performs is not deconstructable to its original components. To view a building as an assembly of components is to distort and oversimplify the functioning building.

The pragmatic difficulty with accounting for building performance through its components is, simply, that many problems in buildings emerge not at the "solids," but in the voids, meaning, among other places, at the joints and seams. Buildings deteriorate for a multitude of reasons, and the failure of an individual component is only one of those reasons. By conducting an inventory of a building component by component, we presume that somewhere in the investigation the inspector will discover a flaw in a component. The discovery of that flaw will lead the inspector on an investigative trail, which will eventually end at the source of the flaw, hence to the source of the deterioration. Certainly this is a possibility, but not the only one, and so it is hazardous to rely on the presence of flaws to alert us to deterioration and/or failures.

5.4.1   The Systems Approach to Building Assessment

Annotating flaws is a common approach to building assessment, and there is little harm in the process. It is, however, quite time-consuming and costly. Unless it results in the identification of an active deterioration mechanism, the process yields little more than an annotated list of the routine manifestations of aging. The disconnection between failure and aging is profound: failures occur for reasons other than aging, and aging is not a failure. We have, therefore, adopted a somewhat more active approach to assessment and diagnostics. The systems approach begins not with a flaw or the perception of a flaw as indicative of failure, but goes directly to the inability of a system to perform the necessary and essential function for which it was intended.

If failure is anything less than imminent, the system is performing, and any perceived flaws are either cosmetic or, at worst, indicators of potentially threatening defects. This is, in some respects, a decidedly more relaxed standard of assessment than is customary in the field, as it places far less emphasis on defects and far more emphasis on performance. An example may help make the distinction clearer. Suppose that the floor of an old building is significantly deflected, let's say exceeding the customary standard for new design of $\frac{1}{360}$ of the span. In the flaw-identification approach to assessment, the deflection is a diagnosable deficiency. Because the identification of deficiency is interpretable primarily in terms of digression from the norm—in this case meaning deflections less the $\frac{1}{360}$— the observation of a deflection outside the norm is, in and of itself, grounds for intervention.

In contrast, using the systems approach to assessment, the deflection is not by itself of any significance. Intervention is not warranted until the deflection is determined to signal a failure, meaning in this example that very shortly the floor will not be able to perform the structural function it was designed to perform. The flaw, in other words, is not a failure, and intervention is probably unwarranted until such time as the member cannot perform. Deflection, even gross deflection, does not determine failure. If we elect to intervene in a component that is otherwise performing, that is a decision based on something other than the reasonable criteria of necessity.

One pragmatic consequence of the flaw-identification system of assessment is that buildings and their components are overdiagnosed for problems. The second pragmatic reason is just the opposite, and goes to the observation that the component approach to assessment tends to emphasize, as a result of being fabric-oriented, what can be seen, and tends to minimize what cannot be seen. This has, in turn, two prongs: things that are physical and hidden from view, and things that are at interfaces, gaps, and joints, which are the voids between components. Another example may help here. Imagine a building with a built-in gutter. The built-in gutter collects water at the lower reach of a sloping roof panel. It was formed by depressing the gutter into or, more specifically, below the plane of the roof inboard of the extreme edge of the roof. Such gutters are common because they are not visually intrusive from the exterior, which is the most common objection to the outboard slung gutter. Thus the gutter is embedded into the attic area, and the bottom of the gutter is not easy to reach for inspection. Depending on the overhang of the eave and the exact location of the gutter relative to the extreme edge of the roof, the centerline of the gutter may be outside of the wall line, over the wall line, or inside of the wall line.

The built-in gutter is a rather elegant design solution to the problem of water collection at a roof edge; however, there are definite problems associated with these devices. When built-in gutters fail, it is noticeable only *after* the rupture of the gutter or the gutter liner. Built-in gutters typically are constructed of seamed sheet metal, so the specific site of failure is at the soldered joint of a seam. Partly due to debonding between the solder and the sheet metal, partly because of differential expansion and contraction of the solder and the sheet metal, and partly due to inherent fabrication flaws, the joint fails; usually, the bond between the solder and the sheet metal fractures, and a fine-line crack forms between the exposed surface of the gutter liner and the underside of the gutter, which is precisely the point on the gutter that is not readily visible. (Of course, the fractured seam is not the only way that sheet metal gutters fail. They also fail because of pinhole corrosion, physical damage, and other reasons; the fractured seam failure is used to illustrate the point.) This is not to say that such gutters are not reasonable designs. Built-in gutters, particularly those made of copper or lead-coated copper, and where the soldered joints are properly fused and expansion joints are properly designed and installed, have in some instances served for over a hundred years.

The motive force that will move water across the crack is a combination of hydrostatic pressure of water in the gutter and/or differential pressure between the exterior of the gutter and the underside of the sheet metal. Once the necessary and sufficient conditions are satisfied, a leak initiates. In point of fact, the formation and progression of leaks in metal built-in gutters is not different from leaks in outboard slung metal gutters. They all leak, because of their materials and fabrication, not because they are outboard or built-in. The difference in the consequences are, however, quite profound and the reason for using this particular example. When any gutter develops a leak, the magnitude of the leak is initially rather small. If the leak continues without intervention, the originating conditions combined with water now moving through the pathway will increase the width of the pathway, thereby permitting further increase in the volume of water flowing through the crack. Recall Poiseuille's Law, which states that the volume of water that can flow through an orifice is a function of the diameter of the "pipe," taken to the fourth power. In short, the exponential increase in volume of water through an ever increasing flaw is extremely rapid. If the leak is from the underside of an outboard slung gutter, the identification of the leak may not be immediate, but it is likely to be detected relatively soon after initiation. Probably the owner will wait to call a roofer, and he or she can afford this liberty because the entire mechanism is visible, and the attendant damages such as water collecting at-grade and potentially migrating into basement walls, are readily identifiable for evaluation and assessment.

If the same leak occurs at a built-in gutter, the scenario is totally different. The initial flaw, which will pass little more water than an occasional drip, will descend until it strikes the eave soffit, the top of the wall below, or the interior attic floor or ceiling below, depending on the location of the centerline of the built-in gutter. In any case, the water from the initial leak will collect on materials which, typically, are vulnerable to moisture damage, meaning wood, plaster, and masonry if the exposure is prolonged. The water, in most cases, will be absorbed by these materials, and so the presence of the leak will go undetected; but already deterioration mechanisms have started at those locations and in those materials. The damage resulting from those mechanisms will progress until their deterioration is manifest and the problem is finally detected.

Where the built-in gutter is outside the wall line, the damaged items typically include the soffit itself, the supporting outriggers, the substrate of the gutter, and possibly the exterior surface of the wall below the soffit. Damage is not necessarily restricted to these items but they represent the most likely candidates. If the damage is, in fact, limited to the soffit area, the intervention is fairly direct, and repairs are relatively inexpensive—but still considerably more than to repair of the gutter alone. In other words, the consequential damages far exceed the immediate damage to the gutter by itself. The damages associated with a leaking built-in gutter, the centerline of which is inside the wall line, are similar to the eave soffit condition. The problem will announce itself in the form of damage to the ceiling below the leak and in the occupied zone. Assuming that the area where the damage is manifest is occupied or, at least, visited occasionally, this leak will be detected when the damage is still manageable. This is not to minimize either of these two possibilities; remember that the volume of water flowing through the flaw is accelerating at an exponential rate. By the time that eave soffits or plaster ceilings are damaged enough so that the problem is evident, the mechanism itself is already well developed and accelerating in its damaging potential. Deferring the problem any longer than is required to mobilize a roofer will result in rapidly rising repair costs.

The eave soffit or the plaster ceiling scenarios are serious problems, but they are minor compared to the third possibility: when the centerline of gutter is over the wall itself. When the water passes through the flaw, instead of falling on finished surfaces, the water will wet materials that can absorb significant quantities of water before manifesting problems and that may adversely affect the structural system of the building. If, for example, the wall is a masonry wall with a timber plate at the top, which accepts roof and ceiling rafters, the water from the leaking gutter can rot the rafter ends and a portion of the wall plate before being detected. Water can collect in the upper reaches of the wall in concentrations beyond any ordinary amount but for the failure of the gutter. In many cases, the first sign of a serious gutter problem is the collapse of a portion of the roof, where the rafter ends rotted, or the collapse of part of the ceiling inside, where the ends of the ceiling rafters failed. As the wall plate rots, the wall may become unsupported at its upper edge, hence become unstable and begin to move laterally. In one actual case in Yellow Springs, Pennsylvania, the top of the wall had begun to move so as to form an outward bulge at the top of the wall. The rafters had rotted to the point at which the wall was no longer braced at its top edge. The problem originated with a faulty built-in gutter.

This case, and many others like it, progress so far either because people presume that problems do not exist until they announce their presence or that problems are not knowable until they announce their presence. So it is with the flaw-identification approach, which presumes the proper functioning of the built-in gutter until damage is manifest. Obviously, delayed detection is not desirable, but it is an unavoidable consequence of the flaw-detection approach to assessment and intervention.

Conversely, the systems approach presumes that failure will occur and that our inability to detect failure does not mean that it is not there. The systems approach further presumes that materials and systems are in a constant state of deterioration. By understanding the mechanisms inherent to any given set of materials and assemblies enables us to reasonably predict the location and occasions of many, if not most, failures. The existence of a built-in gutter is enough of a signal to alert the investigator of an impending problem. In summary, the flaw-detection approach presumes performance until problem manifestation; the systems approach presumes active deterioration until proven otherwise.

We delayed the discussion of the differences between these two approaches to building diagnostics until this chapter for two reasons. First, this information is much easier to understand after learning the relevant mechanisms and the consequences of delayed detection. Second, there is no system that so dramatically demonstrates the risks and consequences of delayed discovery as the horizontal closure system. The example of the built-in gutter is but one of many and it is categorically misleading to suspect that such problems are relegated to roofs or roof components.

The fundamental problem with the horizontal closure system that prompts this discussion is that it is one of only two subsystems that *invites* water. (The other subsystem designed specifically for the transportation, collection, distribution of water is the domestic sanitary, sewer, system.) In contrast, the structural system is normally shielded from water and certainly is not functionally involved with water management; the vertical closure system may provide some barrier to water and, therefore, implicitly provide for some contact with water, but its overall design is to minimize contact with water; the environmental modification or climate control system may include some limited water circulation, but the operative term is that it is limited; and the power system is moisture-averse.

It is abundantly clear by now that moisture is a major agent of deterioration; it is a major and persistent element of many sets of necessary and sufficient conditions (see Fig. 5.4). We can consider moisture reduction as a viable approach to timber rot mitigation, but it would be antithetical to

FIGURE 5.4 This is end-stage roof failure. In many jurisdictions this would have been condemned by authorities even earlier, but here it is "protected" by a four foot-high wall beyond.

the function of the horizontal closure system to suggest that we can mitigate deterioration by reducing the amount of water being collected and transported in the system designed to collect and transport water. The horizontal closure system must collect and transport water on demand. Because that very process is potentially damaging, we must presume that the system is in an active state of deterioration at all times.

In addition to being the agent of it own demise, the horizontal closure system tends to execute its collection and transportation function via a series of somewhat elusive and clandestine devices. Although the methods of collection and channelization are at times visible as in surface-mounted leaders and outboard slung gutters, designers as a rule seem to prefer hidden and enclosed devices: built-in gutters, leaders encased in walls or column covers, flat roofs hidden behind parapets, and buried storm water systems. Determining that the components of the system are better when they are not visible or expressed is a major contributing factor to the delayed damage phenomenon. For these reasons, we cannot afford the luxury of responding to systemic deterioration on a reactive basis; the diagnostician must be anticipatory and predictive in his or her assessment. The consequences are too grievous to wait for the mechanism to present itself. Good pathologists are hunters, not trappers.

5.4.2   How Roofs Leak Water

Roofs leak in a limited number of ways, depending on the roof type. Recall that there are three fundamental types: single-layer, multiple-layer, and hybrids. Earlier we introduced and defined them primarily in terms of materials selections. Here we will discuss the specific vulnerabilities of each of these three roof types. We begin by reviewing definitions.

The roof is that part of the horizontal closure system located between the atmosphere and otherwise exposed-grade, relative to descending precipitation. Without question there are elements of selective barrier—hence, with the vertical closure system—associated with roofing, but the functional definition of the horizontal closure system focuses on the roof and the performance of the roof as a water collection and diversion device. The basic approaches to roof construction, though not identical, are analogous to the basic approaches to wall construction to this extent. The necessary and sufficient conditions for water to penetrate a roof or a wall are (1) a pathway traversing the barrier, namely the roof or the wall; (2) water occluding one end of the pathway; and (3) a motive force or pressure differential to move the fluid through the pathway. We categorized walls based on the devices employed to defeat the pressure differential; we can employ a similar analysis to roofs.

To describe walls we used terminology that enabled us to simultaneously generalize the underlying physics and to accommodate the peculiarities of wall systems. We want to do the same thing with roofs: generalize the underlying physical principles and accommodate the peculiarities of roofs. First some recognition of similarities and differences between walls and roofs is in order: Sloped walls are at least in part roofs, and vertical roofs are walls; the elemental difference between roofs and walls is essentially geometric—vertical versus horizontal. The physical significance of that distinction has to do with the source or sources of the pressure differential across the barrier.

In walls, the primary source of pressure differential is the variation in atmospheric air pressure between the inside and the outside of the building. The origin of that pressure differential is often the wind, but it may also be affected by mechanical equipment operation and stacking effect. Capillarity, too, plays an important, though limited, role in the transportation of fluid across the exterior barrier. All of these factors are potentially present at any roof as well. The laws of physics are valid for horizontal as well as vertical surfaces. Water can be pulled through a flaw in a roof as quickly and as surely as it can through a flaw in a wall for precisely the same reasons.

But, the geometric distinction between roofs and walls brings with it an added source of pressure differential, namely hydrostatic pressure. It is true that hydrostatic pressure can develop in walls to the extent that the fluid-filled pathway is inclined downward at the same time that it is leading away from the exterior surface. This element of hydrostatic pressure in walls is not, however, a major element in the overall calculus of water penetration through walls. For roofs, hydrostatic pressure is a major factor in whether, when, and how much they will leak.

Hydrostatic pressure manifests in two primary ways. One is the pressure exerted by water acting under the influence of gravity. The weight of a cubic foot of water is 62.4 pounds (28.2 kg/ft^3 or 996 kg/m^3). When a pool of water is sitting on the roof for whatever reason, for every square inch of area and every inch of depth, that water weighs a bit more than half an ounce. The water at the bottom of the pool is under that same amount of pressure. A half ounce of pressure is not a tremendous force, but it is more than enough to account for a leak measurable in some number of gallons per hour through a hole no larger than the diameter of a pencil lead.

Imagine now that in the plane of the roof is a pool of water resting in a shallow swale, and that the roof is slightly tilted up on one side. The water will flow downhill and, presumably, off the edge of the roof. In the process of traversing the roof, the moving water exerts a downward gravitational force on the surface. The magnitude of that force and, therefore, the attendant hydrostatic pressure perpendicular to the roof plane, decreases as the slope of the roof increases. In other words, the gravitational force on the water, whether it is moving or at rest, is the same, 62.4 pounds per cubic foot (996 kg/m^3), and that vector is always pointing toward the center of the Earth, whether the water is flowing on a slope or at rest on a plane. Relative to any plane on which the water is located the gravitational vector can be divided into two components: one is perpendicular to the surface of the plane and one is parallel to the surface of the plane. When the water is at rest on a horizontal plane, such as a flat roof, the component force parallel to the roof surface is zero, and the component vector perpendicular to the roof is equal to the total force of the water acting under the influence of gravity. If the plane is perfectly vertical—that is, a wall—the force acting on the water parallel to the surface is the full effect of gravity (less the frictional effect of running over the wall surface), and the gravitational force perpendicular to the wall surface is zero. At any position between horizontal and vertical, the respective vectors vary depending on the angle of incline of the roof. One of the consequences of this simple observation is that for the same depth of water, the differential pressure perpendicular to the roof plane due to hydrostatic pressure will be less on a sloped roof than on a comparably sized flat roof; and the magnitude of that pressure decreases as the slope increases.

SINGLE-LAYER SYSTEMS

The idiosyncrasies of single-layer systems relative to the passage of water through the membrane are partly related to deterioration mechanisms associated with materials and partly related to the geometry and configuration of the roof planes and panels. This section concentrates on the issues of geometry and configuration, with occasional digressions on the specific materials as they are warranted and not otherwise covered in other sections.

Recall that all low-slope roofs are necessarily membranes, hence single-layer systems. The use of the term single-layer is related to how the material works. This is worth mentioning because, within the industry itself, modified bitumen (MBM) and built-up roofing (BUR) are considered to be multi-layered because of the way they are installed, but they perform as single-layer mats. The fundamental question associated with any roof failure is: How does the water get from the open environment to the underside of the roofing material? We know that membranes accomplish exclusion of water through perfect impermeability of the material. It is assumed that flat roofs sheathed with perfect membranes will be intermittently flooded with water, however minimal the depth of inundation. The motive force under such conditions is supplied by the gravitational force on the fluid.

We know that any pathway that develops between the exterior and the underside of the roofing and whose exterior orifice is choked with water and that requires no ascent to breach a superior point will result in a leak. If there is a superior point in the pathway, a leak will occur when the pressure differential between the exterior and the underside of the roof is sufficient to lift the water over that high point. Once the water is on the underside of the roofing membrane and the leak is active, rarely will the leak be immediately evident, in, for example, the form of a drip from the point of penetration to the floor of the occupied zone. Were we so fortunate. It is customary to support membranes on substrates, which themselves are planes of relatively impermeable materials. Between the substrate and the roofing membrane there may be other layers of semipermeable or impermeable materials. BURs frequently have other layers of old roofing below the current weathering surface. When the water penetrates the outermost layer, the water may encounter a second layer. If there is no ultimate pathway beyond the penetration of the outer layer, there is no connecting pathway to the interior; so, at worst, the water accumulates in a small pool between the two outer layers of the roof construction.

Even if, as often happens, the water seeps through several layers without actually penetrating to the interior, there is no completed pathway, hence no active leak. In some instances, the presence of water in the interstitial spaces between layers is in itself a source of potential damage. This is, in fact, typical of multiple layers of BUR. The water works laterally between the layers of the new roof, ultimately to encounter the former weather surface of the old roof. The crack in the old roof was very likely to have been sealed when the new roof was applied, so water accumulates between the two roofs until a new pathway emerges. The process continues from layer to layer with the water ever migrating across the width and breadth of the roof. The water will remain in between the layers for prolonged periods because very little of it is likely to evaporate. The water alone can deteriorate the fibers of the felts, if they are organic material. Freezing and thawing will increase the crack network in the chilled, brittle asphalt roofing, also increasing the capacity to hold water. Eventually, the weight of the saturated roofing alone can pose a structural concern. We once observed a roof with over four inches of accumulated asphalt completely saturated; it weighed 15 pounds per square foot, which, by the way, was more than an inch of solid slate on a flat roof designed for a dead load of less than a pound per square foot of roofing. Replacement hot roofs were installed when the old roofs failed—which meant when water was actively penetrating the cumulative roof membrane. Given the lateral migration potential of water in flat-roof construction, probably the new roof was installed over a saturated predecessor. When the molten asphalt hit the old, wet roofing, steam formed below the new roof, raising blisters, which break the bond between the old and the new membranes at the moment of installation. It did not take years for the delamination to occur; it occurred

A Roofing Story

This sequence of photographs comprise a virtual time lapse record of the re-roofing of a residential building using rolled bitumen felts and hot bitumen binder. The lessons to learn from this project have less to do with the technology or theory of roofing and roofing materials than with the process, especially if design professionals are not involved.

In this story, a roofing contractor is reroofing the house with a "hot" roof. The home is a typical Philadelphia rowhouse, three stories high at street level, steeping down to two stories in the rear, hence the ladder from the upper portion to the lower portion (Figure A).

At 7:00 A.M., two trucks arrive at the house, one towing the "pot." Both carry rolled roofing, ladders, rigging, buckets of hot and cold bitumen products. The second truck has a portable breaker bar for folding the sheet metal for cap flashing, edging, and drip caps. The total crew number seven. Between 7:00 and 7:30, the crew erects the ladders and fires up the pot, which is set in a parking space on the street (at the left in the photographs). The crew erects a pulley for hoisting goods and tools, and for lowering debris.

At 7:30, the crew begins removing the top layer of the existing roof. Three workmen are dispersed on the two portions of roof; another one is assigned to a small segment of roof (off the frame to the right). Two are tending the pot, and one is setting up the breaker bar.

By 8:00 four workers are on the roof; three are removing roofing and one is hoisting up rolls and buckets of cold-applied bitumen. Conveniently, the neighbor's roof is available for storing materials and accumulating debris, so this aspect of the work moves rapidly. In only 10 minutes (Figure B), the lower rear portion is nearly stripped. This is done first to get the rear portion ready for application ahead of the front portion.

By 8:20, stripping of the rear portion is closer to completion (Figure C). The old edge caps and drips are removed. At the front portion, the materials stockpile is nearly complete. (Unfortunately the odor of the hot melting bitumen cannot be represented here.)

By 8:40, the rear is completely stripped; the crew begins sweeping the surface to eliminate foreign debris (Figure D). The workman at the front portion,

A. 8:00 A.M.

B. 8:10 A.M.

C. 8:20 A.M.

D. 8:40 A.M.

E. 9:00 A.M.

F. 9:30 A.M.

G. 9:40 A.M.

H. 9:50 A.M.

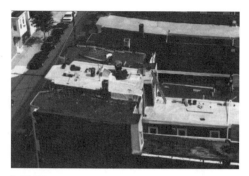

I. 10:10 A.M.

J. 10:30 A.M.

K. 10:50 A.M.

L. 11:10 A.M.

M. 11:30 A.M.

N. 12:00 P.M.

O. 12:30 P.M.

P. 1:00 P.M.

however, seems to be stalled at a shallow swale at the rear of that section. In fact, he is cutting more old roofing back at this location to expose what was most likely the area that leaked giving rise to this project.

Within two hours, the stripped rear portion is almost ready for roofing. All the stripped roofing has been moved to the front portion; the rolls of new material are distributed to the rear portion (Figure E). Shortly before 9:30 (Figure F), three workmen begin mopping the rear portion. Hot bitumen is placed over the clean surface. This proceeds quickly. Figure G shows the rolling bucket of liquid hot bitumen; Figure H shows the echelonned rolls.

At 9:50 (Figure I), it appears the crew has stalled. For the next 20 or 30 minutes (Figure J) a worker installs a scupper box through which to conduct water to the descending leader. By 10:30, (Figure K) the scupper is completely installed; and the rolls extend to the rear edge of the lower rear portion of the roof.

Meanwhile, at the front portion, the lone roofer there has been rolling out sections of felts, measuring them, cutting them to length, rerolling the cut lengths, and lining them up along the far edge of the upper portion of roof. By 10:50 (Figure L), the crew at the lower portion is installing the metal edge and cap flashings. The roofing is cut and rolled at the upper level. Between 10:50 and 11:10 (Figure M), the bucket is hoisted to the upper level. In the interim, all the debris has been bagged, lowered to the street, and loaded onto one of the trucks.

By 11:30, after a short break, the work begins in earnest on the upper level (Figure N). By noon, the upper portion is reroofed, except for the swale and around the scupper (Figure O). In another half hour, the crews are loading equipment. The breaker bar is back on the truck, Trash is removed, and one roofer is applying cold bitumen around the scupper (Figure P).

By 1:00 P.M. the crew is gone and the job is done. Though there were a few violations of roofing etiquette, such as not removing the old roof to the substrate and installing copper scuppers upstream from steel leaders, this accurately describes how the roofing process works. All things considered this roof will function for 15 or more years.

in the first 10 seconds. This is precisely why old roofing should be removed *before* installing a new roof.

Although the lateral migration of water in BUR is legendary, it is by no means restricted to hot roofing. Any membrane roof is at risk of developing a penetration in the outer layer. If the water pools below the outer layer, it can run laterally, following the terrain of the substrate. Although it is possible to slope the structural steel, a common way of developing the angle in low-slope roofs is with tapered insulation or tapered lightweight concrete, both of which may be placed on horizontal decks or slabs. When the water breaches the membrane, it migrates through the tapering material to the surface of the deck or slab. Because this substrate is horizontal, there is the risk that the water will "pond" on the deck, and may accumulate to substantial depth before completing the pathway to a point and in a manner that enables detection.

Ponding

Ponding refers specifically to a phenomenon associated with low-slope roofs. The ponds can be divided into two categories. Some ponding occurs due to depressions in the plane of the roof, essentially forming shallow dishes in the roof that collect water. These ponds fill with water until they overflow, and when the rain stops they empty by evaporation. The dimension of this type of pond is relatively fixed because the low point of the pond remains at much the same elevation relative to the overall plane of the roof. This type of fixed-dimension pond we will call a *static pond*; every time it fills, it fills to the same predictable level (see Fig. 5.5). The second type of pond is variable in dimension. This class of ponding is generally the result of flexure of the structure: the weight of the accumulating water depresses the structure, and the low point descends, resulting in an ever increasing pond depth and diameter. The added weight depresses the deck more, in a potentially devastating spiral of weight and deflection. Roofs supported on lightweight, flexible structures or decks are notable for this class of ponding, which we will dub *dynamic ponding*.

Dynamic ponding can accelerate to the point at which the structure may actually fail. In January 1978, a series of winter storms dumped several inches of freezing rain on the Northeast and Midwest. Because the water froze as it hit building surfaces, even roofs that were adequately sloped, and ordinarily more than stiff enough to drain adequately, began to pond. On January 18, the Hartford (Connecticut) Coliseum roof failed. Three days later, the roof of the auditorium at C.W. Post College on Long Island, New York, collapsed. In between those dates, more than 700 roofs collapsed from ponding caused by the same storms. The specific reasons varied, but

FIGURE 5.5 This bitumen roofing is heavily "alligatored," which by itself does not mean that the roof will leak, but it does indicate age and vulnerability. Of greater immediate concern is the standing water. Note the proximity of the drain, which is not at the actual low point.

the collective cause was extreme accumulation of weight and the accelerating collection of water, much of it in the from of ice, due to ponding.

Points of Entry

Ponding, dangerous though it can be, is not what is infuriating about flat roofs and water. What is infuriating is the lateral migration between layers. It is an axiom of flat-roof leaks that the location of detection is unrelated to the point of entry. When the outer skin of a membrane roof ruptures, the likelihood of the drip appearing on the interior immediately below the point of initial penetration is remote. Not only are the points of entry and detection unrelated, but so is the timing. The pathways may be sufficiently remote and circuitous that a given molecule of water may have to travel several hundred feet before emerging at the point of detection. If the resistance to fluid flow through the pathway is high, the water that actually emerges following a given storm may actually be water that flooded the pathway during the *previous* storm. It may take hours to flood the pathway and days to completely drain it, and not a single foot of the pathway is directly visible for inspection.

Finding the point of entry once the leak is manifest on the interior can be so frustrating that owners often simply have the building reroofed prophylactically rather than try to find the point of entry. When BUR became popular, the cost of reroofing was still quite low, and installing a new roof directly over the old roof was feasible. Now that membranes are a bit more precious and less likely to fail from material deterioration, more attention and time may be expended to find the point of entry. To this end, an interesting and, at times, amusing, array of technical innovations have been attempted to detect water below the membrane surface. Some of these techniques are utter wastes of money; others have some theoretic foundation; still others may even detect water. But the presence of interlaminar water in the roof is not the real issue. We do not need a space-age detector to tell us water is in the roof when all we have to do is go inside and watch the leak. The alleged purpose of such detection devices, whatever their technical merit, is to find the location of *maximum* water concentration, and from that to identify the point of origin. The problem is that the point of maximum interstitial concentration is incidental to the point of entry. Water will concentrate between roofing layers for a number of reasons, but proximity to the point of entry is not one of them. Water ponding below the surface of the roof is absolutely unrelated to how the water got to the pond.

If the delay, or lag time, from when water is applied to the outer surface of the membrane until it appears inside is a reasonable period, such as a few hours, there is an inexpensive diagnostic technique that can save a great deal of time and money searching for the point of entry. It is based on a simple process of elimination:

1. Place rotating garden sprinklers on the roof, and initiate the leak under otherwise dry conditions.
2. Measure the time required to initiate the leak.
3. Place a temporary tarp or other protective membrane over half the roof, then sprinkle the exposed portion for at least as long as was required to initiate the previous leak.

If the leak recurs, the point of entry is on the exposed half; if it does not, it is under the protected half. If the point of entry is under the protected portion, remove the tarp from half of the protected portion to expose a quarter of the roof. Repeat the sprinkling. If the leak recurs, the point of entry is in the newly exposed quadrant; if not, it is still under the tarp. Repeat the process as many times as is economically feasible to repair or replace, whether the point of entry is found or not. In just seven iterations, including the original wetting, the suspect area can be reduced to 1.5 percent of the total roof area. As with many testing techniques, however, the issue is one of time more than of expense; and a technique such as this takes time and vigilance.

Another simple, but far from certain, trick for finding points of entry is to inspect the roof during wet conditions. This means that the inspector has to actually go out into the rain; but, simply, that is the best leaking-hunting condition. If the roof is wet, there is a possibility that when the roof is depressed in the vicinity of the point of entry, bubbles will erupt from the pinhole. In a few cases, we have seen miniature geysers spout from the roof deck. While less than definitive, such techniques are far less expensive than other procedures, and they have the added benefit of placing the inspecting professional at the scene of the event as it occurs. One reason that leaks are at times difficult to detect is that the detective often conducts his or her investigation under dry conditions.

Implicit in this discussion is the assumption that the leak at issue is developing over time as a through-body failure of the roof fabric. That is not necessarily the case. Many leaks occur at roof junctures and seams, particularly when the roof form is flat. Seam and joint failures are typically matters of workmanship and installation; therefore, they tend to activate during the warranty period of the roof. The owner resolves the situation without the intervention of a pathologist by invoking the warranty. The roofer returns to the job; and, assuming he or she can find the fault, he or she repairs the roof. Where roofers look for their own oversights is instructive to anyone who may have a need for the same information after the warranty expires. The most likely locations of workmanship problems are at points of roof penetrations, meaning projections originating from within the building and through the roof plane. The list is long: skylights, stacks and other vent pipes, fans and ventilators, exhaust and supply hoods, chimneys and flues, and plumbing for roof-mounted mechanical systems.

Every time a roof is pierced, the line surrounding the projection must be sealed against moisture penetration. If the roof is a membrane system, the seal can usually be integrated into the same system as the roof to form the seal or flashing around the projection. The general idea is to turn the membrane vertically and run it up the side of the projection to a height well above any likely build-up of water on the roof. The edge of the flashing, which is actually extended roofing, is then clamped or adhered to the face of the projection to prevent water, nominal though it may be, from slipping between the face of the projection and upstanding edge of the flashing. This particular type of flashing condition is so vulnerable to separation from the projecting face that another cover is applied over the exposed edge; this cover is typically called the counterflashing. Obviously the counterflashing must end at some point as well, so the same basic termination edge condition must be resolved sooner or later.

The flashing problem is often dealt with in two stages to allow the flashing to adapt to and integrate well with the roof, whereas counterflashing is more successful when it is an extension of and integrated with the projecting material. It is a common detailing problem among junior designers who know that there are two pieces to the typical flashing solution, but who try to make both the flashing and the counterflashing out of roofing material and roofing technology. One location where this situation is particularly demonstrable is where the flat roof meets a projecting parapet. (We first mentioned parapets in connection with walls, but deferred the in-depth discussion until the section on roofs. As promised, a later section addresses the peculiarities of parapets. For now we will concentrate on the subjects of flashing and counterflashing.)

The intersection of the flat roof plane and the vertical face of any projection is an inherent problem because of the exposure of the incised roofing edge. If water passes between the edge of the roofing and the face of the vertical projection, the water can collect on the underside of the roofing until a pathway develops to the interior. No intersection is more fraught with such risk than that of the roof plane with the inboard face of the parapet. Despite the continued use of the term flat roof, the intersection of the roof and the parapet are usually not orthogonal. In fact, there are three basic conditions of parapet/roof intersection: along high lines, along low lines, and along traversing lines.

High lines and low lines are horizontal, and are the easiest to visualize. The slightly sloping roof intersects with the vertical face of the parapet, and the roof plane is either sloping slightly away from the parapet or slightly toward the parapet. If the roof plane is sloping downward away from the parapet, that is a high line, meaning, simply, that relative to surrounding points on the roof plane, the line along this particular intersection with the parapet is a line of higher elevation. Any water that runs down the back face of the parapet or strikes the roof at the precise location of the intersection is directed away from the parapet itself. The low line is the exact inverse: the slope is toward the parapet, so any water washing down the back of the parapet, plus any water following the slope of the roof panel, will be directed toward the intersection of the roof and the parapet. The high line is a form of ridge, whereas the low line is a form of gutter. That these two conditions are typically dealt with using the same detailing approach is quite remarkable.

There is a risk at the high line that the roofing edge will disengage from the parapet wall, resulting in a gap through which water can pass. But the amount of water that will, in fact, collect at the high line is limited to the water collected on the back of the parapet. This is not negligible, but it is a limited amount of water. If the roofing runs up the back of the parapet, and the edge is protected by a counterflashing, there is little more required at the high line. A threat in all flashing situations is that the roof will flood, and that the level of the water will crest the top of the flashing. This threat is minimal at the high line because water will have already crested all the other flashings on the roof before it jeopardizes the high line, precisely because it is the high point of the roof. The counterflashing at the parapet intersection is relatively easily regletted or otherwise integrated into the parapet, so the exposed edge is protected (see below regarding parapet pathology). Because the elevations of the edge of the flashing and counterflashing are constant, the integration of the counterflashing into the parapet is done to produce a horizontal line across the face of the back of the parapet. Given the construction of most parapets, the development of a horizontal line across this face is not a difficult detail.

The low line is an altogether different condition, despite the tendency of designers to detail the condition similar, if not identical, to that at the high line. In this case, the collection of water behind the parapet is precisely the intent, and the risk of water cresting the upper edge of the flashing is much greater than can ever be the case at a high line. Another factor having to do with the membrane itself is particularly manifest at the intersection of the roof and the parapet at the low line, namely thermal expansion and contraction of the membrane itself.

Lateral movement of the roof membrane is a major factor in flat-roof durability. The materials are subject to thermal expansion; the range of temperature differential are very high; the rates of thermal shift are high; the emissivity of the materials is often high. Every environment factor that can damage materials and systems is harsher at the roof. It is amazing that roofs ever perform for more than a year! In this particular application, the issue is that the membrane moves due to thermal expansion and contraction. The movement places considerable shear stress on the bond of the roofing membrane with the substrate. In those cases, such as BUR, where the composition of the roof is laminar, these stresses also occur across the section of the roofing material itself. The roofing may, therefore, split from the substrate or along the plane of lamination. It is precisely the development of these delamination planes that allow the water to travel laterally once it breaches the outer skin of the roof. This assumes, obviously, that the membrane is adhered to the substrate. But, not all membrane roofs are adhered; some are laid loose on the substrate. Metal roofs are typically cleated to the substrate; but from a practical standpoint, there is sufficient play in these systems such that they expand and contract with little lateral restraint.

If and when the lateral movement of the roofing is unrestrained, either by design or by adhesion failure, the roof begins to move. Restraint of that movement will occur wherever the roof is bonded to a rigid fixture. Any vertical penetration of the roof plane where the roofing is flashed to the vertical face of the projection is a potential line or point of restraint; certainly, the intersection with any parapet qualifies as such a condition. If the parapet is rigid relative to the expansion and contraction of the roofing membrane, with every temperature shift, the roofing will either expand toward or contract away from the line of intersection. At the line of intersection with a vertical plane, the roofing typically folds up at that face. This means that the fold in the membrane is opening or closing with each change in temperature. Of these two options, the opening condition is the more threatening.

Despite the fact that low lines are seldom treated differently from the standpoint of detailing, it is an observable fact of building pathology that low lines are far more vulnerable to water intrusion and, therefore, to secondary deterioration mechanisms from water intrusion than other locations on flat roofs. The low line behind a parapet is more vulnerable than other low-line conditions only because the angle of the intersection between the roof and the parapet face is more acute than the low-line fold of a roof with interior drains, where the low points and lines are removed from close proximity to the parapets.

The traversing lines on the back faces of parapets are those flashing conditions on the cross-slope of the roof. The plane of the roof at the low lines and the high lines is basically horizontal. The development of the intersection details, therefore, occurs along horizontal lines, which means that the integration with the parapet itself occurs along horizontal lines. At traversing lines, the roof slopes relative to the parapet; and the presumption is typically that the top edge of the flashing and counterflashing will follow this slope. This means that the counterflashing must be let into the back of the parapet on an angle relative to the face of the parapet. The implications of this presumption are profound, and ordinarily work to the detriment of the entire parapet and roof. Designers commonly set the width of the flashing and counterflashing to be the same at all locations around the perimeter of a roof. A far more rational approach is to strike a line for the flashing and counterflashing at the high line, which is convenient for installation and fabrication, and to carry that elevation all the way around the roof, and at roof penetrations as well. This puts the maximum depth of flashing at the low line, where it belongs, and keeps all the counterflashing integration lines on the same horizontal plane.

Not all roof edges resolve at parapets; however, the issues and principles remain the same. Another common roof edge detail is the drip edge or drip cap detail. The detail is simply a piece of folded sheet metal that, simultaneously, clamps the edge of the roof, restrains the ballast from rolling off the roof, and throws water away from the facia below. The prevailing detail that dominates the roof edge is the low-line detail. Presumably, if the detail satisfies that condition, it is satisfactory for the high line or the traversing line. Unlike the parapet detail, which is typically driven by the high-line resolution, the drip edge is driven by the low-line detail, which as it turns out is, generally, appropriate and economical in this case.

Ballast

Mentioned in connection with the drip edge was the term *ballast*. A few items are worth discussing regarding ballast and ballasted roofs. One of the risks associated with flat roofs is that air moving over the edge of a roof creates a vacuum at the surface of the roof. The negative air pressure can lift the membrane off the roof. One way to counteract the negative pressure is to adhere the membrane to the substrate of the roof; however, the bond between the membrane and the substrate eventually ages and weakens, and the bond ruptures, at which point the membrane may be pealed off the roof. We can mechanically fasten the roof to the substrate, but mechanical fasteners are often antithetical to the development and maintenance of a perfect membrane. Metal roofs, in particular, have effective cleat systems for holding down the roof.

Another way is to charge the roof with enough weight to counteract the negative pressure and to hold down the roof. The counterweight material is called ballast, and it is usually in the form of aggregate. The ballast not only provides the necessary weight, normally on the order of 10 pounds per square foot, it also provides physical protection from projectiles and ultraviolet radiation. The typical problems associated with ballast are attributable to the shape of the gravel particles. If the gravel is crushed rock, the pieces may be sharp-edged and so underfoot traffic may puncture the membrane. If the ballast is smooth river gravel, the puncturing potential is mitigated, but the gravel tends to roll around the roof. The movement of water or of wind may be enough to dislodge and relocate the gravel. On a hot roof, the ballast may be cast while the bitumin flood coat is still tacky, then at least one layer of the gravel will adhere to the roofing reasonably well. In time, all ballast tends to collect at the low points of the roof, but it can be respread or more can be applied.

During much of the twentieth century, slag was used as ballast. Slag, as defined earlier, is a by-product of iron and steel production, particularly the former. In places such as Pennsylvania where smokestack-style iron production was concentrated, the operators of mills were constantly looking for ways to dispose of the slag. It was often dumped into the closest streambed, but it was also dumped just about anywhere that can be imagined. It was also used as roof ballast. Slag ballast had two problems: it was rather light, and it contained sulfur. The density of slag varies considerably, but 75 to 90 pounds per cubic foot (1200 to 1440 kg/m^3) is common. This meant that, as roofing ballast, it was particularly vulnerable to dislocation.

The sulfur was more problematic. Sulfur is a common element in coal, particularly bituminous coal. When the coal is burned, the sulfur is not heated to a high enough temperature to combust, but it is high enough to concentrate the elemental sulfur. When the sulfur in the slag is subsequently exposed to atmospheric conditions, the sulfur combines with the rain water to form sulfuric acid, which, in turn, attacks and destroys any metals with which the runoff comes in contact. Particularly vulnerable are where the water concentration is high, such as at valleys and gutters. Sulfur-enriched water rolling over a drip edge can eat through the metal in remarkably short order.

A relatively recent development in the realm of ballasted roofs is the protected membrane roof, the PMR for short. The key to PMR is the

Classic Installation Problems, Metal Single-Layer Systems

FIGURE 5.6A This is an example of a recently installed standing-seam, 20-ounce copper roof. The original roof was a symmetrical, double-hip design. The protrusion at the right is a recently installed skylight.

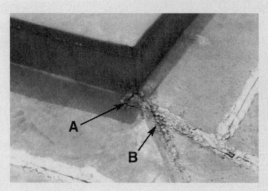

FIGURE 5.6B Both of these locations are at the upslope side of the skylight protrusion. Location A shows that the metal is actually fractured; B shows an optical-level crack in the solder.

Classic Installation Problems, Metal Single-Layer Systems (*continued*)

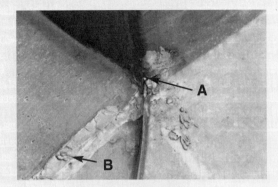

FIGURE 5.6C Similar conditions at the opposite corner. Location A shows a large fracture in metal. Location B shows excessive solder application, indicating a failure of the roofer to understand how solder works.

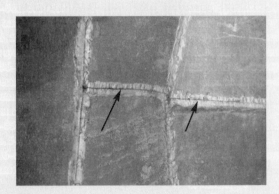

FIGURE 5.6D The solder is cracked, as shown by arrows. While theoretically possible for solder to fill the seam joint beyond, in service, this joint has no more water integrity than one with no solder at all.

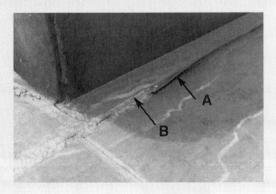

FIGURE 5.6E This shows a seam folded upslope (A) and an open flat seam (B), also facing upslope.

development of insulation, which can remain effective even after it is wetted. The development of extruded rigid foam products allowed builders to put the insulation on top of the membrane. The benefits are considerable: the membrane is physically protected and shielded from damaging ultraviolet radiation by the replaceable and relatively inexpensive insulation; more importantly, the membrane can be brought within the thermal envelope of the interior of the building. This means that the roof is thermally part of the inside of the building, where temperatures are held within a fairly narrow range. The reduction in the thermal range of the membrane means that expansion and contraction are reduced dramatically, thus the roof survives considerably longer.

Because the extruded styrenes are rather buoyant, unless restrained, they may float right off the roof. The insulation can be adhered to the membrane, but one of the preferred ways is to ballast the PMR; but in this case, at least two inches of gravel are required to do the job, double the amount commonly used for other ballasted roofs. Because of the weight requirement sometimes loose-laid paver squares are used in lieu of the gravel ballast. The load can have structural implications and costs associated with the specification of ballasted PMR. If a leak does occur, the removal process to expose the membrane to inspection requires moving 20 pounds per square foot of ballast, which means it must be placed to one side without overloading the structure. PMR roofs tend to be somewhat more expensive than exposed membrane roofs, but their durability may more than compensate for the initial costs.

MULTIPLE-LAYER SYSTEMS

Multiple-layer systems are pressure-equalized systems, of two basic types. In the first type, the attic as a whole serves the pressure equalization chamber; in the second, the space between the back of the shingle and the outer face of the sheathing is the pressure equalization chamber. Both are effective and can work well, as long as the requirements of each are recognized and respected. The principles of pressure equalization were explained in detail in Chapter 4, so we will not repeat that material here. We reiterate only that the effectiveness of the chamber to equalize pressure depends in large part on how quickly equalization can occur; and that, in turn, is a function of the cross-sectional area of the vents to the chamber (see Fig. 5.7).

Where the entire attic is the pressure chamber, obviously, the volume of the chamber is large when compared to a two-inch-wide wall chamber. This means that the attic must be connected to the outside via large openings. Eave and ridge vents, in addition to ventilating the attic, can provide substantial area. Unfortunately, these same vents also allow currents to develop in the attic. In fact, this is precisely the reason they were installed. It

FIGURE 5.7 This hipped dome is, ostensibly, a thin, single-layer system. It has, however, an extensive system of small gutters and leaders around every individual light, to collect and divert water along a path to weeps and back outside. Thus, the system is, in fact, a drained cavity system, more as a wall than a roof.

is part of the ventilating scheme that air will enter at the eave line and, through convection, exit at the ridge. But as a pressure equalization scheme, this is not ideal. Freely moving air, as with freely moving water, cannot develop the requisite pressure to equalize the excess positive pressure outside the attic. Even when the vent area is large, the chamber is open-ended, and there is not effective pressurization of the chamber.

Another way the attic can be pressurized is through a louver or other similar opening in a gable end wall. Large, circular and half-round windows in the gables of attics serve as pressure equalization vents for the attic that is acting as a pressure equalization chamber. The louvered gable opening can be effective as long as the occupants know that the louver or window must remain open for the system to work properly. When it does not, the results can be disastrous. We cite the actual case of a National Historic Landmark church in Philadelphia which was essentially free of roof leaks until the energy crisis of 1973. During that time, the large circular window in the east gable was sealed shut in the interest of conserving heat lost through the open attic window. Needless to say, the pressure chamber was defeated; equalization did not develop, and water was pulled into the attic. Subsequently, the shingle roof was replaced with a standing-seam copper roof, based on the incorrect belief that the shingles had failed. The new copper roof also began to leak, because the new membrane was less than perfect, and the relatively negative pressure of the attic was sufficient to pull water up and through the standing seams, particularly as the water accumulated near the eaves.

As with eave and ridge vents used to generate pressure-defeating currents, it is possible to be too generous with gable vents as well. If the attic has louvers on opposite ends, the two vents may provide the channel through which an air stream can flow without developing the requisite pressure. The solution is to simply keep one—only one—of the gable windows closed.

In a similar way, it is possible to defeat an attic pressure chamber by indiscriminately connecting the attic with the interior of the building below. When the air chamber is the attic as a whole, the air barrier is necessarily at the ceiling line of the room below. The ceiling line is also the attic floor level. If this plane is pierced with overly generous stairs or other openings, the air barrier may be compromised to the point at which pressure cannot accumulate at the requisite rate or magnitude. The church just described had a large grilled opening in the ceiling of the sanctuary connecting the nave with the attic. This was a way of ventilating the nave, pulling air through open windows, up through the grilled opening, across the attic, and out the gable windows. As a ventilating device, its probably worked well enough; on rainy days, however, the grill had to be closed to ensure that the pressure chamber was a cell with a single point of entry.

Another example of how the attic chamber can be defeated is via the so-called cathedral ceiling. Occasionally, a design calls for cutting out the top-floor ceiling and the ceiling rafters and reaffixing the ceiling surface to the undersides of the inclined roof rafters. The effect is intended to be a dramatic increase in volume and height. If this is done without an awareness of how the system works as a whole, an otherwise properly functioning roof, when deprived of the well-vented attic, will develop a pressure differential across the roof plane as a whole and will, eventually, leak.

When the attic is effectively developed as the pressure chamber, the shingles can be placed on spaced boards because the water is not moved up and over the shingle. Air may move around and through the irregularities of the shingles which can be a an effective means of changing air pressure—that is, air moving directly through the shingle plane. If the volume is great enough, there may be no need for more open area. Irregular slate shingles can provide enough vent area without other supplemental sources of air pressure equalization. Cedar shingles, on the other hand, are fitted tightly enough to minimize the amount of air moving through the roof plane; therefore, typically, wood-shingle roofs need a supplemental air exchange source if they are to perform effectively as pressure-equalized systems.

If the attic does not perform as the pressure chamber and the roof is designed to perform as a multiple-layer system, the chamber is restricted to the small lens of space behind the shingle itself and the outer face of the roof sheathing. It is possible to make such roofs function as pressure-equalized systems, as long as it recognized that this is the system, and the components are allowed to perform their respective functions. If the relatively small sliver of space behind and below the shingle is to act as the pressure equalization chamber, the sheathing plane is the air barrier. This means that the sheathing must be developed as an effective impermeable layer. This can be relatively easily accomplished by the installation of a membrane on the sheathing before the shingles are applied. It would improve the performance of the air barrier if the lap seams of this membrane were sealed, but more often they are loose-laid because the membrane below the shingles is not recognized an air barrier. Usually, the installer views the membrane below the shingles as a water barrier. The membrane may, in fact, be called upon to repel water, but that is not inconsistent with performing first as an effective air barrier.

The second requirement is that there be an adequate vent path from the front side of the shingle to the back side. For materials such a slate shingles, this is not normally an issue. The slate is irregular enough that the fit between layers is sufficiently imprecise to reliably provide gaps between slates that can act as air vents. Because the chamber itself is quite small, the volumetric change to satisfy the requirements for pressure equalization are also small. The smaller the chamber, the smaller the required opening area to connect it to the outside.

Answering how small the chamber can be and still operate as an effective chamber is not easy. If the chamber is quite small, and the vent path is correspondingly small, what little water may accumulate in the chamber or in the material cannot dry out from occasional water entry or condensation. As we will see with hybrid systems, the possibility exists that if the chamber space is too narrow, water may be trapped between the back of the slate and the face of the membrane. For this reason it is recommended that for shingles laid on solid sheathing a layer of nailed stripes be installed on top of the sheathing and membrane to provide a slightly larger chamber. The larger chamber will still be relatively small, and pressure will equalize quickly, but it will provide a true ventilation capacity for drying.

Shingle Systems

A couple of examples of small chamber systems are worth discussing. One is actually a wall system, namely the clapboard wall. Clapboard is a vertical shingle system wherein the shape of the shingle is exaggerated in one dimension. Clapboard is not a membrane; it never was intended as such. The clapboard can operate as a pressure-equalized system in two possible ways. The pressure chamber may be the small gap immediately behind the clapboard; or the chamber is the stud space, because the sheathing is so open to the stud space that segregation is not feasible. Obviously, if the pressure chamber is the full stud space, the vent area must be correspondingly larger than if the chamber were the small lens behind the clapboard and in front of the sheathing.

A surprising number of houses from the early nineteenth century have the original clapboard, and they are still shedding water. The primary way to defeat a clapboard wall is to convert is from a vertical, pressure-equalized shingle system into a membrane system. This is usually done inadvertently, and the agent of the conversion is an improper paint job. The gap between the individual clapboard strips is the vent path though which air enters and exits the pressure chamber. If that gap is sealed, as can occur when the clapboard is painted and the paint fills the gap and clogs the vent path, the chamber is defeated. For a brief period of time, the paint will act as a perfect membrane, to exclude water. Over time, however, the paint between the clapboards may crack, producing a fine-line pathway across a plane, with differing pressures on either side. All that is needed in order to initiate a leak is for water to run down the clapboard and occlude the outside end of the pathway.

The proper way to paint a clapboard house is to shim the clapboards, to exaggerate the gap until the paint dries. This maintains the air space between the individual clapboards. Though it requires added work to install, remove, and touch up the locations of the shims, if this procedure is observed, there is no reason a clapboard house cannot retain its original fabric indefinitely. All that said, the important point here is that the clapboard works as a pressure-equalized shingle system, which happens to be oriented vertically.

Another less obvious example of a pressure-equalized shingle system is thatch. Thatch roofs are constructed of reeds bundled in such a way and to a depth that ensures there are always several layers of reeds between the exterior and interior of the building. The individual "shingles" are shaped like sticks, and they are sloped. Because the water tends to adhere to an individual reed, as water descends down the reed, it also traverses the span of the room such that, as the water drips from reed to reed, by the time the water might otherwise enter the occupied zone it is outside the perimeter of the building. For this cascading of water and adhesion to the surface of the reed to work as a lateral transportation mechanism, the reeds must remain aloft, meaning that they must remain fluffed up. If the reed is more or less a sloping cantilever, the air pressure is equal all around the reed. Other than gravity, there is no motive force to pull the water through the thatch into the interior of the building. If the thatch becomes matted and compacted, as it is inclined to do as the individual reeds weather and lose their individual stiffness, the roof will convert from a pressure-equalized system of sticklike shingles to a plate of compacted fibers acting as a poorly performing membrane; ultimately the thatch roof will leak.

Animal fur is essentially a thatch shingle system. Individual strands of fur are like reeds, and work similar to thatch to shed water. Likewise, the fur works better when the individual hair strands are separated and aloft. If the fur becomes matted, the coat becomes a poorly performing membrane and water will soak the animal to the skin. We bring up animal fur as a shingle system to point out that the materials that comprise the multiple-layer system have less to do with whether the system works than the recognition that *all* shingle types must adhere to the basic principles of pressure equalization in order to exclude water.

With that said, in addition to pressure equalization, we must also take into account the fundamental method of cascading of water that characterizes shingles and shingle systems. In short, we also must understand how shingles are designed to work. As water trickles down the shingles, it cascades from one shingle to the next until it reaches the gutter. If we look closer at the shingle pattern, however, we will notice that as the water rolls off the edge of one shingle it is likely to drip in the joint between two shingles, not directly onto the exposed surface of another shingle immediately below. This is because installers deliberately stagger the joints between shingles as they progress up the slope. The water, therefore, drips partly onto the shingles immediately below the first shingle and partly between the two shingles that are overlapped by the first shingle.

The water that cascades directly onto the exposed surface of an inferior shingle will conjoin with more water, and potentially continue cascading from one exposed shingle to the next all the way down the roof. As the depth of the now-accumulated water increases, there is a natural tendency of that water to seek the lowest depth at any instant even as it is flowing generally downslope. This is why water flowing down a slope tends to fan out. On a shingled roof, a certain amount of the water, as it fans out, will roll over the side of the shingle into the joint between the two adjacent shingles, not off the butt end of the shingle. This water is joined by the modest percentage of water from the superior shingle, which is shedding directly into this same joint. This is not a problem, because there is always a third shingle exactly aligned with the first shingle—which is to say that it is centered on the very joint into which water has collected. This third shingle collects the water from the joint of the two overlapping shingles above, even as the water continues to cascade down the slope. If, at any point on a shingle roof, we drilled a hole perpendicular to the roof, we would drill through at least three shingles. (The only exception to that rule is at the extreme downslope edge, where there may be only two layers.

If we take a section through a lateral joint between two adjacent shingles, the joint itself forms a small trough, which is filled, let us say, with water from the shingle above and some water which has fanned out over the edge of the two laterally adjacent shingles. Looking directly up slope, the water in this small trough or groove will be coming directly toward us. The trough or groove is formed by the side edges of the two adjacent upper shingles and the horizontal surface of the lower shingle that covers the gap between the two upper shingles. If we look closely to either side of the groove, we will see a fine-line gap between the bottom surface of the upper shingles and the upper surface of the lower shingle. As water fills the trough, even though it is moving downslope toward us, a low degree of hydrostatic pressure is pushing water laterally into this fine-line gap.

If the water in this fine-line gap were to move laterally all the way to the edge of the lower shingle, there is at least a geometric possibility that it will roll over the edge of the lower shingle. If there is a third layer, this laterally migrating water will be intercepted again, but in precisely the same way as it was before. The water migrates laterally to the edge of this third shingle and drops to the shingle below that if, there is yet another shingle. If there is no other shingle, the water will penetrate through the shingle system altogether, after which it will be free to penetrate further into the roof structure. All this lateral migration through the fine-line gaps between shingles assumes, of course, that there are no other forces acting on the water.

In a healthy, functioning shingle roof, water will get into the fine-line gap between shingles for the same reason and in the exact manner we just described. But in such a roof this is not a problem because even as the water migrates laterally, it moves downslope in the same fine-line gaps, and in very short order emerges from under the shingle onto the exposed surface of the next shingle downslope and back into the mainstream of water flow. This is what is meant by describing shingles as "self-flashing back to the exterior." Water reenters the main current of water as it traverses the roof as a whole.

To this point, however, we have described cascading and self-flashing as if there were no pressure differential acting across the section of the roof. When there is, and it is negative between the outer surface of the shingles and the underside of the shingles, a motive force is working contrary to the gravity force that is otherwise accelerating the water down the slope. A negative pressure differential can, and generally will, pull water through the fine-line gap around, more than up and over, the downslope shingles.

Shingle Materials

With the principles of shingling defined, we can inventory some of the more common shingle materials and the peculiarities associated with them. The characteristics of the shingle cannot be discussed completely without also describing and analyzing the attachment detail. The material that comprises the individual shingle is ideally impermeable, durable, and not particularly heavy. Unfortunately, not many shingles fit the ideal, and the primary devices for compensating for the inadequacies are the designs of the substrate and of the attachment detail. Thus, the shingle and the attachment detail are integral issues.

In this section we will discuss three classes of conventional shingling materials: stone, clay, and timber. Two other general classes of shingles will be discussed in the following section, namely asphalt shingles and shingles made of synthetic and recycled materials. In truth, almost any material can be formed into a shingle, and at one time or another many have been tried as roofing. The thinking is that if a material is at all impermeable, it can be used to make a shingle roof, because the nature of shingle construction allows for gross imperfections in the manufacture and alignment of the pieces while still producing an effective shelter from rain. Therefore, shingle systems range from primitive constructions of grasses and leaves to very sophisticated molded and interlocking glazed tiles. Fortunately, it is not necessary to inventory and discuss every single shingle type and material, because what causes shingle roofs to fail has far less to do with the idiosyncracies of the fabric than with adhering to basic principles of construction. This is true even of the three classes of economically significant materials we will discuss here as examples of the class as a whole.

When we refer to stone shingles, we are for the most part talking about slate, but slate is not the only stone shingle material. The Druids used cleft field stones as shingles. Slabs of granite also have been used as massive shingles; and sandstones have occasionally been used. But slate is the preferred material for shingles, and for good reasons. Slate is nearly impermeable; it can be split into relatively thin sections, which makes it possible to retain the impermeability properties without incurring excessive weight. The slate itself is very stable in an open environment. On the negative side, not all slates are created equal. Some, notably some Pennsylvania slates, weather rather rapidly. When slate weathers, it tends to delaminate along the bedding planes. Thin flakes or wafers of decomposed slate lift from the body of the slate. In extreme cases, the slate can weather completely through the shingle, although typically the slate will fracture and fall prior to weathering to this extreme.

Because slate is inherently brittle and thin, it is vulnerable to cracking and fracturing. The single most common reason that can cause slate to crack is ice, more specifically, the cumulative weight of ice and snow bearing on the surface of the slate. When slate is laid, it is supported on the upslope end directly on the substrate or nailer. On the downslope end, it is supported on the edge of the slate below. This creates a void below the slate, which spans the distance. When a load is applied at the surface of a shingle of slate, it is put into a state of bending. If the magnitude of the load is great enough, the shingle will fracture. This scenario is exacerbated by the effects of cold weather, when the conditions that produce the bending situation are generally present.

Unless slate is attached carefully, another source of force can come from ice that forms between slates. Water can come from many sources, not the least of which is condensation. If the water caught between slates freezes, it will heave the top slate upward. It the attachment detail is such that the slate is rigidly fixed at its upper end or attachment point, chances are, the lifting action of the ice will snap the shingle. To prevent this, the fasteners should not clamp the shingle down so it is fixed. The shingle has to be able to "flap" or "rock" on its attachment detail.

When nails are used to attach the slate to the nailer or substrate below, the risk is that the nail will be over-driven, thereby clamping the slate; or the final blow delivered to the nail may impact the slate itself. The nail hole must be countersunk, so that the nail head does not stand proud of the surface of the shingle. If the head protrudes above the surface of the shingle, the next shingle laid will come to rest on the point of a nail head, not on the edge of the slate below. When a load is applied to the slate, the nail will act as a point support, which may be more than enough of a stress concentration to fracture the slate.

A traditional method of attaching shingles generally and slate shingles in particular is to tie them to space boards using wire to effect the tiedown. It is virtually impossible to overdrive the wire tie; and the desirable movement in the slate itself is easy to achieve. The disadvantage of wire ties is similar to that presented by the proud nail head: that is, the next shingle bears on the wire ties as though they are point supports, which has a similar effect as the shingle bearing on the head of a nail.

A method for attaching slate which is popular in Europe is to use one or two hooks to carry the slate along the bottom edge of the slate. The wire hook is then attached to the substrate. A similar hook is used to replace slate in a field of shingles; it is a sound item to use as an original attachment detail. That said, care must be taken to ensure that the hook nestles between the slates below, to prevent the point-load situation, which is a risk with nails and wires. On rare occasions, slates have been attached to the nailer or substrate using screws instead of nails. This limits the risk of over-driving, but at a considerably higher cost. Historically, the fastener material of choice for slate has been copper, either as nails or as wire ties. Today, stainless steel is a viable alternative. In either case, the added cost of the fastener is a value-added, because otherwise the shingle would outlive the fastener in terms of material deterioration.

When a single fastener fails, the slate can slip out of the field. With the shingle removed, a strap can be installed along the centerline of the re-moved slate. When the slate is slipped back into position, the protruding end of the strap can be folded up and over the edge of the slate to form a hook detail. If the attachment is a double fastener detail, which is typically the case when nails are used, probably one fastener will fail before the other. The shingle may rotate slightly about the surviving fastener, but it will not slip out. To effect a repair, the second fastener must be released. This is normally accomplished with a tool that has a flat strap or thin bar of steel with a sharpened slot in one end and an upstanding crooked han-dle on the other. The strap is slipped under the appropriate shingle so that the sharpened slot engages the surviving fastener. The roofer then hits the opposite end with a mallet until the fastener is sheared off, and the shingle slips out.

The standard sizes of slate shingles has changed over the years. Al-though slates can be custom-cut to almost any dimension, precut slates come bundled, and are sold in bulk based on stock or standard sizes. Size is not an issue when designing a new building, but it can be a significant problem when a roof is being partially resurfaced, thereby mixing old and new slates. Color matching is difficult enough, because older slates tend to fade; but what may come as a surprise is that what was considered a stan-dard size when the building was originally erected is now a custom size, which, in plain language, means that it costs more. The "ideal" thickness of a slate shingle is a topic of ongoing debate. Although thicker is arguably better, thicker is also heavier and more expensive. The most common thickness is three-eighths of an inch (technically 9.5 mm, but more likely found as 10 mm) and one-fourth of an inch (6.35 mm, more likely 6.5 mm).

As to color, slate comes in many shades of gray, green, purple, red, and, of course, black. Shingles also can be shaped on the exposure side to form half-round, diamond, or half-octagon ends. In short, the mixing and matching possibilities are endless. Compatibility issues are rare, except in highly blended installations, where it is possible to have some slates that will last 150 years blended with others that are good for only 50 years. And because prices often reflect the durability, there is the prospect of paying for 150-year material, which will be driven by an adjacent material of sig-nificantly shorter lifespan.

When wood is used as a shingle, the term shingle itself takes on a more idiosyncratic meaning, because there are two methods for producing wood roofing materials. If the shingle is sawn, it is called, simply, a shingle. A sawn wood shingle, unlike a slate shingle, is tapered. The thick end points downward, and is called the *butt*. The thickness of the butt varies slightly, but is usually on the order of three-eighths of an inch (9.5 mm). The width is approximately four inches (10 cm), although any width is available from custom cutters. Lengths are in approximate increments of four inches. The designations—royal and double royal, for example—refer to terms associated with length, which is also related to the number of overlaps. Length is an important consideration less because of how much of the shingle will be exposed and more because of how much of the shingle will be overlain by subsequent layers. Traditionally, the theory has been that the greater the number of layers or laps, the better the roof. Actually, the number of overlaps relates more to the pitch of the roof: the shallower the roof pitch, the more overlaps must be employed.

The other method of forming wood roofing is to split the piece from a timber solid or round. When this method is employed, the resulting piece is somewhat more irregular in dimension and shape, and is called a *shake*. Shakes take on a natural taper from the way annual tree rings form—they become thinner as the tree ascends. There is little technical distinction between shingles and shake, although the advocates of each would make it seem there were.

Timber roofing is always nailed into place; and because several layers overlap, each shingle may have four to six nails going through it into the substrate. Wood roofing is tightly clamped into position. If water collects between layers, the wood will flex and absorb the deformation without fracturing—at least the majority of the time.

Wood shakes and shingles can be formed from almost any species of wood, but it is hardly worth the effort unless the species is inherently resistant to fungal rot. Given the exposure to water and the prospect that water will be trapped between shingles particularly, the risk of rot is high. In fact, timber roofing is the one of the few installations that is seriously threatened by surface rot. Normally, surface rot is dismissed as an interesting, but insignificant, nuisance. But in the case of wood roofing, though it may take 20 to 40 years to develop, surface rot can reduce the thickness of the roofing to the point at which a shingle may very well snap.

Wood species that have the requisite inherent resistance are limited. White oak, for example, might seem to make an excellent roofing material, but because it is so dense and heavy, roofers would have to work twice as hard as with most soft woods to get it up and installed on the roof. Moreover, the acid in oak is hard on fasteners, and the cost is prohibitive. Among the less dense, more affordable species, the one which dominates the market is western red cedar. It comes bundled in pallets good for 100 square feet of coverage (one square of roofing), and it comes in three grades. The top grade is number 1, also identifiable by a blue or purple label on the bundle. Blue label cedar is clear grade, 100 percent heartwood with straight grain, and free of knots. It is the only grade recommended for use where there is direct exposure to weather. Number 2, or red label cedar, has a certain percentage of sapwood, and the grain is not as true. It is approved only for sheltered exterior use, such as under a covered porch. Number 3, black label, is fit only for interior decorative use or as kindling.

In recent years, number 1 grade has been shipped with a water-soluble stain applied to the wood. The roofing looks better initially; it is not as bright as unstained cedar. The idea is that as the rain leaches out the stain, the wood is weathering slightly, so that by the time the stain is gone, the wood looks good again. As long as the stain goes directly into the gutter and away from the building, no harm is done. If, however, the stain gets loose and stains the house, the homeowner will not be happy.

Fastening the shakes or shingles to the substrate is typically accomplished with metal nails. Because the life expectancy even of a good wood roof is on the order of 40 to 50 years, the fasteners are usually not an issue. If the wood roof lasted 80 to 100 years, fasteners would be more of an issue, as is the case with slate roofing. Although the wood brings with it some acid that is antithetical to metal fasteners, the corrosion product that accumulates on the fastener is seldom a problem in a material as relatively elastic as wood. What is occasionally a problem is the staining, which can issue from a corroding nail; other than that, fasteners are seldom an issue with wood roofing.

More problematic than the fasteners are the metal flashings, particularly valleys. The valley is always vulnerable; and because acid from the wood roofing concentrates there, the valleys are good candidates for synthetic flashing materials, as opposed to metal of any type.

In temperate zones, north-facing wood roofs are vulnerable to moss, algae, and other forms of biological growth. These growths themselves are not particularly damaging to the wood, but they are indicators of fungal decay and hydrolysis of the cellulose. These sorts of growths are often combated with dilute solutions of chlorine bleach but that does not change the fact that their presence is a sign of far greater problems. Wood roofing is still wood, and to many species on this planet, dead wood is food. The natural resistance of some wood, notably western red cedar heartwood, is fugitive, meaning that it does not last forever. As the chemicals that are the source of that resistance leach out of the wood, the omnipresent spores and larvae are waiting for the threshold of safe ingestion.

Another consideration when dealing with wood roofing is that it is a fire hazard. In many jurisdictions, wood roofing is prohibited or severely limited for that reason. Wood can be treated by pressure injection with fire retardant chemicals, but this process has serious drawbacks. First, the injection process strips out of the wood any residual chemicals that might offer any resistance to decay or infestation. Second, the chemical injected is even more attractive as food to the species that prey on wood in the first place. The upshot is that wood treated for fire retardation is subject to accelerated deterioration, from rot in particular.

The terra-cotta tile roof is, at least technically, the ultimate roof. The material is highly inert; will not corrode, burn, or rot. It can expand and contract without consequence. It is manufactured under controlled conditions, it offers excellent quality assurance as to flaws and cracks. There are only two economically significant threats to terra-cotta roofing: fracture and anchorage failure. Tiles crack and fracture from the usual impact and accidental reasons, such as thrown or dropped objects and the misplaced footfall of an inspector or roofer. It is also possible, although not common, for a clay tile to crack from freeze-thaw cycling or other naturally occurring, environmentally induced deterioration mechanisms.

A tile roof may outlast the rest of the building, and it is almost the only roofing system about which that can be said. It is far more likely that the anchor holding the tile in place will fail before the tile itself does. Depending on the geometry of the tile, the anchors or fasteners typically go through the tile and into a batten or nailer strip of some sort. In common barrel tiles, the batten typically runs up and down the slope of the roof. This offers the advantage that occasional water, such as condensation, will have an unobstructed path down the slope, and so will not collect behind a nailer placed laterally on the roof substrate.

The alternative is to place nailers across the plane of the roof. Lateral nailers are more common with pan tiles or some other shape than the half-round family. Whenever lateral nailers are used on a roof, whether they are combined with tiles or other shingle material, provisions must be made to prevent occasional water from collecting on the upslope side of the nailer. If water is allowed to accumulate in the upslope crotch of the nailer strip, the water will corrode the anchors and rot the wood nailer. There are several details that can be applied to lateral nailers to mitigate the risk of trapped upslope water. The strips can be slotted on their backsides so that the slot provides a pathway under and through the strip. Or the strips can be angled so that collecting water is pitched downslope. At some appropriate location, the nailer is broken so that the collected water is channeled through and past the nailer line. It is also possible to lay a double grid of strips, the lower one running up- and downslope, the other running laterally across the lower strips and, obviously, separated from the plane of the substrate. However the objective is accomplished is not as important as how to keep the anchor strip from deteriorating. This is a particular issue with tile roofs, because the first and only time that the attachment strips will ever get inspected is after they have begun to fail and tiles are sliding off the roof.

Although we have consistently referred to the strips as nailer strips, because of the extremely long life of the tile, the attachment device may be more than a common roofing nail. Tile roofs offer one of the few occasions in shingle roofing when the attachment detail may be deleted. Because the tile is relatively heavy, and because the tile can be molded to any form and shape, occasionally a cleat or lug is integrated into the fabrication of the tile itself. The lug then hooks over a lateral nailer strip, and the weight of superior tiles holds the entire system in place. Today, this is more likely to be found in concrete tiles, but examples of lugged clay tiles exist. The roof of the Philadelphia Museum of Art is an exquisite example of a glazed terra cotta tile system, held in place by the gravitation force of superior tiles and restrained from sliding by lugs.

When tiles do displace, for whatever reason, the repair technique is the same as described for slate shingles. The tile is removed and a strap is installed. The tile is put back into position, and the end of the strap is folded up and over the butt end of the tile. This is satisfactory as long as the item that failed was the anchor. And in tile roofs, it is probable that the anchor failed, because the attachment strip rotted out and the anchor released. The repair of an attachment strip may require removal of the entire roof, which in the case of tile is not infeasible, but neither is it easy. Nailer strip repair can be approached piecemeal, but the first sign of failure of the attachment strips is certain evidence that all the nailers are approaching the end of viable life.

HYBRID SYSTEMS

Throughout this discussion about roofs and roofing materials, we have scarcely mentioned what are arguably the most widespread of all roofing systems and materials—those that travel under the label of asphalt shingle roofing. Ironically asphalt shingles have been the object of some derision in the architectural community. As a roofing material, asphalt shingling started as an extension of the hot roofing industry. The cotton lint felts were impregnated with asphalt, then cut for nailing on a sloped roof, rather than for being adhered on a flat roof. The gritty or sandy texture we associate with asphalt shingles was originally intended as solar protection, to prevent the amorphous solid asphalt from bleeding out of the shingle and running down the roof. More recently, we have come to appreciate grit because it reflects UV rays, which is the deterioration mechanism that will defeat most asphalt shingles (see Fig. 5.8).

The market soon replaced cotton lint felts with wood fiber and, later, mineral fiber felts. Today, the fiber of choice for asphalt shingles is fiberglass. The asphalt itself is often chemically modified to improve resistance to heat and, particularly, to UV rays. With these improvements, asphalt shingle roofing today is one of the best buys on the market.

FIGURE 5.8 This is end-stage asphalt shingle failure. The shingles are so brittle at this point that they fracture on contact even when they are warm.

Asphalt shingle systems are neither true membranes nor true shingles. The absence of true membrane action is obvious. The way they are laid makes no pretense that they meld into a homogeneous, impermeable membrane. On the other hand, they cannot be laid on spaced boards, because they will collapse under the heat of the sun, into the voids between the boards. Therefore, they must be placed on solid sheathing, which means that pressure equalization of the attic as a whole is ineffective, at least to the extent equalization can be achieved through the roof plane. This still leaves the option of pressure equalization at the local level, except that the space between the back of the asphalt shingle and the sheathing is minimal, if it exists at all. Even if there is a small void, the air path to the outside is ineffective. That is why most asphalt shingles now come with strips of heat-activated adhesive. When the sun heats the surface, the shingles become glued to the strip below, forming, if not a membrane, at least a continuous sheet of asphalt shingling.

According to the conceptual basis of this book, however, if the system is not a membrane and pressure equalization is not feasible, there is nothing left. Clearly, we need to amend the model, hence the notion of the hybrid systems. In fact, asphalt shingles comprise just one member—albeit the dominant member—of a growing class of roofing. Whereas single-ply membranes dominated industry attention relative to changes in roofing

FIGURE 5.9 This roof is 75 years old and still working, albeit with some maintenance. The shingles are asbestos cement, firmly attached and adequately pitched, and have outlived the flashings.

over the past 30 or so years, there are several new entries to the market that are variations on the hybrid approach.

In the abstract, hybrid roofing works as an outer layer that is an imperfect membrane backed up by a second membrane that is structurally inferior but less permeable. Asphalt shingles can be applied directly to the substrate without a second moisture barrier; but that is not the standard. The outer shingles defeat heat and UV rays and shed the vast bulk of the water. If any moisture gets under the asphalt shingle, it is intercepted by a second membrane. As the water descends the slope behind the outer shingles, it is effectively flashed back to the exterior at the head of every line of shingles. Any water that is trapped behind the shingles and freezes is unlikely to damage the highly flexible asphalt shingle.

If there is a current, widespread weakness in the use and installation of asphalt shingles, it is the relative inattention paid to the backup membrane. The material currently used for this purpose is typically 15-pound construction paper. While this is a highly economical choice, it is easily damaged in construction and offers no self-sealing effect to the piercing caused by the roofing nails or staples.

But for the age and nostalgic value associated with the material, most modern wood shingle roofing, technically, is closer to a hybrid system than to true shingle systems because it is more often than not installed with a membrane between the roofing and the sheathing. Despite its prevalence, this is not a recommended way of installing wood shingle, because of the consequences of water trapped between the shingles. Water behind an asphalt shingle is not desirable, but water trapped behind a wood shingle is potentially devastating.

For a number of years, manufacturers have tried, with mixed results, to sell fake slate roofing. The objections to date have been either that they simply did not work or they looked just like what they were, namely perfect, monotonous replicas of the same molded item. That is changing. The fake slate business is expanding the range of simulated slates in terms of colors, shapes, textures, sizes, and mold patterns. Technically, they will operate in the realm of hybrid systems. As such, installers and designers are cautioned to recognize the functional importance of the second membrane, and to ensure the quality and protection of that which is arguably the real roof, and treat the simulated shingles as the physical protection of that roof.

GUTTER, LEADER, AND DRAIN PROBLEMS

Imagine a hollow circular cylinder, capped on both ends, full of water and lying on its side. The maximum pressure of the water is at the point of maximum depth of the water, which is to say at the bottom of the pipe as it is lying horizontally. If the pipe is, say, six inches in diameter, the hydro-

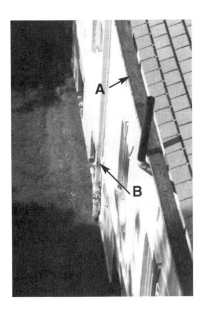

FIGURE 5.10 This is a new roof and gutter. Note the heavy accumulation of debris from surrounding trees in the gutter (A). This debris has clogged the leader, causing water to back up in the leader, which is sealed at the boot. Freezing of the mixture of debris and water has split the metal leader at location B.

static pressure at bottom, dead center, is six inches water times the density of water, which is 62.4 pounds per cubic foot; thus, the pressure at the bottom of the pipe is approximately $3\frac{1}{2}$ ounces per square inch (1.5×10^3 Pa). If there is a small pinhole in the bottom of the cylinder, the pressure of the water will push the water through the hole.

Now suppose that the cylinder is, in fact, a pipe, and that water is at rest inside an otherwise perfectly horizontal section of the pipe. The situation is the same; the only pressure is the gravitational pressure of the water on the pipe. If the water inside the pipe is released to flow out of the pipe, the gravitational pressure will decrease as the pipe drains. If the water is replaced as rapidly as it drains out, such that water flows through the pipe but the depth of water inside the horizontal pipe remains constant, the pressure still is no greater than the same gravitational pressure as when the water was at rest.

If the pipe is inclined, and the water flows out of the pipe without any restriction at the outlet, and the released water is replaced precisely at the rate of outflow, the velocity of the water moving in the pipe will increase as the slope increases. This is because the gravitational vector in the direction of the pipe increases; therefore, the accelerating force on the water in the pipe increases, and the force vector perpendicular to the bottom of the pipe decreases as the pipe is inclined further. When the pipe is perfectly vertical, the water accelerates through the pipe to a velocity based on the frictional drag of the pipe wall on the fluid and the length of the pipe. This is very similar to the situation described earlier in the context of roof planes.

One simple but, as it turns out, very important difference between the roof and the pipe is that the pipe is enclosed and the roof, being a flat plane, is, obviously, open. When the roof plane is tilted into a vertical position, there is still no restraint opposite the roof plane. No matter how much water sheets down the face of a vertical plane, the force vector perpendicular to the plane is zero. To demonstrate the difference, imagine the pipe filled with water and standing vertically. The pressure at the bottom of the pipe is equal to the density of the water times the height of the water in the pipe; this called the *head*. If the water inside the pipe is twelve inches deep, the pressure at the bottom of the pipe is about seven ounces per square inch (3.0×10^3 Pa). Because the water is a fluid, the pressure at the bottom of the vertical pipe is exerted not only downward but in all directions, meaning laterally against the wall of the pipe as well as the bottom.

If the lower end of the pipe is sealed, the pressure against the bottom will be the same as the pressure against the wall of the pipe at the same level, namely seven ounces per square inch (3.0×10^3 Pa). If the bottom suddenly opens, making the entire cross-sectional area of the pipe unobstructed, the water, obviously, will flow out of the pipe. Hence, the confinement of the fluid will release completely; the pressure against the bottom will be zero because there is no bottom; and the pressure against the wall of the pipe will drop to zero. The lateral pressure of unrestricted fluid flowing inside the pipe is zero. When the bottom was closed, the pressure was the same as the head of water; when the pipe is completely open, the lateral pressure is zero.

If water passing through a vertical pipe is sufficient to fill the pipe, and if there is no obstruction to the flow of that water, the pipe will not leak even if there is a fine-line crack in the wall of the pipe because there is no motive force to push or to pull the water through the crack. If the pipe is inclined even slightly, there is now an identifiable "bottom" to the pipe, with an associated force vector caused by gravity. If the crack is located such that the water opposite the crack is subject to even the slightest gravitational force, the crack will leak, if water flows in the pipe long enough to force water through the capillary. If as the water descends in the pipe, there is a partial obstruction to the flow of water, and the pipe is full above the obstruction, the water above the partial obstruction will be arrested, even if only slightly. The water column will not be able to fall freely, and the water above the obstruction will exert lateral force on the enclosing pipe.

Often, water passes through pipes at volumes significantly below the full capacity of those pipes. If the volume of water flowing through a pipe is not enough to fill the pipe, the fact that there is no lateral force on the wall of the pipe is intuitive. If, under such conditions, the pipe is partially obstructed, back pressure above the obstruction will not occur until and unless the obstruction is great enough to fill the pipe at the location of the obstruction. If the water in the pipe at the location of the obstruction is flowing at maximum free capacity, and any more water is added to the pipe, water will back up in the pipe above the obstruction, and lateral pressure will occur.

The water that collects in various forms of gutters and leaders must pass through a series of open troughs and pipes before appropriately discharging at a location off the site. Each of these conduits is vulnerable to developing an open pathway through which uncontrolled water can flow. We have discussed the risk of a corrosion pinhole developing in metals. Such flaws are certainly common enough, regardless of the metal of choice. Common wisdom says to select copper for such occasions, but there are many examples of corroded copper gutters and leaders, particularly when they are located in the vicinity of acidic deciduous trees. The leaves collect in the gutters and release tannic acid, which is corrosive to any metal, including copper. Lead-coated copper is an excellent material; but, for example, the leaders removed from Berkeley College at Yale University often had pin-holes through the leaders—though, admittedly, they were 65 years old. The point is, all metals are vulnerable to such corrosion. Other gutter and leader materials may also display specific material-related vulnerabilities. Rigid PVC, despite its appearance limitations, has become a popular leader material precisely because it is not subject to corrosion. On the other hand, it is extremely brittle when it is chilled, and often cracks on impact, particularly in cold weather.

Drains, leaders, and pipes, generally, can develop fine-line cracks and pinholes, but neither the presence of water nor the combined presence of water in a pipe and a flaw in that same pipe means that the pipe will leak. There must be a motive force; and water flowing freely through a pipe without obstruction does not exert lateral pressure on the wall of the pipe. Water flowing in a pipe that is even partially obstructed, however, is subject to the development of lateral hydrostatic pressure, even if the water continues to flow. In other words, a pipe does not have to be completely clogged to develop hydrostatic pressure.

If an eight-inch (20 cm)-diameter leader from a roof drain is 30 feet (9.14 m) long and is completely clogged at the very bottom, the pressure at the bottom of the pipe is approximately 13 pounds per square inch (8.96×10^3 Pa). If the leader is constructed from an eight-inch standard pipe, the hoop stress in the wall of the pipe is approximately 160 psi (1.1×10^6 Pa). In the scheme of stresses and materials for a steel pipe, this stress is almost trivial. If the same conduit were made of 20-gauge (0.91 mm) sheet metal, the stress in the metal would be almost 1,500 psi (1.03×10^7 Pa) which is not trivial. More important, the sheet metal conduit is joined by forming a seam along the length of the tube. The force across that seam at the bottom of the sheet metal leader would be 52 pounds per linear inch (9.26 kg. per linear cm.) of seam, which is more than enough to rupture the seam. As long as the water flows freely in the leader, the lateral pressure is zero, so the hoop stress in the conduit is zero; but as soon as the leader clogs, the forces across the joint in the sheet metal will climb.

This is one reason why many sheet-metal leaders split along the seam. They may also rupture because the water in the leader freezes; but the hydrostatic pressure alone is enough to rupture the seam if a long leader is clogged. Exacerbating the situation is the industry preference to turn such seams toward the wall. This is probably done to ensure that water spewing from the ruptured joint does not have to search for a wall to saturate. Such a target will always be straight ahead of the rupture and completely opposite any point from which it might be easily detected. Are sheet metal seams so unattractive that we must align them in a way that guarantees maximum damage when they fail?

With steel pipe, the risk of fracture from freezing water is real. In the same situation, if the water freezes, the stress in the pipe wall increases to as much as 40,000 psi (2.76×10^8 Pa) if the temperature of the ice reaches as low as 18° F (−7.8° C). Although a standard steel pipe is probably capable of withstanding such a stress under ordinary conditions, meaning at higher temperatures, the elevated stress of the freezing water condition occurs at exactly when the brittleness of the metal is increased. For seamless steel pipe, even this is not a particularly threatening condition, but for cast iron pipes and boots, it definitely is a threat.

A simple detail that can overcome this problem is to allow successive segments of pipe to be loosely, rather than tightly or rigidly, joined. When the steel leader meets the cast iron boot, simply leave the packing material out of the joint. If the boot clogs, for example, water is forced up and out through the joint between the segments. Pressure is relieved and the clog is well announced. Even if ice forms in the boot below the joint, there is the prospect that the pressure will force at least some of the ice vertically, thereby relieving some of the lateral pressure. This assumes, of course, that an overflowing splice joint is even momentarily acceptable, and this is not always the case. Sound maintenance theory says that an overflowing boot will attract attention and, therefore, will be rectified. Experience, however, suggests otherwise.

Buildings with rainwater conductors running in enclosed chases in the exterior walls, interior column covers, or similar enclosures are particularly vulnerable to failed leader joints. Those in exterior walls are also vulnerable to freezing conditions, but it is the sealed joint that is the subject here. The design presumption of a built-in leader is that, (a) it will never clog up, and (b), if it ever does clog up, it will not leak. The fact is, either a leader

FIGURE 5.11 When it began leaking several years ago, the built-in gutter was filled in. Now the water rolls off the edge of the building. What does not wash the wall immediately below, splashes at-grade and saturates the base of the wall.

will or will not clog. If it never clogs, sealing the joints is a waste of money because we know that freely flowing, unobstructed water does not exert lateral pressure. The joints, therefore, will never leak. If, however, the leader does clog, the pressure along the perimeter seal between the pipe and the boot of our previous example is 52 psi (3.6×10^5 Pa). This is not an excessive pressure across a well-sweated or taped threaded joint. It may, however, be enough to blow out an aging oakum joint.

Oakum is hemp rope, traditionally saturated with pine pitch, although tar and asphalt were also used. The oakum was wedged into an overlapping plumbing joint with more pitch or tar added on top of the oakum. A similar technique had been used for centuries to make watertight sailing vessels. The problem with old oakum joints is that the pitch hardened and shrank, creating fine-line cracks between the oakum and the plumbing joint. The same thing can happen with unthreaded solder joints, particularly in large castings, less so with sweated copper, and far less so with threaded joints.

Where built-in leaders and sheet metal leaders with joints turned toward a porous wall are at issue, it is simply incorrect to assume that constructed items will not fail; they will. When systems fail, it is imperative to detect the failure as soon as possible, not wait for secondary and tertiary damage to make the announcement. Burying sealed-jointed leaders in inaccessible wall cavities is bad design, even when a cleanout is located adjacent to a likely point of failure. Cleanouts, simply, while they may be required by plumbing codes, are crutches for bad designers.

SUMMARY ON ROOFS

The information presented in the first part of this chapter on roofing materials and configurations is not, by any means, a complete account of this topic. The intent was to describe classes of material types, along with their associated anchorage problems and general approaches to intervention. Unfortunately, this type of overview may lead to the conclusion that problems associated with multiple-layer systems are concentrated at the individual shingle, the individual anchor of the shingle, or the attachment point. Those are certainly common sources of problems, they are secondary matters to the vulnerabilities and problems that beset the folds and crenulations of the roof plane. Problems at folds on any roof are more problematic, generally, than problems in the field, for two fundamental reasons

- Problems at folds are more likely to be the result of workmanship, and because of their configuration are not readily available for inspection.
- When a failure occurs at a roof fold, probably it will become apparent only after the consequential damage has progressed to the point that it is identifiable; for example, valley leaks are rarely announced by the original leak itself.

Citing workmanship above is not to point fingers at roofers or any other trade as the causes of roof leaks. In fact, it is the generally high quality of the work of tradespeople that accounts for buildings not all failing the day they open. Rather, the point is, that despite the best efforts of tradespeople, the laps and seams at the edges and folds of roofs are areas where

FIGURE 5.12 This leader drops to grade at a low point. The concentration of water at the base of the wall is sufficient to generate damage at the entire corner.

workmanship can make the difference between a long-lasting detail and a premature failure. As a rule, the handiwork is concealed as it is installed, so it is simply not feasible for every piece of stepped flashing or valley side lap to be inspected, much less tested, before it is covered by subsequent work.

For some trades, this is more of an issue than for others. Electricians and plumbers can produce a substantial quantity of work before closure of the wall or riser is necessary. Moreover, today it is common practice, for good reason, to leave electrical and plumbing work exposed until inspectors and engineers have seen and approved the work. The same is true for structural steel and rough framing, rod setting, and most of the "finish trades." The work is visible as it progresses; and if the installed work is found to be unacceptable, timely identification can be followed by timely removal and reinstallation.

For other trades, this is not as feasible. Though the exposed surface of a masonry or brick wall is visible, and the exposed surfaces of such walls typically receive close scrutiny, the backup block is less likely to be inspected or even reasonably available for inspection if the face brick or stone goes up at the same time as the backup. Today, the block may go all the way up the wall before the face veneer is even started because, frequently, there must be a pause between layers to allow for installation of rigid insulation and/or dampproofing in the cavity or pressure chamber. Still, there are many instances where work must be enclosed or covered as it progresses, rendering inspection impossible. In no place is this restriction likely to occur than with roofing and, especially, flashing work.

Consequently, the trade must be self-supervising—which is not necessarily an improper or imprudent practice. The problem is, however, that only one pair of eyes will see the stepped flashing or the cricket before it rains, putting it to the test. If the roof is, in fact, faulty, the first notification of the fault may be weeks or months and thousands of dollars of secondary or consequential damage after the initial failure. Thus, the point is not that workmanship is inherently faulty, but that the more work is field-crafted—particularly with reduced quality control—the more vulnerable it is to the vagaries of chance and risk.

This leads to a few simple rules about roofs and roofing that relate to water exclusion:

- *Steeper is better.* The faster water moves off the roof and into the lower portion of the drainage system, the less chance there is that leaks will occur; and when they do occur, they probably will be detected earlier, hence with less collateral damage.
- *Simpler is better.* The more a roof plane twists and turns, the more linear feet there will be of flashing, crickets, and folds, the locations of increased long-term vulnerability and maintenance cost. And when edges and/or folds fail, the repair process raises the risk of introducing flaws at the same time the original flaw is fixed.
- *Shingles are better than membranes.* When leaks occur, they are far easier to track and repair in systems composed of discrete pieces than those of monolithic pieces.
- *All roofing problems are greater near low points than at high points.* Therefore, a good design maximizes the linear feet of the high line and minimizes the number of linear feet of low lines and points. Water should not be allowed to accumulate more than two inches in depth at

a failed drain before overflowing in some sort of failsafe backup, such as an emergency overflow scupper or an adjacent drainage panel.
- *Exposed gutters and drains are better than anything built in that conducts water on a routine basis.* Remember the military adage: It's not what you expect that works; it is what you inspect.

All of the examples and explorations of multiple-layer systems assume the water is in liquid form. This is not without justification; however, snow and ice, too, are important issues particularly on sloped roofs. Flat roofs are inherently membranes, so the primary issue associated with them and solid forms of water is structural capacity. But when the roof is not a membrane, and is sloped, the issues associated with water in a solid state become significant.

To launch this discussion, we must cite the universal interplay between roof design and configuration and climate. The variables that ultimately determine roof design are not numerous, but the degrees of influence are subtle in the extreme. In short, there is more to consider than rain and snow. As will become apparent when we address snow guards, it may be desirable, on balance, to encourage snow accumulation. Roofs may also serve as ventilation stacks or as cisterns. Roofs may have to endure strong winds. Rain may come torrentially or, seemingly, as perpetual drizzle. Moreover, material availability and costs influence roof design, as does durability and what's occurring under the roof. Roofs may need to be non-combustible or reflective.

Roofs in tropical locales may be thermally light, for heat rejection, generally with steep pitches and deep overhangs for moving large quantities of water rapidly away from the enclosed areas, while ventilating the volume. Roofs in the far north may be flatter and structurally heavier, to hold snow to take advantage of its insulating effect. Roofs in arid climates may be flat-bottomed cisterns for catching and collecting water, also flatter and whiter, to minimize emissivity and exposure area. Only in temperate zones with complicated weather patterns is roof design indeterminate.

When we don't consider climate before building roofs, we court disaster. Nothing demonstrates this more clearly than the phenomenon known as the *ice dam*, a condition that can occur at or near the outboard edges of steep-sloped roofs. It is related to a broader issue called the *cold roof principle*. We will first discuss ice dams briefly, followed by the broader principle, then return to the specific issue.

Typically, the first sign of an ice dam is that the building, usually a residence, displays signs of leaks at or near the perimeter of the building, during cold weather. Snow and/or ice have accumulated on the roof, which is, necessarily, a sloped, shingled roof or a sloped, hybrid roof. There may or

FIGURE 5.13A   The upper end of the leader is disengaged from the scupper catch box, and. . . .

FIGURE 5.13B   The deposition at the base of the wall has led to the deterioration of the stucco and the substrate.

may not be any liquid precipitation at the time of announcement. Inspection of the attic, assuming it is available for inspection, will reveal saturated sheathing at the outer perimeter. The sheathing at the north side of the building, generally, will be the wettest; the south side will be the least wet, although that is not a definitive sign. To the extent that it is possible to tell, the shingles themselves will appear to be intact and sound, although evaluation may be limited by the ice and/or snow. Inspection of construction details will probably reveal that there is no underlying membrane, or that what is working as a membrane is construction paper or a slip sheet. The house is probably less than five years old, and a detailed inspection of weather records will reveal that this is the first winter the building has endured this amount of ice accumulation. What is happening?

This is where the cold roof principle comes in. Snow falls on a sloped shingle roof and accumulates. The sun comes out and warms the snow pack from the exterior inward. At the same time, a limited amount of heat migrates from the interior of the building into the attic, and warms the roof from the underside. The combination of the two heat sources melts some, if not all, of the snow in a single day. During the meltdown, water liquified by the sun percolates through the snow and joins with the water that accumulates on the shingle surface as a result of the warmth of the attic. Together, these two sources of water tend to percolate at the boundary between the bottom of the snow pack and the shingle surface heading downslope. In the ideal situation—which is to say on a perfectly cold roof—the water continues down the slope until it enters the drainage portion of the horizontal closure system and is evacuated.

When the sun goes down, obviously, that source of warmth is interrupted. The warmth from the interior of the building may be enough by itself to continue the meltdown. If it is not, the water that has saturated the snow at the boundary with the shingles may refreeze during the night, to resume the melting process the following day.

Recall that water moves down and, to some extent, through a shingle roof; and keep in mind the cascading and self-flashing phenomenon that characterizes shingle roofing; finally, imagine the water migrating through the fine-line gap between shingles as the overall water migrates downslope. In the ice dam scenario, as the water flows downslope in the fine-line gaps between shingles, it will reach a line across the breadth of the roof where all the subsequent downslope gaps are sealed by ice between the shingles. Any water caught in the fine-line gaps cannot progress downslope any farther. Water, therefore, accumulates in the gaps and exerts hydrostatic pressure such that the water can only move laterally. The result? It will roll off the immediate shingle edge and collect on the one below. But the downslope edge of that shingle is also frozen shut by ice, so the progression will continue until the water is released on the backside of the shingle system altogether.

FIGURE 5.14 Water from the downspout is actually directed upslope and under the shingles.

The ice acting as a seal is formed from water migrating through the fine-line cracks that encountered freezing conditions during the descent down the slope of the roof. This means that the temperature that was causing the thawing at the upper reaches of the roof, thereby inducing thawing of the snow and ice, goes down toward the edges. If the temperature of the entire roof is above freezing, ice will not form in the fine-line gaps. If the temperature of the entire roof is below freezing, water will not melt more at the upper zone than the lower, and ice dams will not form. If, for whatever reason, the roof temperature is higher at the edges than toward the ridge, water will descend through the warmer portion, and ice dams will not form. For ice dams to form, the upper portion must be warmer, to thaw water. That water passes through a region of the lower roof that is below freezing.

One of the primary heat sources that can cause this unusual set of conditions is that from the interior occupied zone, through the ceiling of the attic and into the triangular region of the attic itself. The warmer air rises toward the ridge, resulting in a warmer upper region and a relatively cold lower region. The collected heat at the ridge, by itself, may be enough to trigger thawing, even as the outside temperatures remain well below freezing. This sets the stage for ice dam formation. It is possible to defeat ice dams by keeping the entire attic and roof at the same temperature, although that may require keeping the attic as cold as the outside. Obviously, such an attic will be uninhabitable during the winter, but at least the roof will be of a single uniform temperature and ice dams will not form. The key to accomplishing this is install substantial insulation at the attic floor level and to ventilate the attic to the outside air. This will both protect pressure equalization and ensure a cold roof.

One of the more recent culprits responsible for the formation of ice dams is mechanical equipment located in the attic. Too often, people view the attic volume as wasted space if it remains unoccupied. But if we treat the attic as part of the occupied interior, we generally must push the insulation into the zone between the attic ceiling and the underside of the roof, to accommodate, for example, heat pumps, hot water heaters, air handlers, exhaust fans, and other modern environmental modification equipment. In this way, the attic becomes home to energy-consuming, heat-producing devices, and the insulation is in the wrong place.

Fortunately, ice dams are a rare occurrence in regions such as New England that are most prone to their formation because builders and residents have been contending with the phenomenon for generations. A relatively common and simple device to defeat ice dams is the slip sheet or skirt. This a band of exposed sheet metal roughly three feet wide placed along the edges of the building. The width is approximately that of the band of maximum vulnerability to ice dams. The sheet metal introduces a band of sloped membrane roofing into the system but it also offers some offsetting benefits. First, it provides a highly heat-conductive material where the ice dams are likely to occur. More heat is conducted toward the eaves, keeping the surface and the zone generally warmer, thereby mitigating the conditions conducive to ice formation. Second, if and when ice does form on the metal, it will release sooner and sheet off the roof faster. Third, it eliminates the overlapping construction of shingles where frozen interlaminar water would otherwise cause a problem. A variation on the theme is to put the membrane below the shingles in the same location and of the same material, but carry the shingles over the slip sheet to the edge of the roof. This will mitigate the condition, but it sacrifices the quick-release aspect of the exposed metal.

The options for intervention after the fact depend almost entirely on money. Certainly, it is possible to tear off the roof and replace it with a superior design, keeping in mind that ice dams are most likely to form when the building is less than five years old. This is not because roofs improve with age, but because the weather conditions that result in ice dams may not occur every year. It is possible to ventilate the attic to induce the cold attic principle even if air has to be pumped in. Heat can be mitigated by wrapping any equipment in insulation, providing that doing so will not cause overheating or other problems with the equipment itself.

It is possible to add heat to the eaves and lower reaches of the roof, although this is in some ways less reliable than dealing directly with the source of the problem. Power fans, for example, can be added to connect the eaves and attic extremities with the occupied zone. This will put heat at the desired location, but it also relies on the proper performance of equipment; furthermore, there is a permanent hole between the occupied zone and the attic, through which unconditioned air can get back into the occupied zone. More troubling is that the occupied zone is now physically connected to the attic, making pressure segregation extremely difficult, if not impossible, to maintain.

Another technique is to install tapes along the downslope edges, generally, under the shingles themselves. These low-voltage resistance heating tapes are connected to temperature and moisture detecting controls. When

FIGURE 5.15 The original leader location was on the opposite side of the tower. When it was relocated, the additional charge overloaded the drain, so water spills over the edge of the gutter and down the face of the wall.

the conditions are such that it is possible for ice to form at the eaves, a relay is thrown; and a low-voltage DC current flows through the heating tapes. This approach, at least as a remedial measure, avoids the physical connection of the occupied zone and the attic, and it puts heat precisely where it is most desirable. The downside is that, again, the success of the technique depends on a mechanical device to compensate, rather than to actually mitigate or eliminate, the offending source.

A relatively minor option, but one that attracts considerable attention, is the snow guard. The term is technically accurate, but our interpretation is faulty. We have come to believe that the device will protect or guard us from the uncontrolled descent of ice and/or snow from a sloping roof, by intercepting sheets of sliding ice and snow pack and breaking them up into small pieces, which, presumably, are less life threatening than the unbroken slabs. But, in fact, snow guards guard the snow, not people.

The primary function of snow guards is to keep the snow piled on top of the roof, where it acts as an insulating factor when the outside air temperature is below freezing. If the outside air temperature is precisely 32° F (0° C), in a thermally static state, the air, the snow, and the roof surface are all at the same temperature, 32° F. If the air temperature rises above 32° F, the snow has a chilling effect on the roof surface, because the temperature of snow cannot rise above freezing as long as any crystalline water is present. In a thermally static condition, when snow is on a roof and the ambient outside air temperature is below freezing, the snow acts as a blanket on the top of the building. Heat generated from within the building is conserved with snow on the roof. The obvious trade-off to using snow guards is that the structure has to be designed to accommodate the accumulated snow load, a practice modern structural engineers do not encourage.

5.4.2 Parapets

The subject of parapets may seem more logically included as part of the discussion of the wall, but experience shows the connection of the parapet to the wall, while physically obvious, is functionally more part of the roof, as, arguably, are cornices. The parapet does not provide a barrier, either selective or otherwise, so it is not functionally a wall. Furthermore, the problems that beset parapets are different from those that afflict walls, and are more akin to those that afflict roofs.

Parapets are somewhat bewildering design elements. We suspect even that they are built from habit as opposed to being dictated by some design imperative. As guardrails, they are exorbitantly expensive. As water collectors and diverters, they are costly and prone to failure. As wall caps, they are remarkably inept. The parapet is, simply, a construction extension of the masonry of the wall below. It is built of the same materials and with the same or similar construction technology. If the designer is paying attention, the parapet is debonded from the wall below to facilitate the differences in movement due to thermal expansion, among other things.

The environment surrounding a parapet is vastly different from that of an exterior wall. The interior of a wall is relatively stable in terms of moisture and temperature, whereas the parapet is exposed to the elements on both faces without mitigation. The top edge or surface of a parapet is one of the most vulnerable sections of an entire building. If water penetrates the coping of the parapet, from there it is often a freefall to the window head of the top floor of the occupied zone. Thereafter it is a direct and unobstructed line around the frame to levels below and/or the interior of the building. Detailing a coping as though it will perform as a perfect membrane is, simply, daft.

There are only three possible relationships between the parapet and the roofing: the intersection is a high line; the intersection is a low line; or the intersection is a transverse or sloping intersection. If the intersection is a high line, water will drain away from the relatively vulnerable intersection with the wall; however, that advantage is at the expense of directing the water back toward the interior of the building, which, of course, is contrary to the object of draining water from a roof to begin with. If the water is directed away from the interior of the building and toward the perimeter, it is also directed against the back of the parapet, and is, therefore, concentrated and collected at precisely the vulnerable detail of the parapet. This is, essentially, a form of built-in gutter, hence is fraught with all the attributes of that problematic detail.

By virtue of its exposure, the parapet is in the direct stream of air as it moves over the profile of the building as a whole. It is not unusual for air velocities over a parapet to be twice the ambient air velocity. At the leading edge of the building, the air moving up and over the parapet creates a compression of the air mass at the very top of the parapet and a corresponding zone of negative pressure at the leeward side of the parapet. The pressure differential across the thickness of the parapet is significant. Because of the relative compression of the air mass at the parapet, as compared to the mid-height of the wall, the pressure differential across the thickness of the parapet may be two or three times greater than the sum of the positive and negative pressures across the section of the building as a whole.

Depending on the exact profile of the coping and the direction of the wind, the effective pressure at the coping may be absolutely positive or negative. The absolute pressure at the coping is not nearly as important, however, as the differential between the pressure at the coping and the pressure at other locations on the interior of the parapet. For example, if the pressure at the coping is +15 psf (7.2×10^2 Pa) and the pressure at the exterior windward surface is +20 psf (9.6×10^2 Pa), the relative pressure of the exterior windward surface is greater than at the coping surface; therefore, actual migration of air and, potentially, any water trapped in the intermediate capillaries will be from the exterior windward surface toward the coping, even though the absolute pressure at the coping is positive.

To fully understand the relative movements of gases and liquids inside a porous mass such a parapet, it helps to look at the section of the construction as though it were a weather map, and to plot the lines of equal pressure. Lines of equal pressure are called *isobars*, and they provide a two-dimensional picture of the pressure "terrain." Migration of gases and/or fluids will move from higher pressure to lower pressure regardless of the absolute values of the pressure. Fluid will move from a location of +20 psf to a location of +15 psf because the fluid will flow from the location of relatively greater pressure. In our example of the parapet coping, the fact that the pressure is an absolutely positive value is not enough information to determine whether the coping will be pulling fluid into the joints or blowing fluid out of the joints. Having mapped the isobars on the section of the parapet it is entirely possible that the pressure gradients cause water to be pulled in at the leading edge of the coping and blown out the back edge of the same joint.

Some designers attempt to keep water out of the parapet using a drained cavity or a pressure-equalized chamber; in other words, they extend the wall section into and through the parapet, with the logic that if the section is good enough for a wall it is probably good enough for the parapet. This, unfortunately, is not sound reasoning. Because water at some time or other will be on either the front or the back surface of the parapet, the necessary and sufficient conditions for water penetration into the fabric will be present either on one surface or the other. The presence of a cavity at the core of a parapet wall will not alter the risk of developing a pressure differential between that cavity and one of the exterior surfaces. In short, cavities inside parapets are virtually guaranteed to accumulate water; therefore, they are not worth the effort to construct.

Another common response to the exigencies of the parapet is to completely flash the backside with a membrane. As with any membrane, the efficacy of the detail depends in large part on the perfection of the membrane. If the membrane flashing has been reasonably installed, there is at least a pressure differential-defeating barrier on one side of the parapet. If that same membrane is extended across the section of the parapet immediately below the coping, there is an added advantage relative to penetration through the coping itself. Such a detail, while generally prudent, does not make up for all the deficiencies of parapet design, but it does provide a viable starting point, at least with regard to moisture penetration.

The danger, of course, is that water not only will saturate the parapet material but will migrate into the wall below. This is easily avoided by physically separating the parapet from the wall below with through-wall flashing at the base of the parapet so that there is no reasonable path between the two regions. Unfortunately, often this is not done, much to the distress of the owners of the leaking building.

Water penetration is not the only interference to the performance of a parapet. Parapets are also outside the thermal envelope, which is polite way of saying that they are abandoned construction. The parapet is not alone is this regard, but there are not many projections of buildings completely outside the thermal envelope. Parapets, cornices, cold roofs, some buttresses, and overhangs all are treated as construction exceptions. They are not exceptions. They are all items of construction that are effectively freestanding pieces. From the standpoint of weathering, they may as well be items of lawn sculpture, for they are exposed to the elements without much, if any, benefit of temperature or moisture mitigation from the occupied zone. Whatever the ambient temperature is, in a relatively short period of time, that will be the through-body temperature of any of these items.

The long-term benefit of occupancy and, therefore, of internal environmental modification, meaning primarily heat, cannot be overemphasized. In terms of weathering, the only difference between a building and a comparable stack of unassembled building materials in an open field is occupancy; and the only difference in terms of durability between a stack of building materials in an open field and a comparable pile of raw materials that go into those building materials is that the raw materials are probably more stable and, therefore, more durable in their unaltered state. Occupancy, however, has certain detriments, primarily that people do damage to materials, surfaces, and structures. Still, surveillance, maintenance, and environment modification are all that stand between a building and a deteriorating pile of construction materials. When a building is abandoned, all the benefits of occupancy are removed, and the building passively responds to the elements the same as a rock outcropping, but with less stability.

The effects of unmitigated temperature changes, combined with hostile assault by water and pollutants, and amplified by wind loads are particularly telling at parapets. One of the characteristic problems associated with parapets is called *ride-off* or *parapet jacking*. If we stand back and observe a masonry parapet from a distance, there is a high probability that we will see a diagonal crack, starting at the corner of the building at the base of the parapet, and running up and laterally across the face of the parapet, until it intersects the coping some distance away from the corner. If we move 90 degrees around this same corner so that we are looking at right angles to our original position, we will also probably see a similar crack on that face. These two cracks converge at the corner. The lateral movement of a parapet is significantly greater than a comparable section of the field of the wall below because the temperature conditions are more extreme at the parapet. Remember, there is no modulation of those extremes by a more temperate interior, and the design of the parapet reinforcement does not take into account the first two items. The parapet, in the course of expanding and contracting, is constrained by the friction connection with the building below.

Once this characteristic fracture occurs, the continuing alternation of expansion and contraction has the effect of pushing the corner of the parapet off the corner of the building. Even though the crack forms for the same reasons and in the same manner as wall cracks, generally, one other detail accounts for the progressive migration of the corner off the building. Basically, the crack forms as the materials contract from being chilled. Once formed, the fact that the crack is on the bias means that there is mass bearing on the horizontal component of the crack, providing friction at the crack plane. There are also bits of crushed debris in the void of the crack. When the materials reexpand upon warming, the combination of the friction and the debris preclude the smooth closure of the crack to the original relative positions of the parapet segments. Instead, the friction and the debris act as obstructions to closure, and the expanding parapet pushes the corner wedge further toward the corner. Of course, the same action is occurring at right angles to the primary façade, so the combination of the two vectors tends to push the parapet off the corner of the building entirely.

Effective prevention of ride-off requires the preemptive recognition that the conditions at the parapet are significantly more severe than at the field of the wall, hence that the detailing approach must be quite different than for the wall. One approach is to dowel the parapet to the wall below with reinforcing rods grouted vertically through the interior of the parapet and into the wall. This is far more effective if the common horizontal wall reinforcement is also doubled in density, particularly within the vicinity of the corner of the parapet. This approach is one of resisting the forces that tend to push the parapet off the building with enough counteracting material to overcome the rupture of the fabric. Of course, this means that potentially corroding metal is now encapsulated in the parapet, which may itself become a source of deterioration in the parapet.

The other approach to any differential expansion problem is to build in releases. In parapets, this means constructing vertical slots in the parapet to ensure that any movement is not constrained. The slots may be filled or left open. If they are filled, of course, the material used as a filler is subject to its own set of aging problems. If the joint is left open, there is the risk that debris will foul the joint, providing the rigid obstruction necessary for pushing the corner off the building. And an open slot makes it easy for water to work its way down to the base of the parapet. In summary, neither of these two basic approaches is better than the other; however, either is preferable to doing nothing.

We also have to address the fact that not all parapets are at building edges. One such projection above the roof line of a building typically occurs at party walls of adjoining buildings. Party wall projections above the roof, unlike parapets at the perimeter of the building are of unequivocal theoretic benefit. The party wall projection of noncombustible material is an excellent device to restrict a fire on one side of the party wall from burning across the roofing and down into the adjacent building. Unfortunately, because of the characteristic water penetration problems associated with the party wall projection, the fire division aspect is often circumvented.

Typically, the party wall projection is constructed of the same materials as the party wall itself. In the classic urban rowhouse, the party wall projection and the party wall below are constructed of eight inches of brick. In suburban townhouse developments, the party wall and the party wall projection are often constructed of studs and fire-rated gypsum drywall. Neither was developed specifically for exterior conditions. When the party wall breaks the plane of the roof, the flashing condition at the party wall is typically a transverse condition; and to some extent, the face of the party wall projection is exposed to the elements on both side of the projection. This means that a party wall projection has flashing conditions and weather exposure on both sides; thus it was not a construction developed for an exterior condition. The party wall projection is vulnerable from the sides; and it is exposed at the top edge, which means that there will be a coping detail, which in this case must also be noncombustible and must be maintained by two different owners. This is a setup for problems.

We won't elaborate here on all the sources and reasons why this detail will leak. They have been covered in other sections in this book. In summary, the flashings, the faces of the projection, and the coping are all vulnerable to their peculiar problems, all at a location that must be maintained by two different parties who are not necessarily bound by mutual agreement or covenant. Although many owners in such locations do cooperate in the maintenance of party wall projections, many do not. And those who do cooperate, often do so without the awareness of the long-term implications or the violations of local fire code ordinances.

The most expeditious solution to a leaking party wall projection is to continue the roofing up and over the projection so that it is covered by the same roof membrane that covers the two adjacent properties. If the basic roofing is watertight, the exposed sides and coping will be protected by an equally competent covering. Because this solution is easy to execute, inexpensive, pervasive, and beyond the reach of casual inspection, it is routinely the option of choice among rowhouse occupants. The difficulty is that, even today, the roofing material most likely to be used in such remediations is BUR, and BUR is combustible. The rest of the story is obvious.

5.4.3 Decks and Terraces

Decks and terraces are special roofing conditions that are disproportionately problematic. They occur when the building designer or the occupant attempt to merge the concepts of roof and traffic, either pedestrian or vehicular or both. The results are usually unsatisfactory. The fundamental reason for the failure of trafficked roofs is the belief that the two functions are compatible. While they may be brought into peaceful coexistence, it is a tense peace at best and perpetually fragile. Many designers and building owners regard a flat roof as an unexploited terrace, the area below an exterior plaza as a below-grade room or building gone unexcavated, and a porch as something that can carry traffic above and an occupied volume below. Whether a roof is converted to a deck or an exterior paved area is converted to a roof or a combination of roof and deck, the problem is essentially the same. The temptation is to collapse the roof and deck construction into a single package, to save depth and cost.

PAVING OVER UNOCCUPIED AREAS

To further demonstrate the problems of combining roofing and traffic surfaces, we will digress for a moment to discuss road construction. There are two general approaches to designing a road. One is to build an impermeable mat; the other is to build a permeable mat. The standard asphalt or concrete highway exemplify two variations on the first approach; a gravel road or railroad bed, the latter. There are obvious limits on gravel roads, but there are obvious costs associated with impermeable highways. We could build roads out of anything and in any manner we wished but for the fact that they must endure rolling loads, which means that they will be subjected to cyclic compression.

The essential problem, then, is to prevent wheel or foot traffic from pressing the paving into the soil until the road eventually subsides into the ambient terrain. Soil is a plastic material, and as such can be deformed and does not resume its original shape. A wheel or foot traversing unconsolidated soil of sufficiently low-bearing capacity will sink into that soil. The resistance to further movement is a function of the degree to which the wheel or foot becomes mired in the terrain. It is not that we cannot move through a swamp; we can, but only very slowly and with very light loads. The purpose of any road or traffic surface is to reduce the subsidence of the point load into the terrain, to facilitate efficient traverse, by expending more energy moving laterally and less overcoming the resistance of the soil. This is not difficult to achieve. The solutions depend on three primary variables: the magnitude of the point load of the impact; the capacity of the soil to support that load without deformation; and the capability of the road to endure weather. Certainly, there are other considerations in the design of a road, such as cost and maintainability, but they are not determinative of the essential concept. In particular, today we are more cognizant of the environmental consequences of laying a road in a given location, but environmental compromise is another form of cost; it does not determine the detailed design of the road itself.

The mechanics of a point load on the surface of soil are, simply, a matter of the load sinking into the bearing material. The soil is partially consolidated and partially displaced. If the soil particles are loosely sorted (that is, they are not compacted into the densest possible arrangement), there are voids filled with either air or water, which under pressure can be collapsed. The process of squeezing the air or water out of loosely packed soil particles is called *consolidation*. When, for example, sand is loosely packed, the mechanism is to vibrate the particles until they nest into a denser arrangement. Ironically, if the sand is saturated with water, and the entire mass is vibrated, a point load will actually sink into the sand. A patented pile-driving technique pumps water into a vibrating, perforated pipe. The water pushes through the hole in the hollow pile and saturates the surrounding sand. The oscillation of a weight that is eccentric relative to the inside the pile causes the entire pile to vibrate. As it does so, it sinks into the sand, which is the purpose of the exercise.

For surface loads on soil, even if the effect is to consolidate the soil beneath the load, there is some subsidence of the point load into the soil. If the form of the load is a wheel, as the soil below the wheel is consolidated, and the wheel subsides relative to the surrounding soil, some of that soil will fall to the side of the wheel, some of it in front of the wheel, and some, to a lesser extent because the soil there is now consolidated, behind the wheel. As the wheel rolls forward, it must either climb back up on top of the soil in front of it or plow through the soil mounded in front of the wheel. Visualizing this from consolidation relative to the surrounding soil, we can imagine the compounding effect of the direct displacement of soil as the wheel pushes or plows through it. To overcome the accumulated soil in front of the wheel, the vehicle is constantly climbing uphill even as the net motion is horizontal. A similar action occurs with locomotives on steel rails. The rolling wheel of the locomotive creates a "wave" in the rail in front of the advancing wheel. The amplitude of the wave and the stiffness of the rail are significant determinants in the efficiency of the train/rail interaction. Much of the engineering of ties and rails goes to minimizing the rolling wave in front of the wheels.

When the load is a foot or a hoof, the direction of motion is much more up and down, at least relative to the soil. The forward motion executes even when the foot or hoof is stationary relative to the ground. The loss of energy from the properties of the soil is the result of a number of factors, which are different from those of the rolling load of a wheel. For one, the foot or hoof may have to overcome suction at the contact plane, and the push required to propel the body may dislodge or shear the soil behind, and lose traction. Actually, this happens with wheeled vehicles, too, as anyone who has been stuck in snow or mud can attest.

In the construction of roads that are little more than improvements to the basic soil, the objective is to gain load distribution without incurring displacement of the material. In gravel roads, larger granules are used, which tend to interlock with less sliding than finer-grained soils, thereby reducing lateral displacement of the bearing material. The depth of the gravel is adjusted to reduce the contact pressure between the subgrade and the gravel to minimize the tendency of the gravel to press into and become part of the soil mass. The general tendency of the gravel to blend with the soil below is called contamination of the ballast, and it is the primary cause of granular road failure.

As the fines of the soil contaminate the ballast, the mixture becomes more plastic, and subsequent rolling loads can more easily displace the gravel laterally. Eventually, the road develops a rut, or linear depression, in the ballast; then, instead of rolling on the surface of the gravel, the wheel comes in direct contact with the soil/gravel mixture. If the material is dry, there may be no consequence, but if the mixture is wet, the fines in the soil will behave almost as a lubricant relative to the gravel. One way to enhance gravel ballast construction is to confine the gravel layer from dispersing laterally. This is accomplished using what amounts to above-grade curbs, or "cribs," which contain the gravel between two retaining structures. When the wheel load compresses the gravel laterally, the curbs confine it, preventing displacement. Another enhancement is to use crushed stone, instead of river gravel, whose sharp edges tend to interlock and prevent lateral movement. Of course, crushed stone is much harder on tires. The depth of the ballast is an important factor in reducing the bearing pressure at the contact plane of the ballast and the subgrade. Lowering that value becomes significant in the reduction of contamination.

A direct outgrowth of the search for a better, inexpensive road material was what has come to be known as *filter fabric* or *geotech fabric*. Originally developed by the Celanese Corporation, this material consisted of a compressed mat of polypropylene fiber. The fibers were extruded into a deep pile of mounded polypropylene, which was then simply smashed into a mat. The mat stayed together without rebounding to the original depth of the loose fibers through an ingenious invention: perpendicular to the surface of the compressing plate were needles with barbed ends. When the plates compressed the polypropylene, the needles passed through the mat and through holes in the opposing plate. When the plates were retracted, the barbs snagged a convenient polypropylene thread and pulled it back through the mat as the plate retreated. The snagging and pulling of these occasional loops of fiber was enough to stitch the mat together. Much of the filter fabric on the market today is stitched in much the same way.

The history of the development of this material is interesting. The engineers at Celanese were approached by the Georgia-Pacific Company, along with other pulp lumber interests, to come up with a system that would enable them to get logging trucks loaded with pulp wood without sinking axle deep in the swamps of the southeastern United States. The logging roads were not precious enough to warrant all-weather hard surfaces, but the usual composition of temporary roads were sinking into the swamps. In retrospect, the solution seems simple, as so often is the case with great inventions. It was well known that the depth of gravel required to disperse the wheel loads on swamp soils was considerable, three feet or more. Putting the gravel in place was no problem, but as soon a load, even nominal, was applied, the fines immediately began contaminating the ballast. The filter fabric of compressed polypropylene fibers allowed water to pass through, and the mesh of the fabric prevented the fines from the soil from pumping back up into the ballast. Because the ballast remained clean, it continued to provide durable bearing capacity.

Though it took several years before engineers recognized the full merits of filter fabrics, once they did, they started using it wherever possible. Now filter fabric is routinely used to restrict erosion in railroad beds, under highways generally, in drainage layers and dry wells, below walkways, and around construction sites. However, the polypropylene was discovered to be UV-vulnerable, so more recently, other less vulnerable alternatives, such as polyester and nylon meshes, were developed for exposed conditions.

Filter fabrics notwithstanding, there are still limits on the viability of gravel roads. For faster, more efficient travel with heavier wheel loads, loose granular systems were not the answer. Long before the invention of filter fabric, another approach was implemented to overcome the vulnerabilities of the granular road, and that was the hard surface road. The hard surface road comes in two varieties: the stiff road and the flexible road. Both operate on the same two basic principles: first, to put a tough durable material between the wheel and the ballast so that the ballast is not dispersed laterally; second, to keep as much water as possible from percolating through the surface so that the subgrade is not saturated and, therefore, does not contaminate the ballast from below. This is a very different approach from that used in the gravel road and all its cousins, including cobblestones and Belgian blocks, which allow water through the paving.

The Romans built roads of multiple layers of fitted stones, which accomplished the objectives of the all-weather road, but by the eighteenth century, this approach was seen as far too expensive and tedious. Even before railroads, there was increasing pressure to move bulk goods more rapidly and reliably than gravel and dirt roads could support. One early attempt to seal a road against water was developed by a Scotsman, John L. McAdam (1756–1836), a surveyor, who experimented with binding the gravel into a water-repellant and cohesive surface. The system was so successful it became known simply as macadam; and roads treated in this manner were said to have been macadamized. The binder of choice was often tar, which when contracted with the word macadam became *tarmac*, a term that survives to this day, although it now commonly refers to the area where airplanes park and taxi.

But macadam was considered a luxury road surface in the nineteenth century, so several streets were paved with the less expensive and supposedly temporary Belgian block. Macadam survives today as the often maligned asphalt paving.

Asphalt paving is an excellent compromise of durability and cost, but as we discussed earlier, asphalt is inherently UV-sensitive. Any road paved with asphalt will become brittle and eventually crack. When that happens, what occurs is similar to the process of frost heaves and road failure, as the result of cyclic wheel loads, water, ballast contaminations and the rest of the roadbed failure scenario. Once the paving is fractured, water enters through the paving into and through the ballast to the subgrade, and is then pumped into the ballast. The rest should be familiar by now.

Historically, one of the problems with asphalt and gravel laid directly on ballast was flexibility. Two events in the 1950s occurred to counteract the flexibility problem. One was the growth of the concrete industry, courtesy of federal investment during World War II. The second was the enactment of a national commitment to a federally subsidized interstate highway system. The road that had moved the nation for 30 years was a reinforced concrete slab on compacted sand. The current generation of interstate highways is likely to have filter fabric between the sand and the subgrade, and they will almost certainly be reinforced with epoxy-coated reinforcement. Most of these reinforced concrete slabs have asphalt-wearing surfaces on top of the concrete. The asphalt is relatively easy to remove and replace, and it helps keep water out of the concrete as well. Another appropriation of the polypropylene mat was developed by Phillips Petroleum, called Petromat. Phillips took the common filter fabric, saturated it with asphalt, and embedded the mat between the topping and the concrete base and/or between layers of asphalt. It turns out that the same fabric that was good at filtering soil fines is also good at providing tensile reinforcement to asphalt paving.

Back to the subject at hand. Pedestrian paving and exterior decks and porches on-grade fortunately are not required to support truck loads; nevertheless, they are not immune to the problems associated with paving generally. Typically, for reasons associated with costs, paved areas designed for pedestrian loads are not prepared as well as the average roadbed. When a highway is built, considerable time and attention are given to the compaction and consolidation of the subgrade materials below the ballast, because in most common roadbeds, the existing subgrade is inadequate for long-term durability of the road itself. Therefore, soil is removed and replaced with clean, compacted fill. No organic material is allowed in the subgrade; excavation continues until all residual organic material is excavated. The clean fill is compacted with rollers, then inspected by geotechnical engineers for density before the ballast is laid.

In contrast, with pedestrian paving, three presumptions are at work: one, that the loads will never be of a magnitude to justify the expense required for controlled subgrade development; two, that the consequence of a depression in a sidewalk is minor—that someone may someday step into a puddle formed in the depression; three, that money is considered better spent to fix a problem than to prevent it. Thus, pedestrian paving is generally poorly designed and constructed. Unless an owner, after a number of bad experiences, insists on a higher standard, designers will operate on these three presumptions and install inferior sidewalks and paving.

Pedestrian paving fails for the same basic reason that any paving fails: water enters the subgrade. The source of the initial wetting may be through lateral migration, even condensation. In the presence of water, the subgrade, which is often poorly compacted, may consolidate by self-weight. This results in a void below the paving, which the paving often bridges because of its relative structural capacity. If the void is nominal, it may persist for years without incident. If the void continues to grow, most significantly, in diameter, the paving will have to span longer and longer distances. At some indeterminate span, the self-weight of the slab will induce cracking. Probably though, long before the void expands to a critical dimension, some extraordinary event such as a passing truck will overload the spot and crack the paving. Regardless, the avenue of water entry is now advanced several orders of magnitude before the crack.

Once the water enters through the crack, it saturates the subgrade; the cracked paving subsides into the depression below the crack location. Any subsequent load, even footfall, pumps the water/soil mixture through the crack and laterally under the paving. The depression increases due to lost material; the bearing capacity of the subgrade decreases because of the saturation. The only significant difference between the failure mechanism of a heavy traffic slab and a light traffic slab is the prospect of the development of the initial void below the slab, which is far less likely in the case of a road.

One effort to prevent void formation and slab failure is to lay the paving in precracked segments. Loose pavers may be set in a bed of sand so that water will intentionally and quickly run through the cracks in the paving pattern and drain from under the slab by moving laterally through the granular layer to the edges of the slab. The concept of loose pavers for light traffic paving is valid; but, again, generally for reasons of economy, not all the necessary pieces are installed to ensure long-term performance. The concept of allowing or, at least, not resisting, the passage of water through the wearing surface is not that far removed from the gravel road. It can work, and it offers the prospect of easy maintenance. If a paver or stone becomes misaligned, the maintenance procedure is to simply lift the stone or paver, toss in some more sand, and tamp the paver back into place.

For loose paving to work, and work well, however, it is not enough to accept that water will percolate through the wearing surface. As with any water management problem, the water must be tracked all the way to where it is appropriately discharged from the site. Anything less rigorous is an invitation to premature systemic failure. The first step in the design is to assess the effects of water coming into contact with the paving material. The next step is to assess the effects of water passing into and through the joints between the pavers. If the intent is to allow water to pass, the joint material, if there is any at all, is presumably porous. The joint material of choice is often the same sand as the bedding material. If the joint material is different from the bedding material, chances are, one will intermingle with, and potentially contaminate, the other.

The porosity of the bedding material is even more critical than the porosity of the joint material. The intention of loose-laid paving is that water *may* pass through the joints; there is no requirement or invitation for water to pass through the joints. Once water reaches the bedding material, however, it *must* pass through the bedding material, and so it is desirable that it do so completely and quickly. The material most commonly used for the bedding layer is sand, whose properties make it the logical and economical choice. It does, however, have its limitations. If the sand is to provide even and firm support to the pavers, it must be compacted to at least 90 percent of its maximum dry density. If the sand is not adequately compacted, it will liquify when saturated under load, it will migrate, and/or it will compact. Any one of these behaviors will cause the pavers to subside, resulting in unevenness.

When the sand is compacted to provide adequate support, the percolation rate can be inhibited, especially when the sand is well-graded (which means that the distribution of grain sizes covers a wide range). If the sand will be used to produce mortar or concrete, well-graded sand is definitely preferred. When the intended use is as a drainage layer, however, well-graded sand is not a virtue; in fact, for this use, the narrower the grain size distribution, the better for drainage. Even when uniformly graded sand is well compacted, a substantial percentage is void; and that is desirable for maximum fluid flow through the medium.

Gravel, particularly small-diameter gravel, is preferable to sand in many respects, but it raises the issue of the compatibility of the joint material and the drainage layer. If the joints are filled with sand, and the drainage layer is gravel, water will carry the sand grains into the gravel, thereby eliminating the joint material and contaminating the gravel. This problem is, however, so easily remedied that it is probably not the reason gravel is not used more often. The remedy is to place a layer of filter fabric between the bottom of the pavers and the top of the gravel. (This is probably a good idea even if the bedding material is sand.) The primary reason sand is preferred over gravel is that sand is easier for the paving installers to work, to compact by hand, and to level.

Below the bedding layer or drainage layer is the subgrade. The better prepared the subgrade is, the more durable the paving. Subgrade preparation is to paving what wall preparation is to painting. The best paving materials on earth cannot compensate for poor subgrade preparation. The subgrade must be free of organic material, which, otherwise, will ultimately rot and produce voids under the drainage layer. The subgrade must be compacted, even, and free of hard and soft spots. It is also good practice to put a layer of filter fabric between the subgrade and the drainage layer to minimize contamination. When all this is done, the water will, indeed, collect in the drainage layer, and not pump fines into the bedding material.

But there's more, because the water has not left the site; the water is still under the slab. It is precisely at this point that even competent designers often stumble. The water must go somewhere. If it is not tended, the water will either accumulate or it will move of its own accord in one of three general directions. If it accumulates, the drainage layer will become saturated, and pressure on the top surface will pump water back up through the joints, flood the surface, and potentially push the joint filler up out of the joint.

As to directions, the water may move downward through the drainage layer, through the filter fabric if it is installed, and into the subgrade. If the subgrade can percolate the water at an acceptable rate, and if filter fabric is installed, this is a reasonable accommodation. If filter fabric is not installed, and the subgrade is not itself granular, the accumulation of water in the drainage layer will result in the subgrade being pumped into the drainage layer, leading to contamination. The water may move longitudinally under the slab until is reaches a low point in the undulation of the drainage layer, where the water will accumulate instead of at the point of entry. This is no consolation if the paving fails at a point removed from the source of the water. Finally, the water may move laterally toward the edges of the paving. Although this may not seem to be significantly different from moving directly into the subgrade below, it does offer an opportunity for detailing advantage. The edge of the paving is also where water from the top of the paving surface is likely to move when the designer has deliberately pitched the surface to one side or the other. The subgrade at the edge of the paving can be trenched, lined with filter fabric, and filled with gravel to provide essentially a linear dry well. Even more positive devices such as porous pipe can be placed in the trench. The filled drainage ditch need only be on one side of the pathway if the pitch above- and below-grade is sufficiently positive to that side.

Once the water has been collected, it must be transported off the site in a reasonable and appropriate manner. Appropriate water disposal is not an engineering or a technical term; it is a legal term and, on a broader level, an environmental term. Water disposal is one of the older issues addressed by property law. The basic rules are fairly simple: Rule One: If water flows from your land onto your neighbor's land in its natural bed, you are not responsible for the consequences to your neighbor; you are not responsible, for example, for the flood damage emanating from a stream that crosses your land. Rule Two: If water flows from your land onto your neighbor's land as a consequence of your alteration of the course of a natural stream, you are liable for the negative consequences; if, say, you accumulate water from a paved surface and dump it onto your neighbor's property, you are liable for the negative consequences. In sum, appropriate water disposition means that you eliminate water from your property in a way that precludes your liability to damages and is not harmful to the environment.

The relevant point of all this to road and paving construction relates to the problems of paving over unexcavated terrain. The paving must endure the wear and tear of traffic; it can be flexible or stiff, and it can repel water at its surface or allow water to percolate through the body of the paving. In fact, there are many ways to solve the exterior paving at-grade problem, but they all have inherent technical problems. To overcome those problems requires technical sophistication, attention to details, and keen appreciation of the deterioration mechanisms that affect paving.

PAVING OVER OCCUPIED AREAS

The conservative approach to designing an exterior deck over an occupied zone is to (1) design a completely competent roof, absent any traffic, and (2) then, and only then, design a competent, structurally supported traffic surface, surmounted on the roof, leaving ample space between the two constructions for unobstructed drainage. This means that the designer must deal with the duplication of construction depths, and the owner must deal with the duplication of construction costs. Designed and executed in this manner, there is no fundamental reason why the roof should not endure in a reasonable manner and for a reasonable period; it will not endure forever, and there is the compounding problem of servicing and maintaining a roof surmounted by a deck. Within limits there are reasonable ways to compress the section of a deck or plaza with an occupied area below; but every inch that the section is reduced complicates the construction and exponentially increases the risks and consequences of failure.

One of the more successful light-traffic deck solutions is to use solid pavers on a protected membrane roof (PMR) section. This is accomplished effectively by replacing the recommended two-inch ballast with a two-inch-thick paver block. One problem with this approach is that the pavers are loosely laid, and are, therefore, easily removed. On the other hand, the advantage of easy removal is that repairs are facilitated; the disadvantage is that pavers tend to disappear. When a PMR is over lain with pavers, and the roof is accessible only on a limited basis and for limited loads, this is, however, an inexpensive adaptation. It is not satisfactory at exterior entrances where prying fingers can remove the pavers, nor is it satisfactory for heavy wheel loads, which can stress the pavers eccentrically, crack them, and potentially puncture the membrane and/or crush the insulation.

The conceptual problem with trafficked roofs is that the exposed surface must be a wearing surface. This is typically viewed not as a problem but as an opportunity. The wearing surface is dense, impermeable, and tough, desirable properties in a roof as well as a traffic surface. The logic is that what makes a good wearing surface must make a good roof. To begin our examination of this theory, we begin with a hypothetical roof composition that has been adapted as an exterior deck. It has a reinforced concrete slab as the structural substrate. On the concrete slab is placed a liquid applied membrane. On top of the membrane is a mortar bed, into which are set terra cotta tile pavers. The tile set in mortar is a good wearing surface; the liquid-applied membrane on concrete is, at least, a reasonable solution to water exclusion, given the protection of the pavers. While this composition has some prospect of success, it will fail sooner rather than later, and the failure will be virtually impossible to remedy short of demolishing the entire wearing surface. The primary reason that the roof will fail is because, sooner or later, a crack will emanate in the concrete, and particularly in cold weather the crack will reflect through the membrane. Even though the membrane is flexible at the time of installation, it will embrittle. The bond between the membrane and the concrete will not arrest crack progression, and the crack tip will bridge from the concrete to the membrane as though it were a homogeneous material.

While the advancing crack is perfecting the pathway, water is delaminating the tile from the setting bed or the setting bed from the membrane, primarily through freeze-thaw action. Eventually, the pathway from surface to surface is completed, and differential pressure will supply the motive force to pull water through the paving and the crack to the interior of the volume. Searching for the point of entry is futile, and largely irrelevant, because there are probably multiple points of entry, any and all of which can supply the flaw in the membrane. Efforts to seal the top of the wearing surface, too, are doomed to failure, so the only reliable fix is to find and repair the flaw in the membrane. As with flat roof construction generally, finding the flaw is virtually impossible. Finding where the water is accumulating is not the same as finding the flaw. Sealing the underside of the concrete slab is not even conceptually close to a fix, because the pathway and differential pressure are still present. The best that an inside repair can reasonably accomplish is a momentary respite from the leak. Eventually, the owner will have to demolish the entire wearing surface, reapply the membrane, and relay the wearing surface. Nevertheless the crack in the concrete will still be present, and as soon as it flexes, it presents a risk of propagating across the bond and through the membrane again.

Based on this little tale of horror, it is tempting to blame the flaw and the failure on the composition of the deck, membrane, and slab; but additional examples will tell a similar tale as long as the materials are porous, rigid, brittle solids bonded to either side of an adhered membrane. Cracks will initiate in the wearing material and in the structural support. When a

FIGURE 5.16A This skylight was added on top of the original version to provide a side panel for ventilation.

FIGURE 5.16B With the cover removed, the profile of the original pointed-head skylight is visible. It is very difficult to glaze a panel into a frame and not develop a fine-line crack at the boundary plane.

flaw develops in the membrane, detection will be difficult; access is limited; and repairs almost invariably result in the demolition of the topping and complete reconstitution of the membrane.

5.4.4   Site Drainage Problems

Site drainage comprises a class of problems. Those of greater pathological consequences are associated with terrain and soil, more so than underground storm water and sanitation sewers. The difference is in the persistence of distress. A clogged sewer that backs up and floods can usually be repaired, and the incident is over, whereas a poorly graded exterior perimeter may direct water against an exterior wall every time it rains.

PLANT DAMAGE

It need not be said that plant roots migrate to moisture. The ability of fine root filaments to enter tight crevices is common knowledge. We also know that roots, particularly those of the woody, stemmed varieties can exert lateral pressure at the point of penetration to expand the root diameter. Where the underground conduit is a clay tile pipe, the root may not be able to initiate a crack, but, being hydrophilic, it will find a fine-line crack and exploit it. When a breach is made through the conduit wall or, more likely, through a joint, the root will grow a filagree of small, branching roots inside the pipe. The intermittent nature of water flow through the pipe is a perfect culture for root development. (If water were to flood the conduit all the time, many roots could not survive for lack of oxygen, but the intermittent nature of storm water is perfect for immersion followed by aeration.) Of course, sewers are even more attractive to plants because they provide water, plus nutrients.

The aggregation of roots inside a storm or sanitation sewer will continue as long as the supply of water continues, and at some point the root mass will begin to inhibit flow. As explained previously, even a minor obstruction to flow can precipitate a reduction in velocity and, therefore, lateral pressure upstream. Sewers may then leak from the topsides and other inactive cracks unless there is a back up of water in the pipe. These new sources of moisture also attract plants, and the process accelerates.

A clogged underground sewer makes itself known when water backs up into the occupied area, and floods the basement or some other low point in the building. If the sewer is a sanitation line, the situation becomes anything but sanitary. The response of the building owner or operator in the face of such an emergency is to engage a service that will ream the line with a device that consists of a spade bit at the end on a rotating, flexible cable. Unfortunately the power of this device is such that the solution can be as detrimental as the problem, particularly for clay tile lines. Once the basement is flooding, however, time is of the essence, and while there are severe drawbacks to the rotating reaming approach, it is quick, and it clears the line.

A less traumatic, but admittedly less reliable approach is to employ less radical means of lines purging. For one, the lines can be deliberately flooded in a controlled manner to allow the application of herbicides to the roots. Another alternative is to occasionally, but routinely, schedule reaming with lighter and slower bits to cut young rootlets without digesting the conduit walls. These preventive maintenance measures are, however, often set aside in favor of other operating priorities.

SITE-WATER DAMAGE

There are two basic classes of site-water-related problems. One is caused when water collected from around the building is improperly or ineffectively managed; the other is caused by water that falls to grade, away from the building, and is subsequently directed against the perimeter of the building. The first problem, mismanaged drainage from the building, has been directly and indirectly addressed intermittently throughout this book: We discussed perimeter water problems in Chapter 3 relating to hydrostatic pressure, and faulty leaders and drains earlier in this chapter.

The general mismanagement problem arises when concentrations of water are deposited at or near the base of the building. Sometimes this action is a conscious, but misguided or unwise, decision. For example, it is not unusual to see leaders coming off the roof and depositing the water into a splash block, which often is little more than a three- or four-foot-long trough directing the water from the leader away from the base of the building by its length. If splash blocks have any merit, it is as erosion-control items. When the water reaches grade after descending in an unobstructed leader, it may be moving 20, 40, or more feet per second. Even if the leader ends in an "elbow" that redirects the vector to a somewhat more horizontal direction, when the water actually strikes the soil, the velocity will still be more than enough to erode the soil at the point of contact. When the water hits the splash block, the velocity is quickly dissipated, and subsequent erosion or cavitation is minimized. Thus, used specifically for erosion control, splash blocks make sense.

As collection items, however, splash blocks leave the water only three or so feet away from the building perimeter. From the standpoint of building pathology, splash blocks are virtually useless. Where the building is constructed on piers, with a well-ventilated crawl space on granular soil— as may be seen in residential construction in some coastal areas—the consequences may be negligible, and for buildings constructed on slopes, splash blocks on the downslope side of the structure may not be harmful. But in almost all other locations, splash blocks are an illusion of competent water management. When water is deposited at or near the base of a building, the probability is that the water will saturate the soil at that point, and, therefore, wet the surrounding area, including the foundation of the building.

Water collected from the roof of a building must be led directly away from a building. If the designer wants to percolate that water back through the ground to the water table, that is possible, but it cannot be accomplished by assuming that any water deposited on the ground will seep into the soil without consequences to the building. Unfortunately, this is a very common design error.

The rate at which water will pass through the soil is called the *percolation rate*, or the perk rate; it is expressed in terms of gallons per square foot of exposed area per hour or minute (liters per square meter per minute or hour). Another way of expressing the same value is in terms of the number of inches of depth of water that will percolate through an area in an hour. If the soil is of a pure granular composition, such as gravel or sand, the perk rate is high; conversely, if the soil contains a large percentage of clay and or silt, the perk rate will be low. In either case, water will not pass through soil in a straight column. The water disperses in a pattern more or less resembling a cone or wedge as it passes through the soil. If the perk rate is high, and the water passes through rapidly, the slope of the cone is fairly steep. If the water passes through slowly, the angle of the dispersal pattern is shallow, but the affected area is greater. The precise rate of percolation on a given site is a value that can be tested relatively inexpensively. For a property owner planning to dispose of storm water in bulk on his or her land, the test probably will be required by local ordinance if not by state statute. For a property owner planning to dispose of water collected off the building, a perk test probably will not be required, though the information would be valuable to have.

Designing for Water Disposal

By itself, the perk rate is not enough information to formulate a plan of water disposal. The designer must also know the quantity of water to be dispersed in a given period. Suppose that the building for which the drainage system is being designed has 1,000 square feet (92.9 m²) of roof area, and that the design storm for the area is 10 inches (25.4 cm) of water in an hour. If the perk rate is two inches (5.08 cm) per hour, to keep up with the rain on a minute-to-minute basis, a percolation field five times the area of the building, in this case 5,000 square feet (464.7 m²) will be re-

FIGURE 5.17 This area drain is almost choked by a pollonia tree, which though it has been cut and poisoned, still returns. The bulge at the arrow is the remnant of the trunk. The root system in this case has filled the drain line. The removal cost alone could run almost $1,000.

quired, because the design storm is five times the percolation rate of the soil. A major caveat to that statement is that the calculation assumes that the percolation area is available exclusively for building runoff. In fact, probably at the same time we want the area to be absorbing the water from the building, the percolation area will be coping with the water being deposited directly on it by the same storm. This water we will call the *surcharge water.*

There are two general methods for designing around the surcharge water. One is to calculate the surcharge into the percolation problem; the second is to dispose of the surcharge in another manner, which, typically, means allowing the surcharge water to run off as surface drainage. In most jurisdictions, this is permissible unless the surface has been modified to reduce the permeability at the surface. In other words, it is not permissible to build a percolation field under a paved parking lot and declare the runoff from the parking lot as uncontrolled surface water. On the other hand, suppose that, in an area of moderate percolation immediately adjacent to a natural water course, we install a substantial dry well two or three feet below-grade, and we reconstitute the soil above the well with the original soil. In this design, it is permissible to calculate at least some the surcharge as runoff to the natural water because we have not created an area less permeable than what was there before.

Whether the surcharge is part of the percolation design or is disposed of as surface runoff, we are still left with the basic problem of calculating the area of the percolation field. We know that where the perk rate is lower than the design storm rate we will need an area greater than the area of the building dedicated to the purpose. If such a large area is available, and surcharge is not a problem, the design of the dry well will resemble a large tile field. The water will be directed either through pipe or through granular-filled trenches to an area large enough to cope with the collected water, at a rate equal to its accumulation. Unfortunately, this is a luxury seldom enjoyed in the real world.

Far more reasonable to assume is that the available area for percolation will be less than that required for coincidental collection and percolation. The storm will deposit water faster than it can be removed, in which case the excess can be stored temporarily and dispersed later. To continue our example, let's now say the building area and the perk areas are equal, 1,000 square feet, but that the storm rate is still five time greater than the perk rate. The differential is eight inches (20.3 cm) over 100 square feet (9.3 m^2) or 667 cubic feet (18.9 m^3); and that is the quantity of water that must be stored until after the storm. If the same 1,000-square-foot perk area can be used for storage, one solution is simply to allow eight inches of water to accumulate over the area of the percolation field; within four hours after the storm has passed, the pond will percolate. If the surcharge must be added to the area, the depth of the pond at the end of the storm will be ten inches each from the field and from the building, less the two inches that perked during the storm, or 18 inches. Thus, the total volume of retained water is:

$$1,000 \text{ ft}^2 \times 1.5 \text{ ft} = 1,500 \text{ ft}^3 \ (42.47 \text{ m}^3)$$

If this amount of water, that is, 1,500 cubic feet, were to be held in a cylindrical tank rather than on the ground, to hold that much water its size would have to be 15 feet high and almost 12 feet in diameter, not a small tank. One way to construct such a "tank" without actually building a true tank is to create an equivalent void underground and fill it with gravel. This is the aforementioned *dry well*. The gravel should be relatively large and of the same size to maximize the void. But even doing so, achieving better than 35 to 40 percent void is difficult. If the void volume is roughly 35 percent, the size of a single dry well would be 15 feet deep and 20 feet in diameter, and that is one big hole to fill with gravel—about eight large dump trucks worth, as a matter of fact. Therefore, a dry well of this volume is probably better constructed as several dry wells, interconnected with conduit. Still, that does not solve the fundamental problem of accounting for water.

Let us review for a moment how we got into the underground tank business. The problem is composed of three factors: the intensity of the storm for which we design, the percolation rate, and the terrain available to discharge the water. If the design storm is smaller, the problem will be less pronounced, nevertheless the designer always has to include an overflow provision even when the system is sized to the design load. Seldom is availability of terrain an issue—either it is owned or it is not. That said, many owners think that they have more land available than they do. For example, if the anticipated recharge or dry well area is in a wet land zone or a flood plane (particularly, a five- or ten-year flood plane), local ordinances and/or state statutes may restrict the use of flood planes and wet lands as retention or recharge areas on the theory that when owners need them for disposing of roof drainage, those same areas will be doing the same thing for the flood water.

The variable about which the least is known, or even is reasonably knowable, is the percolation rate. The perk rate is not something a layperson or even a practiced engineer can determine simply by looking at the ground. Qualitative information of how well the surrounding area drains water is not enough, a fact the county engineer will mention when the request for a permit is filed. A test conducted within reasonable proximity may suffice, but the preferred method is to conduct the test at the location of the proposed drainage area. The test is simple: dig a hole; fill it with water; time how long it takes to drain in total or in part; divide the time by the available percolation area, taking into account that some of the water migrated through the walls of the hole not just the bottom. The results of such a test are reasonably reliable, and they eliminate the guesswork. Some owners resist spending money on site testing, but this is some of the best money they can spend on any project.

There are countless variations on the configuration of dry wells. We can dig a great big hole and fill it with gravel, or we can dig a series of smaller holes and fill them with gravel. We can sink conventional perforated pipe, as would be used for a wet well, but backflood it instead. The contemporary wet well is a pipe driven into the ground by a device closely related to a pile driver. The pipe is perforated along its length so that when the driller reaches a saturated strata, water will flow into and fill the pipe. In most cases, the pipe itself can serve as a reservoir. An eight-inch diameter pipe can hold over 2.6 gallons per linear foot. If the well is driven or drilled into the saturated strata 40 feet, the residual storage capacity of the pipe alone is over a hundred gallons. If the percolation rate at the well is high, and the pump is reasonably robust, the well may recharge almost as fast as the pump evacuates the pipe. In that case, an above-grade reservoir is hardly necessary. In most cases, however, the well owner will also have an above-grade water storage tank or reservoir, and when the storage tank gets low, engage the pump to refill the tank, rather than supply the end user directly.

The point here is that this same device can be filled with water above the water table, then recharged to an area as if it were a dry well, but relying more on the vertical surface of the pipe as a percolation surface rather than the area of the bottom of the well, which is, after all, negligible. This same sort of pipe may be laid more or less horizontally, instead of drilled vertically, and accomplish essentially the same function.

The term *French drain* is given to a device similar in principle to the dry well. A hole is filled with granular material that serves as a holding tank while the water percolates out the bottom. The French drain is normally considered to be a gravel- or boulder-filled trench or ditch. At reduced volumes, the trench acts as a simple conduit; at higher volumes, the trench acts as a retention device; and at even higher volumes, the trench, if it connects in the end to a natural water course, can serve as the overflow conduit as well. Ingenious uses have been made of French drains below pathways and sidewalks. The same granular layer recommended below paving is converted to a trench that is filled with the same material. In this way, it acts as a French drain and as ballast for the paving at the same time.

The preceding comments segues to one of the detriments of dry wells and French drains: they become contaminated. Therefore, routinely now dry wells are lined with filter fabric below where they are charged with granular material. Because the direction of fluid flow is from the well into the soil, the risk of mass movement of fines into the gravel from the surrounding soil is not high.

The primary source of dry well contamination is from the runoff itself. The storm water carries with it bits of debris, leaves, bugs, and dirt to the well. Over time, these solids accumulate in the voids of the granular materials, and foul the well. When the top of the well is exposed at its surface (typically the case with window wells), we may need only to purge the top several inches of the granular material. But for large below-grade dry wells, we usually must unearth the well, remove the gravel, and recharge the well with fresh, clean granular fill. This process is, however, more than most homeowners want to tackle, so often it is deferred until the feeder lines back up and overflow.

Let us now suppose that the property is conveniently located near a municipal storm water sewer or a natural water course. Do not assume that runoff can be directed to the street or to a convenient stream. Linking to an existing storm water system usually requires a permit; and often, today, it will be determined that the system cannot accommodate additional water. Not so long ago, typically, the sanitation and storm water systems operated on a combined system. In larger and older cities, many systems still are not completely separated. Where the disposal of storm water is combined with sanitation water, the concern is that the storm water will carry untreated effluent into a natural water course during a major storm, because the treatment facility cannot handle the combined water. Even where the systems are separated, storm water has been found to be not contaminant-free. One way to regulate the contaminants is to limit the runoff, or at least that is the theory.

The same general concern may impact an owner's options for discharging to a convenient natural water course even when that stream or creek may be wholly within the owner's property. Thus, the proximity of a municipal storm water sewer or a natural water course is no guarantee that the property owner will not have to manage storm water. The difference between retaining water for purposes of percolation and for purposes of subsequent discharge into a sewer is minor. If the purpose of the retention pond or dry well is to hold water until it percolates into the soil, the surface area of the percolation area is important; and the only outlet from the retention structure is for occasional overflow, the location of which is at the top of the retention structure. If the purpose of the retention structure is to slow down discharge, the percolation area is more or less irrelevant, and the structure may even be lined to preclude percolation. There may still be an overflow outlet at the top of the structure, but the primary device for discharge is a regulated outlet at the bottom of the structure. Operation of the lower port is how the rate of discharge out of the retention structure is regulated.

WATER COLLECTION AND CONVEYANCE

The second major class of site water management is the collection and conveyance of water that does not originate from the buildings or paved surfaces. A good deal of water enters the site either from direct precipitation onto the open area of the site or from off the site altogether. The source of the water must be of interest to us because we may want to mitigate its impact on the building at issue. As made clear throughout this book, water is a necessary condition of many deterioration mechanisms. So, when water is present on a site in uncontrolled and, therefore, unpredictable quantities, moving unpredictably, it should concern us. Static water will saturate surrounding soil, and to the extent that the underground water will migrate toward a building, it may enter the building at- or below-grade. If water is moving across the site, even if the building is not currently in its direct path, there is always a risk that the direction will change or that the depth will increase enough to endanger the building.

When it comes to site drainage, the investigator must assess site water management based on the drainage pattern as well as on how that pattern might change over time and/or how it will alter when the system is overloaded. In other words, site drainage systems are under constant threat of erosion and flooding; it follows then that site water is a constant actual or incipient threat to a building.

Water moves toward a building for two basic reasons. The first is the most common and the most obvious, that the building is simply in the path of water. The second major reason is that a building is often over a hole or void in the ground. It, therefore, represents a point of relatively lower pressure into which water will flow. The first of these two reasons is a function of moving water; the second is a function of static water.

Moving water problems can be classified either as surface water problems or below-surface water problems. Surface water problems as they relate to buildings are very simply a result of the building being at a lower elevation than the source of the water. Whether the source of the water is from off or on the site, the physics of the situation are the same. The fluid flows from a higher elevation to a lower elevation, and a building gets in the way. It is not surprising that one of the more obvious and common sites of water problems is where the building is constructed on a slope. By definition, a building on a slope has a high side and a low side, as well as a continuing upward slope above the high side.

The consequences of building on a slope are easy to explain. Water runs downhill. If that water meets an obstruction, the momentum of the fluid is interrupted; pressure is, therefore, exerted against the obstruction. The consequential forces on the structure are of two general types. First is the overall structural load on the obstruction: the moving mass of the water will effect a translation force on the structure, resulting in the tendency of the building to follow the same path as the water down the slope. To a lesser extent, the vertical location of that translation force may also exert a rotating force on the building, resulting in a tendency for the building to slide, even tumble, down the hill. The two factors that combine to produce this force are the mass of the water and the velocity of that mass at the time of impact. Opposing the external force is the rigidity of the building and the degree to which the rigid structure is attached to an unyielding base.

The water at issue can move on or through the ground, but the velocity of moving water is likely to be far greater when it is moving over the ground, as in a flood. The mass of the water can also be extremely great in a flood, but this fact should not overshadow the considerable impact of moving ground water, whose velocity is low, but whose mass is considerable. Because the velocity of moving ground water is so low, for analytical reasons, it is often considered to be static, and the pressures resulting from it are ordinarily assigned to the second category of forces, namely hydrostatic forces. This is understandable, however, because except in, for example, mudslides in southern California, this phenomenon is rarely dramatic. More commonly, the buildings gradually yield to these forces, manifesting in the form of deformations and distortions, rather than gross and catastrophic translations, particularly when combined with hydrostatic pressure.

The available approaches to intervening in moving water situations is an excellent opportunity to apply the intervention matrix because there are, generally, several general and specific intervention options (see Table 5.4). The necessary and sufficient conditions for dynamic impact of water are (1) water, (2) a gradient for the water to traverse, (3) an obstructing surface or object, and (4) direct impact of the moving water with the

Table 5.4 INTERVENTION MATRIX: DYNAMIC IMPACT OF WATER

Necessary and Sufficient Conditions	Intervention Approaches					
	Abstention	Mitigation	Reconstitution	Substitution	Circumvention	Acceleration
General Mechanism	Accept the mechanism and the consequences.					Abandon the site.
1 Water		Dig diverter trenches and/or dry wells and retention ponds.			Redirect the channel around the property.	
2 Gradient		Level the terrain.			Build a dam; conduct the water across the site in a conduit.	Dig a channel through the terrain.
3 Obstruction		Streamline the impact profile.			Build an intercepting obstruction.	Remove the obstruction; demolish the building.
4 Direct impact		Build moats around base of the building.		Reinforce the structure to withstand the impact.	Backflood the system.	

obstruction. This last condition may seem minor, but its inclusion illustrates some important points about the assembly of the list of necessary and sufficient conditions.

Let us review the options in the matrix on the basis of their individual merits and feasibility to the topic at hand: moving water. The abstention, "do nothing," option is always worthy of consideration because it is both the cheapest and easiest option, and it provides a baseline against which to measure the efficacy of other options. In this case, there is an element associated with the mechanism that we have heretofore not developed: The nature of dynamic impact of water is that it is sporadic, in contrast to most deterioration mechanisms, which though they operate more or less in this sporadic fashion, not with the obviousness of a spring flood. Thus, what we decide to do may amount to temporary measures. A perfectly viable response to a moderate site water problem may be simply to rent a sump pump when the event occurs, and clear the basement after the flood—even if that means doing so every spring. Likewise, it may make sense to pile sandbags along a levee every 15 years or so to combat a flood.

Where water is concerned, one other element must be considered before taking the nonintervention option if the building at hand is of historic value. When the fabric of the structure itself is deemed to possess unique and intrinsic value, it cannot be replaced. No insurance policy can reimburse the owner for its loss.

The more tangible options in the matrix of Table 5.4 are several mitigative options, more so than with other deterioration mechanisms, though as with all mitigative options, the basic process is to reduce the damage of any single event, to prolong the tenure of the structure.

Opting to construct dry wells and retention structures on the property presumes that the necessary area is available. The physical limitation of this approach is that it tends to be totally effective at eliminating impact up to the holding capacity of the wells and ponds, then the structure absorbs the full brunt of events that exceed that capacity. The building is completely protected from the pesky little floods so it can be at full strength when the big ones hit. Consequently, this approach has very little technical merit relative to the durability of the structure because the damage from one major flood will be more than the sum all of the small floods on record. Thus, what is being mitigated are the costs of cleaning up after the flood, not the structural integrity of the building.

Terrain modification can be a very effective method of interdicting and redirecting flood water, but it is rather expensive, perhaps with some costs hidden. Constructing levees and dikes, for example, are classic modifications that are, indeed, effective at mitigating the damage of raging rivers. A levee is a raised embankment along the edge of the water course, behind which surface water accumulates, keeping it from running directly into the water course. The trade-off is that land adjacent to the river is less well drained because of the presence of the levee. And when the levee is constructed "naturally," meaning that the immediate bank of a river is higher than the terrain inland from the river, water can accumulate behind the natural levee and conform to a water course independent of, but parallel to, the major water tributary. This is called the Yazoo effect, named after the Yazoo River in Mississippi. The Mississippi River has thrown up enough sediment along its banks that the area inland is actually lower than the top of the bank itself. Water accumulates behind the bank and begins to flow behind the naturally occurring levee. If that flow line becomes large enough to classify as a permanent tributary, then that is a Yazoo.

Leveling the terrain on one segment of land usually compounds the problem somewhere else, however, most likely downstream. The benefit of the approach is that as the gradient decreases, velocity decreases; as velocity decreases, momentum and impact decrease, and the building endures less stress. But this benefit has subsequent inescapable geometric consequences to the land.

Leveling a segment of land along what is otherwise an unbroken slope requires either cutting, filling, or both to form a single earth-moving operation. The line that represents the slope is the trajectory of the unaltered path of the water. As water descends the slope, the velocity increases, due to the influence of gravity, at the same time it is restrained by frictional contact with the surface. If the line is leveled out for a short distance, part of the line segment is depressed below the original plane, or it extends above the original plane, or some of both. There will be three bends, or changes in direction, for total cut and for total fill, and four bends for cut and fill. Every bend represents a change in the force vector acting on the water. Wherever the slope angle steepens, the water accelerates; where it flattens, the water decelerates. When the velocity is minimized across the flat segment, the depth of the water increases. So, we may manage to slow down the water across the site, but at the expense of increasing the depth. Furthermore, some section of the hillside is at the base of a veritable waterfall, which means increased scouring and erosion. If the plan is to just cut, the zone at risk is at the upslope side of the cut. If the plan is just fill, the scouring occurs at the downslope side. The point is, cuts and fills may mitigate conditions at the building site, but the modification may essentially transfer the problem up- or downslope.

Shaping the building to mitigate the impact of moving water may seem a bit far-fetched, but it is certainly possible. If the problem is known beforehand, and is unavoidable, a fairly simple siting option may be to rotate the plan of the building to point a corner of the building in the direction of the approach of the water. The corner acts as a prow, and the building "sails" through the flood. Such an orientation, obviously, has to be balanced against other considerations, but there are distinct structural advantages to this approach.

Digging a moat or trench at the base of a building, particularly at the base of the upslope side, can relieve pressure from moderate amounts of moving water. Much of the resistance to the dynamic force of the water is effectively transferred to the upslope retaining wall, thereby relieving some pressure from the building. Unfortunately, as with wells and ponds, the device is effective only as long as it is not filled. Once the moat is flooded, the upslope retaining wall is obviated because water is now on both sides of that structure, applying full force against the base of the building.

Addressing the building to more successfully withstand the pressure of the water is arguably a form of substitution. How substitution or reinforcement is accomplished depends on the form of the distortion or deformation manifested by an unreinforced specimen. If the problem is one of sliding, the intervention may be some form of deeper or repeated anchorage. If the problem is overturning, bracing and/or tying may be appropriate. The weakness of all substitution options and reinforcement principles, generally, is that they are valid only up to a limit, then they fail. If that limit is high enough and economically viable at the same time, resisting forces can work. Where freeze-thaw cycling is involved, we can fairly accurately predict a worst-case scenario and design accordingly. But when it comes to flooding, there is always a distinct statistical probability that the water rise will be higher than almost any reasonable design assumption. And design costs rise concomitantly for every added foot of water depth and/or velocity increase. Simply put, building a structure that can withstand any environmental condition is not practical.

The options for circumventing the problem of moving water include both some of the more commonly used and the most far-fetched of the entire matrix. The redirection channel is an effort to accomplish in plan what cutting and/or filling accomplishes in section. Lateral redirection of water on a site can be as simple and subtle as a simple swale or as complicated as an artificial river bed. The idea is the same: allow the water onto and across the property, but not near the building. If that is not enough coercion, we might dam up the water and pipe it across the site.

A simple, inexpensive, and remarkably effective device is a chevron trench, a slit in the ground upslope from the building. Both the surface runoff and the shallow ground water flow into the trench. The trench is cut on the bias relative to the slope so that as the water descends the slope of the trench, it is also moving laterally on the site. If the building is located in the lee of the chevron, much of the water that may otherwise have impinged the building will be carried beyond the lateral limits of the building. When the water emerges from the trench, it will be far enough from the lateral limits of the building to pose no threat. Elaborations on this approach include lining the trench with filter fabric and filling it with gravel, and/or installing a porous pipe drain in the bottom of the trench and leading it beyond the building. As with many of the site alteration options, the upslope chevron trench presumes dominion over and occupation of the site surrounding the target. This is certainly not always the case.

When immediate control of land, particularly the upslope side of the property, is not possible, and the actual dimension of the property available on the upslope side is severely constrained, it may be necessary to consider other forms of land interest to accomplish the preferred intervention. One of the older and still effective interests in land is an easement. An easement is provided by the owner of the upslope property in exchange for consideration, normally money. Once the easement is perfected, it travels with the title to the encumbered land. If, for example, the installation of a chevron trench on a neighbor's land causes little or no disruption to that neighbor, the price of the easement may be modest, but the benefit to the easement holder is as great as if he or she held title to the adjacent property.

A somewhat exotic but interesting option, which was discussed in substantive detail relative to hydrostatic pressure in Chapter 3, is to backflood the structure. Though the building is as flooded as if the water had come down the slope, the cleanup is measurably less difficult, and the long-term damage to finishes is significantly reduced.

In addition to dynamic impact problems, there are always static pressure problems. While the source of the hydrostatic pressure was not discussed in Chapter 3, all of the consequences and intervention approaches were. Here, it is important to emphasize that hydrostatic water pressure is the source of numerous and major deterioration mechanisms.

6 The Active Systems

We have relegated all of the active systems to a single chapter, for reasons more pragmatic that conceptual. Each active system is a potential book in it own right, and it is beyond the scope of this one to do more than comment on the active systems. But the comments that comprise this chapter reflect the salient observations of 30 years of practice, and raise issues not commonly discussed in other books on the topic. Thus, the sections here are not as conceptual as those dealing with the passive systems, and are considerably more anecdotal.

The active systems include:

- Utilities systems
- Environmental modification systems
- Protection systems

6.1 Utilities Systems

The power and the hydraulic systems are the most pervasive of the utilities systems, but they are not the only ones found in buildings. In addition to power, meaning electrical service and distribution, and water, other utilities may include compressed air, vacuum air, natural gas, acetylene, butane, propane, oxygen, nitrogen, and carbon dioxide.

All utilities have as a common analytical basis three elements:

· Supply and distribution
· Application
· Collection and disposal

Failure of any utility can occur at any one of the three portions along the pathway of the utility. When failures occur at the point of application, the problems are usually easy to identify and correct. When failure occurs in either the supply or collection branches, it may develop slowly enough that it does not disrupt the utility, and consequently, either may precipitate considerable damage prior to detection or potential damage in the form of explosions and fires.

When dealing with any utility, it is extremely important to assume that failures *will* occur, hence to be vigilant. All utilities can be pressure-tested for even trace leaks. Electrical circuits can be tested for stray currents, which are the electrical equivalent of leaks. In a recent case involving the re-installation of a gas-fired boiler in an aging church, the gas company found that at least 5 percent of the gas passing through the gas meter was leaking out of the pipes and into the building. In defense of the gas company, the last date the system had been pressure-tested was the last time any significant work had been done on a section of the gas system, more than 20 years earlier. In the interim, fittings had opened and seals had cracked. The leaks were not, however, severe enough to attract attention, and the building was sufficiently—and luckily—ventilated to prevent an accumulation in any one area. Aside from the danger posed to the occupants and the building, the parish had been paying 5 percent more for gas than was being used.

A rule of thumb for utilities systems is that the greater the risk to people and property, the more frequently the utilities need to be tested. If the building is an empty warehouse of no commercial or real estate value in an unpopulated location the utilities may not need to be tested regularly. Every other building in the inventory will benefit from routine testing of the utilities.

6.1.1 Power Supply

The power system of choice today in residential construction is 110/220-volt AC service. Larger buildings with heavier demands have service connections with considerably higher voltage. If the building is nonresidential, and older than 40 years, the electrical distribution system is probably a replacement installation (see Fig. 6.1). If it is more than 60 years old, the service and panel box are probably also replacements. If the building is 100 years old or older, the current electrical system is at least third-and proba-

FIGURE 6.1   Note open junction box with splices exposed and dangling from the joist above. This a common item on buildings, and represents a significant potential source of ignition.

bly fourth-generation service, distribution, branch protection, and/or wiring. Before the first electrical wiring, the power source may have been gas, and before that, open combustion.

One lesson from the preceding comments is that power systems change with some regularity. In the nineteenth century, the primary changes came in the basic power source. In the twentieth century, the changes reflected requirements for better reliability and safety, both to people and property. In the twenty-first century, changes are occurring to meet increased power demands in the branch circuits. Residential power consumption is increasing, but total consumption is less of an issue than peak circuit demand.

In a contemporary dwelling, total power consumption is increasing for several reasons, only two of which are primary. First, the dwelling has more systems that operate with electricity, and they do so whether the unit is occupied or not. The sum of this increase is relatively small, however. The largest consumers of ongoing use are heating and cooling, refrigeration, clocks and other timing equipment, security and emergency systems, telephones and other communications equipment and humidification equipment. Though the devices in this category are growing, the power that these systems use is not the major cause of increases in power consumption.

A big jump in power consumption we can attribute to so-called discretionary power use. The number of devices in the category is growing, but more important is that the power draw on the specific end items is increasing. Though some items are becoming more efficient, such as the central processing units of computers, others, such as the monitor on the same system may now draw 1,000 watts. Refrigerators today are twice as efficient as in 1973, but people have more of them. We now have lightbulbs that are three or four times as efficient as their counterparts 30 years ago, but we install four times more of them in residential structures. And as the median age of the population goes up, the demand for increased illumination will continue to rise as well.

The major source of end-item power consumption, however, is a collective category—the growing number of items that operate on electricity. Everything from hedge trimmers to toothbrushes are electrically powered. Electric irons now consume up to 2,000 watts per use; hair dryers, 1,800-plus; sound systems, many thousands of watts; big-screen televisions, 2,000 watts and climbing. The kitchen alone is becoming a major power drain. The family entertainment center is becoming numerous individual entertainment centers. Recharging of portal electronics is beginning to be a measurable point of power consumption.

The only mitigating force on this increased power consumption is that the average family size is declining, for fewer end users per unit, but that is minimal on balance. And an even more costly prospect is that branch wiring will have to be replaced, even wiring in good condition because of the potential for a single user to load a single circuit with tens of thousands of watts.

Not only is the quantity of consumption high and climbing higher, but most of that power is going through one or two circuits. Ten-ampere breakers on a bedroom outlet is not going to be enough in the very near future. The outlet, the wire to the outlet, and the circuit breaker will all be replaced before 2020, and probably again every 20 years for the foreseeable future. As long as power is cheap, we will increase our use of it, so the distribution systems will have to be replaced.

This pattern of power change, to upgrade, has been going on for 200 years, earlier at intervals of 50 or so years, then accelerating to every 20 years (see Fig. 6.2). This probably will result in a cumulative assault to the

FIGURE 6.2 These panel box covers were not installed 15 years ago, which provides a notable source of risk in itself. Far more common at panel boxes is upsizing of replacement breakers because the "faulty" ones constantly were popping. Another abuse is to tape the breaker switch down so it will not pop.

materials and the structure of buildings. The building that has endured four, five, or six changes in power systems has had to bear that many assaults to the fabric of the structure, with each set of scars only cosmetically disguised. The holes and slots drilled and cut into the building are still there, whether we can see them or not, and the residual damage can be serious, even building-threatening.

6.1.2 Hydraulic Supply

Ironically, after dedicating much of this book to the problems associated with the entry of water into the fabric of a building and into the building proper, we now address a system whose objective is the deliberate introduction of water, and lots of it, into and through the building. Water is a necessity of life, and so easy access to it is imperative to our quality of life. But bringing water into the domicile is a relatively new concept. Until 200 years ago, indoor running water was a luxury available only to the wealthy. It became routinely available to the middle class in cities in the nineteenth century; even in some rural areas today in this country we find the well outside the premises.

As much as we expect it today, however, the truth is, buildings were, on the whole, better off when the water source was outside the perimeter of the walls. Aside from the convenience, there are of course very good reasons to bring the water supply inside. It is less vulnerable to freezing, which is a severe detriment to pump operation. It is less vulnerable to pilferage and contamination.

But having brought the water inside means that we also must collect and return the waste water to the outside. Every point of application, meaning every water outlet, requires a supply line, a control device, and a drain. All three elements of the complete loop are vulnerable to deterioration and failure. The obvious problem is containment. As long as the water stays within the prescribed conduits and containers, there is no problem.

Water as we know is a primary necessary condition for timber rot, for corrosion, for freeze-thaw cycling, for ionic dissolution, for hygroscopic expansion, and others. As emphasized throughout the book, the mere presence of unconfined water inside a building should be cause for concern.

The hydraulic system is a major cause of damage to the fabric of the building. Any flaw or failure in the system may increase the magnitude of the flaw, because the precipitating mechanism is progressive, and contaminate other systems, thereby potentially jeopardizing the integrity of the entire building. In one notable example in North Philadelphia, a house was abandoned but the water was not turned off. Subsequent inspection revealed that a water supply line in the basement had burst, probably from freezing. The water was spraying upward onto the first-floor joists. Later, the building was in a state of collapse because the floor joists had rotted and failed, resulting in an unbraced party wall that buckled and collapsed, bringing down the second floor and roof.

Another issue of hydraulic systems is that major sections of them are installed in concealed areas. This means that failures may go unnoticed until it become clear from secondary or collateral damage. Some years ago, a shower liner developed a pinhole leak that went undetected until it had caused a thousand dollars worth of ceiling damage and another thousand dollars worth of tile work to get to the leak. All three types of damage from concealed lines is exemplified in this example: the damage itself, in this case a failed shower liner; the collateral damage of the ceiling below; and the secondary damage caused by excavating the failed item.

Moreover, because so much of the hydraulic system is concealed, it is difficult to inspect. The slightest signals of problems, therefore, warrant concern and quick response.

6.2 Environmental Modification Systems

As stated at the beginning of this chapter, space here precludes comprehensive coverage of the active systems inherent to buildings. This limitation may be the most apparent in regard to the environmental modification systems. Other books, however, give this topic its just due. Victor Olgyay wrote *Design with Climate, Bioclimatic Approach to Architectural Regionalism*, in 1963. Though out of print, it is worth seeking out. Baruch Givoni has extended many of Olgyay's concepts in two more recent works: *Passive and Low Energy Cooling of Buildings* (1994), and *Climate Considerations in Building and Urban Design* (1998). For a complete inventory and discussion of how all the pieces of these systems work, there is Benjamin Stein's and J.S. Reynolds', eighth edition of *Mechanical and Electrical Equipment for Buildings*. In this section of this book we can only begin to touch on the subject at hand, and recommend further exploration elsewhere.

Environmental modification to buildings is ordinarily done in the name of human health and comfort. All of the books just cited are based upon the premise that the human occupant is at the center of the inquiry into the various conditions and systems. Olgyay and Givoni articulate the ways we can achieve health and comfort through the use of the site, the landscape, and architectural form and expression. Stein and Reynolds inventory the machinery and devices we can use to modify and control the interior environmental factors, again, with the occupant as the object of the exercise. All of these writers treat the building as a tool, which in conjunction with the site, the landscape, and the systems are used to accomplish the prime directives of occupant health and comfort.

Our perspective here is slightly different from that of from Olgyay, Givoni, and Stein and Reynolds in that we focus on the building not only as shelter, but as object and objective in its own right. We are not suggesting that we should subordinate human health, safety, and comfort to the welfare of the building; we are stating that the interests of the occupants and the interests of the building are not mutually exclusive. When the building is seen as nothing other than as shelter for the contents, it becomes a sort of machine, to be pushed to its limits, and sacrificed if necessary, to achieve the prerogatives of the occupants.

There are two general scenarios upon which the message of this book and others written in the same vein come to bear. First, there is the prospect that the total costs to the fabric of a building for a given activity are not understood and, therefore, are not taken into account relative to the value of the activity. This is essentially the scenario of the mismanaged, purely functional "container" building. Second, there is the possibility that the building comes to be valued for other than its original purpose. For example, a building originally in service as a warehouse becomes worth more divided into apartments. The new value may be a function of historic significance. When a building accrues values beyond its market value, the paradigm for maintenance and use of the structure shifts. Nowhere is this more apparent than in determining how a building's environmental modification systems operate and to what ends.

For the purpose of this discussion, we can divide buildings into three classes: buildings operated as short-run containers; buildings operated as long-run containers; and buildings operated as objects. The pathological consequences of these three basic classifications are reasonably predictable: short-run containers are consumed; long-run containers are maintained; and objects are preserved. The first two are financial imperatives of investment and rent. Buildings that make money are well maintained; buildings that lose money are not.

There are two major exceptions to the market-driven situations. One is the third category mentioned, namely those buildings that have metamorphosed into *objets d'art* or, as are more often the case, *objets trouvés*. The other exception emerges from the vast inventory of buildings that either are not subject to market forces or which are subject to markets other than the real estate market. Let us deal with the latter case first.

The classic institutional building is difficult to analyze because the school, the church, and the town hall do not come into being by following conventional rules of real estate, and they are not normally operated in accordance with axioms of market analysis. The maintenance and operation of institutional buildings is derived from preferences and priorities several degrees of separation from the condition and/or deterioration of the building.

The *objet* building can occur within an institutional setting or grow from a property on the open market. In either case, the maintenance of the building can take on a life of its own. It is possible to actually overmaintain such a building. The most likely candidates for this behavior are buildings that have surpassed the normal considerations associated with utility and value. For example, the maintenance budget for Independence Hall in Philadelphia is enough to build a new comparably sized and designed structure every five years. That is not a criticism; it is simply proof that the value of the object building cannot be evaluated according to customary standards.

6.2.1   Heating and Cooling

Buildings can be crudely considered either as caves or tents. The cave is a massive construction, which, structurally, is a form of barrel vault. It is relatively stable thermally, easy to defend, and noncombustible. On the other hand, it is immobile, poorly ventilated, and has limited egress. The tent, though structurally more sophisticated, is such that all the members are in a state of axial tension and compression. The tent is well ventilated, readily transportable, and has multiple means of egress. The tent is, however, vulnerable to attack, and it is usually combustible. Humans have been working within these rough boundaries of habitation for as long as we have decided to come in out of the rain. For most of our existence, the structures we have found or constructed to stand between us and the elements have done little more than to serve the function of shelter. The type and form of construction tended to depend on the stability of food supplies, weather, and available materials. When food supplies were nonmigratory and multiseasonal, we built heavier more permanent constructions. When food supplies were migratory in conjunction with fluctuating water supplies, and/or the absence of suitable hard materials, we opted for more flexible shelters. In both cases, our choices were made for survival purposes.

The emergence of wealth as a factor to these choices has a fascinating relationship with architecture and construction. Historically wealth came from control of food and water sources. Particularly, in nonmigratory habitations, those in temperate climates, the domicile became heavy and permanent. Permanence brought shape and efficiency; ventilation was added. Wealth began to be seen relative to the dwelling itself. But concurrent with these developments emerged the problems inherent to matching human interests with nonhuman materials.

FIGURE 6.3 This boiler was originally coal-fired, and was later converted to an oil-fired burner. The entire boiler is encased in amphibole asbestos.

Heating and cooling systems are a prime example of where the principles of human interests versus fabric practicality can be at odds. When the temperature changes outside, there is a delay before the interior surfaces and interior air of a building are affected. That delay reflects the thermal mass of the structure. A structure can be "tuned" to some extent, to play off of its thermal mass so that the ambient interior temperature remains fairly stable. This may or may not coincide with the perceived ideal temperature for people, but it almost invariably works to the benefit of the structure.

The primary reason for this advantage was discussed in principle in Chapter 4 as it related to thermal gradients and internal stresses derivative of temperature changes. Recall that much of the internal stress that occurs within the fabric is a function of the magnitude of the temperature shift and the rate or length of time over which the temperature change occurs. The extreme outer skin of the building can be shaded or otherwise protected, but in the end, the weather will change according to its own rules with or without our intervention. Relative to whole buildings, the thermal gradient from whatever the extreme outside surface temperature is to the core of the wall through to the interior is very much a function of the thermal mass of the envelope. The higher the thermal mass, the longer the time required to dissipate or to saturate the thermal mass; therefore, the longer it will take to affect the temperature of the occupied zone. The advantage to the fabric of the building of such a situation is that the gradients tend to be less steep, and internal stresses to the materials tend to be lower. At the extreme outer surface, the stresses and the gradients remain the same, but beyond the immediate skin, the gradients drop sharply for a high thermal mass structure.

Thermally lighter structures act more as reflectors than as absorbers; the thermal mass is negligible. If the thermal mass is all that stands between the occupants and an unhealthy temperature condition, the occupants relocate themselves—and the tent as it were in some cases. In others, they find methods and devices to ventilate the lighter mass, to modify the inherent lack of thermal lag in the structure. The nature of thermally light structures is that they are subject to large degrees of thermal expansion, and those movements occur over relatively short periods of time. To absorb these movements, it follows logically that thermally light structures should either be highly elastic or highly fragmented. The former will result in low internal stress accumulation; the latter will relieve those stresses with reduced effect.

If the mechanical systems are designed to provide only for the comfort of the occupants, without regard to the nature and movement of the envelope, commonly, the selection of envelope materials and design details is antithetical to the environmental conditions acting on the envelope (see Fig. 6.3). The effects of these inappropriate decisions are mitigated relative to the occupants because the mechanical systems operate to the benefit of the occupants, but not necessarily to the benefit of the envelope. Insulation, for example, is implicitly viewed as a substitution for high thermal mass. If the costs of a thermally massive structure are deemed as too high, we build a lighter and cheaper structure, and regain the thermal properties of the more massive structure with insulation. But this, simply, is bad science. Thermal mass is a function of specific heat, which means the capacity of materials to gain or surrender energy. If the effect of high thermal mass is that the occupant remains cool during a blistering hot day, it is because the specific heat and mass of the assembled materials between the exterior and the interior were sufficient to absorb the radiant energy without apparent transfer of that energy to the interior. Insulation combined with a light structure has very low thermal mass, but it does offer significant resistance to thermal transmission. The net thermal gain to the interior may seem to be the same to the occupant, but the effects on the envelope are radically different.

6.2.2 Humidity Control

Similar effects accrue to a building's materials relative to moisture and humidity. There is an analogue to thermal mass which we will call moisture mass. Though not a widely accepted term, we can reasonably perceive its character and value. If the materials of the envelope have a higher than average capacity to absorb and retain moisture, we may say that such capacity is analogous to specific heat; and we will call this property the specific moisture of the material. There are, however, distinct mathematical limitations to this analogy, so we will want to press the issue beyond the qualitative level, but we do not need to, to make our point here.

If materials differ in their specific moisture, then the assembly of various materials will result in varying moisture masses, meaning that the total capacity of a wall, for example, to absorb and hold moisture will vary. If the exterior conditions suddenly become quite dry, even uncomfortably dry, a structure with a high moisture mass will surrender that moisture from all evaporative surfaces, including those on the interior. To the extent that the interior volume is enclosed, the humidity of the interior volume will drop far more slowly than the climatically driven humidity of the exterior. In effect, the high moisture mass of the structure has a mitigating effect on the moisture content of the interior, hence on the ambient-interior relative humidity, comfort of the occupants, and the moisture content of the finishes, furnishing, and fixtures.

Installing mechanical devices to stabilize the humidity of the interior air volume allows for the reduction in thermal mass of the envelope. But as the thermal mass of the envelope drops, it becomes increasingly important to maintain stricter segregation of the interior air volume and the ambient exterior environment. The lower the moisture mass, therefore, the higher the requirements for impermeability of the envelope.

The combination of light thermal mass and light moisture mass pushes the design of the envelope in the direction of the tent, specifically, in the direction of thin, perfect membranes. When drained cavities and pressure-equalized chambers are employed with light thermal and moisture mass concepts, there is often a design disconnection. The concepts are not antithetical, but they are not totally compatible either. Antithetical is the concept of low thermal and moisture mass and thick membrane construction.

Designers do not often consider the effects of the interior environmental modification in terms of envelope compatibility, consequently wind up unnecessarily sacrificing an otherwise viable envelope to the creature comfort of the occupants and benefit of the contents.

6.2.3 Ventilation

For thousands of years, the methods for managing the interior air mass of habitations included external combustion of fossil fuels for heat, the introduction of water features into the occupied zone for humidification, and various penetrations through the envelope for ventilation. In temperate zones, the need for humidification was seasonal at most, and often negligible. In damp zones, whether temperate or tropical, the use of deliberate humidification was not a priority. In humid zones, therefore, environmental modification was with heat and ventilation.

The available means of cooling meant primarily sensible cooling because latent cooling, meaning reducing humidity through induced means was not feasible. The basic approach to interior humidity control was to accept the ambient air moisture content as a baseline, and to the extent possible use ventilation to prevent accumulations of humidity in the occupied zone above ambient levels. This meant, of course, that some ambient conditions were hot and humid.

Ventilation as a humidity and temperature control device to this day generally benefits building materials and indoor air quality, but at the expense of close tolerance of temperature and humidity levels. For close tolerance control, mechanical devices are essential, but only if close tolerance is required. In recent years, the notion that we and our objects are better off with close tolerance conditions has directed interior environmental design.

For humans, the established desired set point is around 70° F (21° C) and 60 percent Rh. When energy is inexpensive the set point bifurcates, such that we actually prefer a higher temperature in cold weather and a lower temperature in hot weather. In fact, individual temperature and humidity preferences are as variable as for music and food. Likewise, the set point or points for objects is variable, based on changing forces and trends. The current theory is based less on set points, which are fragile and expensive to maintain, and more on modulation of the rate of change in temperature and moisture content. There is increasing recognition that the rate of change may be more damaging to fabric than the magnitude of the change. If that is true, then the mixing of air masses through ventilation as ambient conditions change may be economical, easier to maintain, and less vulnerable to failure without unwarrantedly jeopardizing objects and finishes.

6.3 Protection Systems

Protection systems follow the same track as mechanical systems, characterized by an increased reliance on sophisticated and costly equipment. We address protection systems in a book about building pathology because they fit the definition of other systems relative to deterioration. The failure of protection systems will contaminate and jeopardize other systems and, thereby, the building as a whole. The degradation of protection systems follows parallel patterns of small-scale deterioration, leading to gross deterioration, leading to systemic failure.

What distinguishes the failure of protective systems is what distinguishes the failure of many of the active systems: that systemic failure is signaled by a catastrophic event. As with the power systems, the signaling event may occur separate from the actual failure of the system. The contacts corrode long before the fire breaks out.

This book cannot begin to detail all the hardware and software related to emergency systems, a rapidly growing field. The purpose of this discussion is to focus on basic principles and common sense.

6.3.1 Security Systems

Physical security is a legitimate concern of any property owner. Physical security may include personnel security, property security, and content security. Each of these areas has different parameters, applications, and operational imperatives. First and foremost, it is important for property owners or operators to be clear in their mind what they perceive as the security risk before beginning to identify the security threat. We emphasize this point because property managers tend to rely on vendors of security devices and services for guidance, which, for obvious reasons, is a faulty approach. Far and away the easiest way to sell security is to scare the consumer. Threats exist; that is fact. Not all threats, however, are present at all times, nor are they directed toward all aspects of all the potential targets all the time. If the threat to property is exclusively at night, for example, then purchasing round-the-clock property security is a waste of resources.

This book primarily addresses threat to property and, specifically, buildings. Buildings are vulnerable to arson, which has two components: the incursion and the fire; vandalism, including from defacement, invasion by human and non-human intruders; and accidental invasion or damage from all sources. Other security risks tend to be threats to persons or objects.

Before addressing how to defeat each of these threats, it is appropriate to assess the building in terms of value and to factor in the role insurance plays in the overall security picture. Threat assessment is difficult to do log-

FIGURE 6.4 The archaic fire alarm is heat-activated, whereupon it theoretically releases compressed air, which will prompt an air horn alarm.

ically, because decisions in this area are often driven by fear, not reality. Still, the threat can be measured on two scales: frequency, and form of assault or damage. But in general, if the risk of threat is low, and the potential damage is also low, an appropriate response may be to purchase a casualty policy against the event and forgo active security altogether.

Moreover, property owners often have misguided ideas about what is and is not a legally supportable defense of property. This brings us to the second point here, regarding the development of a physical security plan. Any such plan should be reviewed by legal counsel. The local police department, too, should be consulted regarding realistic threats and responses to threats.

The plan itself should be based on a terrain map of the site, which identifies various points of threat and avenues of approach for each threat vector. Three aspects of each threat should be quantified: the deterrence potential, the detection potential, and the response potential. The map should show the specific avenues, or lines, of entry and exit. Different threat maps may be in order for different times of the day and/or different days of the week.

If the deterrence plan includes obstacles, such as fences, it is important to remember that no obstacle is more effective than the observation of the obstacle. Furthermore, it is imperative to recognize that all threats can not be defeated at the perimeter. Once an intruder breaches the outer perimeter, he or she must be further channelized into points of detection and response.

6.3.2 Emergency Systems

Physical security and emergencies systems are interrelated. The only difference between physical security and emergency preparedness is the vitality of the threat. Emergency threats normally include fire, flood, wind, and earthquake, and, depending on the terrain, mudslides or avalanches. In this category, similar methods of analysis of the threats are used relative to magnitude and frequency, direction, vulnerability, insurability, deterrence, detection, and response. As with physical security, there is a trend to substitute mechanical devices for human observation. But spending more money on deterrence devices does not obviate the need for detection and response.

Nowhere is this axiom more applicable than fire emergencies. Among those properties on the National Register of Historic Places that are destroyed each year, more will be lost to fires than all other causes combined. Fire is a constant threat from without, in the form of arsonists and natural causes, and from within due to human error and equipment failure. The necessary and sufficient conditions for a fire are: fuel, oxygen, and ignition. Eliminate any one and you eliminate the fire. Properties can be assessed and analyzed in terms of these three components. Given a fire event, we want to detect it and respond to it in the shortest time practical. Having responded, we want to ensure evacuation and suppression. To do this successfully, it is necessary to divide the event into as many elements as reasonable, and develop a comprehensive plan for each of those events.

Epilogue

··

T here are, as has been stated throughout this book, two fundamental
methods of solving building pathology problems. One is to match
the apparent signs and symptoms of the current condition with
those signs and symptoms of a known condition. The second method
traces the same set of signs and symptoms to underlying principles.

The first method has certain advantages and decided disadvantages.
The advantage of the comparative method is relative speed and, therefore,
relative costs of obtaining a diagnosis. If the signs and symptoms are distinctive and reliable and if the correlation with the known solution is very
strong, then the comparative method may be both accurate and efficient.
This method also carries with it a simplified approach to prescriptive interventions, because the intervention of choice will be whatever worked
with the known or control problem. If Problem x = Problem y, which is
known, and Solution A worked for y, then A will work for x. Diagnostics
and intervention are reduced to a decision tree with the characteristics of
the manifestation arrayed in list form. If the unknown condition x matches
all of the characteristics of condition y, then $x = y$. If A was, in fact, effective as an intervention in y, then it is not unwarranted to expect the same
solution to apply to the identical set of characteristics.

Manuals abound which assemble the lists of known problems in ways
which make it as simple and mindless as possible to solve the puzzle. At the
end of the description of the typical condition is THE prescription. Possibly the prescription includes a few variations which take into account a
corresponding number of subthemes. Certainly one advantage of this sort
of problem solving by comparative reference is that thinking is minimized.
Matching characteristics may not be mindless, but it is also not higher

order analysis. A second advantage and one not to be discounted in our society is that solutions to problems by comparative reference is a strong defense against negligence claims. The argument is that in using a reference manual we have exercised reasonable care, and unless we have grossly mismatched the characteristics, we have met the standard of care common to the respective branch of practice.

The disadvantages of the comparative method of diagnostics and intervention are fairly much the basis of this book. There are sources of mismatching the evidence which are fairly obvious. If the characteristics are misread or incomplete, the accuracy of the match is degraded. If the diagnosis is degraded, the prescription is degraded. Another and a not so obvious limitation of comparative methods is the prospect that the pre-existing list of characteristics leads the observer to rationalize the subject at hand to match the pre-existing condition. In other words, the investigator will find the characteristics of the known solution in the problem at hand precisely because he or she will look for those characteristics and discount countervailing evidence. The power of suggestion being what it is, there is a strong incentive to make a match by working rather more from the known to the unknown rather than the other way round.

Reality has an annoying way of being more complex than any set of lists will ever be. If the subtle divergences between the problem at hand and the model problem were always trivial, then the prescriptive solution, even if the fit is loose, would be reasonably accurate. The reality of field conditions is, however, that the variations between a general classification of the problem and the site specific problem is enough to invalidate the entire methodology. In other words, even if we identify the class of problem as predominantly one of rising damp, the range of appropriate solutions within that class is as great or greater than an overlapping range of options had we mistakenly diagnosed the problem as freeze-thaw cycling.

It is also the case that much of what we do as preservers and conservers, despite our sincere efforts to the contrary, is often not reversible. Once we commit to a solution, that solution is permanent. We cannot reverse ethyl silicate consolidation; we cannot reasonably reverse the insertion of a damp proof course. When the intervention is not on target, the solution may and often does become a secondary problem.

The alternative to the comparative method of diagnostics is to resort to first principles and build the story of the mechanism from properties and environmental conditions. No pieces can be discounted and solutions are inherently site specific. The advantage is that the solution is necessarily in harmony with the existing conditions. The disadvantages are that the data collection may be time consuming, expensive, and/or inaccessible.

Having collected the information alone is not necessarily the same thing as arriving at the diagnosis. Advocates of the first principles approach often operate under the illusion that the solution lives in the data. It follows, therefore, that when we cannot discern a solution, it can only be because we have not collected enough data or the data is inaccurate.

These are not trivial concerns; however, the risks of arriving at inappropriate solutions are less than the comparative method, albeit at the price of having to think and accept the personal responsibility for being wrong. Ultimately, the point of being a professional is, however, being willing to state a position with the full knowledge we could be wrong and accept that responsibility willingly. This is written for those among the professions willing to take that risk.

Glossary

absorption The process by which a liquid is drawn into and tends to fill permeable pores in a porous solid material.

abutment A surface or mass provided to withstand thrust.

adsorption The action of a material in extracting a substance from the atmosphere and gathering it on the surface in a condensed layer.

auger A handheld carpenter's tool for boring holes in wood.

autolysis The breakdown of plant or animal tissue by the action of enzymes contained in the affected tissue.

ballast Coarse stone, gravel, or slag used as an underlayer for poured concrete.

beam Structural member whose prime function is to carry transverse loads.

Bernouilli's Law States that the work per unit volume of fluid is equal to the sum of the changes in kinetic and potential energies per unit volume that occur during the flow.

blow count The number of blows required to drive an object into soil.

brazed connection Bonding of two pieces of metal using a second metal.

brown rot A fungus that destroys wood cellulose, leaving a brown powdery residue behind.

caisson A watertight structure or chamber within which work is carried on in building foundations of structures below water level.

cantilever A beam, girder, or truss that projects beyond its supporting wall or column.

capillary action (capillarity)   The movement of a liquid in the interstices of porous material as a result of surface tension.

catenary   The curve formed by a flexible cord hung between two points of support.

Charpy V-notch test   A single-blow impact test utilizing a falling pendulum, which breaks a specimen, usually notched, supported at both ends.

cleat   A small block or strip of wood nailed on a member or on a surface to support a brace or to hold a member or object in place temporarily.

coefficient of friction   The ratio of the force causing a body to slide along a plane to the normal force pressing the two surfaces together.

coefficient of thermal expansion   The change in dimension of a material per unit of dimension per degree change in temperature.

collar joint   The joint between a roof rafter and a horizontal member, which ties together two opposite common rafters.

colloid   A substance made up of a system of particles dispersed in a continuous gaseous, liquid, or solid medium.

compression   The state of being shortened by a force.

conduction   Transfer of heat by direct contact of one material with another.

consolidation   The process whereby particles are packed more closely by the application of continued pressure.

convection   The transfer of heat via a heated gas or fluid.

corbel   In masonry, a projection or one of a series of projections, each stepped progressively farther forward in height.

corrosion   The deterioration of metal or concrete by chemical or electro-chemical reaction, caused by exposure to weathering, moisture, chemicals, or other agents in the environment.

creep   The permanent and continuing dimensional deformation of a material under a sustained load, following the initial instantaneous elastic deformation.

deflection   Any displacement in a body from its static position.

dehumidification   Condensation of water vapor from air by cooling below the dew point.

delamination   Failure in a laminated structure, characterized by the separation or loss of adhesion between plies, as in built-up roofing or glue-laminated timber.

derrick   A hoisting machine for heavy loads.

differential settlement   Relative movement of different parts of a structure caused by uneven sinking of the structure.

dry rot   Decay of seasoned wood, caused by fungi capable of carrying water into the wood they infest.

ductile   Capability to be stretched or deformed without fracturing.

elasticity   A material property that will enable the substance to deform under load and will return to its original shape when the load is removed.

elastomeric   A macromolecular material that returns rapidly to approximately its original dimensions and shape after substantial deformation by a weak stress and release of the stress.

end-bearing pile   A pile principally supported at its point, which rests on or is embedded in a bearing stratum.

extractives   Substances in wood, such as colorants, oils, tannins, or resin, that are not an integral part of the cell structure and can be removed with solvents.

flange   A projecting collar, edge, rib, rim, or ring on a pipe or shaft.

flitch beam   A beam built of structural timbers bolted together with a steel plate that is sandwiched between them.

footing   The portion of the foundation of a structure that transmits loads directly to the soil.

friable   Easily crumbled or pulverized.

gambrel   A roof with two pitches on each side.

heaving   Upward movement of soil caused by expansion or displacement resulting from phenomena such as moisture absorption, the driving of piles, the action of frost.

homeostasis   Tendency of a system to maintain internal stability, owing to the coordinated response of its parts to any situation or stimulus that tends to disturb its normal condition or function.

hydrolysis   Chemical decomposition in which a compound is split into other compounds by reacting with water.

hydrophobic   Having little or no affinity for water.

hydrophyllic   A high capability to absorb water.

hydrostatic pressure   Pressure equivalent to that exerted on a surface by a column of water of a given height.

hygroscopic expansion   Expansion resulting from readily absorbing and retaining moisture from the air.

impulse   Force multiplied by time.

isolation joint   Joint between two adjacent structures that are not in physical contact.

isotropic   A material that has the same physical properties in all directions.

knee brace   A corner brace; a diagonal member placed across the angle between two members that are joined.

lintel   A horizontal architectural member supporting the weight above an opening.

liquefaction   To make or become a liquid.

malleability   The property of a metal that permits mechanical deformation without fracturing.

mat foundation   A continuous foundation under the full extent of a structure.

modulus of elasticity   The ratio of the unit stress to the corresponding unit of strain in an elastic material that has been subject to strain below its elastic limit.

modulus of rupture   Measure of the ultimate load-carrying capacity of a beam.

Mohr's circle   The sets of shears and axial forces that resolve to a single pair of principle stresses on the perimeter of a circle.

moment of inertia   Of a body around an axis, the sum of the products obtained by multiplying each element of mass by the square of its distance from the axis.

momentum   Of a moving body, the product of the mass of a body and its velocity.

parging   In masonry construction, a coat of cement mortar on the face of rough masonry.

peen   The end of a hammer head with a hemispherical, wedge, or other shape; used to bend, indent, or cut.

permeability   The property of a porous material that permits the passage of water vapor through it.

photosynthesis   Production of complex organic materials from carbon dioxide, water, and inorganic salts, using sunlight as a source of energy and with the aid of chlorophyll and associated pigments.

pile   A concrete, steel, or wood column usually less than two feet in diameter, which is driven into the soil to carry vertical load or to provide lateral support.

plasticity   A material property that enables a substance to deform under load but not to return to its original shape or position when the load is removed.

Poiseuille's Law   States that the volume rate of flow is inversely proportional to viscosity, directly proportional to the pressure gradient, and varies as the fourth power of the radius of the pipe through which the fluid is flowing.

polymerization   A chemical reaction in which the molecular weight of the molecules formed is a multiple of that of the original substances.

polystyrene   A foamed plastic weighing about one pound per cubic foot; low in cost and high in thermal insulation value.

porosity   A ratio, usually expressed as a percentage of the volume of voids in a material to the total volume of the material, including the voids. The voids permit gases or liquids to pass through the material.

radiation   The transfer of heat from one surface to another by means of electromagnetic waves.

section modulus   The moment of inertia of the area of the cross-section of a structural member divided by the distance from the center of gravity to the farthest point of the section; a measure of the flexural strength of the beam.

shear   A deformation in which parallel planes slide relative to each other so as to remain parallel.

sheet piling   A barrier used to retain soil or to keep water out of a foundation.

shrinkage crack   A crack caused by restraint of shrinkage.

soffit   The exposed undersurface of any overhead component of a building, such as an arch, balcony, beam, cornice, lintel, or vault.

soldering   Joining metals by fusion using an alloy with a lead or tin base.

spall   A small fragment or chip removed from the face of a stone or masonry unit by a blow or by action of the elements.

spandrel   In a multistory building, a wall panel that fills the space between the top of the window in one story and the sill of the window in the story above.

strain gauge   A very fine wire or thin foil that exhibits a change in resistance proportional to the mechanical strain imposed on it; used in the experimental determination of stresses.

strongback   A frame attached to the back of a concrete form to stiffen or reinforce it.

superstructure   Those structural forms and elements above the foundations of grade.

surfactant   Any surface-active substance.

tensile stress   The stress per square unit area of the original cross-section of a material that resists its elongation.

tension   The state or condition of being pulled or stretched.

thermal emissivity   The ratio of the rate of radiant heat energy emitted by a body at a given temperature to the rate of radiant heat energy emitted by a blackbody at the same temperature in the same surroundings.

thermoplastic   A material with a linear macromolecular structure that will repeatedly soften when heated and harden when cooled, e.g., styrene, acrylics.

torque   The product of a force and a lever arm that tends to twist a body.

torsion   The twisting of a structural member about its longitudinal axis by two equal and opposite torques.

translation   A linear displacement; in kinematics, a motion of a body such that a set of rectangular axes, fixed in the body, remains parallel to a set of axes fixed in space.

viscosity   The internal frictional resistance exhibited by a fluid in resisting a force that tends to cause the liquid to flow.

wall ties   In masonry, a type of anchor used to secure facing to a back-up wall.

white rot   A type of decay in wood caused by a fungus that leaves a white residue.

Index

U.W.E.L. LEARNING RESOURCES